HANDBOOK ON THE GEOGRAPHIES OF GLOBALIZATION

Handbook on the Geographies of Globalization

Edited by

Robert C. Kloosterman

Virginie Mamadouh

Pieter Terhorst

Department of Geography, Planning and International Development Studies (GPIO) Centre for Urban Studies, University of Amsterdam, the Netherlands

EE Edward Elgar
PUBLISHING

Cheltenham, UK • Northampton, MA, USA

Published by
Edward Elgar Publishing Limited
The Lypiatts
15 Lansdown Road
Cheltenham
Glos GL50 2JA
UK

Edward Elgar Publishing, Inc.
William Pratt House
9 Dewey Court
Northampton
Massachusetts 01060
USA

Paperback edition 2020

A catalogue record for this book
is available from the British Library

Library of Congress Control Number: 2018945836

This book is available electronically in the Elgaronline
Social and Political Science subject collection
DOI 10.4337/9781785363849

Printed on elemental chlorine free (ECF)
recycled paper containing 30% Post-Consumer Waste

ISBN 978 1 78536 383 2 (cased)
ISBN 978 1 78536 384 9 (eBook)
ISBN 978 1 80037 239 9 (paperback)

Typeset by Columns Design XML Ltd, Reading
Printed and bound in the USA

Contents

PART III   GEOGRAPHIES OF FLOWS

PART IV   GEOGRAPHIES OF PLACES

Figures

Tables

Contributors

Paul C. Adams is Professor in the Department of Geography and the Environment at the University of Texas, where he also serves on the Graduate Studies Committee of the Center for Identity. He is the author of *Geographies of Media and Communication* (Wiley-Blackwell, 2009), co-editor of *Research Companion to Media Geography* (Ashgate Press, 2014), and co-author of *Communications/Media/Geographies* (Routledge, 2016). His research addresses theoretical issues at the intersection of media theory and human geography, including place representation, biosemiotics, digital simulation models, critical geopolitics, and cartographic discourses. He is a co-investigator on the research project 'Responsible Adoption of Visual Surveillance Technologies in the News Media (ViSmedia)' funded by the Research Council of Norway through the University of Bergen. He has held visiting research positions at Karlstad University (Sweden), University of Bergen (Norway), McGill University (Canada) and University of Montreal (Canada).

Anne-Laure Amilhat Szary, PhD, full Professor at Grenoble-Alpes University, France and a member of the Institut universitaire de France. She is the Chair of the CNRS *PACTE* research unit, a pluridisciplinary social sciences research centre (https://www.pacte-grenoble.fr/en). A political geographer dedicated to critical border studies, she is studying comparatively American and European borders. She co-edited *Borderities, the Politics of the Contemporary Mobile Border* (Palgrave Macmillan, 2015) with Frédéric Giraut and *Après la frontière, avec les frontières: dynamiques transfrontalières en Europe* (Édition de l'Aube, 2006) with Marie-Christine Fourny, and is the author of *Qu'est ce qu'une frontière aujourd'hui?* (PUF, 2015), as well as of 70 papers and chapters, and over 15 other edited books or special issues of international journals. Her latest research concerns the interrelations between space and art, in and about contested places. She is a founding member of the 'antiAtlas of borders' collective (http://www.antiatlas.net/en/), a science-art project.

Dennis Arnold is Assistant Professor at the University of Amsterdam, Department of Human Geography, Planning and International Development. Dennis publishes and teaches on labour, migration and citizenship; global production network analysis; and geo-economics. He does research in Southeast Asia and Southern Italy. His email address is D.L.Arnold@uva.nl, and his publications are available at: https://uva.academia.edu/DennisArnold.

David Bassens, a human geographer and social and cultural anthropologist, is Assistant Professor of Economic Geography at the Geography Department of Vrije Universiteit Brussels where he acts as co-director of Cosmopolis: Centre for Urban Research. He doubles as Associate Director of Financialization of the Globalization and World Cities (GaWC) research network and is Executive Committee member and Treasurer of

FINGEO, the Global Financial Geography network. David Bassens' PhD research studied processes of world city-formation through the lens of the geographies of Islamic finance. His current work focuses on understanding processes of world city-formation under conditions of financialized globalization, with an empirical focus on continental Europe and emerging markets. His work has been published in leading journals in the field of human geography.

Soyoon Choo is a PhD Candidate in Urban Planning and Development at the University of Southern California. Her research interests revolve around the idea of 'development with dignity' and mutual accommodation in an urban context, including equity and community-based planning processes, grassroots activism, and creative/artistic urban practices in envisioning alternative futures. Specifically, she looks into how artists, community organizations, and residents build trust and create productive, collaborative relationships to reconstruct norms/values around urban development and quality of life. She received her Masters in Public Administration at the University of Southern California and her BA in International Studies at Ewha Womans University (South Korea).

Kevin R. Cox is Distinguished Professor Emeritus of Geography at the Ohio State University. He has been a Guggenheim Fellow. His most recent books are *Making Human Geography* (Guilford Press, 2014) and *The Politics of Urban and Regional Development and the American Exception* (Syracuse University, 2016).

Elizabeth Currid-Halkett is the James Irvine Chair in Urban and Regional Planning and Professor of Public Policy at the Price School of Public Policy at the University of Southern California. Her research focuses on the arts and culture and most recently, the American consumer economy. She is the author of three books – *The Warhol Economy: How Fashion, Art and Music Drive New York City* (Princeton University Press, 2007), *Starstruck: The Business of Celebrity* (Farrar, Straus and Giroux, 2010) and *The Sum of Small Things: A Theory of the Aspirational Class* (Princeton University Press, 2017).

Simon Dalby is Professor of Geography and Environmental Studies at Wilfrid Laurier University, Waterloo, Ontario. Simon Dalby was educated at Trinity College Dublin, the University of Victoria and holds a PhD from Simon Fraser University. He is the author of *Environmental Security* (University of Minnesota Press, 2002) and *Security and Environmental Change* (Polity, 2009) and recently coedited (with Shannon O'Lear at the University of Kansas) *Reframing Climate Change: Constructing Ecological Geopolitics* (Routledge, 2016). His current research focuses on globalization, the cultural politics of climate change and Anthropocene geopolitics.

Elena dell'Agnese is full Professor of Geography at the University of Milano-Bicocca, where she is also Director of the Center for Visual Research. In 2014, she was elected as vice president of the International Geographical Union. Her research is focused mainly on the cultural aspects of political geography, but she has also been working on gender issues from a geographical point of view and on Critical Tourism Studies. On these topics, she published a dozen books and more than 120 contributions, including journal articles and book chapters, in Italian, English, French, Spanish, Portuguese, and Croatian. She has also been working extensively on the cultural geographies of food, starting with the analysis of the 'Mediterranean Diet' as a geopolitical discourse (1998), and getting to the study of 'pizza' in the world as a symbolic combination between global trends and local claims (2015).

Ben Derudder is a Professor of Urban Geography at Ghent University's Department of Geography, and an Associate Director of the Globalization and World Cities (GaWC) research network. He has been a Marie Curie Research Fellow and a Visiting Professor at the Chinese Academy of Sciences (Nanjing). His research focuses on the conceptualization and empirical analysis of world city networks in general, and their transportation and production components in particular. His work has been published in leading academic journals, and he has co-authored the second edition of *World City Network: A Global Urban Analysis* (Routledge, 2016, with Peter Taylor).

Tatiana Fogelman is an Associate Professor of Mobility and Urban Studies at Roskilde University in Denmark. Her overarching intellectual interest concerns the relationship between difference and citizenship. Her research, published until 2017 under the surname Matejskova, focuses on both the governance of transnational migrants as a component of state-making on the one hand, and immigrant subjectivity and experiences of settlement and mobility on the other. Her publications examining issues such as the role of integration projects in urban encounters across difference, landscapes of belonging in marginalized neighbourhoods, immigrant understanding of integration, the impact of the integrationist turn on citizenship, and the national-urban lens in migration studies, have appeared in such journals as *Social & Cultural Geography*, *Antipode: Journal of Radical Geography*, *Urban Geography*, *Migration Studies* and *Ethnicities*. Together with Dr Marco Antonsich she co-edited the edited volume *Governing through Diversity: Migration Societies in Post-Multiculturalist Times*, published by Palgrave Macmillan's Global Diversities series in 2015.

Christopher Gaffney is a Clinical Associate Professor in the School of Professional Studies at New York University. He is an urban geographer whose research focuses on the urban and social impacts of sports mega-events. His recent publications examine the ways in which event coalitions produce changes in the political economy of urban regions. He is the author of *Temples of the Earthbound Gods* (University of Texas Press, 2008) and editor of the *Journal of Latin American Geography*.

Joyeeta Gupta is Professor of Environment and Development in the Global South at the Governance and Inclusive Development programme group of the Amsterdam Institute for Social Science Research of the University of Amsterdam, with a special focus on the Law and Policy in Water Resources and Environment at IHE Institute for Water Management, Delft.

Markus Hesse is a Professor of Urban Studies at the University of Luxembourg. With an academic background in urban and economic geography and spatial planning, his research interest is to explore the relationship between cities/regions and flows. A current research project deals with the emergence of 'relational cities' in Europe and South East Asia. Also, a particular emphasis is placed on the science-policy interface in urban planning and governance.

Rory Horner is a Lecturer in Globalization and Political Economy of Development, ESRC Future Research Leader and Hallsworth Research Fellow at the University of Manchester's Global Development Institute. He is an economic geographer by training, with research interests in the political economy of globalization, industrial development, trade, India and Africa. He holds a BA from Trinity College, Dublin and a MA and PhD in Geography from Clark University. His current research is focused on South–South trade, and involves a major project on India's pharmaceutical industry and local pharmaceutical manufacturing in sub-Saharan Africa.

Shirlena Huang is Associate Professor of Geography at the Faculty of Arts and Social Sciences, National University of Singapore. She is also a member of the Steering Committee of the Faculty's Migration Cluster. Her research focuses mainly on issues at the intersection of transnational migration, gender and family, with a particular focus on the themes of care labour migration and transnational families within the Asia-Pacific region.

Arne Isaksen is a Professor at the Department of Working Life and Innovation at the University of Agder, Norway. He has a PhD in economic geography from the University of Oslo. His research interest is theoretical and empirical studies of regional industrial development, focusing on the development of regional clusters, innovation systems and path development, companies' innovation mode, and regional innovation policy and policy lessons.

Andrew E.G. Jonas is Professor of Human Geography at the University of Hull in the UK. Andy's research interests address the politics of urban and regional development, focusing on the USA, Europe and, more latterly, China. His recent textbook, *Urban Geography: A Critical Introduction* (Wiley-Blackwell, 2015), is co-authored with Eugene McCann and Mary Thomas. Andy's co-edited books include *Handbook on Spaces of Urban Politics* (Routledge, in press, with Kevin Ward, Byron Miller and David Wilson), *The Urban Growth Machine: Critical Perspectives Two Decades Later* (SUNY Press, 1999, with David Wilson), *Interrogating Alterity* (Ashgate, 2010, with Duncan Fuller and Roger Lee) and *Territory, the State and Urban Politics* (Ashgate, 2012, with Andy Wood).

Alun Jones is Chair of Geography at University College Dublin. Prior to joining UCD he was Professor of Human Geography at the University of Leicester, and before that Associate Professor of Human Geography at University College London. He has also held visiting posts at the Universities of Cambridge, Oxford, and Bonn. He was a recipient of a senior Leverhulme Research Fellowship in 2002–2004, and in 2004–2005 presented with the Edward Heath Award by the Royal Geographical Society (with the Institute of British Geographers) (RGS/IBG) for his contribution to research on EU governance. In 2006 he was made an Academician of the Learned Societies for the

Social Sciences in the UK. In 2011 he was elected to the Royal Irish Academy (MRIA) and to the Academia Europaea. He is currently Chair of the Geosciences and Geographical Sciences committee of the Royal Irish Academy (RIA), and National contact point for Ireland for the International Geographical Union (IGU).

Jana M. Kleibert is a Visiting Professor in Economic Geography at the Goethe University Frankfurt am Main and Honorary Research Fellow of the University of Manchester. Previously, she was a postdoctoral Research Fellow at the Leibniz Institute for Research on Society and Space. She holds a PhD in Geography and an MSc in International Relations from the University of Amsterdam. Her research interests are economic and urban geographies of globalization, global production networks, special economic zones, and the Philippines. She is co-editor of *Globalisation and Services-driven Economic Growth: Perspectives from the Global North and South* (Routledge, 2017).

Robert C. Kloosterman is Professor of Economic Geography and Planning at the University of Amsterdam. He is head of the research group Geographies of Globalisations and was director of the Amsterdam Institute of Metropolitan and International Development Studies. He has advised the Dutch national Social-Economic Council and the OECD on the issue of migrant entrepreneurship. His research is guided by questions about how the social, economic and cultural transition of advanced urban economies that gathered pace after 1980 has affected cities and why different outcomes have emerged. He has published extensively in English language journals on urban issues such as labour market developments in urban areas, migrant entrepreneurship, and more recently on cultural industries, especially music and architectural design, and planning issues related to cultural amenities. He is a member of the Editorial Board of *Urban Geography* and of the Editorial Board of *Built Environment*.

Rosa Koetsenruijter holds a Research Master's degree in Urban Studies from the University of Amsterdam. She also studied at the University of British Columbia in Vancouver, Canada. She has published on processes of globalization of cultural industries. Her Master thesis, *Changing of the Seasons: Investigating the Role of Fashion Weeks and Film Festivals in Amsterdam and Vancouver*, highlighted the importance of temporary gatherings of workers in the cultural industries in facilitating crucial social networks as well as in fostering innovation. She is co-founder and editor of the online platform on the Urban Commons of Culture. She is currently working on a study which focuses on the impact of the recession in the wake of the credit crisis on architectural practices in the Netherlands.

Theodora Lam is a postdoctoral Fellow at Asia Research Institute, National University of Singapore (NUS). She obtained her PhD in Geography from NUS and her dissertation focused on understanding changing gender subjectivities, the web of care and relationships within the family in the wake of transnational labour migration. Her research interests cover transnational migration, children's geographies and gender studies in Southeast Asia, and she has also published on themes relating to migration, citizenship and education.

Juho Luukkonen works as a postdoctoral Researcher at the University of Helsinki, Finland. Juho's research interests concern the political geographies of European integration, spatial planning and development, politics of scale and space as well as the state territorial processes. His research has been published in such journals as, for instance, *European Planning Studies, Environment and Planning* and *Planning Theory*.

Virginie Mamadouh is Associate Professor of Political and Cultural Geography at the Department of Geography, Planning and International Development Studies of the University of Amsterdam and affiliated to the Centre for Urban Studies and ACCESS EUROPE. She is an elected member of Academia Europaea. Her research interests pertain to (critical) geopolitics, political culture, European integration, (urban) social movements, transnationalism, and multilingualism. She published mainly in English, Dutch and French. For the past 25 years her work on the political and cultural geographies of languages has been focused on Europe and the European Union, ranging from multilingualism in the internal dynamics of the European Parliament, through the link between preferred language regimes and conceptions of national identities and transnational networks across monolingual national communication settings (especially those linked to migration, grassroots movements and new media) to the linguistic diversity in European cities. She is an editor of the international academic journal *Geopolitics* and Chair of the Commission on Political Geography of the International Geographical Union (IGU). She recently co-edited with John Agnew, Anna Secor and Joanne Sharp, *The Wiley Blackwell Companion to Political Geography* (Wiley-Blackwell, 2015), and with Anne van Wageningen, of *Urban Europe: Fifty Tales of the City* (Amsterdam University Press, 2016). Full profile at: http://virginie mamadouh.socsci.uva.nl.

Valentina Mazzucato is Professor of Globalization and Development at Maastricht University. She specializes in migration from a transnational perspective, which focuses on the linkages that migrants forge and maintain between their societies of origin and destination. She studies the consequences of such linkages for migrants, their families and communities both at home and abroad. Her research is characterized by multi-sited research designs and mixed-method approaches combining survey and ethnographic methods and working in interdisciplinary teams. She heads several international research projects on transnational families that live between Africa and Europe. She currently heads a five-year project on the Mobility Trajectories of Young Lives in Global South and North (MO-TRAYL) funded under the ERC Consolidator Grant scheme. Before joining the Faculty of Arts and Social Sciences in Maastricht, Professor Mazzucato lived and worked in and on Africa for over twenty years, focusing on West Africa (Niger, Burkina Faso, Mali, Ivory Coast, Ghana).

Evan McDonough is an urban geographer from Toronto who has published research on transportation infrastructures in London, Zurich and Madrid. His research interest lies at the intersections of urban–regional and transport geographies, and the socio-spatial experience and local negotiation of extending and externally oriented transportation flows.

Michiel Van Meeteren is postdoctoral Researcher at Cosmopolis, Vrije Universiteit Brussel. He holds a research Master in the social sciences from the University of Amsterdam and a PhD in Geography from Ghent University. At Cosmopolis, he currently studies the historical geography of financial integration in the EU and the contemporary geography of financial technology. His work covers a broad scope, alternating between urban, economic, financial and political issues. A recurring focus in this work is the bridging of different theoretical and methodological traditions combined with a strong historical-disciplinary awareness.

Byron Miller received his PhD (Geography) from the University of Minnesota in 1995, MA (Geography) from Arizona State University in 1984, and BSc (Geography) from The Pennsylvania State University in 1978. He worked as an urban planner for the city of Scottsdale, Arizona, in the early 1980s, spent three years living and studying in Freiburg, Germany, in the late 1980s, and taught at the University of Cincinnati (1993–2000) before taking his current position at the University of Calgary as Coordinator of the Urban Studies Program. In Calgary he teaches courses on urbanization and urban planning, urban social geography, urban politics and governance, globalization, and field courses on urban sustainability in Europe as well as a seminar on the history and philosophy of geography. In 2000 Miller was a summer fellow at the Institute for Advanced Study in the Behavioral Sciences at Stanford University. He has served as a board member of the Urban Geography Specialty Group (1999–2001) and the European Geography Specialty Group (2009–2011) of the American Association of Geographers, as an editorial board member of *Geography Compass* (2008–present) and *Spaces and Flows* (2012–present), and as a board member of Sustainable Calgary (2008–present). Byron Miller's recent work focuses on the spatial constitution of social movements, urban governance and governmentality, neighbourhood change and inequality, the politics of urban and regional sustainability, and the social and environmental implications of smart cities. He is the author of *Geography and Social Movements* (University of Minnesota Press, 2000), co-editor (with Walter Nicholls and Justin Beaumont) of *Spaces of Contention: Spatialities and Social Movements* (Ashgate, 2013), and co-editor (with Kevin Ward, Andrew Jonas, and David Wilson) of the *Handbook on the Spaces of Urban Politics* (Routledge, 2018).

Sami Moisio is Professor of Spatial Planning and Politics in the Department of Geosciences and Geography at the University of Helsinki, Finland. His recent research interests include political geographies of Europeanization, state spatial transformation and the geopolitics of the knowledge-based economy. He has published papers in journals such as *New Political Economy*, *Political Geography*, *Environment and Planning C*, *Environment and Planning A*, *Geopolitics*, *Global Networks*, *Progress in Human Geography* and *Territory, Politics, Governance*.

Martin Müller is Swiss National Science Foundation Professor at the University of Lausanne. He is a human geographer working on the planning and impacts of mega-events and contributing to conceptual debates around actor-network theory (ANT) and assemblage thinking. His recent publications have developed the concepts of 'hazardous planning' and 'improvisation' in the context of planning mega-events.

His most influential publications include 'The Mega-Event Syndrome' (*Journal of the American Planning Association*) and 'What Makes an Event a Mega-Event?' (*Leisure Studies*). See also www.martin-muller.net.

Barbara Oomen is a Professor in the Sociology of Human Rights at Utrecht University, and teaches at University College Roosevelt, UU's Liberal Arts and Sciences Honors College in Middelburg, the Netherlands. She has a background in law, anthropology and political science and has published extensively on the interplay between law, culture and human rights, in fields ranging from customary law and transitional justice to the legitimacy of human rights. In the academic year 2016–2017 she was a Fernand Braudel fellow at the European University Institute in Florence. Her most recent research concerns local authorities and international human rights law, on which she co-edited a book (*Global Urban Justice: The Rise of Human Rights Cities*, CUP, 2016) together with Martha F. Davis and Michele Grigolo. In 2017, she received a Vici grant from the Netherlands Organization for Scientific research for a five-year project titled 'Cities of refuge: local governments and the human rights of refugees in Europe'.

Soul Park is a Lecturer in International Relations at the National University of Singapore. He received his BA from Seoul National University, MSc from the University of Wales, Aberystwyth, and MA and PhD from the University of Notre Dame. His current research focuses on international security and foreign-policy analysis, particularly on the political decisions relating to the use of force.

Giacomo Pettenati is a post-doctorate researcher in Political and Economic Geography in the Department of Cultures, Politics and Society at the University of Turin. In 2014 he obtained a PhD in Spatial Planning and Local Development at the Polytechnic University of Turin. He currently teaches Didactics of Geography and Geography at the University of Turin. His main research topics are food geographies, landscape heritage, mountain regions development and participatory mapping. His latest research activities focus on Alpine foodways, regional food systems and alternative food networks. He has published more than sixty contributions, including journal articles and book chapters, in Italian, English, French and Spanish. He is part of the interuniversity research team 'Turin Food Atlas' and he is on the board of the research association, Dislivelli, working on mountain areas.

Mark Rosenberg is a Professor of Geography in the Department of Geography and Planning and cross-appointed as a Professor in the Department of Public Health Sciences at Queen's University in Kingston, Ontario, Canada. He is the Tier 1 Canada Research Chair in Development Studies. Professor Rosenberg received his undergraduate training at the University of Toronto and his graduate training at the London School of Economics and Political Science (LSE). In 2012, Professor Rosenberg received a Queen Elizabeth Diamond Jubilee Medal for his research on Canada's aging population, contributions to gerontology and service to older Aboriginal and non-Aboriginal populations. In 2017, Professor Rosenberg was presented with the 2017 American Association of Geographers Health and Medical Geography Specialty Group

Melinda Meade Distinguished Scholar Award for his contributions to medical and health geography. Publications from his research can be found in the leading journals of geography, gerontology, social science and medicine.

James W. Scott is Research Professor of Regional and Border Studies at the Karelian Institute of the University of Eastern Finland. Professor Scott obtained his Habilitation (2006), PhD (1990) and MA (1986) at the Free University of Berlin and his BSc at the University of California Berkeley (1979). His principal fields of research include: urban and regional geography, borders, border regions, geopolitics, regional and urban governance, Cohesion Policy and Central European studies. Professor Scott has participated in the establishment of the BRIT (Border Regions in Transition) network. Since 1999 he has coordinated several medium-sized and large research consortia focusing on border studies and supported by the EU's Framework Programmes, the European Science Foundation, the Finnish Academy and other sources. Presently he is coordinator of the GLASE project (Multilayered Borders of Global Security), funded by the Academy of Finland and is scientific coordinator of the Horizons 2020 project RELOCAL that investigates the role of the local level and local strategies in Cohesion and Territorial Development.

Matthew Sparke is Professor of Politics at the University of California Santa Cruz. He is the author of *Introducing Globalization: Ties, Tensions and Uneven Integration* (Wiley-Blackwell, 2013), and *In the Space of Theory: Post-foundational Geographies of the Nation-State* (University of Minnesota Press, 2005); and has also published widely in leading social science journals and edited books on global studies, and the changing geographies of citizenship and sub-citizenship amidst globalization.

Pieter Terhorst is a retired Assistant Professor at the University of Amsterdam (Department of Geography, Planning and International Development Studies) and an informal member of the research group 'Geographies of Globalization'. His main research areas are political-economic geography, variety of capitalism, rescaling of economy and state, the economic geography of tourism cities, and, more recently, the urban restaurant industry.

Kees Terlouw is Assistant Professor at the Department of Human Geography and Planning at the Utrecht University in the Netherlands, where he teaches political geography. Situated at the intersections between geography, sociology and political science, his research has several focuses. After writing a PhD thesis and many other publications on the world-systems approach and regional development, he now concentrates more on the relations through which regions are formed. He has published on how different regional stakeholders shape cross-border relations. The main focus of his current research is the changing relations between identities and politics. He is especially interested in the relation between local and regional identity discourses and how these can help or hinder specific policies, like municipal amalgamation or the promotion of the competitiveness of metropolitan regions.

Franz Tödtling is a retired Professor at the Institute for Multi-level Governance and Development, Vienna University of Economics and Business. His main research areas are regional development, clusters, innovation systems, and the knowledge economy from a spatial perspective. He has been involved in a number of international research projects and has contributed a large number of articles in professional journals.

Michaela Trippl is a Professor of Economic Geography at the Department of Geography and Regional Research at the University of Vienna, Austria. The main focus of her research is on regional innovation dynamics, long-term regional structural change and innovation policies.

Peer Vries studied economic and social history at the University of Leiden, where he also defended his PhD and worked for many years. From 2007 to 2016 he was Professor of Global Economic History at the University of Vienna. Currently he is honorary fellow at the International Institute of Social History in Amsterdam. His major recent publications are *Escaping Poverty. The Origins of Modern Economic Growth* (V&R unipress, 2013) and *State, Economy and the Great Divergence. Great Britain and China, 1680s–1850s* (Bloomsbury, 2015). He is now writing a book on Japan and the Great Divergence, that will be published in 2018. His main fields of interest are the Great Divergence and the role of the state in economic development, with a focus on the early modern era.

Lauren Wagner is Assistant Professor in Globalization and Development at Maastricht University. Her research focuses on how diasporic practices evolve in face-to-face interactions and emerge as global, international, and transnational collectivities and social networks. She investigates practice through microanalysis of everyday encounters, using ethnomethodological parameters to analyze linguistic recorded data as well as observed embodied practices in materialist atmospheres. As an interdisciplinary mobilities researcher, she participates in geography, anthropology, sociology and sociolinguistic discussions. More extended linguistic analyses are now available in her book *Becoming Diasporically Moroccan* from Multilingual Matters (2017). Her next project examines the value of caring for home as a problem of mobilities and financialization of homeownership across the globe. Her educated opinions can be found at www.medium.com/LBWagner and her current projects and publications are updated regularly on ORCID (ID: 0000-0002-4778-7408).

Wu Yue-fang is an Associate Researcher in the Department of Finance and Management at Foshan Polytechnic. She obtained her PhD degree from Sun Yat-sen University. Her PhD dissertation is about lifestyle mobilities in China. Her research interests include second home, lifestyle mobility, intangible heritage and folk culture in tourism.

Xu Hong-gang obtained a PhD from Asian Institute of Technology in 1999. Then she joined Sun Yat-sen University. She has been involved in sustainable tourism development research, teaching and practice. She is Vice Chairman of the Chinese Tourism Geography committee since 2013. She is the PI of several National Tourism Accreditation programmes and over 20 tourism consulting projects. Professor Xu is the key member of sustainable tourism monitoring centre UN World Tourism Organization which carries out the monitoring work of 10 important tourism destinations in China, including Huangshan, Zhangjiajie, Kanas, and Xishuangbannan. Professor Xu has substantial international experience and has been invited by universities from America, Thailand, Spain, Japan and France to cooperate in the research and deliver graduate courses. She has published more than 100 journal papers, including the most internationally prestigious tourism journals.

Takashi Yamazaki is Professor of Geography at Osaka City University, Japan. His research interests concern political geography, critical geopolitics, and Okinawa studies. He has published many articles both in Japanese and English journals including *Political Geography* and *Geopolitics*. His book *Space, Place, and Politics: Towards a Geography of Politics 2nd Edition* (*Seiji, kūkan, basho: seiji no chirigaku ni mukete* 政治・空間・場所—「政治の地理学」にむけて [改訂版], Kyoto: Nakanishiya, 2013) is the latest textbook on political geography in Japan.

Brenda S.A. Yeoh is Raffles Professor of Social Sciences at the Department of Geography as well as Research Leader of the Asian Migration Cluster at the Asia Research Institute, National University of Singapore. Her research interests include the politics of space in colonial and postcolonial cities, and she has considerable experience working on a wide range of migration research in Asia, including key themes such as cosmopolitanism and highly skilled talent migration; gender, social reproduction and care migration; migration, national identity and citizenship issues; globalizing universities and international student mobilities; and cultural politics and the transnational family.

Acknowledgements

As members of the research group *Geographies of Globalizations* (GoG) of the Amsterdam Institute for Social Science Research (AISSR) of the University of Amsterdam (UvA), we have been looking at different spatial manifestations of processes of globalization for quite some time. We were already very much aware that we only covered a small and selective set of issues related to this near-ubiquitous phenomenon. The contributions to this *Handbook* have only strengthened our sense of modesty as they present a rich panorama of the many different ways in which globalization is articulated. A project all the more important in times where the globalization as natural phenomenon discourse (hegemonic since Mrs Thatcher's 'There-is-no-alternative' in the 1980s) has suddenly been replaced by renationalization rhetoric in many regions of the world.

We would first like to express our deep appreciation to all the contributors to this *Handbook* for sharing with us their expertise on globalization processes, and for contributing to this collection. Thanks to them, the chapters cover a wide range of topics, approaches and insights related to geographies of globalization.

We thank all who have contributed one way or the other to the making of the *Handbook*, especially our colleagues at GoG for their comments on our early plans for this volume, the team at Edward Elgar for their support and patience, and Joanna van der Leun for the design of the cover.

Finally, we are particularly grateful to Rosa Koetsenruijter who was responsible for much of the communication with the authors as well as assisting us with the editing of the chapters and, moreover, has done a great job in keeping us on track for the completion of this *Handbook*.

Robert C. Kloosterman
Virginie Mamadouh
Pieter Terhorst
Department of Geography, Planning and International Development Studies
Amsterdam Institute for Social Science Studies (AISSR)
Centre for Urban Studies (CUS)
University of Amsterdam
Amsterdam, the Netherlands

PART I

INTRODUCTORY CHAPTERS

1. Introducing geographies of globalization: genealogies of the concept, existing views on globalization inside and outside geography

Robert C. Kloosterman, Virginie Mamadouh and Pieter Terhorst

Globalization is more often than not associated with remote places and large-scale perspectives while it is pervasive of our most personal and mundane activities. As we take our breakfast we might be conscious of the importance of globalization when we read the daily news, but we might easily overlook that the cutlery, microwave or smart phone we use as well as what we eat and drink are shaped by globalization and have shaped it. Christian Grataloup – a French geographer specializing in the *géohistoire* of globalization (Grataloup, 2009, 2010, 2017) – has looked in his book *Le monde dans nos tasses* (Grataloup, 2017) behind the seemingly self-evident façade of our breakfast. As the geographer David Harvey notes, we can 'consume our meal without the slightest knowledge of the intricate geography of production and the myriad of social relationships embedded in the system that puts it upon our table' (Harvey, 1990: 422–423). Grataloup, however, has shown how this everyday experience is the product of particular paths of globalization – according to the sociologist Anthony Giddens (1990: 64) 'the intensification of worldwide social relations which link distant realities in such a way that local happenings are shaped by events occurring many miles away and vice versa'. In particular he reminds us in his analysis of the world in our breakfast, that this specific meal is a rather recent phenomenon. It emerged in Europe in the seventeenth century, replacing a meal that was quite similar to the other meals of the day – which it still is nowadays among groups less affected by globalization around the world. Indeed the new fashion spread from the urban bourgeoisie in Amsterdam and London to the European aristocracy, then to popular classes in Europe and beyond Europe among urbanites and well-to-do classes. The drinks associated with breakfast – tea, coffee and chocolate – and the sugar used by many to sweeten them, were originally very expensive and demonstrated economic prosperity and social distinction. Nowadays, they are not only travelling great distances to arrive on our breakfast tables, their very production is a quintessential component in the history of globalization of the past centuries.

The control of the production and trade of these by now common breakfast staples was one of the main drivers of the early-modern European expeditions across the world seas. With this urge to control came claims on territories and sea routes, the founding of colonies, the establishment of plantations in these colonies, the transplantation of crops (Crosby, 2003, 2015), and the movement of enslaved and forced labour across continents. Contemporary areas of production are now often far removed from those of origin: we tend to associate cacao more with the current main producers Ivory Coast and Ghana than with Mexico, its region of origin. The same can be said with regard to coffee: it is Brazil, Colombia and Vietnam rather than Ethiopia that now springs to mind. Sugar cane is more linked to Brazil than to India, its country of origin. In addition, while China has remained the main producer of tea, former British colonies – where tea was introduced by the British (India, Kenya, Sri Lanka) in the nineteenth century – are the next most important producers. With his unpacking of the typical contemporary breakfast, Grataloup thus has chosen a strategic window to explain globalization to a general public. It reveals the many temporal and spatial scales of globalization, its material and ideational dimensions, as well as its economic, political, cultural and social facets which usually remain unobserved.

There are many ways to look at processes of globalization, reflecting their manifold expressions and multi-dimensional character of globalization. Social scientists from various disciplinary backgrounds have looked at globalization through their own particular lenses, using various conceptual tools and methodologies. This has resulted in a wide range of accounts with each covering a particular, inevitably highly selective, part of these nearly all-encompassing and interrelated processes of globalization. Like the parable from Jainism in which the blind men touch different parts of the elephant each experiencing/conceptualizing a different thing, scholars of globalization have difficulties in grasping the larger picture it represents. The essays in this volume are no exception, but the coverage is so extensive that at least some of the broader contours of this, well, global phenomenon come into view.

Although an inherently spatial phenomenon, one could argue that the *geography* in globalization is frequently underplayed – sometimes to the amazement of geographers (cf. Yeung, 2002 and Sheppard, 2016 regarding the need for a geographical analysis of economic globalization to counteract accounts by economists). In this *Handbook on the Geographies of Globalization*, the emphasis is explicitly on how these processes are articulated in space and, in turn, how space shapes these processes. The spatial dimension of globalization is, hence, in each of the following 32 chapters, the guiding and structuring perspective. We thus approach globalization as a phenomenon that cannot be grasped without a proper understanding of its spatial dynamics, while highlighting the diversity of geographical approaches to processes of globalization.

The first overarching aim of this *Handbook*, then, is to highlight the myriad of ways in which a great variety of cross-border flows of people, goods, services, capital, information, pollution and cultures have (re)shaped concrete places across the globe and how these places, in turn, shape those flows. The second aim is to position globalization in a broader historical perspective and indicate long-term continuities and ruptures in the development of cross-border linkages. Third, we want to present a variety of geographical perspectives on how to grasp these processes of globalization.

Before we briefly sketch some of these broader contours of globalization, we discuss the concept 'globalization' itself and take a look at its surprisingly rapid emergence in the 1990s in both academic and public debates. In the next section, we turn to the plurality of geographical expressions of and geographical approaches to globalization, hence the plural in the title of this *Handbook*. After that we introduce the contents of this volume.

GENEALOGIES OF A CONTESTED CONCEPT

Globalization, then, refers to many different things. The concept is contested and contestable. This is often the case with key concepts in the social sciences and especially when they refer to large-scale developments in combination with being widely used in public debates. The concept of globalization is, however, unique regarding the speed at which it has emerged and became a buzz word, not only in academia but also in everyday language (James and Steger, 2014). At the same time, it is notable that despite its short history, no one can be named as inventor of the term, in the way the parent of a neologism can be identified (such as the concept geopolitics can be traced back to the Swedish political scientists Rudolf Kjellèn), moreover very little has been written about the origin and the development of the concept.

Paul James and Manfred Steger (2014) have written 'A genealogy of "globalization"' which has been published as an introduction to *Globalization: The Career of a Concept*, a special issue of the academic journal *Globalizations* (established in 2004 and dedicated to opening the widest possible space for discussion of alternatives to narrow understandings of global processes and conditions) – and it is very much the exception. The special issue further consists of interviews with 'crucial contributors to the rise of this keyword' and other key figures in the emerging field of Global Studies: George Modelski, Roland Robertson, Saskia Sassen, Joseph Stiglitz, Arjun Appadurai, David Held, Jan Aart Scholte, Jonathan Friedman, Nayan Chanda, Mark Juergensmeyer, James Mittelman and Barry Gills. Strikingly, only Robertson explicitly remembered his first encounter with the term. Most, however, recalled that they started using it in the early 1990s, in the period after the fall of the Berlin Wall and the end of the Cold War.

According to Sandu Cuterela (2012: 137), the *Oxford English Dictionary* states that '[o]ne of the earliest uses of the term "globalization", as known, was in 1930 – in a publication entitled *Towards New Education* – to designate an overview of the human experience in education'. This, however, remained an isolated occurrence. In textbooks and articles we sometimes find a sketchy genealogy of the term globalization, but these histories are typically highly contingent on the disciplinary, linguistic and national backgrounds. In its obituary of Theodore Levitt in 2006 (Feder, 2006), the *New York Times* first endorsed then retracted the common misperception among economists that this American economist had coined the term globalization in an article entitled 'Globalization of Markets' published in *The Harvard Business Review* in 1983 (Levitt, 1983). French economists, by contrast, might refer to the French economist François Perroux and his article 'L'économie planétaire' (the planetary economy) published in *Tiers-Monde* in 1964 and in the second edition of *L'économie du XXe siècle* (Perroux, 1964, 1969). Sociologists, in their turn, may bring forward the Canadian communications theorist Marshall McLuhan who coined the notion of global village in his book *The Gutenberg Galaxy* (McLuhan, 1962) to describe the extension of the village mindset to the whole planet thanks to radio and the television, the rapidly spreading mass media in those days. Scholars of International Relations, on the other hand, might refer to George Modelski's article on 'Communism and the globalization of politics' in *International Studies Quarterly* in 1968 (Modelski, 1968). James and Steger (2014) also found earlier unrelated occurrences of notions of globalization (as in the field of education with the global reading method or the intriguing writing about the global-ization of 'the US negro question' triggered by the experiences of African-American soldiers overseas during the Second World War).

The awareness of interconnectedness on a global scale evidently preceded the use of the term globalization. After the closure of the European 'Age of Discoveries', when much of the planet (except the almost inhospitable Antarctica) had been mapped, a sense of global history emerged (Mackinder, 1904; Heffernan, 1998). Expressions of global consciousness (James and Steger, 2014: 422) in the first half of the twentieth century can also be seen in the use of 'globe' in the title of newspapers as well as in the appearance of globes in the logos of film studios and related activities (e.g. the creation of the Golden Globe Awards in 1944). This global imaginary strengthened after the Second World War. It is, for example, addressed in the writings of Hannah Arendt (1958) in her analysis of the *Human Condition*, and in Marshall McLuhan's 'global village' (1962). The pictures of the Earth as the home of human kind ('Spaceship Earth') which became available in the late 1960s and the early 1970s with the exploration of space constituted a very visual representation of the global scale. The colour picture of the Earth as a blue marble taken by the crew from Apollo 17 in December 1972 even became an iconic image (see Cosgrove, 1994, 2001). Only quite recently has this been challenged by satellite pictures a few clicks away for every Internet user in the current age of Google Earth.

Notwithstanding these earlier understandings and representations of global inter-connectedness, the widespread use of globalization to label far-flung, cross-border linkages only took off in the late 1980s. This sharp rise in the popularity of the concept holds for English, but also for other main languages. N grams produced from the collection of books digitalized by Google show that the term was hardly used before the mid-1980s. After that, in the second half of the 1980s, a slow increase occurred with a very sharp increase in the 1990s and a subsequent stabilization in the 2000s (NB: the collection ends in 2008). In other European languages, we can observe rather similar trends, but with earlier occurrences and, remarkably, even drops after the turn of the century: in German after 2001, and after 2002 in Spanish and French (this is also the case for *globalisation* and *mondialisation*). By contrast in simplified Chinese (i.e. books published in the People's Republic) the rise in the use accelerated strongly in the 2000s (see for counts and mappings of the use of the term globalization: Murray, 2006: 18; Lévy, 2008; Sidaway et al., 2016: 6; James and Steger, 2014: 419).

Interpretations of the concept globalization are (partly) contingent on the linguistic context. In France, for example, globalization is typically seen as a borrowing from English. In French, therefore, the terms 'globalisation' and 'mondialisation' are frequently used, sometimes as synonyms of globalization, sometimes with slightly different connotations, based on the etymology (the former referring to the shape of the Earth or to the three dimensional model representing it, the latter to the world in terms of human civilization). Moreover, the use of these terms in French also foregrounds more explicitly conflicting visions of globalization and often entails discussing the possibilities of alternative forms and projects of globalization. In the English language academic discussions, these tensions are usually underplayed.

The concept globalization is now widely used in many different societal, disciplinary, geographical and linguistic contexts. Unsurprisingly, many definitions abound but most, if not all of them share the reference to linkages between different regions across the globe. A concise and very influential summary of the conceptual connotations of globalization has been proposed by David Held and his co-authors at the Open University (Held et al., 2000). They emphasize four distinctive features of globalization as process:

1. The stretching of social relations over larger areas across the borders of the nation states (they thereby implicitly assume that the international system of modern territorial states is the 'natural' basis organization of human kind) including flows of people, capital, information, licit and illicit goods, germs, and pollution.
2. The intensification of interaction and interconnectedness through these flows.
3. The increasing interpenetration and locally increasing cultural diversity.
4. The building of a global infrastructure consisting of material and ideational, formal and informal arrangements which enable an intensification of flows within and between globalized networks. This ranges from the submarine fibre-optic cables and Internet Exchange Points to agreements and protocols regarding Internet routing, air travel, money transfers, human rights regimes or the rise of English as global langue (i.e. language of global communication).

All kinds of combinations of these features figure prominently in many academic contributions to debates on globalization from various disciplinary backgrounds. What, then, characterizes geographies of globalization?

GEOGRAPHY AND GLOBALIZATION

Globalization obviously is about space. This *Handbook on the Geographies of Globalization* demonstrates that geographical perspectives can generate highly relevant and important views on the multifaceted processes of globalization. The *Handbook* also makes clear that there is not one dominant geographical perspective on globalization but, instead, a plurality of viewpoints anchored in different theoretical frameworks and using different research methodologies. Before we outline the main arguments of the ensuing chapters and show the many ways in which the relationship between globalization and geography are now being explored, we briefly sketch why geographers initially were not very much involved in debates on globalization.

Exploring spatial relationships evidently forms the core business of geographers. One would, then, expect that they would have taken the lead in studying globalization. The economic geographer Peter Dicken (1986) and the political geographer Peter Taylor (1985) were indeed in the vanguard and published influential books in the early phase of debates on globalization. Yet, one cannot say that geographers have dominated the early phases of globalization studies. On the contrary, geographers, with the exception of the sociologist/geographer Saskia Sassen, are markedly absent in the above-mentioned *Global Studies* list of key figures in globalization studies. The near-absence of an explicit geographic approach can be traced back to a combination of an initial isolation of the discipline of geography and a neglect of space by other social science disciplines.

For a long time, the relationship between geography and other social sciences such as economics, sociology, and political science was mainly asymmetrical. Until the 1950s, geography was very descriptive, quite similar to the discipline of history. It was focused on the unique qualities of places, not very much theoretically informed and without much ambition to see more general patterns (cf. Scott, 2000). Geographers, then, stood mainly with their backs towards other social sciences (Soja, 1989). Only in the second half of the twentieth century can we observe a strong drive to make the discipline more scientific and, hence, less descriptive and more theoretical. Doing so mainly by extensively borrowing concepts and theories from other social sciences – notably economics, sociology, political science and international relations – but this remained for a long time very much a one-way street.

Other social scientists were not very interested in space and when they were, they typically ignored the work of geographers. Neither economists like Weber, Christaller, Lösch and Myrdal nor for example the philosopher/sociologist Lefebvre, who each took space seriously, were significantly influenced by geographers. The relatively low status of geography among the ranks of social sciences was the result of the until recently firmly established subordination of space to time in social theory. Key figures of early modern social science – Marx, Weber, and Durkheim – paid little attention to space because they prioritized time and history over space and geography in their abstract theoretical work. If they paid any attention to space and geography at all, they tended to view these rather unproblematically as a stable context or site for historical action (Soja, 1989). A crucial consequence of this prioritizing of time and history over space and geography is that spatial differences are interpreted as different stages in a single temporal development. This is reflected in grand narratives of unilinear progress, modernization, the sequence of stages of production, the transition from industrial to post-industrial society, and in unilinear development paths of cities.

According to Doreen Massey (2005), this interpretation of spatial differences has led to an *a-spatial* view of globalization, in which globalization is seen as an all-embracing movement that spreads from the West to the 'rest' of the world (it is a Western narrative too). From this perspective of a-spatial globalization – which at a first glance seems a *contradictio in terminis* – national states, which have been typically seen as the crucial spatial units of analysis, are different because they have not yet become modernized and are still behind on the same path of development. This view has been dominant, particularly among mainstream economists who preach 'best practices' to attain high rates of economic growth, as well as in many policy circles (Raworth, 2017). States with low levels of GDP per capita, the essential yardstick in this view, were apparently not able to fully adapt to the logic of capitalism. They were seen as lacking in free markets with firms stifled by regulation, failing to display good governance and their key actors not being adequately entrepreneurial, rational, and self-interested. The Washington consensus supported by the World Bank and the IMF, which became the dominant development paradigm in the 1980s, for instance, proposed in essence one-size-fits-all neoliberal policies of deregulation, privatization and liberalization to move up on the one and only road to economic development (Baldwin, 2016; Raworth, 2017).

Space, however, has become ever more salient from the 1980s onwards when globalization gained momentum. This was not just in the sense of more cross-border flows, it also meant 'the ascendance of other spatial units and scales' other than that of the nation state (Sassen, 2005: 28). Notably cross-border linkages between so-called global cities and global city-regions (Sassen, 2001; Scott, 2002; Hall and Pain, 2006) have gained in importance. Against this background, geographers have, predictably, highlighted the importance of space in understanding contemporary societal development at the local, national and global scale. Harvey (1985) and Soja (1989) have even argued that space should be included in social theory right from the start, to grasp both the drivers and the impacts of contemporary social change.

At the same time, a number of leading sociologists, political economists and political scientists – such as Giddens (1990), Urry (1985), Featherstone et al. (1995), Boyer and Rogers Hollingsworth (1997), Castells (1996/2011) and Jessop (2002, 2009) among others – began to take space and processes of spatial differentiation seriously. It seems, then, that the dividing line between geography and other social sciences has become less sharp and that time and history have become less prioritized over space and geography than before. Social dynamics are increasingly seen as not just a matter of time, but also as having an inherently spatial dimension. As a result, time and space, history and geography and have gained a more equal status than before, and the relation of geography to other social sciences has become less asymmetrical, which has led to more mutual interaction between the various disciplines and a greater plurality of approaches to globalization. Research on aspects of globalization has thus become, on the one hand, more fundamentally spatial, and, on the other, more multidisciplinary.

Research from various disciplinary backgrounds has meanwhile covered many aspects of current processes of globalization. It has, often following the lead of Held and his co-authors, looked at the increase, spread, and speed of global flows of all kinds of goods, services, capital, workers, tourists, refugees, messages, ideas, images, animal and plant species, and diseases. Studies have revealed that, although these flows in many cases run more or less parallel to each other, they tend to have their own geography and history. In addition, these flows display their own characteristic dynamic patterns. Some of them are flow specific, but there are also more general factors at work fostering global flows such as the overall growth of the world economy, declining costs of transport and communication, liberalization (including the opening up of China and the former Communist countries), increases in diaspora and expat populations in cities across the globe, and the growing importance of supranational organizations like the EU creating more level playing fields for trade.

Studies have also shown that the sharp acceleration in the size, coverage and speed of global flows since the 1980s has not led to the flat world predicted by Thomas Friedman (2005). Globalization, unmistakably, comprises homogenizing forces – witness the near-ubiquity of McDonald's, Apple or, more recently, the emergence of hipster spaces selling latte macchiato and vegan food in many cities. Spatial differentiation, however, can also be viewed as part and parcel of globalization. In more general terms, we can say that spatial differentiation is a precondition for the global space of flows. Geographers have stressed that the processes of spatial homogenization or deterritorialization are intimately intertwined with spatial differentiation or re-territorialization and uneven development at various scales (Smith, 1984; Brenner, 2004). New, complex and dynamic spatial divisions of labour have emerged thereby creating a mosaic of interlinked cities and regions with concentrations of different sets of place-bound resources which are essential for global economic activities (Sassen, 2005). Financial services in New York, London, Tokyo, and Singapore; film production in Los Angeles and Mumbai; and fashion design in Paris and Milan are just a few examples of such concentrations of specific resources and assets. The space of global flows is thus inevitably linked to these *spatial-temporal fixes* in the form of the built environment, infrastructure, social, political and economic institutions, and culture (Harvey, 1985). For Marxist geographers like Neil Smith (1984), David Harvey (2005) and Eric Sheppard (2016), the concept of uneven development has an even deeper meaning than growing differences between national states, regions, and places. They argue that uneven spatial development is a precondition for the reproduction of globalizing capitalism and therefore also inherently about power struggles.

Reterritorialization and uneven development and the ever-changing geography of flows mutually influence each other. Globalization obviously impacts on the development of places. Some places suffer from declining exports and disinvestments due to increasing global competition – the fate of former industrial powerhouses as Detroit, Sheffield, Roubaix, and Essen are testimony to this. Places, however, do not only suffer from globalization, they may also succeed to strengthen their position in the global economy. In doing so, they simultaneously actively 'produce' globalization by creating and accumulating the assets and the linkages that are necessary for global flows of various kinds (Sassen, 2005). Cities, particularly global cities like London, New York, Hong Kong and Shanghai are the places in and through which globalization is, to a large extent, being produced. They are the 'actors' in globalization. Or, more precisely, actors within these cities are actively producing globalization. Firms, for example, that increase their exports to far-away markets, create complex, spatially dispersed value chains, or that expand by taking over firms located in other countries. Universities and museums that set up branch institutes in the Middle East or the Far East. Low-cost airlines in conjunction with local citizens which offer their homes on Airbnb and thus enable ever more tourists to visit other countries. These are all actors producing globalization.

'Production' of globalization may also be aimed at creating a unique quality of place. Just as private firms try to escape from exacerbated global competition by innovations or creating oligopolies and monopolies (Schumpeter, 1934), we can observe in many cities combinations of collective and private actors that seek to carve out a vantage point by creating place-bound assets that cannot easily be imitated elsewhere, such as specific modes of governance (e.g. low taxes), the revitalization of historical districts in order to attract more tourists, the establishment of large cultural amenities often in combination with buildings designed by 'starchitects' such as the Guggenheim in Bilbao or the new design museum in Shenzhen (Kloosterman, 2014), or huge infrastructure investments to strengthen the competitive position of the trade and distribution sectors.

A large body of research has thus shown that space is a crucial dimension when looking at globalization. Space, moreover, has to be unpacked into different spatial scales as nation states, regions, cities and urban neighbourhoods may all be involved differently in processes of globalization. In addition, many studies have emphasized the dialectic character of the relationship between specific places and processes of globalization: places (re)produce globalization and globalization, in its turn, shapes places in many ways. These processes are anything but 'blind' or 'natural': they require actors, both private and collective, which in a myriad of ways interact with each other, are inserted in configurations of power, and are, typically, locally embedded.

GEOGRAPHIC LENSES: FOCUS AND WIDE-ANGLE

It is hard to find a place on earth nowadays which has not been touched by globalization one way or the other. Climate change is obviously indeed a truly global phenomenon, but as the contributions to this *Handbook* amply show, globalization in its many expressions affects numerous places in many, often unexpected ways. Its omnipresence as well as its (potential and actual) significant impact on the lives of people in all kinds of environments make globalization an important field of study for society at large, policymakers, and, evidently for academics. Its multifaceted character, far-flung and, typically less than transparent, linkages in combination with its complex causal dynamics, however, also make processes of globalization hard to grasp from a theoretical, conceptual and empirical point of view.

. With this *Handbook*, we aim to present a pluralistic overview of geographic approaches to globalization. The plurality is reflected in the range of topics, the different analytical and methodological viewpoints. To be able to disentangle the myriad of processes of globalization in a meaningful way, the chapters in this *Handbook* look at particular aspects while stressing the interconnections – the scalar dimensions – of these processes offering both more focused, in-depth analyses as well as more wide-angle, broader pictures of the complex relationships between space and globalization. As mentioned above, many of these contributions depart from the widely accepted definition of Held et al. (2000), thereby emphasizing the dimensions of extension, intensity, volatility, interpenetration of cultures, and the construction of infrastructures (both tangible and intangible) to enable global flows. Geographic approaches to globalization, then, are firmly embedded in the broader field of social sciences allowing both for cross-disciplinary debates and collaborations.

Although the chapters represent a wide variety of topics and approaches, we can observe a few common elements, some of them shared with accounts of processes of globalization in other social sciences, but also some which are characteristic for applying a geographic lens. First, many chapters explicitly point to a rather recent acceleration of globalization. Different starting points are suggested – from the late 1970s to the early 1990s – different labels are used – for example, space of flows, second globalization, hyper globalization, second unbundling. This acceleration is usually linked to a combination of technological developments in transport and communication and changes in the regulatory framework exemplified by the opening up of China, the fall of the Berlin Wall and the drive towards trade liberalization. The capitalist drive towards accumulation is usually seen as the ultimate driver of these processes, but geopolitical motives (e.g. China) are also acknowledged.

Second, this acceleration in globalization covers, in principle, all the dimensions of the definition by Held et al. (2000): cross-border flows display a greater spatial reach; cover more people, goods, services, capital, pollution, culture and ideas; have become more intense and more volatile; lead to more interpenetration and hybridization, and require an ever more elaborate infrastructure of, for instance, airports and digital communication networks.

Third, this new phase also entails the emergence of new global division of labour in the sense that what was once seen a 'world system' with selected Western countries as the core and most of Asia, Africa and Latin America as (semi-)periphery has now been replaced by a much more polycentric system with the rapid rise of notably China. The concomitant erosion of the hegemonic position of the United States will have consequences which transcend the economic and political realm by far as, for instance, already testified by the emergence of globally cultural industries outside the former core.

Fourth, though globalization is anything but a carefully planned comprehensive attempt to change the world, it is still a human force. Collective actors as TNCs, supranational institutions, national governments, NGOs and terrorist organizations, together with individual actors such as expats, migrants, refugees and tourists, may all be involved in accelerating processes of globalization. Complex configurations of different kinds of actors, prone to power struggles, interact in often unforeseen ways. Globalization is, hence, not a set of smooth processes, but subject to oblique or often open contestation. The outcomes of these man-made processes and complex struggles, therefore, can be rather different than intended by the actors involved. Emergent effects are, then, crucial elements of globalization. Given that globalization is man-made also implies that this, in principle, could be reversed. The period between the two world wars is a clear example of such a reversal after the globalization era of the end of the nineteenth century. The recent rise of both right- and left-wing populism might in a similar way trigger a race between nation states to shut their borders to global flows.

These key elements themselves are more or less shared by other social science accounts of globalization. Using a geographic lens, however, as the contributions to this *Handbook* amply reveal, stresses the salience of spatial scales. Processes of globalization are embedded and articulated at various scales in different ways. This holds for the key dimensions of extensity, intensity, volatility, hybridization, and the necessary infrastructure. Neighbourhoods, cities, regions, nation states and supranational regions may all be very differently affected by processes of globalization. These divergent spatial articulations are related to the inherent spatiality of a particular process – for example, the clustering of the film industry or the impact of climate change have rather different spatial footprints. But the spatial articulation is also related to the uniqueness of places: their path-dependent legacies of both in terms of the built environment and the less tangible sociocultural and institutional make-up at different scales (Kloosterman, 2010). A much more fine-grained map of globalization is thus offered with a more precise understanding of the emergence of socio-spatial cleavages from within cities to between countries, a wider range of actors from the local to the supranational, and, hopefully, providing a sounder base for policies aimed at a more equal and sustainable world.

MORE GEOGRAPHIES OF GLOBALIZATION

We can recognize many of the elements just mentioned in much more detail in the contributions to this *Handbook*. Each chapter is a stand-alone contribution opening up a particular perspective on, or field of research on, geographies of globalization. Many of them contain concrete cases to illuminate how globalization can proceed and how it may affect places and people. They all have a list of references which are relevant to their particular approach. Together, we claim, they form a rich panorama of geographies of globalization, touching on a wide range of topics and referring to many different strands of literature, theories and related methodologies. The *Handbook* thus reflects not just the diversity of the processes of globalization itself, but also shows the plurality of contemporary approaches to grasp these processes and especially to understand their spatial articulations.

There are six sections. The first two sections offer general accounts: the first features a set of introductory chapters broadly contextualizing globalization, the second introduction of globalization in specific geographical fields. The next two sections present more specific chapters. Although they all deal with the relations between cross-border flows and places, they are grouped according to their main entry points: flows in Part III, places in Part IV. The fifth section turns to issues of governance of globalization and its politics, including resistance to it. The final section addresses specific challenges globalization poses for fieldwork and for teaching.

For *Part I – Introductory Chapters* we have invited contributors to position processes of globalization in broader temporal and spatial contexts, addressing globalization as a specific period or as a specific scale and a globalized as specific space (i.e. the borderless world). Chapter 2 provides a long-term, historical overview of globalization by Peer Vries that contextualizes the present phase of globalization in a longer process of economic globalization since Columbus 'discovered' America. In Chapter 3 Kevin R. Cox to some extent starts where Peer Vries stops, namely at the onset of what he calls the 'second globalization' which began sometime during the downturn in the 1970s. He discusses the global scale and delves into the relationship between globalization and spatial scales. In Chapter 4 James W. Scott questions the idea that globalization brings about a borderless world. He shows how borders are an essential component of regulatory spaces. Paradoxically, with globalization the study of borders has become more important as flows cross ever more formal and informal borders, each with different meanings and different actors and Critical Border Studies have been blossoming.

For *Part II – Globalized Geographical Perspectives* we have commissioned chapters that elaborate distinct geographical approaches to globalization. Each chapter introduces a specific geographical approach or field and discusses how they address globalization and how globalization has impacted that sub-discipline. In Chapter 5 Kees Terlouw presents world-systems analysis, the hugely influential contribution to the debates on globalization made by Immanuel Wallerstein with his notion of the world economy as a singular economy combined with a plurality of states and the three layers structure necessary to stabilize power relations (core, semi-periphery and periphery). The other chapters introduced four subfields of (human) geography: international development studies, economic geography, cultural geography and political geography. In Chapter 6, Joyeeta Gupta considers globalization in the light of international development studies and deals with the relationship between globalization, environment and development, which are increasingly seen as being intertwined as in the Sustainable Development Goals (SDGs) promoted by the United Nations. In Chapter 7 Robert

C. Kloosterman and Pieter Terhorst highlight the distinct approach of economic geography to globalization that builds upon a layered ontology which acknowledges 'rich' places and 'rich' actors (in contradistinction to economic approaches to globalization). Next, Soyoon Choo and Elizabeth Currid-Halkett examine in Chapter 8 how cultural and media geographies deal with the globalization of culture and especially how the recent phase of globalization enabled by modern communication technology can be grasped. The approach of political geography is expounded in Chapter 9 by Sami Moisio, Juho Luukkonen and Andrew E.G. Jonas. They provide a thorough geographical analysis of how political entities as well as content of policies are related to processes of globalization.

The following two sections present a wide collection of thematic chapters. We have a cluster focusing on flows and another focusing on places. In *Part III – Geographies of Flows* each chapter deals with specific cross-border flows. The first set of chapters covers different aspects related to flows of people. In Chapter 10 Anne-Laure Amilhat Szary captures migration as a particular form of spatial mobility. In Chapter 11 Tatiana Fogelman looks at cross-border flows from a different angle by exploring how contemporary migration alters conceptions of citizenship in both sending and receiving countries. Brenda S.A. Yeoh, Shirlena Huang and Theodora Lam focus in Chapter 12 in another aspect of migration, that of transnational families and households. They show how the macro-level of migration flows is related to the micro-level of families in Asia. Dennis Arnold turns in Chapter 13 to labour geographies and shows how the evolving global division of labour is intertwined with global migration flows. Migrants tend to be overrepresented in the precariat and are thus part of the drive towards flexibilization. Finally people may also move across borders as tourists and Hong-gang Xu and Yue-fang Wu dissect how tourism is affected by globalization in Chapter 14.

Globalization is, evidently, also about other cross-border flows: the flows of information and knowledge, the flows of goods and services, the flows of capital. Languages are affected worldwide by migration, trade and other exchanges. In Chapter 15 Virginie Mamadouh examines how globalization has shaped new linguistic geographies and notably has boosted the role of English as *the* global language of communication (sometimes labeled Global English or Globish) but also the linguistic diversity in almost every locality while also challenging conventional ideas about languages and their relationship with territories. The most impressive flows boosted by globalization processes concern goods, investments, services, and bits of information. In Chapter 16 Jana M. Kleibert and Rory Horner provide a comprehensive framework for analyzing complex, spatially dispersed production chains, the global production networks. In Chapter 17 Elena dell'Agnese and Giacomo Pettenati focus on food. They describe the historical process which has led to a globalization of food stuffs resulting in a high rate of homogenization of the supply of food in many places in the world, but, at the same time, may contribute to localized *foodscapes* through hybridization with existing dishes and practices. In Chapter 18 David Bassens and Michiel Van Meeteren show how financial flows constitute a global financial system with its own particular geography and traceable spatial footprint while simultaneously fostering globalization by searching for profitable outlets across the world. In Chapter 19 Mark Rosenberg unpacks the complex relationship between health and globalization. This relationship is not just about germs crossing borders, but also about changes in the global environment (e.g. climate change), the movement of people and the development of health care as a global industry. In Chapter 20 Paul C. Adams highlights the role of digital media in enabling flows of information and how these new forms of communication are becoming integrated into daily practices, not just in developed but also in developing countries. In Chapter 21, Robert C. Kloosterman and Rosa Koetsenruijter turn to cultural industries, and link the emergence of a polycentric geography of cultural nodes of production, increasingly encompassing Asian countries, to an erosion of the Western and in particular American cultural hegemony. Finally, in Chapter 22, Martin Müller and Christopher Gaffney position mega-events such as the Olympic Games firmly within a globalization perspective by emphasizing the key dimensions of extensity, intensity, and velocity. This chapter functions as a transition to the next section which is more centred on places as entry points (here the cities hosting such mega-events).

Part IV – Geographies of Places looks first and foremost at how global flows affect concrete territorial entities. The selected chapters foreground places at different scales. For Chapter 23 Simon Dalby discusses Gaia (i.e. the Earth) as a place and uses the concept of the Anthropocene to describe an age in which human activities have such a profound impact on the globe (i.e. influencing climate change) that this warrants the declaring of a new geological epoch. In Chapter 24 Soul Park analyzes how a global marketplace for security services has emerged while nation states are still crucial actors in the globalized security environment. In Chapter 25 Franz Tödtling, Arne Isaksen and Michaela Trippl foreground subnational regions and economic clusters, and explain why the regional scale has become more important as globalization enhances the need for innovation in subnational economic clusters. As the capacities to innovate differ between regions, we can also observe divergent trajectories of regional development. This is also true for localities that are dealt with in the last two chapters of Part IV. Ben Derudder shows how a particular group of well-connected cities (the so-called world cities) are inserted in processes of globalization, both as drivers and as recipients (Chapter 26), while Markus Hesse and Evan McDonough centre their contribution on the impact of increasing flows of goods on port cities and their infrastructural facilities (Chapter 27).

Part V – Geographies of Governance aims at capturing how globalization is related to shifts in governance. The chapters in this section address specifically global governance challenges, such as human rights, macroregional integration, maritime trade and sea lanes, and global social justice. In Chapter 28 Barbara Oomen looks at the way in which the globalization of human rights has been conceptualized and how this has been institutionalized. In Chapter 29 Alun Jones turns to macroregional integration and examines a concrete example of supranational governance – the European Union – and its dual response to globalization, intensifying its internal market while protecting its external borders. In Chapter 30 Takashi Yamazaki explores how forms of supranational cooperation have been crucial to the governance and protection of maritime trade, a key component of both historical and contemporary globalization processes. His contribution foregrounds the case of the Indian Ocean as Japan's sea lane. Finally, in Chapter 31, Byron Miller examines how the uneven distribution of costs and benefits of globalization is contested and what alternatives are envisioned. Responses include left- and right-wing anti-globalization movements seeking re-nationalization, but also alter-globalization initiatives which seek global engagement and exchange on a basis that protects and advances values of social, economic and environmental justice.

The final section of the book takes a slightly different tack. It is about doing geographies of globalization, and its challenge for geographers (or spatially sensitive social researchers), as researchers and as teachers. Indeed *Part VI – Researching and Teaching Geographies of Globalization*, covers more practical questions: how can we actually explore and teach the multifaceted, multi-scalar, interrelated transformations we call globalization from a geographical perspective? In Chapter 32 Valentina Mazzucato and Lauren Wagner discuss a research design typically associated with the empirical research of global flows: the multi-sited fieldwork. They assess the potential and the difficulties of doing multi-sited research – examining 'global' relationships by actually doing empirical fieldwork in geographically distant spaces which are linked through followable connections. Chapter 33 is devoted to teaching: Matthew Sparke emphasizes the need to highlight the layered character, the man-made nature, and the spatiotemporal dimensions of actually existing globalizations, stressing the plurality of the phenomenon. This chapter can be read as a conclusion, since it presents the main insights from the field of geographies of globalization and presents them to students.

REFERENCES

Arendt, H. (1958), *The Human Condition*, Chicago: University of Chicago Press.
Baldwin, R. (2016), *The Great Convergence; Information Technology and the New Globalization*, Cambridge, MA/London: The Belknap Press of Harvard University Press.
Boyer, R. and J. Rogers Hollingsworth (1997), 'From national embeddedness to spatial and institutional nestedness', in: J. Rogers Hollingsworth and R. Boyer (eds) *Contemporary Capitalism. The Embeddedness of Institutions*, Cambridge: Cambridge University Press, 433–484.
Brenner, N. (2004), *New state spaces: urban governance and the rescaling of statehood*, Oxford: University Press.
Castells, M. (1996/2011), *The Rise of the Network Society*, New York: John Wiley & Sons.
Cosgrove, D. (1994), 'Contested global visions: one-world, whole-earth, and the Apollo space photographs', *Annals of the Association of American Geographers*, **84**(2), 270–294.
Cosgrove, D. (2001), *Apollo's Eye: A Cartographic Genealogy of the Earth in the Western Imagination*, Baltimore: Johns Hopkins University Press.
Crosby, A.W. (2003), *The Columbian Exchange: Biological and Cultural Consequences of 1492* (Vol. 2), Westport: Greenwood Publishing Group.
Crosby, A.W. (2015), *Ecological Imperialism*, Cambridge: Cambridge University Press.
Cuterela, S. (2012), 'Globalization: definition, processes and concepts', *Revista Română de Statistică – Supliment Trim IV*, 137–146.
Dicken, P. (1986), *Global Shift. Mapping the Changing Contours of the World Economy*, London: Paul Chapman.
Feder, B.J. (2006), 'Theodore Levitt, 81, who coined the term "globalization", is dead', *New York Times*, 6 July 2006.
Friedman, T.L. (2005), *The World is Flat; A Brief History of the Globalized World in the 21st Century*, London: Allen Lane.
Giddens, A. (1990), *The Consequences of Modernity*, Cambridge: Polity Press.
Grataloup, C. (2009), *L'invention des continents: Comment l'Europe a découpé le monde*, Paris: Larousse.
Grataloup, C. (2010), *Géohistoire de la mondialisation*, Paris: Armand Colin.
Grataloup, C. (2017), *Le monde dans nos tasses. Trois siècles de petits déjeuners*, Paris: Armand Colin.

Hall, P.G. and K. Pain (eds) (2006), *The Polycentric Metropolis: Learning from Mega-city Regions in Europe*, London: Routledge.

Harvey, D. (1985), 'The geopolitics of capitalism', in: D. Gregory and J. Urry (eds) *Social Relations and Spatial Structures*, Basingstoke: Macmillan, 128–163.

Harvey, D. (1990), 'Between space and time: reflections on the geographical imagination'. *Annals of the Association of American Geographers*, **80**(3), 418–434.

Harvey, D. (2005), *Spaces of Neoliberalization: Towards a Theory of Uneven Geographical Development*, Wiesbaden: Franz Steiner Verlag.

Heffernan, M. (1998), *The Meaning of Europe, Geography and Geopolitics*, London: Arnold.

Held, D., A. McGrew, D. Goldblatt and J. Perraton (2000), 'Global transformations: politics, economics and culture', in: C. Pierson and S. Tormey (eds) *Politics at the Edge*, London: Palgrave Macmillan, 14–28.

James, P. and M.B. Steger (2014), 'A genealogy of "globalization": the career of a concept', *Globalizations*, **11**(4), 417–434.

Jessop, B. (2002), *The Future of the Capitalist State*, Cambridge: Polity Press.

Jessop, B. (2009), 'Avoiding traps, rescaling states, governing Europe', in: R. Keil and R. Mahon (eds) *Leviathan Undone? Towards a Political Economy of Scale*, Vancouver, Toronto: UBC Press, 87–104.

Kloosterman, R.C. (2010), 'This is not America: embedding the cognitive-cultural urban economy', *Geografiska Annaler: Series B, Human Geography*, **92**(2), 131–143.

Kloosterman, R.C. (2014). 'Cultural amenities: large and small, mainstream and niche: a conceptual framework for cultural planning in an age of austerity', *European Planning Studies*, **22**(12), 2510–2525.

Levitt, T. (1983), 'The globalization of markets', *Harvard Business Review*, **61**(3): 92–102.

Lévy, J. (2008), *L'invention du monde: Une géographie de la mondialisation*, Paris: Presses de la fondation nationale des sciences politiques.

Mackinder, H.J. (1904), 'The geographical pivot of history', *Geographical Journal*, **23**(4): 421–437.

McLuhan, M. (1962), *The Gutenberg Galaxy: The Making of Typographic Man*, Toronto: Toronto University Press.

Massey, D. (2005), *For Space*, London, Thousand Oaks, CA and New Delhi: Sage.

Modelski, G. (1968), 'Communism and the globalization of politics', *International Studies Quarterly*, **12**(4): 380–393.

Murray, W.E. (2006), *Geographies of Globalization*, London: Routledge.

Perroux, F. (1964), 'L'économie planétaire', *Tiers-Monde*, **5**(20): 843–853.

Perroux, F. (1969), *L'économie du XXe siècle*. Paris: Presses universitaires de France.

Raworth, K. (2017), *Doughnut Economics; Seven Ways to Think Like a 21st-Century Economist*, London: Random House Books.

Sassen, S. (2001), *The Global City: New York, London, Tokyo*, New York, Princeton, NJ: Princeton University Press.

Sassen, S. (2005), 'The global city: introducing a concept', *The Brown Journal of World Affairs*, **11**(2), 27–43.

Schumpeter, J.A. (1934), *The Theory of Economic Development. An Inquiry into Profits, Capital, Credit, Interest, and the Business Cycle*, London, Oxford and New York: Oxford University Press.

Scott, A.J. (2000), 'Economic geography: the great half-century', *Cambridge Journal of Economics*, **24**(4), 483–504.

Scott, A.J. (ed.) (2002), *Global City-Regions: Trends, Theory, Policy*, Oxford: Oxford University Press.

Sheppard, E. (2016), *Limits to Globalization. Disruptive Geographies of Capitalist Development*, Oxford: Oxford University Press.

Sidaway, J., M. Bradshaw, P. Daniels, T. Hall and D. Shaw (2016), 'Introduction geography: finding your way in the world', in: P. Daniels, M. Bradshaw, D. Shaw, J. Sidaway and T. Hall (eds) *An Introduction to Human Geography, Issues for the 21st Century* (5th edn). Harlow: Prentice Hall (Pearson Education), 1–15.

Smith, N. (1984), *Uneven Development. Nature, Capita Land the Production of Space*, Oxford: Blackwell.

Soja, E.W. (1989), *Postmodern Geographies. The Reassertion of Space in Critical Social Theory*, London/New York: Verso.

Taylor, P.J. (1985), *Political Geography, World-economy, Nation-state and Locality*, Harlow: Longman.

Urry, J. (1985), 'Social relations, space and time', in: D. Gregory and J. Urry (eds) *Social Relations and Spatial Structures*, Basingstoke: Macmillan, 20–48.

Yeung, H.W.C. (2002), 'The limits to globalization theory: a geographic perspective on global economic change', *Economic Geography*, **78**(3), 285–305.

2.   A very brief history of economic globalization since Columbus

Peer Vries

INTRODUCTION

This chapter provides a historical overview and analysis of the process of economic globalization over the period from the Late Middle Ages till roughly the last quarter of the twentieth century when the global economy again went through fundamental changes for various reasons. I focus on *economic* globalization. The period of roughly the last half-century will be discussed extensively and repeatedly in other contributions in this volume. Economic globalization will be – admittedly quite loosely – defined as an increase in extensity, intensity and velocity of (mostly power-laden) intercontinental exchanges with a clear and direct economic impact (Held et al., 1999).

Any discussion of globalization in global history has to begin with pointing out the immense difficulty of any kind of intercontinental contact and exchange until quite recently, due to the state of the existing means of transport and communication. It was only in the nineteenth century that major breakthroughs occurred in this respect. The limits set by technology impacted differently on the exchange of commodities, to some extent depending on whether they were for ordinary subsistence or luxury, of people or of information, but from our contemporary perspective they were very strict. The limits of what was impossible and possible were to a large extent defined by geography (Braudel, 1981; Hugill, 1993). Where one lived to a large extent determined how one lived. Trade was often confined to importing what one simply could not do without or what one was wealthy enough to afford anyhow and to exporting surpluses.

As a rule, transport over water was faster, less troublesome and cheaper than over land. Considering the fact that it largely depended on wind and currents it was far from predictable. Of the populated regions in the world, Europe was best endowed with waterways. Almost without exception large cities worldwide had to be near a waterway for their provisioning. When transport over water was not available, land use often was in accordance with the model that the German economist Von Thünen (1783–1850) created in his *Der isolierter Staat in Beziehung auf Landwirtschaft und Nationaloe-konomie* (Von Thünen, 1826). Intercontinental overseas trading was almost entirely done on European ships. But even the total tonnage of the European fleet was low. It has been estimated at about 1 000 000 tonnes in 1600 and about 1 500 000 tonnes in 1670. At the very end of the eighteenth century it still was less than 4 000 000 tonnes. Ships were small: The average tonnage of ships going to Asia from Europe during the early modern era was 600 to 700 tonnes while oil tankers nowadays can measure as much as 500 000 tonnes. Additionally, transport on such ships was slow. A round trip from Spain to the Philippines could last some five years in the 1570s and during the eighteenth century, it took a ship of the Dutch East India Company on average 235 days to go from the Netherlands to Batavia and 225 days to Canton.

Intercontinental contacts were first and foremost a maritime and therefore coastal affair until the transport revolution of the nineteenth century. In that period the global economy primarily was a thin coastal crust surrounding a vast number of more or less self-enclosed economic regions that were not directly integrated into a world market. There was a big difference between port cities and coastal regions on the one hand and the countryside and hinterland on the other. It would be an exaggeration, though, to regard them as disconnected. The global economy was slowly developing as a network but it was a network with many holes. The size and importance of those holes were, in all likelihood, smallest in Europe. But even there, until far into the nineteenth century there was a clear difference between coastal, lowland regions and landlocked, mountainous regions. European grain markets until the end of the eighteenth century still consisted of two basic zones: a coastal zone and an inland zone (Studer, 2015). Inland zones only began to become integral parts of transcontinental networks from the late nineteenth century onwards with the impact of railroads, more and better roads and automobiles and canals (Bateman, 2012; Shiue and Keller, 2007; Li et al., 2013; Studer, 2015). But even in the economically most advanced industrializing regions of the world at that time in Europe, geography, for example the location of coal or iron ore, continued to have a huge impact (see for example Pollard, 1981).

A look at the map of Africa, the Americas and the enormous, expanding Russian Empire suffices to see that integrating them economically was all but impossible with the existing primitive means of transport. The exchange of goods, with other continents or just to other countries, was problematic not just because of physical circumstances. Intervention by rulers was the norm. Such exchange might be prohibited, restricted, promoted, taxed, or concentrated in specific places. Rulers might give licences or monopolies, intervene in exchange or let it free, close markets or, on the contrary, open or even literally conquer them. One of the main goals of this text is to show that there have always existed inseparable connections between economic globalization and political developments and decisions.

SOME COMMENTS ON PRE-COLUMBIAN GLOBALIZATION

With Columbus and Magellan, intercontinental exchange evidently entered a new phase. It is, however, widely debated how fundamental changes were. To a large extent, this is due to the fact that there is no consensus regarding which indicators should be used. Claims that globalization would be much older are therefore not lacking. Andre Gunder Frank and Barry Gills contend that the connected world in which we live stretches back at least 5 000 years. According to them, capital accumulation, centre–periphery relations, the alternation between hegemony and rivalry and economic cycles with alternating ascending and descending phases, are not fairly recent phenomena but go back thousands of years in world history (Gunder Frank and Gills, 1996). There has even been talk of a Palaeolithic world-system (see for example Chase-Dunn and Hall, 1997). According to John Hobson, globalization existed as early as the sixth century 'as significant flows of goods, resources, currencies, capital, institutions, ideas, technologies and peoples, flowed across regions to such an extent that they impacted upon, and led to the transformation of, societies across much of the globe' (Hobson, 2004: 34). Until the nineteenth century the centre of global trade never lay in the West. For Samuel Adshead a continuous world history began with the creation of the largest contiguous land empire in history, the Mongolian Empire, that existed during the thirteenth and fourteenth century (Adshead, 1993: 4–6). In another publication he suggests the world already had one history at the time the Tang dynasty ruled China, that is, 618–907 AD (Adshead, 2004). Peter Frankopan is very keen on making the claim, in his hugely popular book on the silk roads, that there already was a connected 'world' during what Westerners used to call the Middle Ages and that its centre was in Central Asia (Frankopan, 2015). None of these scholars is very specific and concrete when it comes to actually defining what 'globalization' means, apart from 'intense' and 'relevant' contacts.

Janet Abu-Lughod proceeds more systematically and assumes the existence of a network of globally interconnected trade already *before* the early modern age. That network, however, disintegrated in the crisis of the Late Middle Ages. She admits that compared to the contemporary epoch, thirteenth-century international trade and the production associated with it was neither large nor technologically advanced and that its exchanges were minuscule. But, they were different from anything the world had ever known before, more complex in organization, greater in volume and more sophisticated in execution (Abu-Lughod, 1989). Her world-system, in contrast to Wallerstein's, which we will discuss later on, actually was a polycentric trade circuit. It was not hierarchically structured and lacked a clear hegemon. It, moreover, did not include Western Africa and the Americas, and knew no overseas trans-Cape connection with Asia, all of which were of fundamental importance for Wallerstein's later emerging world-system. These 'omissions' certainly qualify any claim about the existence of globalization before the sixteenth century.

THE EARLY MODERN ERA, 1450s–1850s

One may wonder whether it makes sense to debate when globalization 'began' (see for example Bordo et al., 2003; Findlay and O'Rourke, 2007; Hopkins, 2002; Lang, 2006; Osterhammel and Petersson, 2009). There certainly was a pre-history, but many historians, probably the majority of them, are convinced that Columbus, with his discovery of the Americas, and Magellan, with the first circumnavigation of the world, opened a new chapter in the history of globalization.[1] David Held et al. (1999), to refer to non-historians, in their *Global transformations* discuss the impact of globalization in the early modern era based on its extensity, intensity and velocity (see also Gunn, 2003; Parker, 2010; Reinhard, 2016). They consider its extensity to have been high and its intensity and velocity both low, and conclude somewhat surprisingly that its impact was high, a claim that Adam Smith and Karl Marx would endorse. Adam Smith wrote: 'The

---

[1]   Magellan himself actually did not complete the circumnavigation. He died in 1521 on the Philippines.

discovery of America, and that of a passage to the East Indies by the Cape of Good Hope, are the two greatest and most important events recorded in the history of mankind' (Smith et al., 1981: 626). Whereas for Marx: 'World trade and the world market date from the sixteenth century, and from then on the modern history of capitalism starts to unfold' (Marx, 1976: 247). Some scholars are even more specific. Dennis Flynn and Arturo Giráldez claim that globalization began in 1571 when Manila became a Spanish stronghold connecting the Atlantic and the Pacific (Flynn and Giráldez, 2006). They focus strongly on global trade flows, in particular of bullion. Several renowned scholars before them had already used those flows and their impact on price levels as signs of globalization. In the work of the Chaunus much effort is put into trying to prove the existence of a Spanish Atlantic conjuncture that shows clear parallels to what was happening in the Manila galleon trade in the Pacific, the Brazilian Atlantic sugar trade and the Baltic trade of Amsterdam and Lübeck (Chaunu and Chaunu, 1955–1959, 8(2): 19). They suggest that between 1500 and 1650 the Carrera de las Indias, the sea routes over the Atlantic Ocean that connected the different parts of the Spanish Empire, established 'the first outline, however rough, of a world-economy' that set 'a world rhythm' (Chaunu and Chaunu, 1955–1959, 8(1): 8, 28). Fernand Braudel and several others agree and cautiously hint at global rhythms, worldwide ups and downs in production, trade flows and prices in the early modern era (Braudel, 1981). For the eighteenth century, Fischer is less reticent and writes about its 'price-revolution' as 'truly a world event' (Fischer, 1996: 121). What is causing these global rhythms and waves – if indeed they are global – is hotly debated. Bullion production and bullion flows will certainly have played a role. Many scholars also refer to demographic developments. Fernand Braudel, for example, postulates a general worldwide increase in population over the period 1300–1800, which he suggests might in turn be caused by global climate change (Braudel, 1981: 31–51; Parker, 2013).

Several scholars have found 'Eurasian parallels' in the early modern world that encompass more than just economic developments. Jack Goldstone refers to what he regards as periodic waves of state breakdown in early modern Eurasia that he considers as part and parcel of one single basic process (Goldstone, 1991). Joseph Fletcher points at parallels in population growth, the growth of 'regional' cities and towns, the rise of urban commercial classes, religious revival and missionary movements, rural unrest and the decline of nomadism (Fletcher, 1995). Victor Lieberman focusses on territorial consolidation, administrative centralization, cultural and ethnic integration and commercial intensification (Lieberman, 2003, 2009; see also Richards, 1997). In particular Fletcher and Lieberman are much better in describing their parallels than in explaining them.

THE EARLY MODERN ERA, 1450s–1850s: THE IDEAS OF IMMANUEL WALLERSTEIN AND ANDRE GUNDER FRANK

Immanuel Wallerstein's work has started major, also more theoretical, debates on economic globalization. According to him a European, capitalist world-economy 'the modern world-system', came into existence during the late fifteenth and early sixteenth century (Wallerstein, 1974). This world-economy was distinctly modern and capitalist, which means it thrived on ceaseless accumulation of capital. It is a 'world'-system, not because it encompasses the whole world but because it is larger than any juridical defined political unit, and a 'system' because it is largely self-contained and the dynamics of its development are largely internal. He calls it a 'world-*economy*' because the basic linkage between the parts of the system is economic. This world-system is characterized by an extensive functional (occupational) and geographical division of labour with a clear hierarchy.

Wallerstein distinguishes three constituent parts. First there is the core, the part of the system where free, often skilled labour produces goods with a high added value in strong states. Then there is the 'opposite', the periphery, which is that part of the system where production is primarily of lower-ranking goods that have less added value but are essential for daily use in the core and that are produced predominantly by unfree, unskilled, less well remunerated labour. The polities in the periphery tend to be weak and have a strong export orientation. Then finally there is a third category, the semi-periphery, that part of the system that is in between the core and the periphery on a series of dimensions.

For centuries, port cities and their mercantile and entrepreneurial elites played a pivotal role in the economic and, very strikingly, the political life of Europe's core regions. They continued to do so to a lesser extent when the territorial, national state became more important. Those elites secured well-protected property rights and political leverage in their 'city-republics' that long held a fairly independent political status. Such cities and such independent economic elites were unknown not only in peripheral regions of the European world-system but also in the big empires of, for example, the Ottomans, the Mughals in India or the Ming and Qing in China when they were still outside that system. The protection and political status of citizens as privileged inhabitants of autonomous cities, that in core-regions was later transferred to all citizens of core states, in combination with the fact that those cities were densely populated centres of exchange, communication and migration, provided favourable conditions for innovation (Pearson, 1991). The central city of the core provided the ideal place to find what one was looking for. What is striking is that the central position in the European core region was not fixed but held by different cities in succession – in the European world-economy of the early modern era Venice, Antwerp, Genoa, Amsterdam and then London, and that those cities all tended to become primarily centres that focused on services (Braudel, 1992; Kindleberger, 1996).

The European world-economy was an economic system that, next to its core in Western Europe, in the sixteenth and seventeenth century included parts of Central and Eastern Europe, parts of the Americas and the West Indies, parts of Africa, and some tiny footholds in Asia. This last continent, by far the biggest economy in the early modern world, was to a large extent external to the European world-system. That means, still according to Wallerstein, that its trade with Europe was based primarily on the exchange of preciosities, which did not have a big impact on existing modes of production. Asia's primary production certainly was not (yet) an integral part of a Europe-centred European division of labour and neither did the Europeans dominate Asia. In as far as they had military superiority it was only at sea. It was only in the eighteenth and even nineteenth centuries that India, the Ottoman Empire, China and Japan were opened to the West, on conditions stipulated by the West.

Western Europe tended to extend its relations of unequal exchange, if not downright exploitation, with peripheries outside, and not to forget inside Europe. Eventually, it would become the core region of the entire world. In this process of extending itself, it was driven by an internal dynamism and competition, due to the fact that Europe consisted of a uniquely fragmented state-system (Arrighi, 2007). According to Wallerstein and his followers, the division of labour in the modern world-system to all intents and purposes was advantageous for the core regions that developed at the cost of their (semi-)periphery that they underdeveloped. This view has been criticized on several grounds. Patrick O'Brien has stated that the trade between core and periphery simply was not big enough at the time to have the enormous global historical impact that Wallerstein attributes to it (O'Brien, 1982). Wallerstein's approach that sees the roots of the Great Divergence already in the sixteenth century is emphatically rejected by Andre Gunder Frank (1998 – i.e. in roughly the last decade of his career; before that, he was one of the main proponents of the so-called dependency theory that shows strong resemblances with Wallerstein's approach). According to Frank, as indicated before, a global economy encompassing the entire globe and functioning as a fully integrated, autonomous system, with a logic of its own, predated the early modern era. He analyses the workings of this economy during the early modern era focusing on trade, in particular the flows of precious metals (Gunder Frank, 1998). Looking at those flows, Frank concludes that at the time Asia, and more in particular China, was the global economy's centre, while it functioned as the big depository of precious metals. China's massive imports of silver, in his view, showed that its balance of trade was permanently favourable. He takes the fact that – at least in his view – the incoming silver did not cause any inflation, as proof that China's economy was dynamic and productive. The role of Europe in the global economy in his view, hence, was marginal. The European and even Atlantic economies were no more than backwaters in the world economy of the early modern era that already encompassed the entire globe and had Asia, in particular China, as its centre and not Europe.

Asia was evidently the biggest economy in the world in the early modern era, containing highly developed regions producing textiles, porcelain or lacquer ware which had no equal in the world. Frank's claim that Asia's advanced regions were as rich as or even richer than their (North-Western) European counterparts has now been refuted although recent research has also shown that most of Europe was not wealthier than Asia's most wealthy regions. Frank in particular emphasizes that Asia was the big global silver sink. Is that indeed a sign of major economic strength? The amount of goods exchanged with other continents, predominantly Europe, was so small that it seems farfetched to claim their impact was of fundamental importance for the big economies involved in it. The per capita amount of bullion available to Asians, most clearly in the case of China, continued to be smaller than in Europe. In China its value was tiny in terms of GDP. Besides, Europeans voluntarily traded with Asia. They bought goods there that they did not have at home or that were relatively cheap and re-sold them at far higher prices. That was simply good business. Why would it indicate weakness? It might have been even better business if the Chinese would have been willing to accept something else than silver. But still, the bulk of the value added accrued to Europeans who also profited from the fact that silver had a much higher value compared to gold in China than in Europe. They did so by using silver to buy gold there and then sell it for a much higher silver price back in Europe. The claim that China must have had a strong economy because it imported huge amounts of silver is not convincing. To call Asia the centre of the global economy at the time is also not very convincing considering the fact that Asians in intercontinental exchanges as a rule passively stayed at home and let the Europeans add most of the value to their exports. Europeans, primarily Britons, sold Chinese tea to Europeans at a manifold of the original price in China. Europeans sold Indian textiles in Europe and Africa not Indians. Indian traders did not directly deal with these export markets. The Japanese traded far less than they could have done. The Ottomans voluntarily left it to Europeans or other 'foreigners' to handle much of their exports and imports. It is striking that already before industrialization took off in Europe, Asia's exports had become more peripheral in the sense that manufactured products had become less important. The Asian economy in any case does not seem to have profited a lot from its (assumed) centrality. The silver imports into China – in combination with the introduction of maize, sweet potatoes and peanuts from the Americas – stimulated the economy, led to an increase in population, declining wages and rising prices of resources. Over time this caused major economic problems. Europe could easily have done without China's tea, porcelain or silk. When China no longer received a massive inflow of silver via European traders, that, for whatever exact reasons, had a clearly negative effect on its economy.

THE EARLY MODERN ERA, 1450s–1850s: THE COLUMBIAN EXCHANGE, SLAVES AND MIGRANTS

Not all scholars agree that globalization began with Columbus, Magellan or even earlier. Kevin O'Rourke and Jeffrey Williamson are the most explicit and most cited exponents of a fundamentally different view. What went on during the early modern era undoubtedly may have had a major impact but it, in their view, should not be called globalization. For them the only good indicator of globalization would be the existence of an intercontinental convergence of prices as a consequence of a global integration of markets. They claim these only emerged a couple of decades into the nineteenth century and then primarily in the Atlantic (O'Rourke and Williamson, 2002; see also De Vries, 2010). They, however, use a very limited economic interpretation of globalization and look at economic development as a fully separate sphere with its own logic (Flynn and Giráldez, 2004; O'Rourke and Williamson, 2004; see also Bayly, 2002). Most historians, even amongst those who would agree that a global integrated economy with clearly converging prices only emerged in the nineteenth century, want to study economic globalization in a broader context and from a broader perspective. They point out that in all probability the so-called Columbian Exchange, the exchange of crops, livestock, and diseases between the Old World and the New World, was the most important global phenomenon of the early modern world in terms of economic impact. The introduction of Old World diseases had a disastrous impact on the indigenous population of the Americas, whereas, for example, American crops became very important as foodstuffs for Europeans (the potato and, to a lesser extent, maize) and Chinese (the sweet potato, maize and peanuts) (see Crosby, 1972, 1986; Mann, 2011; Nunn and Qian, 2010).

There is also the intercontinental trade in slaves. The most notorious case is the transatlantic slave trade. Between 1460 and 1870 between eleven and twelve million enslaved Africans arrived in the Americas. We are dealing with a real trade here and real markets where prices were not, as is often suggested, one-sidedly set by buyers, but the result of supply and demand. Africans were also sold as slaves in the Trans-Saharan slave trade, which is less well known and less well studied but in which during roughly the same period of time at the very least an estimated 2.6 million enslaved Africans were traded. Much higher estimates exist (Vries, 2013). In total some 600 000 Muslims were brought to Italy as slaves and some 375 000 Africans to the Iberian Peninsula over the entire early modern period. Europeans could also be enslaved. Some three million East Europeans were sold on slave markets in the Middle East. North African Muslims in the Mediterranean, between 1500 and 1800, enslaved an estimated 1 to 1.25 million Christians in raids (Lucassen, 2016; Davis, 2004).

There were also other forms of transcontinental migration. By far the majority of transcontinental migrants in the early modern era were Europeans. Their total number was surprisingly low. During the period between 1492 and the beginning of the nineteenth century an estimated three million Europeans went to the Americas and some two million to Asia. The number of Europeans going to Africa during that period has to be counted in the tens of thousands. The geography of that continent was far too inhospitable for them at the time. There were also significant migration flows elsewhere. At the end of the eighteenth century some 900 000 Russians lived in Siberia. Although the numbers of migrants were quite small compared to the global population, regional and local impacts on labour markets could be significant. What is striking, is that in this intercontinental migration too, unfree labour (convicts, exiles, people pressed into army or navy, but also many indentured labourers) played such a huge role.

THE EARLY MODERN ERA, 1450s–1850s: THE POLITICAL ECONOMY OF (NON-)GLOBALIZATION

Early modern economic globalization – just like all processes of globalization – was a process driven by agency and embedded in a much broader context that one easily ignores when focusing only on market forces. Globalization tends to be associated with the undermining of state power and over time it indeed might have had this effect. That, however, should not obliterate the fact that from its very beginning it has also been driven strongly by states, most clearly in their efforts to build formal or informal empires. For most of human history major polities were organized as empires (Burbank and Cooper, 2010; Darwin, 2007). All empires are, amongst other things, also mechanisms for economic redistribution. Up until the emergence of the modern European world-system, this redistribution between imperial cores and peripheries occurred via tribute including taxes or monopolistic advantages. The early modern world knew several huge empires that were quite successful in terms of extending their territory and their range of tributaries. Most conspicuous, however, at the time was

West European overseas expansion. In particular in the case of Portugal and Spain that certainly had tributary aspects but was different because Western Europeans built their empires overseas and because, in particular in the case of the Dutch, French and English, it was driven by a capitalist logic as described by Wallerstein. In their empires a large part of redistribution occurred via trade and interference in production. In the nineteenth and early twentieth century this 'modern' empire building was successfully continued and also taken up by new imperialist contenders like Germany, Japan, Italy and the United States (Abernethy, 2000). More generally, empire building always led to the construction of larger economic zones thereby boosting globalization and having redistributive economic effects.

Politically as well as economically, Western imperial expansion would have been unthinkable without initiating, coordinating or at least supporting states. In the so-called chartered companies, the most famous of which are the English and the Dutch East India Companies, the public and the private all but completely merged. In this uniquely European construct, private firms were endowed by their state with elements of sovereignty in their territories. They for example had their own armies and navies, could wage war and conduct treaties (Stern, 2011). They, strikingly enough, increasingly turned into territorial powers that ruled over substantial territories. Wealthy merchants certainly were not a uniquely European phenomenon. The fact that in certain Western states they could have institutionalized political leverage, however, was quite exceptional.

In Western European state- and (overseas) empire-building, foreign trade played a far more important role than in the large land-based empires. In Europe, in general, rulers intervened frequently to strengthen the economy on behalf of the state. In the highly competitive European states-system rulers needed ever-increasing amounts of money to pay for their armed forces. Their policies, actually a fairly improvised set of measures, have often been referred to as mercantilism. They included measures to promote exports and reduce imports and a focus on the production, and then export of commodities with a high added value. Policies stimulating economic activities considered essential to power were also popular as were efforts to create integrated and free domestic markets. In foreign trade, in contrast, monopolies were long preferred to avoid 'wasteful' competition amongst citizens, keep out foreigners and ease tax collection. Mercantilist states as a rule were keen on disciplining the labour force in their realm (Rössner, 2016; Chang, 2002; Reinert, 2007).

In China under the Ming (1368–1644) but certainly under the Qing (1644/1683–1911) dynasty, rulers considered it their main goal to make their huge agrarian population live in security and wealth. Their policy has been described as agrarian paternalism. Merchants, even more so when they were foreigners, were never fully trusted. Rulers were convinced that they must never let them become socially powerful. When they thought the merchants' behaviour would endanger social harmony they intervened in trade, which they normally did not. They, on several occasions, installed a policy to close the country and from 1750 onwards tried to concentrate all contacts with Europeans in Canton. From the 1630s onwards, Japan increasingly tried to control foreign trade and to protect the country against unwarranted foreign influences or goods. Although the country was never entirely closed off, it definitely became much less open than Western European countries. Here too, contact with foreigners (the

Dutch, Chinese, and Koreans) became concentrated in specific ports. Many of the fiefs of which the country consisted, behaved like semi-independent states and pursued mercantilist policies. Korea, from roughly the 1640s to the end of the nineteenth century, was almost entirely closed off to foreigners and foreign trade, apart from very restricted contacts with Japan and China. In the Mughal Empire in India, the central government almost exclusively confined itself to collecting taxes. The Mughals in that sense did not really have an economic policy. The economic policies of the rulers of the Ottoman Empire had three fundamental traits. The first one is 'provisionism', the principle that the government had to see to it that, especially the army and the capital, but also the population at large, could be sure they would not be short of their necessary means. The second one is 'traditionalism'. Changes in the social status quo were opposed. Competition was not expected to create victims and should not become abnormal. Merchants active in international trade could do as they pleased but only as long as their behaviour did not cause social upheaval. The third principle, 'fiscalism', implied that economic activities were primarily valued according to their fiscal benefits and disadvantages. Rulers even at the heydays of the empire often found it unproblematic to give foreigners all sorts of trade privileges in order that they take care of imports. For traditional empires that unlike European states were not part of a system of competing polities foreign trade normally was not of vital interest (Vries, 2002).

In the West the inseparable connection between power, politics and the economy showed most glaringly in the prominent role of violence in what we would now consider economic relationships. To quote Ronald Findlay and Kevin O'Rourke: 'no history of international trade can ignore the causes or the implications of military exploits'. According to them the greatest expansions of world trade, have tended 'to come ... from the barrel of a Maxim gun, the edge of a scimitar, or the ferocity of nomadic horsemen' (Findlay and O'Rourke, 2007: xix, xviii). At sea, in many regions the difference between trade, privateering or even piracy was anything but clear-cut, certainly in the early modern era. For the Dutch East India Company military expenditure in the East during the period 1613–1792 was about 30 per cent of all its overseas investments. Without protection trade often was simply impossible, especially long-distance trade. In this context it is striking that in Europe wars increasingly also could be trade wars.

THE EARLY MODERN ERA, 1450s–1850s: INTERCONTINENTAL COMMERCIAL EXCHANGE

After Columbus, a slow but clear relative decline of overland routes set in. Intercontinental trade increasingly became overseas trade. In the seventeenth century the importance, for example, of the caravan routes of Africa and the Silk Roads in Asia had clearly diminished. Geography was crucial as, typically, apart from bullion, goods were traded because the supplying region had them and the demanding region lacked them, and not because of differences in prices. In early modern global trade, geography played its role also in another respect. Atlantic trade had a different logic for Europeans – the only ones who really actively traded globally – than trans-Cape trade with Asia because voyages to the America's took far less time and required much less capital investment in terms of ships, manpower and freights. In their contacts with Africa the climate and diseases of large parts of this continent were reason for them to confine themselves to visiting the coasts.

When we look at what is actually exchanged, the role of silver and gold is striking. They played a fundamental role in global trade during the early modern era. It is hard to imagine how that trade could have existed without them. The bulk of the silver and gold traded in the world was mined in Central and Latin America, then transported to Spain and Portugal, from where it spread over Europe, that in turn exported large amounts of it (estimates vary from one-third to one-half) to Asia via Russia, the Levant or along the Cape. American silver was also sent over the Pacific on board of the so-called Manila Galleons that sailed from Acapulco to Manila, from where most of it went to China and India (Barrett, 1990). It was almost impossible for Westerners to buy anything in Asia except when they paid with bullion. Some 85 per cent of European exports to Asia over the entire early modern period consisted of it. Demand for bullion in Asia was enormous. In China, home of between roughly one-quarter and at the end of the early modern period one-third of the entire global population, demand was huge anyhow and growing because of the increasing monetization of the economy and the government's decision to collect more taxes in silver. The country, up until the end of the seventeenth century, also imported silver from Japan, but that nevertheless continued to be relatively scarce. The fact that European traders had so much bullion opened many doors for them even when Asian rulers preferred to keep them out. Bullion travelled over the globe for various reasons: differences in general price level due to differences in production and productivity between regions, differences in the overall availability of bullion or differences in the relative values of bullion. In China for example, European traders could buy much more gold with their silver than at home. That of course was a good reason to take silver to China, exchange it for gold and then take that gold back home (Flynn and Giráldez, 2004, 2006; Giráldez, 2015).

Intercontinental trade during the early modern era still was not very substantial. According to Pierre and Huguette Chaunu the value of goods supplied by international trade in the sixteenth and seventeenth century in Europe amounted to no more than 2 per cent of its agricultural production (Chaunu et al., 1955–1959). For 1820, commodity exports are estimated to still have been less than 1 per cent of global GDP (Findlay and O'Rourke, 2007). Accordingly, the large civilizations to a very large extent still were worlds to themselves. Paul Bairoch estimates that even in Europe at the end of the eighteenth century only some 4 per cent of total production crossed any border (Bairoch, 1976: 79). For the big Asian Empires, transcontinental trade will certainly have been an even more modest percentage of GDP.

Percentages of GDP are not necessarily the best indicators of the importance of intercontinental exchange. Some differentiation when it comes to kind of products time and place are in order. Halfway through the eighteenth century, according to Niels Steensgaard, 'As articles of consumption, distributed among perhaps 120 million Europeans, the quantities are ridiculous. There may have been three pounds of sugar for every European, half a pound of tobacco, an ounce of tea, the same quantity of pepper, and perhaps cotton material for a handkerchief' (Steensgaard, 1990: 13). Berkeley historian Jan de Vries claims that only 'a slender thread' connected the economies of early modern Europe and Asia. In the 1770s the Cape-route trade only sufficed to provide Western and Central Europe with about one pound of Asian goods per capita. At their arrival in Europe they would cost about six grams of silver, less than the daily wage of a manual labourer in Holland or England, two or three daily wages for manual workers in Central, Eastern or Southern Europe. For Asia, with a population that was a manifold of that of Europe at the time, the impact of such exports on GDP of course cannot have been big. He points at the fact for Western Europe, trans-Atlantic trade – again in the 1770s – in terms of money was three times as important as trade with Asia (De Vries, 2003: 91–93, 96). But even for the Atlantic world early modern globalization has been described as a myth (Emmer, 2003; Eltis, 1999).

In studies on the history of early modern Africa and Latin America we see contrasting approaches (see for Africa, Northrup, 2002; Pearson, 1998; Thornton, 1992; see for Latin America, Grafe and Irigoin, 2012; Benjamin, 2009; Canny and Morgan, 2011). On the one hand there are studies in which the impact of the Europeans and of the modern world-system are toned down and in which the inhabitants of these regions are credited with much more agency, and, on the other hand, there is a continuing line of research that emphasizes the importance of American bullion, the Atlantic slave trade and slave labour for specific developments in other parts of the world, in particular North-western Europe and China (Vries, 2013, 2015; Beckert, 2014; Blackburn, 1998; Inikori, 2002). One, moreover, has to realize that even though, overall, intercontinental exchange increased undeniably over the early modern period, it was a process with ups and downs. The so-called general crisis of the mid-seventeenth century, for example, certainly impacted negatively on it. European interference in that period led to less trade in parts of Southeast Asia at a time when Japan began to close doors on foreigners. Considering the early modern era in its entirety for Africa as well as Latin America, intercontinental trade as compared to the GDP of the countries involved became less important.

THE EARLY MODERN ERA, 1450s–1850s: SPECIFIC EFFECTS IN EUROPE

It would, however, be incorrect to conclude that intercontinental trade was irrelevant, in particular for Western Europe. There it happened to be the economic sector with the highest levels of growth. Moreover, it showed an extraordinary level of dynamism and innovation in that region. Its requirements in terms of logistics, organization, technology, information and financing were entirely different in scale and scope from what was needed for trade at a local, regional or national level. Unsurprisingly the sector was home to many innovations in those fields. On top of that, intercontinental trade was closely entangled with processes of European state building. Its fiscal-military states supported it and profited from it. At the interface of private and public finance, where big merchants played a central role, many financial innovations took place. Governments tended to heavily tax imported exotic goods. In several parts of Europe and North America the wish to buy them triggered a so-called 'industrious revolution' (De Vries, 2008). Changes in consumption, stimulated by the availability of new exotic products, could become so substantial that scholars talk about a consumer revolution

(Trentmann, 2012). That revolution triggered efforts, many of them quite successful, at import substitution (see for example Berg, 2004).

Outside Europe such effects were much less pronounced. Africa's import of textiles, metals, alcohol, and guns as far as we know did not have similar major effects on the mode of consumption or production, although certain states developed that specialized in catching and exporting slaves. In Asia the volume of exports as compared to total production was small. The bulk of imports consisted of bullion, which had no effects similar to those of imports in Europe. That bullion import led to an increase in the consumption of domestically produced goods or, in the case of nineteenth-century China, boosted the import of opium. Even in the nineteenth century, Asian markets were quite resistant to penetration by products from the West. The situation in the Americas was different as inhabitants of colonies were not free to decide with what countries to trade and what to import or export.

The bulk of what Western traders introduced into intercontinental trade actually was not produced by them, unless one wants to call the bullion from Latin America their produce. In that sense their role in global trade was primarily that of middlemen. That, however, does not diminish its importance. On the contrary. In the world of mercantile capitalism before industrialization – and to a large extent also in the industrial and post-industrial world – most value was not added in production but in service activities like transport, financing, and sales, that is, in the activities of the merchant. In this world the 'real' capitalists were the people at the pinnacle of trading activity. They were well-informed and in the position to select what sectors and places they wanted to be active in. They earned much of their profits thanks to the lack of transparency of the markets in which they operated and that they to a large extent manipulated and tried to monopolize. Support from the state was essential in those efforts. Profits in long-distance trade tended to be superior to those from big investments in industry or agriculture (Braudel, 1981). Marx already knew that in mercantile capitalism 'the lion's share' falls to 'the middleman'. It is 'financiers, stock-exchange speculators, merchants and shopkeepers (who) skim the cream …' (Marx, 1976: 907).

That being the case, it is of course fundamental that Western traders already before industrialization had begun to dominate intercontinental overseas trade. The Atlantic Ocean had become a European sea; the Pacific was becoming one. In this respect it is striking that Europeans not only traded so much with but also increasingly for the world. In the eighteenth century, for example, the British and the Dutch increasingly acted as trade intermediaries between Asian countries. Where possible – which as a rule meant in their peripheries – European traders also interfered in the way certain commodities were produced in order to increase their profits. They often transferred products or labour to other regions and introduced forms of forced labour, as shows most clearly in the many plantations they created. The importance of information in long-distance trade and colonization was paramount. The Europeans who explored, charted and subdued so much of the entire globe, helped by their governments that also were eager information gatherers, had a major advantage over others in this respect (Black, 2014; Marks, 2016). The bulk of capitalist capital accumulation from the very beginning primarily and predominantly took place in the service sector, not in the sphere of commodity production. In that respect post-industrial, service economies are less of a novelty than many people think, although of course the scale of the service sector and the amounts of money circulating in it have become unprecedented (see Cain and Hopkins, 2016; Lee, 1986).

THE LONG NINETEENTH CENTURY: GLOBALIZATION'S FIRST GOLDEN AGE

A new phase began in the nineteenth century. In contrast to the period before, there now are clear indicators of market integration and price convergence in the economies bordering the Atlantic and parts of Asia from roughly the 1820s onwards. Intercontinental contacts became more extensive, more intense, faster, and their impact increased. The (intercontinental) exchange of flora was now undertaken more systematically. Tea, for example, was transplanted from China to India and Ceylon, rubber from Brazil to several regions in Asia and Africa, and the Cinchona tree, producing quinine, from Latin America to for example India and Java.

Intercontinental migratory flows after 1800 were of an entirely new order of magnitude. Intercontinental migration from Europe in particular to the Americas, and other white settler colonies reached an unprecedented level. Many Asians too went to the Americas to work. The amount of trade in the world increased very fast, even faster than the amount of production, which in itself was increasing very fast (see for example Maddison, 2007: 81). Not only did the amount of trade increase; for the first time in global history bulk goods were massively exchanged between continents, for example, raw materials like cotton, wool, wood, coal or minerals, and foodstuffs like grain or meat. Now the so-called settler colonies, but also parts of Central and Eastern Europe, started to massively provide the European core of the global economy with necessities (Lloyd et al., 2013).

The increase of incomes in the core regions over the nineteenth century will certainly have played some role in the increase of global economic interaction, but the main reason undoubtedly resided in the revolutionary changes in transport (for example railways and steamships) and communication (for example telegraph and telephone), including the opening of the Suez Canal and the Panama Canal that made long-distance exchange so much cheaper, quicker, more predictable and comfortable. Dependency on natural organic materials and thus on geography and location in principle sharply declined with the transition to a mineral-based fossil-fuel economy and the rise of chemical industries that produced dyes and medicines, artificial fertilizer, synthetic fibres, and, later on, plastics.

Again, the importance of politics should not be underestimated. The rise of free trade that began to acquire momentum in Europe in the main trading nations from the 1840s onwards, also played a major role in further integrating the world, the more so as it was combined with the disappearance of many internal barriers in post-Napoleonic European states, the creation of some big new states like Germany and Italy, and the fast growth of population in regions of Western settlement. The worldwide trend towards free trade that ensued was not just the consequence of voluntary arrangements but also of the forceful opening of new markets that has been aptly described as 'imperialism of free trade' (Gallagher and Robinson, 1953). British historian Niall Ferguson even claimed: 'no organization in history has done more to promote the free movement of goods, capital and labour than the British Empire in the nineteenth and early twentieth century' and introduced the term 'Anglobalization' (Ferguson, 2003: xxi, xxiii). India became increasingly integrated in the British Empire and open to the world economy. The Ottoman Empire, China and Japan, to just give the most important examples, all had to sign unequal treaties with Western powers in which they agreed to impose very low tariffs and give their trade partners extraterritoriality and the most favoured clause.

What then emerged has been described as 'the Great Specialization', in which a part of the world produced manufactured goods while the rest produced primary commodities (Robertson, 1938). In this context it is seductively easy to talk about the West versus the rest or, as increasingly became used, the North versus the South, but it would also be misleading as large parts of Europe actually functioned as peripheries in the global economy. The situation was, hence, more complex (Bairoch, 1993, 1997). The Great Specialization overall implied a de-industrialization of the now underdeveloped countries, that is a strong reduction of their manufacturing sector in as far as it could not compete with modern industries abroad. In countries like China and India manufacturing exports admittedly had already started to decrease somewhat before industrialization really took off in the West. With industrialization Europeans began to primarily trade goods they had produced themselves and no longer acted just as middlemen. They now had a huge competitive advantage in manufacturing. As a sequel to the comments on merchant capitalism in a previous paragraph, it is relevant to point out, however, that the industrializing Western world also further extended its dominance in the global service sector. It almost completely dominated intercontinental transports, communication, banking, insurance and, what now for the first time in global history became an important phenomenon, foreign direct investment. Great Britain alone, for example, accounted for about half of the world's fleet and about half of its foreign investment just before the First World War. The mechanism of the gold standard smoothly regulated the existence of fixed exchange rates.

(Neo)classical economics assumes that in a world of ever-closer connections economies would converge. This assumption is based on the expectation that capital would move to poorer countries where it is scarcer and thus would yield more, especially in a low-wage environment, whereas labour would leave countries where it is in ample supply and thus lowly paid, for better wages in a high-wage environment. But actually the nineteenth century was the age of what Pomeranz has famously called 'the Great Divergence', a term indicating the emergence of a huge global gap in wealth, income, growth and development (Pomeranz, 2000; see also Vries, 2013).

What caused this divergence is one of the major questions in global economic history (Vries, 2013, 2016). Most scholars agree that what made the main difference is the fact that the Western world with industrialization started to experience modern economic growth whereas the rest of the world – until Japan's take off at the very end of the nineteenth century – did not. What role might globalization have played in the emergence of this gap?

Global tides were already changing before the West had acquired a distinct economic primacy and there were already signs of a decline of the East (economically and certainly politically and militarily) earlier on. Causes and timing might be different but it was certainly also due to their developing domestic weaknesses that great powers like India, the Ottoman Empire, China and Japan, could be opened and bullied by the West.

That West now included Russia, that entered into several conflicts with China and Japan and was a player in the Great Game in Central Asia, and the USA, that developed interest in China and Japan and became ruler of the Philippines. Japan almost immediately after the Meiji Restoration of 1868 began to build an empire in Asia.

The situation in the Americas changed fundamentally because of the independence of the United States and of the Spanish and Portuguese possessions there. The United States evolved into a major Western power and part of the global core. The former Spanish and Portuguese colonies followed quite differing trajectories but in the end all became part of the informal empire of Great Britain and later the United States. Africa, to make Western hegemony complete, was partitioned. In the entire world the political leverage of Western industrializing countries increased markedly. But most of the (economic) global gap emerged not because the East or 'the Rest' in general *declined* but because the West continuously *rose*. Apparently differences in specialization tended to be accompanied by differences in wealth. Wealthy countries were industrializing and produced end products with high added value whereas poor countries were much less industrialized and produced and exported primary products.

THE LONG NINETEENTH CENTURY: DIVERGENCE, CONVERGENCE AND SPECIALIZATION

Some scholars have suggested that wealthy countries could become so wealthy because of the cheaply imported raw materials and foodstuffs from the rest of the world that in this way provided them with ghost acreage. Georg Borgstrom first introduced the concept in a more restricted meaning in 1965. It refers to the additional land that a country would need from internal sources to provide that net portion of the sustenance of its economy that it actually derives from sources outside its boundaries – including sustenance from the sea – or in case of fossil fuel, from underground sources. It thus refers to a 'phantom carrying capacity' (Borgstrom, 1965). Eric Jones already hinted at its importance for Europe in his *The European Miracle*, where he referred to four ecological zones that provided Europe with very welcome resources: (1) the oceans that provided fisheries, (2) Russia, Scandinavia and the Baltic that provided wood, (3) tropical and subtropical regions that provided sugar, tobacco, cotton, indigo and rice, and (4) North and South America, South Africa, Australia, and Southern Russia that provided grain (Jones, 1981). The importance of ghost acreage, however, has been particularly emphasized and turned into a major cause of the Great Divergence in the highly influential work of Kenneth Pomeranz (Pomeranz, 2000; Beckert, 2014). These ghost acreages certainly were very important, but they were to a large extent created and not simply found. They required substantial investments in money, resources and people to enable their exploitation and these primarily came from core countries. In the nineteenth century, the money for those investments typically came from selling manufactured goods. Therefore, it is hard to see how these ghost acreages caused industrialization; instead they are rather a condition for being able to continue it. The peripheries of industrializing core countries need not be overseas colonies or territories. They could also be – and often were – contiguous frontier regions in Europe, which again emphasizes the importance of European peripheries for the European core (see Webb, 1952; Richards, 2003; Osterhammel, 2014).

The role of the peripheries, though, whatever exact definition one uses, was not substantial enough to explain the now quickly increasing global gap. Internal growth and dynamism of the West continued to be the primary cause of the rise of the West (Vries, 2013). But even so, one may wonder whether there was a logic behind the peculiar division of the world in wealthy and poor countries that relates to what they produced and traded. In the nineteenth century almost the entire globe became divided in a core and (semi-)peripheries linked to each other via unequal exchange as described by the modern world-system approach. To determine to which extent this was the result of manipulation or brute coercion would require an extensive analysis of the history of power politics in international relations which would certainly yield important results as the relations between cores and peripheries – and inside peripheries – often are not just determined by economic and geographic logics but also by power configurations. The focus here, however, is on the extent to which strictly economic and economic geographic logics determined core–periphery relations. That boils down to tackling the question how one might explain the existence of unequal exchange in a situation of free trade and discuss why peripheries might not profit from trade liberalization. Wallerstein does not provide an explicit analysis of the economic mechanism that is supposed to explain this. He basically just claims that the wealthy core primarily produced high-value added goods whereas the poor periphery primarily produced low-value added goods thereby assuming what should be explained.

The question how regions became cores or (semi-)peripheries, can be addressed by using the broader perspective of the dependency theory of which Wallerstein's ideas can be considered a specific variety. Its proponents tend to focus much more on the industrial world than Wallerstein, do not distinguish the category of semi-periphery and often use a somewhat different vocabulary referring to dominant and dependent, metropolitan and satellite, or central and peripheral states. There are different strands, but they share some broad characteristics. The dominant states are the advanced industrial nations. The dependent states are much poorer and rely heavily on the export of specific commodities for foreign exchange earnings thereby depending on the more developed states. The relations between dominant and dependent states tend to reinforce or even intensify the inequality between them. The developed centres under-develop the peripheries. In dependency models, economic and political power is heavily concentrated in the industrialized dominant states while the division of labour and the socio-economic structure in the periphery are also supported by elites in the dependent states collaborating with the centre.

The institutional context also plays a role as wages and profits would be higher in the core thanks to a better organization of labour and more developed commercial institutions. In the periphery, workers and companies would be weaker, unable to obtain better prices for their work or products. Whatever may be the cause of the specific division of labour between centre and periphery – which to some extent still exists – it tends to be considered the ultimate explanation for the periphery's poverty. Greater integration into the global economy, then, is not necessarily beneficial for poor countries. The doctrine of comparative advantage that postulates they would profit from specializing in producing primary goods is considered to be false.

The main strictly economic mechanism one comes across amongst dependency theorists to explain the poverty of the periphery is derived from the thesis of Singer and Prebisch who claimed the existence of a structural deterioration of the terms of trade between industrialized and non-industrialized countries on the basis of trends in the interwar period (Toye and Toye, 2003). In their view peripheral countries continued to be (relatively) impoverished, as they had to permanently export more to get the same amount of industrial products. That usually was not easy, as primary products tend to have a low price elasticity. This would explain that the benefits of international trade mainly accrued to the centre. Whether the terms of trade of primary goods indeed structurally deteriorated in a way and with a chronology that can explain the periphery's predicament is fiercely debated. There certainly is the problem, though, of the volatility of demand for primary materials that can have a negative impact on the economic development of peripheral economies (Williamson, 2011). Countries producing and exporting primary products are liable to suffer more from decreasing returns to scale, from diminishing returns and from the fact that they can trade relatively little with one another and thus strongly depend on the core for their exports (Reinert, 2007).

But then why did countries (continue to) specialize in 'the wrong' products? The Heckscher–Ohlin model predicts patterns of production and commerce based on the factor endowments of a region. Countries will export products that use their abundant and cheap factor(s) of production and import products that use their scarce and expensive factor(s) of production. That being the case, an abundant and cheap labour force can easily become a trap as it does not as such stimulate labour-saving, technological innovations (Allen, 2012), whereas an ample supply of resources can easily cause a so-called resource curse in which the very fact that economic activity tends to become concentrated on the exploitation of amply available resources has negative effects on potential overall growth (Barbier, 2011). The relationships between factor endowment, specialization and economic development are complicated and always mediated by the existing institutions and policies. Much depends on the internal structure of countries and on the international context in which they operate. As a result, quite different paths were taken during the Great Specialization (Findlay and O'Rourke, 2007).

At the beginning of colonization, Latin America was rich in resources, richer than North America. The existing power configuration led to the creation and perpetuation of extractive economic and political institutions. These institutions hampered broad economic development and innovation (Bértola and Ocampo, 2012; Sokoloff and Engerman, 2012; Acemoglu and Robinson, 2012). North America, in particular the USA, took a different path as it did not simply specialize in primary products where its natural comparative advantage lay, but, after independence, developed a conscious policy of upgrading its production and of supporting and protecting new industries. For the white majority there existed an inclusive institutional structure characterized by widespread property- and voting rights and educational opportunities (Cohen and DeLong, 2016). The USA's history is quite exceptional. Economically and politically it could be considered as a periphery until independence, but it had a higher per capita income than its colonizer Great Britain, supposedly the wealthiest country in the world at the time. Independence brought a back-fall. Then there was a recovery, followed by the turbulence of the Civil War, in turn followed by a period of fast development and high growth rates. Already before the First World War the country had become the most wealthy, industrialized and modern part of the world (Lindert and Williamson, 2016).

In Canada, Argentina, Uruguay, Australia, New Zealand and South Africa, the new settlers in their still relatively empty countries could and very often did profit from the rapidly increasing demand for raw materials. With the end of the First World War that changed: whereas production had extended even further, foreign demand for various reasons decreased. It turned out that a strong export orientation was very risky and continuing wealth would be hard to realize without structural change, for example diversification and upgrading of their production. The interwar period would prove to be a very difficult period to accomplish that (Lloyd et al., 2013).

India had already suffered some de-industrialization before mechanization in Great Britain had really taken off and before most of it had effectively been turned into a British colony. The drawbacks of its colonial and peripheral status are obvious but British rule also led to unification and infrastructural improvements. The country's openness to the global economy certainly increased. The question, who profited from that and to what extent, is still open (Parthasarathi, 2011; Chaudhuri et al., 2015). China's path again was very different. During the nineteenth century it became semi-integrated in the global economy and semi-peripheralized. The Opium Wars (1839–1842 and 1856–1860), the war against Japan (1894–1895) and the foreign intervention in the Boxer Rebellion (1900–1901), all had a negative impact. But its economy was huge and the interior of the country hard to penetrate. Semi-integration also offered possibilities. But as compared to the previous century the country and its economy were much weaker. They were hit by over-population, ecological problems and rebellions. The total number of people killed in rebellions in the period 1850–1880 alone has been estimated at 66 million. The state became very weak and the existing resource-endowment did not stimulate labour-saving innovation (Deng, 2012). Japan to a certain extent did specialize according to its factor endowment and used its ample supply of cheap and industrious labour to build a capital stock. Helped by the Meiji government that strongly promoted development and industrialization, its economy went through a process of upgrading (Akamatsu, 1961, 1962). The importance of primary products as compared to manufactured goods in Japanese exports steadily decreased whereas exports as a whole increased steeply (Howe, 1996). Its successful imperialist policies before the First World War and the fact that it did not participate in that war gave a substantial boost to its economy.

DE-GLOBALIZATION BETWEEN THE FIRST WORLD WAR AND THE SECOND WORLD WAR

Overall, the long nineteenth century was a period of increasing globalization. It experienced an unprecedented increase in the amount of international trade although the tempo of that increase slowed down substantially from the 1870s onwards, when many countries in Europe – though not the United Kingdom – again became more protectionist, in particular because of pressure by agricultural interests that felt threatened by imports from the New World and Russia. Surprisingly enough, at least for many mainstream economists, there are no firm indications that at the time protectionism as a rule was bad and free trade good for growth, rather the contrary (Vries, 2013).

The period between the First and the Second World War was a period of de-globalization, a reminder that globalization need not be a continuous process. During the Great Depression of the 1930s, this became patently clear (Crafts and Fearon, 2001; James, 2001). I will confine myself to some comments. First, there is the fact that the Great Depression was all but worldwide (the Soviet Union and Japan were less affected) and showed that global interconnectedness could have global reper-cussions. Second, the depression in its origin, development and the way it was dealt with, again shows the crucial role of politics. Third, it also shows the enormous and unprecedented importance of international finance. More than ever financial ties connected countries all over the globe. This became a huge problem, the more so when the gold standard mechanism no longer (adequately) functioned.

De-globalization means market disintegration. In the case of the Great Depression that was largely a matter of politics. The dismantling of the Habsburg, Ottoman and German Empires was a political decision. Their many, often quite small, successor states all created their protected borders. That the USSR as successor of the Russian Empire was less integrated in the global economy was also, to a large extent, a matter of politics. The decision to combat the depression domestically by protective measures of all sorts was clearly political. The Tariff Act of 1930 in the USA, to just give one example, raised USA tariffs on over 20 000 imported goods to record levels. Several countries strove to become more autarchic. Exclusive trade agreements and inter-national barter became popular. Many measures can well be characterized as neo-mercantilist. States such as Germany, Italy and Japan attempted to build (in)formal empires. Overall, a clear globalization backlash took place.

RE-GLOBALIZATION AFTER THE SECOND WORLD WAR

Although people tend to consider their own lifetimes as very special and interesting, global economic developments in the half-century after the Second World War in which people began to discuss globalization more than ever before, indeed were un-precedented. The world's population increased by 140 per cent. Global GDP per capita nevertheless increased by 185 per cent, which is 2.1 per cent per year. International trade, in that period, grew by almost 8 per cent per year till the oil crisis of 1973. As a consequence of that crisis international trade decreased, but in the 1990s the growth rate was again some 6 per cent (see for example Maddison, 2001).

The process of re-globalization that set in after the de-globalization of the two wars and the interwar period of course did not involve all countries equally, nor was it a linear process. In the first four decades it primarily involved the developed countries in Western Europe, the USA, Canada, Australia and New Zealand, Japan, and some South East Asian regions. Europe's role rapidly changed. In 1945 the continent was severely damaged and traumatized and split by the Iron Curtain. Western European countries started a process of economic integration and during the period 1945–1973, they actually experienced their glorious thirty years of strong growth, largely catching up with the USA. Their empires, which had taken centuries to build, crumbled in a few decades, which, amongst other consequences, led to substantial migration flows to the former colonial powers (Buettner, 2016; Shipway, 2008). The USA in 1945 produced 45 per cent of the world's industrial output and had become an economic superpower that now actually behaved like one and no longer preferred to act isolationist. Japan emerged as a new economic superpower with very high rates of growth.

Beyond these developed countries of the First World, where trade even grew stronger than production, there, for several decades, was far less openness. The communist countries, even though they often were not as united as the word 'bloc' suggested, definitely were distinct and quite separate from the Western developed world. They, to a very large extent, withdrew from the world economy. Economic exchange with the First World was minimal till roughly the 1970s. Even within the bloc it was not very substantial. The developing Third World constituted a rather residual category comprising countries that were poor, less developed and peripheral. Their existence showed that the Great Divergence and the Great Specialization were still facts of life. Latin American countries as a rule were quite protectionist. So were many of the countries that had recently become independent. Thirty-five years after the war the bulk of the world's population still lived in economies which were rather closed. Re-globalization only became a worldwide phenomenon in the 1980s, when the Soviet communist empire collapsed, communist China radically changed its economic policies and many Third World economies, often under the impact of the neoliberal Washington consensus became more open. Even so, however, the majority of world population in the beginning of the 1990s still lived in economies with higher manufacturing tariffs than on the eve of the Great War. Agricultural protectionism was still rife and there existed many non-tariff barriers. It was also in the 1980s that the Great Specialization began to slowly crumble. In the early 1950s, the manufacturing output of the First World was about 90 per cent of the output of the entire world; in 2000 that was less than 80 per cent. When we look at the percentages of the exports of manufactured products the decline of the First World is even clearer – and has continued in the twenty-first century. Not only did the developing world start to export more manufactured products, it increasingly did so to other developing countries. That does not mean geographical specialization disappeared. Developed countries tended to specialize in capital- and skill-intensive production while the developing countries specialized in production that required less capital and skill. As several underdeveloped countries started catching up, the Great Divergence, however, became somewhat less great.

The character of international trade changed fundamentally. Commodity chains have become more complex as ICT enabled a process of second unbundling in which it became increasingly economically efficient to geographically separate manufacturing stages and locate these in different parts of the world including former peripheral parts (Baldwin, 2016). These spatially dispersed production chains were often organized by Trans-National Corporations. Another striking development was the quickly increasing importance of services. In 2000, commercial services amounted to 19 per cent of total of exports i.e. of merchandise exports plus commercial services exports. Finally there was an enormous increase in financial transactions (see Chapter 17 this volume). Foreign direct investment may have only reached the same level as a percentage of global GDP in 1980 as it did just before the First World War, but in 2000 it already amounted to more than 1 000 000 000 000 dollars. Considering the heated debates about the question whether globalization would be 'a good thing', it certainly is not irrelevant to point out with Ronald Findlay and Kevin O'Rourke that just like in the previous phase of globalization the data do not support 'a simple mono-causal relationship between openness and growth' (Findlay and O'Rourke, 2007: 521).

FINAL COMMENT

The world indeed has become more economically integrated over the last decades although the continued prominence of coastal regions in the global economy is striking. But to again quote Findlay and O'Rourke 'globalization is a fragile and easily reversible process' (Findlay and O'Rourke, 2007: 535). A reversal might happen because of strictly material, economic reasons. But more likely and already more real is a new backlash against globalization. Global economic integration depends on agency. The population of countries involved may like in earlier instances easily turn against it as both Brexit and the election of Donald Trump have shown. International tension and conflict can also easily put an end to further integration and even lead to disintegration. One need not be an ingrained pessimist to think that the global economic integration indeed might again be reversed.

REFERENCES

Abernethy, D. (2000), *The dynamics of global dominance. European overseas empires, 1415–1980*, New Haven and London: Yale University Press.

Abu-Lughod, J. (1989), *Before European hegemony. The world system A.D. 1250–1350*, Oxford: Oxford University Press.

Acemoglu, D. and J. Robinson (2012), *Why nations fail. The origins of power, prosperity and poverty*, New York: Crown Business.

Adshead, S. (1993), *Central Asia in world history*, Basingstoke and London: St. Martin's Press.

Adshead, S. (2004), *T'ang China. The rise of the East in world history*, Basingstoke: Palgrave Macmillan.

Akamatsu, K. (1961), 'A theory of unbalanced growth in the world economy', *Weltwirtschaftliches Archiv. Zeitschrift des Instituts für Weltwirtschaft an der Universtät Kiel*, **86**(1), 196–217.

Akamatsu, K. (1962), 'A historical pattern of economic growth in developing countries', *The Developing Economies. The Journal of the Institute of Economics*, **1**, 3–25.

Allen, R. (2012), 'Technology and the Great Divergence: global economic development since 1820', *Explorations in Economic History*, **49**(1), 1–16.

Arrighi, G. (2007), *Adam Smith in Beijing. Lineages of the twenty-first century*, London and New York: Verso.

Bairoch, P. (1976), *Commerce extérieur et développement économique de l'Europe au XIX siècle*, Paris and The Hague: Mouton.

Bairoch, P. (1993), *Economics and world history. Myths and paradoxes*, New York: University of Chicago Press.

Bairoch, P. (1997), *Victoires et déboires. Histoire économique et sociale du monde du XVI siècle à nos jours*, Paris: Gallimard.

Baldwin, R.E. (2016), *The great convergence: Information technology and the new globalization*, Cambridge, MA: The Belknap Press of Harvard University Press.

Barbier, E. (2011), *Scarcity and frontiers. How economies have developed through natural resource exploitation*, Cambridge: Cambridge University Press.

Barrett, W. (1990), 'World bullion flows, 1450–1800', in: J. Tracy (ed.), *The rise of merchant empires. Long-distance trade in the early modern world, 1350–1750*, Cambridge: Cambridge University Press, 224–254.

Bateman, V. (2012), *Markets and growth in early modern Europe*, London: Pickering & Chatto.

Bayly, C.A. (2002), '"Archaic" and "Modern" globalization in the Eurasian and African arena, c. 1750–1850', in: A.G. Hopkins (ed.), *Globalization in world history*, London: Pimlico, 47–73.

Beckert, S. (2014), *Empire of cotton. A new history of global capitalism*, London: Penguin UK.

Benjamin, T. (2009), *The Atlantic world. Europeans, Africans, Indians and their shared history, 1400–1900*, Cambridge: Cambridge University Press.

Berg, M. (2004), 'In pursuit of luxury: global history and British consumer goods in the eighteenth century', *Past and Present*, **182**, 85–142.

Bértola, L. and J.A. Ocampo (2012), *The economic development of Latin America since independence*, Oxford: Oxford University Press.

Black, J. (2014), *The power of knowledge. How information and technology made the modern world*, New Haven and London: Yale University Press.

Blackburn, R. (1998), *The making of New World slavery. From the Baroque to the modern 1492–1800*, London and New York: Verso.

Bordo, M., A. Taylor and J. Williamson (eds) (2003), *Globalization in historical perspective*, Chicago: University of Chicago Press.

Borgstrom, G. (1965), *The hungry planet. The modern world at the edge of famine*, New York: Collier Books.

Braudel, F. (1981), *Civilization and capitalism 15th–18th century. Volume One. The structures of everyday life. The limits of the possible*, London: Collins.

Buettner, E. (2016), *Europe after empire. Decolonization, society and culture*, Cambridge: Cambridge University Press.

Burbank, J. and F. Cooper (2010), *Empires in world history. Power and the politics of difference*, Princeton: Princeton University Press.

Cain, P.J. and A.G. Hopkins (2016), *British imperialism, 1688–2015* (third edition), Abingdon, Oxon and New York: Routledge.

Canny, N. and P. Morgan (eds) (2011), *The Oxford Handbook of the Atlantic World, 1450–1850*, Oxford: Oxford University Press.

Chang, H. (2002), *Kicking away the ladder. Development strategy in historical perspective*, London: Anthem Press.

Chase-Dunn, C. and T.D. Hall (1997), *Rise and demise. Comparing world-systems*, Boulder and Oxford: Westview Press.

Chaudhuri, L., B. Gupta, T. Roy and A. Swamy (eds) (2015), *A new economic history of colonial India*, Abingdon, Oxon and New York: Routledge.

Chaunu, P. and H. Chaunu (1955–1959), *Séville et L'Atlantique (1504–1650)*, Paris: SEVPEN.

Cohen, S. and J.B. DeLong (2016), *Concrete economics. The Hamilton approach to economic growth and policy*, Boston MA: Harvard Business Review Press.

Crafts, N. and P. Fearon (eds) (2001), *The Great Depression of the 1930s. Lessons for today*, Oxford: Oxford University Press.

Crosby, A. (1972), *The Columbian exchange. Biological and cultural consequences of 1492*, Westport: Greenwood Publishing Group.

Crosby, A. (1986), *Ecological imperialism. The biological expansion of empire, 900–1900*, Cambridge: Cambridge University Press.

Darwin, J. (2007), *After Tamerlane. The global history of empire*, London: Bloomsbury Press.

Davis, R. (2004), *Christian slaves, Muslim masters. White slavery in the Mediterranean, the Barbary Coast, and Italy, 1500–1800*, Basingstoke: Palgrave Macmillan.

De Vries, J. (2003), 'Connecting Europe and Asia. A quantitative analysis of the Cape-route trade 1497–1795', in D. Flynn, A. Giráldez and R. von Glahn (eds) *Connecting Europe and Asia. Global connections and monetary history*, Aldershot: Ashgate Publishing, 35–106.

De Vries, J. (2008), *The industrious revolution. Consumer behaviour and the household economy, 1650 to the present*, Cambridge: Cambridge University Press.

De Vries, J. (2010), 'The limits of globalization in the early modern world', *Economic History Review*, **63**(1), 710–733.

Deng, K.G. (ed.) (2012), *China's political economy in modern times. Changes and economic consequences, 1800–2000*, Abingdon, Oxon and New York: Routledge.

Eltis, D. (1999), 'Atlantic history in global perspective', *Itinerario. European Journal of Overseas History*, **22**(2), 141–161.

Emmer, P. (2003), 'The myth of early globalization: the Atlantic economy, 1500–1800', *European Review*, **11**(1), 37–47.

Ferguson, N. (2003), *Empire. How Britain made the modern world*, London: Penguin Group.

Findlay, R. and K. O'Rourke (2007), *Power and plenty. Trade, war, and the world economy in the second millennium*, Princeton and Oxfordshire: Princeton University Press.

Fischer, D.H. (1996), *The great wave. Price revolutions and the rhythm of history*, New York and Oxford: Oxford University Press.

Fletcher, J. (1995), 'Integrative history: parallels and interconnections in the early modern period, 1500–1800', in: B. Forbes Manz (ed.), *Joseph F. Fletcher, studies on Chinese and Islamic inner Asia*, Aldershot: Variorum, 37–57.

Flynn, D. and A. Giráldez (2004), 'Path dependence, time lags and the birth of globalization: a critique of O'Rourke and Williamson', *European Review of Economic History*, **8**(1), 81–108.

Flynn, D. and A. Giráldez (2006), 'Globalization's birth: the case for 1571', in: B. Gills and W. Thompson (eds), *Globalization and global history*, Basingstoke and London: Routledge, 232–247.

Frank, A. (1998), *ReOrient. Global economy in the Asian age*, Berkeley, Los Angeles, London: University of California Press.

Frank, A. and B. Gills (1996), 'The 5,000-year world system. An interdisciplinary introduction', in: Andre Gunder Frank and Barry Gills (eds), *The world system: five hundred years or five thousand?* (second revised edition), Abingdon, Oxon and New York, 3–58.

Frankopan, P. (2015), *The Silk Roads. A new history of the world*, London: Bloomsbury.

Gallagher, J. and R. Robinson (1953), 'The imperialism of free trade', *The Economic History Review*, **6**(1), 1–15.

Giráldez, A. (2015), *The age of trade. The Manila Galleons and the dawn of the global economy*, Lanham: Rowman & Littlefield Publishers.

Goldstone, J.A. (1991), *Revolution and rebellion in the early modern world*, Berkeley, Los Angeles, Oxford: University of California Press.

Grafe, R. and A. Irigoin (2012), 'A stakeholder empire: the political economy of Spanish imperial rule in America', *The Economic History Review*, **65**(2), 609–651.

Gunn, G. (2003), *First globalization. The Eurasian exchange, 1500–1800*, Lanham: Rowman & Littlefield Publishers.

Held, D., A. McGrew, D. Goldblatt and J. Perraton (1999), *Global transformations. Politics, economics and culture*, Stanford: Stanford University Press.

Hobson, J.M. (2004), *The Eastern origins of Western Civilisation*, Cambridge: Cambridge University Press.

Hopkins, A.G. (2002), *Globalization in world history*, London: Pimlico.

Howe, C. (1996), *The origins of Japanese trade supremacy*, London: University of Chicago Press.

Hugill, P. (1993), *World Trade Since 1431. Geography, Technology and Capitalism*, Baltimore and London: Johns Hopkins University Press.

Inikori, J. (2002), *Africans and the Industrial Revolution in England. A study in international trade and economic development*, Cambridge: Cambridge University Press.

James, H. (2001), *The end of globalization: Lessons from the Great Depression*, Harvard: Harvard University Press.

Jones, E. (1981), *The European miracle. Environments, economies and geopolitics in the history of Europe and Asia*, Cambridge: Cambridge University Press.

Kindleberger, C. (1996), *World economic primacy: 1500–1990*, Oxford: Oxford University Press.

Lang, M. (2006), 'Globalization and its history. A review article', *The Journal of Modern History*, **78**(4), 899–931.

Lee, C.H. (1986), *The British economy. A macroeconomic perspective*, Cambridge: Cambridge University Press.

Li, J., D.M. Bernhofen, M. Eberhardt, and S. Morgan (2013), 'Market integration and disintegration in Qing dynasty China: evidence from time-series and panel time-series methods'. Working Paper.

Lieberman, V. (2003), *Strange Parallels: Volume 1, Integration on the Mainland: Southeast Asia in Global Context, c.800–1830 (Studies in Comparative World History)*, Cambridge: Cambridge University Press.

Lieberman, V. (2009), *Strange Parallels: Volume 2, Mainland Mirrors: Europe, Japan, China, South Asia, and the Islands: Southeast Asia in Global Context, c.800–1830 (Studies in Comparative World History)*, Cambridge: Cambridge University Press.

Lindert, P.H. and J.G. Williamson (2016), *Unequal gains. American growth and inequality since 1700*, Princeton and Oxford: Princeton University Press.

Lloyd, C., J. Metzer and R. Such (eds) (2013), *Settler economies in world history*, Leiden and Boston: Brill.

Lucassen, L. (2016), 'Connecting the world: migration and globalization in the second millennium', in: C. Antunes and K, Fatah-Black (eds), *Explorations in history and globalization*, London and New York: Routledge, 19–46.

Maddison, A. (2001), *The world economy. A millennial perspective*, Paris: Organisation for Economic Co-operation and Development.

Maddison, A. (2007), *Contours of the world economy, 1–2030 AD. Essays in macro-economic history*, Oxford: Oxford University Press.

Mann, C. (2011), *1493. How Europe's discovery of the Americas revolutionized trade, ecology and life on earth*, London and New York: Granta Books.

Marks, S.G. (2016), *The information nexus. Global capitalism from the Renaissance to the present*, Cambridge: Cambridge University Press.

Marx, K. (1976) *Capital. A critique of political economy. Volume 1, part two*, Harmondsworth: The Pelican Marx Library.

Northrup, D. (2002), *Africa's discovery of Europe 1450–1850*, Oxford: Oxford University Press.

Nunn, N. and N. Qian (2010), 'The Columbian Exchange: a history of disease, food, and ideas', *Journal of Economic Perspectives*, **24**(2), 163–188.

O'Brien, P. (1982), 'European economic development: the contribution of the periphery', *Economic History Review*, **2**(35), 1–18.

O'Rourke, K. and J. Williamson (2002), 'When did globalization begin?', *European Review of Economic History*, **6**(1), 23–50.

O'Rourke, K. and J. Williamson (2004), 'Once more: when did globalization begin?', *European Review of Economic History*, **8**(1), 109–117.

Osterhammel, J. (2014), *The transformation of the world. A global history of the nineteenth century*, Princeton: Princeton University Press.

Osterhammel, J. and N. Petersson (2009), *Globalization. A short history*, Princeton: Princeton University Press.

Parker, C. (2010), *Global interaction in the early modern age, 1400–1800*, Cambridge: Cambridge University Press.

Parker, G. (2013), *Global crisis. War, climate change and catastrophe in the seventeenth century*, New Haven and London: Yale University Press.

Parthasarathi, P. (2011), *Why Europe grew rich and Asia did not. Global economic divergence, 1600–1850*, Cambridge: Cambridge University Press.

Pearson, M.N. (1991), 'Merchants and states', in: James D. Tracy (ed.), *The political economy of merchant empires. State power and world trade*, Cambridge: Cambridge University Press, 41–116.

Pearson, M.N. (1998), *Port cities and intruders. The Swahili Coast, India, and Portugal in the early modern era*, Baltimore and London: Johns Hopkins University Press.

Pollard, S. (1981), *Peaceful conquest. The industrialization of Europe, 1760–1970*, Oxford: Oxford University Press.

Pomeranz, K. (2000), *The Great Divergence. China, Europe, and the making of the modern world economy*, Princeton: Princeton University Press.

Reinert, E. (2007), *How rich countries got rich ... and why poor countries stay poor*, New York: Public Affairs.

Reinhard, W. (2016), *Die Unterwerfung der Welt. Globalgeschichte der Europäischen Expansion*, Munich: C.H. Beck.

Richards, J.F. (1997), 'Early modern India and world history', *Journal of World History*, **8**(2), 197–209.

Richards, J.F. (2003), *The unending frontier. An environmental history of the early modern world*, Berkeley, Los Angeles, London: University of California Press.

Robertson, D.H. (1938), 'The future of international trade', *The Economic Journal*, **48**(189), 1–14.

Rössner, P. (ed.) (2016), *Economic growth and the origins of modern political economy. Economic reasons of state, 1500–2000*, Abingdon, Oxon and New York: Routledge.

Shipway, M. (2008), *Decolonization and its impact. A comparative approach to the end of the colonial empires*, Oxford: Oxford University Press.

Shiue, C. and W. Keller (2007), 'Markets in China and Europe on the eve of the Industrial Revolution', *American Economic Review*, **97**(4), 1189–1216.

Smith, A., R.H. Campbell, A.S. Skinner and W.B. Todd (1981), *An inquiry into the nature and causes of the wealth of nations*, Indianapolis: Liberty Fund.

Sokoloff, K. and S. Engerman (2012), *Economic development in the Americas since 1500. Endowments and institutions*, Cambridge: Cambridge University Press.

Steensgaard, N. (1990), 'Commodities, bullion and services in intercontinental transactions before 1750', in: H. Pohl (ed.), *The European discovery of the world and its economic effects on pre-industrial society, 1500–1800. Papers of the Tenth International Economic History Congress* [edited on behalf of the International Economic History Association], Stuttgart, 9 23.

Stern, P.J. (2011), *The company-state. Corporate sovereignty and the early modern foundations of the British Empire in India*, Oxford: Oxford University Press.

Studer, R. (2015), *The Great Divergence reconsidered. Europe, India, and the rise to global economic power*, Cambridge: Cambridge University Press.

Thornton, J. (1992), *Africans in the making of the Atlantic world, 1400–1680*, New York: Cambridge University Press.

Toye, J. and R. Toye (2003), 'The origins and interpretation of the Prebisch–Singer thesis', *History of Political Economy*, **35**(3), 437–467.

Trentmann, F. (ed.) (2012), *The Oxford handbook of the history of consumption*, Oxford: Oxford University Press.

Von Thünen, J.H. (1826), *Der isolierte Staat in Beziehung auf Nationalökonomie und Landwirtschaft.* Stuttgart: Gustav Fischer.

Vries, P. (2002), 'Governing growth: a comparative analysis of the role of the state in the rise of the West', *Journal of World History*, **13**(1), 67–138.

Vries, P. (2013), *Escaping poverty. The origins of modern economic growth*, Vienna and Göttingen: V&R Unipress.

Vries, P. (2015), *State, economy and the Great Divergence. Great Britain and China, 1680s–1850s*, London: Bloomsbury Academic.

Vries, P. (2016), 'What we do and do not know about the Great Divergence at the beginning of 2016', *Historische Mitteilungen der Ranke Gesellschaft*, **28**(1), 249–297.

Wallerstein, I. (1974), *The modern world-system. Capitalist agriculture and the origins of the European world-economy in the sixteenth century*, New York: Academic Press.

Webb, W.P. (1952), *The great frontier*, Boston: Houghton Mifflin.

Williamson, J.G. (2011), *Trade and poverty. When the Third World fell behind*, Cambridge, MA and London: The MIT Press.

3. Globalization and the question of scale

Kevin R. Cox

CONTEXT

Under globalization questions of scale have surged to the fore, both in the academy and elsewhere. New spaces of accumulation have emerged at new geographic scales and not just at the global level. Accompanying these have been new regulatory institutions that are inseparable from the problematic of geographic scale as it has been expressed in its most recent more politicized form (Cox, 2014: 139–141).

The more prominent writings on this topic fall under the heading of what might be called the scale restructuring school, notable among whom have been Neil Brenner (2002, 2004), Bob Jessop (2002), and Erik Swyngedouw (1997a, 1997b). This has had several different emphases. If there is an umbrella for them it is what has been referred to as glocalization: a shift of regulatory powers both upward to a more global level and then downward to the regions and localities. The underlying logic is one of competition. Intensified competition at global levels has led either intendedly or by default to a shift of powers to the international; meanwhile, in order to facilitate success in the struggle for global markets, local and regional governments have been empowered. At local levels, again in response to competitive pressures, there have been tendencies to what has been called 'metro-regionalism': the creation of metropolitan-level institutions so as to surmount those contradictions of urbanization impeding the development of the productive forces. Finally, there are what have been referred to as 'new state spaces': as well as decentralizing responsibilities, central states have been especially alert to the need to support new growth areas in terms of both physical infrastructure and regulatory relief. In short, there have been tendencies for central states to choose winners and losers.

The upward shift in state powers and responsibilities is undeniable. This is not just institutional in the narrow sense of organizations like the EU, or coordinative ones like the G7 and G20 and the increasing role of the IMF and the World Bank. It is also the enhanced role of international money in conditioning state monetary policy. The complementary downward shift, though, while providing a nice descriptive symmetry, is less clear. Devolution in the Western European countries has been massively hyped (Cox, 2009) and has never ever approached the sort of decentralized state form characteristic of the US; and that is the second point – the decentralization of the American state is a longstanding feature and has very little to do with globalization.

There have undoubtedly been scalar changes at sub-national levels but they have been much more heterogeneous in their forms. Metropolitan areas are increasingly significant in national space economies, which adds to the pertinence of arguments about metro-regionalism. Stepping outside the state, there has also been some regionalization of production systems and these have often reinforced the significance of metropolitan centres: a major point of assembly in or close to a metro centre that provides residential areas for management and rapid airline or rail connections, surrounded by component suppliers in nearby smaller towns.[1] There have been other dimensions of differentiation, particularly in Western Europe, where regional policy has indeed shifted from its immediate postwar egalitarian ideals to one of emphasizing growth in the most promising areas (Cox, 2016: Chapter 5; Crouch and Le Galès, 2012).

Theoretically there are also important issues. Competition is emphasized but it is like a *deus ex machina* lacking a foundation in other conditions like accumulation and the underlying class relationship. Aside from this, the space over which the processes of scale restructuring occurs is too uniform. There is little or no sense of how uneven development, alongside class tensions, has conditioned processes of globalization. Globalization and its scalar relations is something that has been struggled over: a class stake as well as one for different countries – why, after all, did the US and to a lesser degree, the United Kingdom lead the way, if it was not from a position of relative weakness? And within countries again, some regions have seen advantages in the new scalar dispensation that others have not. In turn, situating scalar struggle in this way suggests the limits of thinking, as is the tendency in the scalar restructuring literature, of 'top-down' effects: as if, that is, the 'top-down' did not in some way and simultaneously, express and represent the 'bottom-up'. One might also remark on the absence of any reference to the role of discourse and discursive formation in scalar restructuring. An understanding in terms of forces of territorialization and de-territorialization, albeit with class consequences, tends to be taken for granted.

Given the significance of accumulation and class struggle when thinking about questions of scale, the first section of this chapter takes up their implications for scalar restructuring. In the second and longest section of the chapter, these ideas are then applied to an understanding of scale and scalar restructurings in what I am calling, and in contrast to the earlier globalization at the end of the nineteenth and beginning of the twentieth centuries, the second globalization. The final section reviews the increasing tensions surrounding globalization and their scalar implications. It should be emphasized at the start that this is decidedly a view from the advanced capitalist countries and to a lesser degree, Anglo-America: how, that is, the wider processes of increased integration stimulated by them at a global level have then been reflected in new scalar struggles that connect the global to the sub-national.

ACCUMULATION AND SPACES OF REGULATION

Crucial to understanding scale and struggles around it is the capitalist form of development; notably the drive to accumulate and the underlying class relation which is

---

[1]   One thinks here of the Japanese auto transplants: Honda/Marysville/Columbus; Toyota/Georgetown/Lexington KY; or Nissan/Smyrna/Nashville; also of the way Silicon Valley reinforced the growth of San Francisco and the relation between hitech in Britain's so-called Sunbelt and London. See also Hancké (2003) on the emergence of regional production systems in France, sometimes focusing on major regional centres.

its necessary condition. Capitalism is central because of the way in which, and in virtue of its contradictions, it becomes *the* transformative agent with respect to all aspects of the social process, and restlessly so: technology, institutions, the relation to nature, ideology, gender, and, of course geography.

Capital assumes diverse concrete forms and circulates through them starting with production, returning to it, and then starting again: labour power and means of production are brought together courtesy of money; the result of the subsequent labour process are new values in the form of products; their sale to consumers and the recycling back of the money so realized into production completes the cycle. It then starts anew but, as a result of the investment of the surplus along with the values originally laid out, on an expanded basis. This circulation of values has a characteristic geographic form that is distance- and network-constrained in a complex way. It is a geography that is unstable and, in accordance with capital's chronic tendency to revolutionizing its relations, gets transformed as the circulation of values through its various forms is displaced. A geography of unevenness emerges as the production points through which capital circulates lose their attraction in favour of new ones elsewhere. Spatially, as elsewhere in its circuit, capital differentiates itself: new spaces of accumulation while existing spaces assume new differentiated forms with the development of the spatial division of labour.

At all points in its circulation capital encounters obstacles that have to be suspended: resistances and tensions that permeate the social process as a whole. Capital is a contradiction-ridden form of production. Wage repression and its tendency to produce a reserve army of the unemployed incite resistance. There are tendencies to over-production as wages fall short of the value of the product, undermining the ability to accumulate. The temptations of gain through deceit and concealment of information inhibits contract formation. And while it is the competition of capitals one with another that forces accumulation on them, the goal is always to achieve monopoly; which in turn, and again, threatens the continuity of the accumulation process.

This has been the background to the growth of a quite massive and sprawling literature on what are variably described as problems of coordination, governance or regulation, according to political taste. The more mainstream literatures have focused on the triad of market, hierarchy, and network (or trust) but it is not difficult to adapt these ideas to more radical purposes as in regulation theory.[2] Regardless, there is an unfortunate technical sense in these literatures: that once the right combination of mechanisms is found for a particular purpose, so long as markets are complemented by appropriate forms of state intervention or networked forms of relationship, then all will be well: harmony will have been restored and the way opened up for a smooth future of accumulation and prosperity.

What this ignores is the fundamentally antagonistic nature of the capitalist form of production; a contradictory nature that is an inevitable aspect of its functioning and that, regardless of governance or regulation, will continue to express itself, if in new ways. Antagonism is then carried through into whatever regulatory regime emerges.

---

2   As in the displacement of the more hierarchical forms of governance characteristic of Fordism by the market and the networked relations underlying flexible production.

It is always regulation or governance to the advantage of those with money power: an expression of a balance of class forces that is by definition unequal.

In the mainstream literature, including its more radical expressions, there is also a methodological ploy that leaves it wanting. It is as if the social body can be broken up into a discrete set of separate parts, some of which then get combined in regulatory processes. It is not just that markets, hierarchies and networks are internally related – markets inevitably have hierarchical and networked aspects. Rather, regulation is everywhere; the social process *is* regulation whether one is talking about particular forms of technology designed with a view to subordinating labour to the rule of capital, an educational system that produces people trained in norms of punctuality, obedience and saving, or a discourse of class unity and 'national interest'. From this standpoint, Harvey's idea of 'structured coherence' (1985a, 1985b) and its later reworking as a set of internally related activity spaces (2010: Chapter 5) has a good deal of merit.

From the standpoint of addressing scale and globalization, what is attractive about this approach is that it restores accumulation and its tensions to the centre of the picture. Contradiction is inevitable as are attempts to suspend it, but it is a process that embraces the social process in its entirety: not something that can be confined to the restricted world of markets, hierarchies and networks, therefore. It is also a capitalist process which means that contradiction enters into the very process of achieving regulatory compromise within a particular space; a compromise more effective for some than for others: a balance between antagonistic social forces that is constantly disturbed and constantly reconstructed in order to facilitate the reproduction of the central class relation.

As Harvey pointed out in his earlier writings, more global forms of capitalist relations of trade, investment, and the like threatened structured coherences already in place. This threat in turn was indissociable from the geographic differentiation of capital and its uneven development. The sense was of a social and physical infrastructure in which firms and workers were embedded and which would be defended against emergent forces of accumulation elsewhere by coalitions of forces that might be cross-class in character, but not necessarily. I would add that elsewhere structured coherences would perform as platforms for respective capitals to invade world markets and undercut rivals.[3] This would then open the way to a territorialization of capital's social relations.

What Harvey does not refer to are the new regulatory practices emerging at more global scales and the equally new structures of relations supporting them: contributing to the formation of an emergent structured coherence at a larger geographic scale which would then be a stake in the defense of more local and geographically unevenly developed social relations. Students of what has been called 'national integration' will recognize the validity of this, but it is clearly a process that builds beyond the national as a putative coherence at international levels and, for example, extending beyond the

---

[3] The German machine tool industry and its successes cannot be understood apart from labour training practices in that country. American success in consumer durables and their penetration of markets elsewhere only makes sense in terms of a large domestic market and the way it facilitated exploitation of economies of scale and long production runs.

Washington Consensus to include norms of diplomacy, the dominance of the English language, international tourism, student exchange and much more.

One result is a reorganization of regulatory divisions of labour; divisions between the more global and the local, that is: a scaled geography of structured coherences which then provide the framework for scalar divisions of labour in the regulatory function. But whatever the particular scalar nexus (Taylor, 2003) arrived at, it will inevitably be superseded as the dynamics of accumulation set up new tensions. These will obviously be around class as more global relations facilitate the increasingly effective realization of the law of value putting downward pressure on labour, its wages and conditions of work and ultimately on the welfare state institutions that can work to reduce capital's leverage. The squeezing of smaller businesses by the forces of industrial concentration can add to the mix, creating opportunities for new class alliances, possibly taking advantage of that social stratification entailed by the division of labour. Likewise, with the changing concrete character of the accumulation process and the geographic implications of, not least, technological and institutional change, local economies will develop unevenly. Those embedded in and dependent on local structured coherences can find themselves more challenged than others so that coalitions of class forces organize themselves territorially in order to contest more global regulatory arrangements. Scalar ensembles then become the terrain of a complex struggle in which the underlying class forces risk being obscured by the claims of a politics of territory; which of course has been one of the major challenges in making sense of the politics of scale that has emerged alongside the most recent round of globalization, and to which we now turn.

THE SECOND GLOBALIZATION

Scalar Shifts

The second globalization began to take shape in the years subsequent to the onset of the 'long downturn' (Brenner, 1998); that is, the sharp deterioration in corporate profitability, rates of productivity increase and employment that became apparent in the early 1970s. What has led to the designation 'globalization' have been increases in trade and foreign direct investment, the emergence of what some have called a 'New International Division of Labour' between the advanced capitalist societies and the so-called Newly Industrialized Countries or NICs and a massive expansion of international financial flows: short-term trading in currencies, loans to international banks and to foreign governments, above all. It is helpful, though, to be more specific about its geography.

Simply put, the global nature of the second globalization has had its limits. Of the increased trade and direct investment the so-called Triad countries of North America, Western Europe and Japan have taken the lion's share (Kiely, 2014: 287–288). Most of the increased foreign investment from those countries has not been in the NICs; rather it has been in each other. In terms of their numbers NICs are a very, very limited bunch: obviously the flying geese of East Asia, along with India, Mexico, Brazil and Turkey, but where else? Most of Latin America, Africa and Western Asia are virtually off the map from the standpoint of globalized production; they may import consumer goods, including luxury ones for ruling classes, but the more or less strong remnants of pre-capitalism along with subordination to neo-colonial impositions have limited them to enclave forms of development. They have borrowed internationally, speculating on a continued increase in revenues from the export of primary products that has failed to materialize, and so found themselves in the tentacles of the IMF. The sort of autonomous capitalist development found in East Asia has been elusive.

Nevertheless, within those limits, and compared to what went before during the Bretton Woods era, or what some designate as Fordist, there is a new scalar architecture. This includes a new global regulatory regime. The growth of international finance and its conditioning of national monetary policies is inseparable from a relaxation of capital controls around the world just as trade has responded to the lowering of barriers, in some localized cases dramatically, as with the EU. These changes have then been supervised and facilitated by international organizations, among which the IMF, World Bank and WTO loom large.

Particularly noticeable within the advanced capitalist countries has been the growth of large cities and their connections with one another: a division of labour between cities, therefore. It is this growth that has generated claims about metro-regionalism, and the mutual girding of loins to promote competitive abilities in the new globalizing world. But getting local governments to cooperate has been more aspiration than reality and a continuing source of contradiction. Politically, class has retreated to be replaced by new senses of social privilege and difference to be protected, among other ways, territorially: exclusionary suburbs, a call to defend green space, even while its preservation will provide very unequal benefits.

Uneven development in the context of the weakening of labour movement agendas around the welfare state, has then led to calls for some decentralization of powers: not the sort of decentralization imagined by the glocalizers but one that sets itself off against the welfare state, working to the advantage of some regions against that of others. This is the story of Italy's Northern League, Belgium's Vlaams Blok, and the attempt in Germany on the part of the growth areas, mainly in the south of the country, to weaken the system of fiscal redistribution between the *länder*. If there has been decentralization and a retreat from central control of fiscal flows, therefore, it has been stimulated by appeals from select regions: part and parcel of a new territorialization of the imaginary. How therefore to explain what has happened?

Class Struggles

Robert Brenner (1998) has argued that the origins of the long downturn lay in excess capacity and its pressures on profitability. This may be so. But as capital sought ways of restoring profitability, labour resisted. To some degree that resistance was bypassed and profitability maintained in a number of different ways.

Easily the most remarked has been the offshoring to the NICs of some manufacturing, including consumer goods and the production of components for final assembly in the advanced capitalist countries. In some instances the more knowledge-intensive pre-fabrication and post-fabrication stages – design, research, marketing, production planning – have remained in the Triad countries while actual production has been decanted to an NIC. Encouraging growth in the NICs has also meant that the market there for the more skills-intensive, capital goods industries of the Triad countries has been enhanced: machine tools, water and sewer systems, transport equipment of all sorts.

The division of labour in the advanced capitalist countries has therefore been reshaped but, and as a result of the nature of what was being offshored, at the expense of the less skilled and with the incorporation of the former socialist countries and their legacy of industrial skills, 'less skilled' has been defined upwards. Meanwhile, as industrial employment has continued to decline, in part due to that offshoring but also to continuing changes in labour processes, the service sector has expanded. In a context of increasing female employment, though, the possibilities for those displaced by plant closures have been limited and even where there has been interest in them as employees, their own gender ideology has often got in the way (McDowell, 2004). The political implications were not immediately evident but they would eventually come to the fore.

The various challenges posed by the long downturn were also expressed in a crisis of the state. In the first place, there was a fiscal crisis. As rates of economic growth slowed down but public expectations of gradually improving and expanding government services built up during the golden years of the 1950s and 1960s failed to adjust, the gap between state revenues and expenditures risked turning to the negative. This was the famous 'fiscal crisis of the state' about which little is now heard, but at the time it attracted a good deal of interest (Gough, 1979; O'Connor, 1973; Offe, 1984).

Second, there was a crisis of monetary policy. In attempts to relaunch economies in the context of the long downturn, governments continued to engage in positive monetary intervention of the Keynesian sort. The problem was, that production did not respond proportionally, resulting in high rates of inflation. The initial attempt to bring inflation down was through controls on prices and incomes; but firms opposed price controls and respective labour movements rejected controls over incomes. The way out would prove to be what was, in effect, a displacement of responsibility for monetary policy from national governments to international money markets. The formation of government monetary policy would henceforth be a process of adjustment to the expectations of wealth holders around the world, something that the easing of capital controls made possible. Governments believed to be engaged in expansionary monetary policy that risked inflation would be punished as wealth holders sold the currency in question, driving down its value, and well below any value that might have been targeted if the purpose of policy had been to devalue so as to achieve an advantage in export markets.

The advantage of this was that governments could no longer be blamed for the shortcomings of respective economies since they could claim that forces beyond their control tied their hands (Burnham, 1997; Clarke, 1990; Notermans, 1997). Henceforth, how a country was viewed by international money would assume priority. Central banks would have to be independent of the government to dispel all suspicion that they might engage in 'irresponsible' policy. Policies long pushed by the labour movement, like public ownership, could likewise be taken off the table. The objective, anonymous power of money would displace democratically informed state restructuring.

This had clear and significant class consequences. The big gainer was and continues to be finance capital since the erosion of its interest payments by inflation was staunched. Meanwhile, under inflation, industrial capital had tended to gain from the gradual devaluation of those same interest payments. It might be argued that the macroeconomic austerity that resulted from yielding to the expectations of international money would have worked to its advantage since it took demand out of the domestic economy and so induced some slack in the labour market. But this also meant a contraction of domestic demand so that international competition for export markets became all the more intense; a regime that Albo (1994) dubbed 'competitive austerity'.

This might be to draw too sharp a distinction between finance and industrial capital. Finance strengthened its grip over industrial capital in other ways, as Dumenil and Levy (2013) and Henwood (1998) have made clear. The relationship between management and stockholders changed under the banner of 'shareholder value' and this after a period during capital's golden years when management was more independent and able to take a longer term viewpoint on investment instead of a fixation on current stock prices and dividends and a short-term focus on current earnings; a regime into which they would be assimilated by fixing management salary increases to dividends or share value increases.

The big losers of the new policy regime, therefore, have been the working class. Even as productivity has continued to increase, wages have stagnated, and particularly for those lacking the skills that qualify them for the new growth industries like financial services, IT, health care, higher education, and transport equipment (excluding the auto industry). Meanwhile, shifting policy emphasis from full employment to curtailing inflation, combined with plant closures as production is transferred to an NIC or to Eastern Europe, has shrunk job markets and put downward pressure on wages.

Seemingly, global imperatives in the form of short-term international money movements and the siren calls of relocation to an NIC or subcontracting to a firm located there have taken over. A new regulatory space has been created supervised by a World Trade Organization bent on freeing up trade, albeit to the advantage of the developed world and then monitoring the arrangements; and by a re-energized IMF to ensure that international creditors get paid back by laggard countries that had been persuaded to engage in orgies of deficit financing. This, alongside the increasing significance of free trade areas and common markets is the globalizing bit of the glocalization argument.

There is also a case for the localizing part. Keynesian economy policy had focused on demand: putting money into the economy to stimulate investment and consumption. But now governments struggled to meet demands for welfare state expenditures. They would never succeed in reducing them; only slowing down expansion. It did though mean a shift in the discursive environment in which governments asked individuals to take responsibility for their own futures and those of their children; to be efficient market calculators and providers. 'Family values' were now emphasized by the right wing of the political spectrum as a means of reducing demand on state services but also as a sop to masculinist anxieties resulting from the rise of the women's movement. A world of centralized state provision was supposed to give way to one of decentralized responsibilities and provisions located in families.

The crisis of the 1970s was also the occasion for the displacement of so-called 'demand side' by 'supply side' policies. These had implications for urbanization and the regions. Urbanization and the decline of rural and small town populations had a long history, of course. New roles called into being by the New International Division of Labour would now lend impetus to the growth of large urban areas (Veltz, 1996). Discursively the virtues of agglomeration were re-discovered, as in Porter's arguments about the competitive advantages of industrial clusters. The same intense division of labour was now seen as providing a crucible for innovation, whether it be in industrial or financial products. The metropolitan area was a condition for the sort of enhanced product competition that globalization in general and the shift to more skills-intensive production demanded. But this in turn meant challenges: a reason, in other words for metro-regionalism as a vehicle for solving housing problems and easing traffic congestion – so many versions of Le Grand Paris, therefore.

A Crisis of Uneven Development

It is significant that the initial moves paving the way for the second globalization were made by the US and to a lesser degree, the United Kingdom. They were the two countries most affected by the long downturn and the increasing severity of international competition subsequent to reconstruction in Western Europe and Japan. In retrospect the closure of the gold window by the US and the decision in 1973 to float the dollar against other currencies was only a matter of time and would be followed very quickly by the floating of the pound sterling. The Reagan and Thatcher governments would then be the poster children for the new globalizing world and the subordination of monetary policy to the judgements of international money.

Both countries stood to gain from this. In the British case it would be via the subsequent growth of financial services which would restore London to the sort of advantages in global finance that it had lost with the period of financial repression during the Bretton Woods years and as a way of making up for the failures of industry. In the American case, and whether it was intended or not, the relaxation of capital controls and a world of floating currencies would give the dollar a vaunted position in international finance.

The dollar would become *the* reserve currency. It was not only that many raw materials, particularly oil were traded in dollars. It was also that because of the absolute volume of American exports, and the greater variety of products exported by the US, there was a greater demand for dollars anyway as opposed to, say, for yen, deutschmarks or pound sterling. In consequence, as the new regime of inflation control set in, for countries guarding the value of their currencies so as not to encourage a run on it, the dollar reserves of central banks assumed an additional importance: they would be used to support the value of the national currency by selling dollars in exchange for it. Meanwhile all the dollars circulating in the world might be parked in American or British banks, facilitating their short-term lending and giving a boost to respective financial services industries, particularly in New York and in London; and this in addition to the fact that given the role of the dollar, most countries prefer to have their trade financing organized through those same banks.

Increasing oil prices during the 1970s would give a further boost to this lending activity. On the one hand, since oil was priced in dollars and increased prices resulted in dollar earnings for the oil producing states of such proportions that they could not be immediately invested, they too were deposited either in American or British banks. On the other, increased oil prices meant trade deficits for importing countries that had to be financed through loans.

To the extent that the debtor countries had difficulty paying back the loans, the emerging scalar dispensation would receive a further boost through what would become one of its major institutions, the IMF. The IMF was a relic of Bretton Woods where it had played the relatively benign role of providing short-term finance for member countries experiencing balance of payments deficits, but without the conditionality for which it would become notorious. It would now be used as a battering ram to force open national economies, particularly those of developing countries, to international trade and to the reduction of those capital controls that were impeding the perfection of dollar hegemony around the world. At the same time, eliminating capital controls enhanced the possibilities of overseas direct investment by American corporations. As the worldwide demand for dollars boosted its value and so made American industrial exports that much more expensive, so this became more important as a way of tapping into foreign markets: easing the rules on repatriation of profits, including taxes; on eliminating those that required some proportion of domestic ownership; opening up strategic sectors hitherto denied to foreign ownership.

Even so, the relaxation of capital controls was very uneven. Among the NICs of East Asia, Japan had been an inspiration for a very different development model. Limiting foreign investment was part of a broader approach emphasizing state intervention and collaboration with banks in the allocation of credit to strategic industries and the like. Foreign investment was seen as a threat to the model and was discouraged. Meanwhile, the NICs, along with Japan, thrived and through their exports were a challenge to American corporations elsewhere in the world, and not just in the US (Wade, 1996). From the American point of view the problem was: How to rein them in, destroy the East Asian model and clear away the obstacles to American investment? The crisis in the region in the late 1990s would be the opportunity for what was termed the Wall Street–IMF–Treasury complex (Wade and Veneroso, 1998). To make up for a reduction in foreign direct investment from Japan, countries had turned to short-term international loans (Bello, 1998). More speculative investments followed by a rise in the interest rate on foreign money then triggered a crisis as banks tried to call in loans to compensate for the non-performing ones. Firms went bankrupt and exports were threatened. The attempt to resolve the crisis was through the IMF, which was in effect the opportunity to open up the countries most affected by the crisis, like South Korea, Indonesia and Thailand to the multinationals and so compromise their planning priorities.

These moves fortified the position of the Wall Street–IMF–Treasury complex. But France and Germany had also dragged their feet. What would subordinate *them* to the new monetary regime and its attendant competitive austerity in international trade would be the EU. One of the earlier arguments for the original common market of six countries had been that of creating a market of a size that could compete with the Americans: facilitate production runs that would lower production costs in the way that it was believed had been the case in the US (Georgiou, 2017). In that respect the various moves towards the perfection of the single market in the form of the Single European Act of 1986 was a landmark. Trade within the EU had already been liberalized via the elimination of quotas and tariff barriers. The goal of the 1986 legislation was to eliminate the barriers posed by production standards which varied in terms of things like hygiene and safety. It also foresaw a process of eliminating limits to the movement of workers between countries. Subsequently, extension of the EU to the east after 1990 to include the former Soviet satellites and the creation of the Eurozone in 1999 would bring the law of value to bear on producers and workers in Western Europe in an altogether more decisive way.

In sum, the US and to a lesser extent the United Kingdom, took the lead in creating the conditions not just for globalization, but for a particular form of globalization. This was, first and foremost, one in which finance would assume the dominant role, as others have noted (Gowan, 1999). This suited the US because of the relative ease with which, in a world where international finance was set free, the dollar would be supreme; and the United Kingdom would benefit because of London's role as an international financial centre that would thrive on doing business in offshore dollars. It was also, though, a distinctly Americanized form of globalization because of the way it expressed the peculiarities of the American social formation. I am not thinking here of consumerism. This is a more general, internally generated feature, of the advanced capitalist societies. Rather it is the quite extraordinary subsumption of labour to capital – economically, culturally, politically – that has long been a feature of that country. The balance of political forces there never even approached that achieved in Western Europe during the golden years. The weakness of the American welfare state is just one result. The US is, from the standpoint of capital, perhaps the ideal society. In Western Europe, the remnants of pre-capitalist forms hindered capital's march to dominance (Cox, 2016: Chapter 8). Globalization, though, whether intended or not, has released forces that would weaken labour, detach it from its old solidarities and pave the way for a more American, and therefore, more capitalist world.

Beyond Institutions

This, and in the spirit of the scale restructuring school, is to emphasize institutions: the role of the IMF, the dismantling of capital controls, free trade agreements, metro-regionalism and the territorial structure of the state. Structured coherences at both national and international levels have been restructured: internally related complexes of institutions, state forms, divisions of labour and employment relations. This, though, is to marginalize the role of discourse. This has both subjectified and disciplined so as to embed a particular imaginary of the global and then work to discourage backtracking.

The imaginary in question has been a territorial one. Notions of territorial competition, business climate and the need to liberalize labour law had already been rolled out in the US. In the 1960s the so-called metropolitan fiscal disparities problem was interpreted through the lens of a competition for tax base. Later, the disparate growth of Coldbelt and Sunbelt would be given a similar interpretation. In order to justify policy changes aimed at making a country more attractive to foreign investment these central ideas would be extended to the international sphere. Supposedly ideologically hide-bound labour movements became the scapegoat for failures to compete in the global marketplace. The axis of interpretation of national political economy shifted from one of class to one of how a country ranked according to international benchmarks of labour productivity, number of days lost to labour disputes, and wages.

Dislodging populations from attachment to the welfare state and to class solidarities and abandoning a collective upward mobility for a more individualized notion of achievement, then freed them up for a territorial discourse that divided people *within* countries as much as between. Within wider metropolitan areas enduring conceptions of territorial competition for tax base (at least in the American case), territorial exploitation, including the effects of residential exclusion, would make metro-regionalism that much more difficult. Localities and regions contested locations and state funding. Subra (2007) writing about the French case lamented the loss of a sense of national interest prevalent during the *Trente Glorieuses* and the blossoming of more local and regional interests and identities; while class interests underwent an eclipse. In Germany the equalizing payments from the favoured *länder* to the less advantaged became an issue. In other words, territory and territorial discourse acquired a new legitimacy. Countries, regions would then be ordered so as to reinforce the message: a world that included at the bottom end 'failed states', 'failed regions', and 'rust belts'.

Globalization in Context

Yet in assessing globalization and its contradictions, it has to be recognized that it has been far from a closed system in its practices and conditions. Rather globalization and its attendant social struggles have overlain or corresponded with other tendencies, mobilized them, been colonized by them, and had effects on them. By no means is it the case that the de-mobilization of class was a straightforward result of globalization. Rather it built on tendencies that had been apparent for a long time. What was being noticed in the late 1950s was a change in dominant family forms: a concentration on the nuclear family and a loosening of wider kin relations. In Britain this was called 'the symmetric family' (Young and Willmott, 1973): symmetric in the sense of a new spousal equality. This was a more child-centred family with a focus on shared consumption: notably the television, the family car and the privately owned home. In part this was associated with increasing differentiation within the working class. The division of labour assumed more intensely divided, hierarchical forms enabling vertical mobility. This would then be the condition for new subjectivities of 'getting on' and competitive display and would lay the basis for the financialization of social relations that began to be noticed from the early 2000s. Discourses of territoriality would help undermine class identities, but the process was already underway.

Changes in the division of labour have lent impetus to this. The decline of employment in those sectors of the economy that were always most unionized – coal, steel, shipbuilding, heavy engineering – has been going on for a very long time and globalization has simply intensified it through the transfer of what remained of those industries to the NICs. Meanwhile, service industries have emerged as the dominant sector, laying the basis for the post-industrial city and a metropolitanization which, again, is far from being a simple result of globalization.

More coincidental with globalization have been two other transformations. The growth in women's wage work has been little remarked in discussions of globalization, unless it is the female-employing light assembly plant in an NIC. But it has been truly dramatic in the advanced capitalist societies. Growth in female participation rates has been particularly strong in North America, Australia, Netherlands, Sweden, and in the

United Kingdom (Costa, 2000: Figures 3 and 4). Male participation rates have tended to decline. What made a crucial difference was the growth of service employment; i.e. those sectors deemed by gender ideology to require personal attributes of a feminine sort, though the way had already been cleared for women to enter wage employment by declining birth rates and the spread of household appliances. This might not sound as if it has much to do with globalization but it has certainly fed into the way in which men would eventually be disproportionately attracted to the recent populist backlash. On the other hand, the dual income family has also been one of the ways in which families have coped with static wage levels over the last three decades, possibly preempting resistance to a restraint on incomes widely attributed in various mediated ways to 'globalization'.

The other coincidental shift has been neo-liberalism. It is certainly the case that globalization made many neo-liberal nostrums politically feasible. Without the subordination of monetary policy to world money and the shucking of responsibility by national governments, the implementation of monetarism as a means of stemming inflation would have been much more of a political hot potato than it was. Likewise, supply-side policies appeared as a possibility once demand-side policies had been made more difficult by the new regime of austerity. Globalization would then facilitate the retreat of the state in other ways. The dollarization of the global economy made the privatization of state enterprises more attractive since, to the extent that the sale was to some American MNC, it could generate dollars. Opening up national economies to foreign investment could serve the same purpose. There was, though, no perfect alignment of neo-liberalism with globalization. To some degree it floated and continues to float freely on its own, as in charter schools, public-private partnerships and a good deal of deregulation, as in the case of the airlines.

CONTRADICTIONS OF THE SECOND GLOBALIZATION

In the claims of the scalar restructuring school conflict is muted: an effect of their emphasis on markets and competition. Glocalization, new state spaces and metro-regionalism are all seen as responses to the new competitive pressures of a globalizing world. This is to minimize the effects of class and territorial contradiction: something that is becoming increasingly evident as they now surge to the fore. Class and geographically uneven development and the tensions that they generate were fundamental to the emergence of the second globalization and its scalar expression. Globalization did not abolish them. Rather they are now expressed in different ways, but this time threatening a retreat from globalization and the neo-liberalism of which it has become one expression.

Class

The first thing to note is that the new neo-liberal regime of financialized globalization never reversed the long downturn in growth rates. Instead, and since the early 1970s, it has been a case of one step up and two steps down.[4] Productivity has continued to increase since 1970 but at a decreasing rate. Meanwhile, inequality, particularly in the US and the UK, has, with some significant exceptions like France and the Netherlands, increased dramatically; in other words, while the propertied classes have taken an increasing share of the surplus, investment in equipment and organizational change so as to increase productivity seems to have lagged.

In the US hourly wage compensation is almost flat from the early 1970s on while productivity continues to increase. In the remainder of the OECD countries this trend sets in later, round about 1990, and while wages are not entirely static, they increasingly lag productivity growth. This has been attributed at least in part to the diminished bargaining power of the working class, resulting in turn from declining union membership, particularly, again, in the US and in the UK; and also to an increased downward pressure on wages as a result of a greatly expanded global labour market, particularly, of course, since the implosion of the socialist bloc. Subsequent plant closures have decimated the labour movement even while declining membership owes something to changes in occupational composition. Wage repression has in turn undermined domestic consumer markets; savings have been run down and home equity drawn on to make up the difference. Meanwhile tax cuts for the wealthy, supposedly to encourage investment, along with strong pressures on behalf of social expenditures – which in turn suggests some mass residual power through the ballot box – have left national budgets adrift. But deficit financing has then created new opportunities for making money for the wealthy; far preferable to seeing their taxes go up so as to bridge the financing gap and giving further impetus to widening income gaps.

One argument is that the defeat of the working class entailed by the second globalization and the neo-liberal turn has been a mixed blessing for capital. In the classic Marxist model, it is worker resistance to lower wages and deteriorating work conditions that provides the incentive for capital to replace workers through mechanization and reorganization of the division of labour, which means the increased productivity of those still employed. Through subsequent overaccumulation of capital and the swelling of the industrial reserve army, this then creates the conditions for the emergence of new lines of production. But if the resistance of the working class is broken, this logic breaks down. As Streeck has argued:

> [c]apitalist progress has by now more or less destroyed any agency that could stabilize it by limiting it; the point being that the stability of capitalism as a socio-economic system depends on its *Eigendynamik* being contained by countervailing forces—by collective interests and institutions subjecting capital accumulation to social checks and balances. (Streeck, 2014: 46–47)

Uneven Development

Everywhere there have been new forms of uneven development while older ones have been deepened. Before the second globalization no-one talked about 'rustbelts' or 'new growth areas' but they are to be found side by side in the US, in Belgium, in Italy and

---

[4]   See, for example, Figure 1 in Streeck (2014).

in Germany. An intensified division between larger cities on the one hand and smaller towns and rural areas on the other has then been superimposed. World cities – and no one talked about them either – have risen on the growth of financial and producer services. They are, though, the pinnacles of a new urban hierarchy, as Peter Taylor and his research group have shown (Taylor et al., 2007); more a matter of variations on the post-industrial city therefore. While numerous small towns have seen their economic base voided by plant closures in favour of greener pastures in China, Eastern Europe and Mexico, expanding service industries in the big cities have sucked in the young.

Implementing metro-regionalism has also been fraught, running into the headwinds of peri- and sub-urban populations determined to hang on to their privileged positions in geographic divisions of consumption. Meanwhile, inner city rustbelts mean that no wider metro area is interested in combining forces with central cities if it means bailing them out financially. It is also the case that emphasis on metro-regionalism marginalizes the metro/small town relation: metro-regionalism, to the extent that it succeeds, will further deepen what was always an unequal relation with surrounding hinterlands simply through the agglomerative power of the metropolitan areas.[5]

The centres of the crisis have been the United States and the United Kingdom: again the two countries that played such a major role in initiating the process of freeing up trade and finance; but also, and significantly, the two that had records, of which they thought that they could be proud, of being hegemonic powers and which has given them a fatal sense of superiority. In both cases again, overlapping contradictions – regional, national, urban–rural, and ones of respective free trade areas exposing economies to migration from far less developed areas – have brought things to a crisis point, resulting in quick succession in populist rejections of the neo-liberal, globalizing agenda.

In the US the crisis of the small town, continuing de-industrialization for which NAFTA is held at least in part responsible, along with the threat to the Caucasian majority of continuing immigration have stimulated the rise of a national populism. The talk is of retreat behind national borders so as to rebuild American industry. The implication is the loss to globalizing forces of its most active, if not aggressive, leader. In the United Kingdom the contradictions have taken the form of a regional question and the country's position within the EU. To some degree they have reinforced one another.

The regional question, opposing London and the Southeast to 'the rest of the country' is of long standing but arguably it is one that has been intensified by membership of the EU. To assign the de-industrialization of the North, Northern Ireland, Scotland and Wales to globalization would be a stretch. They always had a legacy of coal mining, heavy industry and labour-intensive textiles doomed to disappear. And even during the golden years British industry could only fend off the competition of the French, the Germans and the Japanese by staying out of the EU and concentrating on the protected markets of the commonwealth (Georgiou, 2017).

---

[5] As in, and to take two cases, the decimation of small town hospitals and health care more generally by the big urban research hospitals; and the way in which e-commerce is eroding the retail base of the small towns while the compensatory employment in distribution centres goes once more to the larger cities and their suburbs.

Meanwhile the country had always had a global advantage in the financial services headquartered in London and the Southeast and one that had been enhanced by the rise of international money in the 1970s. Once in the EU, and having to confront the enduring weaknesses of British industry, encouraging this sector would be a major plank of British economic policy. On the other hand, membership of the EU also suggested one way in which industrial revival might be approached. Takeovers and new investments by the Americans and the Japanese – always seen as more savvy industrially – would be encouraged by selling the country first as providing access to EU markets; and second, by a business climate emphasizing light touch regulation. The problem was that when, under pressure of technical change and challenge of imports from the NICs, the multinationals rationalized their production in the EU, the United Kingdom, with its lower social costs of plant closure, would be the loser. It was this same light touch that would also result in opening the country up to immigration from Eastern Europe in 2014, and rejecting the permissible seven-year grace period; and immigration would, of course, be a major issue in the Brexit referendum of June 2016 (Georgiou, 2017). Meanwhile inflationary pressures, while not necessarily felt in the rest of the country, would be tackled by an increase in interest rates: something always opposed in the North and in Scotland where it was seen as curtailing the industrial investment that they wanted.[6]

From the standpoint of continuing globalization, at least of the neo-liberal sort, the danger now is that resistance will spread. The spotlight has been on the EU and the Eurozone. Expansion of the EU to the east not only fueled the immigration issue in the United Kingdom; it also created new opportunities for firms to relocate while retaining access to EU markets. France and Germany have been particularly affected. The Eurozone is also now in question. Its constraints on national monetary policy have worked their wonders: no way for Greece to make itself more competitive by devaluing a no longer existing drachma; or for Ireland or Spain to rein in speculative construction booms by increasing interest rates. Rather, in order to finance deficits they had to borrow and when the banks came calling after 2007, the game was up. The tensions are felt more widely. France's withdrawal, unless it can negotiate more generous terms on what is permissible in fiscal deficits, and disruptive as it might be, is only a matter of time. As in the United Kingdom, anxieties have been particularly intense in areas of the east and northeast experiencing de-industrialization: a major hotbed of support for the National Front.

Everywhere, it is those without qualifications, the less skilled who have been most aggrieved. It would be wrong, though, to see this as a direct expression of class. De-industrialization has hit most of all heavy industry and mining: the most masculine sectors, in other words. Through its reliance on the container, globalization then doomed thousands of jobs for dockworkers. The experience of women in wage work has been somewhat different. While still in a minority, their participation rate is now well over 80 per cent of that of men in France, Germany, the United Kingdom and the US. It should be no surprise, therefore, that the vote for Trump was heavily loaded

---

[6]   The US and the United Kingdom also turn out to be the most unequal of the advanced capitalist societies, though this can only in part be laid at the door of a changing division of labour subsequent to the rise of the NICs. Financialization of the firm has also played a part.

towards working class males;[7] the same applies for Marine Le Pen,[8] though by a lower margin and Brexit was more nuanced.[9] The sectoral composition of the economy, in other words, has shifted in a way that in terms of the gender stereotypes is more female-friendly; so to the resentments of class are added those of being masculine in an age of female emancipation (Katz, 2016; Kimmel, 2015). Masculine identity was always associated with being the breadwinner and that status is now seriously in question and in runaway fashion as women opt for late marriage, if at all, and female-headed households multiply.

In this chapter the discussions of scale ignited by the most recent round of globalization have been brought into a relationship with the geohistorical record and reinterpreted through the prism of geographical historical materialism. The class tensions intensified by the initiation of the long downturn, and felt more in some countries than others, would be a major stimulus to a reorganization of regulatory geography: a reorganization of its scalar division of labour in particular. In an unevenly developed world, the nice symmetry imagined by the glocalizers has to be in question. Changes in sub-national scalar geography have been hard to grasp and for this reason. But equally, and as we saw, there has been no straightforward transition to the more global either. Some dragged their feet more than others.

Accordingly, the discussion provides a number of different lessons for future investigations of geographic scale. Class and uneven development have to be to the forefront and this in turn means that what is happening at one scale is internally related to others. The new world of the second globalization was always for capital and for particular countries and, indeed, for those embedded in the more local spaces of accumulation that stood to gain from it, like London, Paris, Southern Germany, and the San Francisco Bay Area. This has meant that the contradictions in the global accumulation process that emerged in the early 1970s never went away and it is to them that we now need to look in order to understand why globalization and its geography are now in question.

REFERENCES

Albo, G. (1994), '"Competitive Austerity" and the Impasse of Capitalist Employment Policy', in: R. Miliband and L. Panitch (eds), *Between Globalism and Nationalism*, London: Merlin Press: 144–170.
Bello, W. (1998), 'East Asia: On the Eve of the Great Transformation?', *Review of International Political Economy*, **5**(3), 424–444.
Brenner, N. (2002), 'Decoding the Newest "Metropolitan Regionalism" in the USA: A Critical Overview', *Cities*, **19**(1), 3–21.

---

7   http://www.pewresearch.org/fact-tank/2016/11/09/behind-trumps-victory-divisions-by-race-gender-education/ (last accessed 12 July 2017).

8   http://tempsreel.nouvelobs.com/politique/elections-regionales-2015/20151208.OBS0971/elections-regionales-qui-a-vote-fn.html and http://www.ipsos.fr/sites/default/files/attachments/les_francais_et_le_fn.pdf (last accessed 12 July 2017).

9   Gender differences were apparent and in the expected direction but, and oddly, difference was hardly apparent among the most working class: https://www.ipsos.com/ipsos-mori/en-uk/how-britain-voted-2016-eu-referendum?language_content_entity=en-uk (last accessed 12 July 2017).

Brenner, N. (2004), *New State Spaces*, Oxford: Oxford University Press.

Brenner, R. (1998), 'The Economics of Global Turbulence', *New Left Review*, **229**, 1–265.

Burnham, P. (1997), 'Globalization, States, Markets and Class Relations', *Historical Materialism*, **1**, 150–160.

Clarke, S. (1990), 'Crisis of Socialism? Or Crisis of the State?', *Capital and Class*, **42**, 19–29.

Costa, D.L. (2000), 'From Mill Town to Board Room: The Rise of Women's Paid Labor', *Journal of Economic Perspectives*, **14**(4), 101–122.

Cox, K.R. (2009), '"Rescaling the State" in Question', *Cambridge Journal of Regions, Economy and Society*, **2**(1), 107–121.

Cox, K.R. (2014), *Making Human Geography*, New York: Guilford Press.

Cox, K.R. (2016), *The Politics of Urban and Regional Development and the American Exception*, Syracuse, NY: Syracuse University Press.

Crouch, C. and P. Le Galès (2012), 'Cities as National Champions', *Journal of European Public Policy*, **19**(3), 405–419.

Dumenil, G. and D. Levy (2013), *The Crisis of Neoliberalism*, Cambridge, MA: Harvard University Press.

Georgiou, C. (2017), 'British Capitalism and European Unification, from Ottawa to the Brexit Referendum', *Historical Materialism*, **25**(1), 90–129.

Gough, I. (1979), *The Political Economy of the Welfare State*, London: Macmillan.

Gowan, P. (1999), *The Global Gamble: Washington's Faustian Bid for World Dominance*, London: Verso.

Hancké, B. (2003), 'Many Roads to Flexibility: How Large Firms Built Autarchic Regional Production Systems in France', *International Journal of Urban and Regional Research*, **27**(3), 510–526.

Harvey, D. (1985a), 'The Geopolitics of Capitalism', in: D. Gregory and J. Urry (eds), *Social Relations and Spatial Structures*, London: Macmillan: 128–163.

Harvey, D. (1985b), 'The Place of Urban Politics in the Geography of Uneven Capitalist Development', in: D. Harvey, *The Urbanization of Capital*, Baltimore: Johns Hopkins University Press: 125–164.

Harvey, D. (2010), *The Enigma of Capital*, London: Profile Books.

Henwood, D. (1998), *Wall Street*, London: Verso.

Jessop, B. (2002), *The Future of the Capitalist State*, Cambridge: Polity Press.

Katz, J. (2016), *Man Enough? Hillary Clinton, Donald Trump and the Politics of Presidential Masculinity*, Northampton MA: Interlink Books.

Kiely, R. (2014), 'Imperialism or globalisation? … Or imperialism and globalisation: Theorising the International after Rosenberg's "Post-Mortem"', *Journal of International Relations and Development*, **17**, 274–300.

Kimmel, M. (2015), *Angry White Men: American Masculinity at the End of an Era*, New York: Nation Books.

McDowell, L. (2004), 'Masculinity, Identity and Labour Market Change: Some Reflections on the Implications of Thinking Relationally about Difference and the Politics of Inclusion', *Geografiska Annaler. Series B, Human Geography*, **86**(1), 45–56.

Notermans, T. (1997), 'Social Democracy and External Constraints', in: K.R. Cox (ed.), *Spaces of Globalization*, New York: Guilford Press: 201–239.

O'Connor, J. (1973), *The Fiscal Crisis of the State*, New Brunswick, NJ: Transaction Publishers.

Offe, C. (1984), *Contradictions of the Welfare State*, Cambridge, MA: MIT Press.

Streeck, W. (2014), 'How Will Capitalism End?', *New Left Review*, **87**, 35–64.

Subra, P. (2007), *Géopolitique de l'aménagement du territoire*, Paris: Colin.

Swyngedouw, E. (1997a), 'Neither Global nor Local: "Glocalization" and the Politics of Scale', In: K.R. Cox (ed.), *Spaces of Globalization*, New York: Guilford: 137–166.

Swyngedouw, E. (1997b), 'Excluding the Other: The Production of Scale and Scaled Politics', In: R. Lee and J. Wills (eds), *Geographies of Economies*, London: Arnold: 167–176.

Taylor, P.J. (2003), 'Global, National and Local', in: R. Johnston and M. Williams (eds), *A Century of British Geography*, Oxford: Oxford University Press: 347–370.

Taylor, P.J., B. Derudder, P. Saey and F. Witlox (eds) (2007), *Cities in Globalization*, Abingdon Park: Routledge.

Veltz, P. (1996), *Mondialisation, villes et territoires,* Paris: PUF.

Wade, R. (1996), 'Japan, the World Bank, and the Art of Paradigm Maintenance: *The East Asian Miracle* in Political Perspective', *New Left Review*, **217**, 3–37.

Wade, R. and F. Veneroso (1998), 'The Asian Crisis: The High Debt Model versus the Wall Street–Treasury–IMF Complex', *New Left Review*, **228**, 3–24.

Young, M. and P. Willmott (1973), *The Symmetrical Family: A Study of Work and Leisure in the London Region*, London: Routledge and Kegan Paul.

4. Globalization and border studies

James W. Scott

INTRODUCTION

The phenomenon of globalization is characterized by a multitude of definitions that emphasize the increasingly international nature of human societies, whereby much attention has been directed to flows, networked global infrastructures and cultural hybridization. In terms of the study of borders, however, it is perhaps understandings of globalization as *interpenetration* and *transformation* that are most salient. Interpenetration relates to transformations of everyday life and of social and economic practices by processes that operate across national boundaries (see Cochrane and Pain, 2000). The transformational nature of globalization is, furthermore, evidenced by iterative processes of wide-scale diffusion and a simultaneous domestication of forces, social, economic and cultural practices and ideas circulating at cross-border and global scales (see Held, 2000).

Importantly for this discussion, globalization has made the notion of a world without borders possible in the sense that socio-economic, financial, resource-based and, ultimately, environmental interdependencies increasingly lay bare the limits of state territoriality and sovereignty (Ceglowski, 2000). Taken to its most extreme expression, in fact, globalization sometimes even implies an end to borders in any politically and economically significant way – a scenario echoed by post-Cold War prophesies of a much less bordered future driven by global technologies, cyberspace, capital flows, political convergence and interstate integration (Ohmae, 1995; Scholte, 1997). Coupled with an equally extreme model of neoliberalism in which the state would no longer have an appreciable regulatory role, these processes might end the need for border studies altogether, except as footnotes in conceptual history. What we instead find is that the study of borders has enjoyed something of a renaissance since the new Millennium, reanimating debates about their social, political, economic and environmental significance (Newman, 2011). In fact, the renaissance of border studies can also be attributed to the emergence of counter-narratives to globalization discourses of the late 1980s and early 1990s, indicating that the 1990s fad of 'borderlessness' was too shallow to be sustained. As will be elaborated in more detail below, globalization has increased border sensitivities and the propensity to engage in border politics for a number of reasons; security and the maintenance of control figuring among the most critical. Given the real and imagined impacts of globalization, and the frequent spectre of threat to ontological and physical security, borders have attained increased significance as a defensive mechanism and guarantor of local stability.

Moreover, as Claude Raffestin (1992) has claimed, political boundaries are bio-ethno-social constants that make societies possible; borders are institutions without which it would be impossible to negotiate an 'inside' and 'outside' of any socially constituted space. Border scholars remind us that there are no inherent contradictions between political, social, cultural borders and globalization for the simple reason that borders fulfill a very basic social need, namely to create and manage recognizable spaces in which everyday life can be conducted. Moreover, globalization has proved a boon for border studies by multiplying the potential socio-spatial contexts within which political, economic, social and cultural processes unfold. According to Chris Rumford (2006: 163):

> A globalizing world is a world of networks, flows and mobility; it is also a world of borders. It can be argued that cosmopolitanism is best understood as an orientation to the world which entails the constant negotiation and crossing of borders … . Borders connect the 'inner mobility' of our lives with both the multiplicity of communities we may elect to become members of and the cross-cutting tendencies of polities to impose their border regimes on us in ways which compromise our mobilities, freedoms, rights, and even identities.

How then, do border studies interact with globalization, both as a concept as well as a process? Spurred on by globalization, border studies reflect continuity and change in scientific thought and thus innumerable contributions to the conceptualization of social space and its workings (Scott, 2012). Thanks to processes of globalization, the study of borders has moved from a dominant concern with formal state frontiers and ethno-cultural areas to the investigation of border-making in diverse socio-spatial contexts and geographical scales. This has elicited a shift to multifaceted processes of border-making and their social consequences. Globalization has also contributed to the breaking down of separations between discrete disciplinary approaches within border research. As a research field, border studies now encompass a wide range of disciplines besides human geography: political science, sociology, anthropology, history, international law as well as the humanities – notably art, media studies and philosophy. One only need browse through recent collections and compendia of border studies research in order to appreciate its post-disciplinary thrust (Donnan and Wilson, 2012; Andersen et al., 2012; Brambilla et al., 2015).

Another important consideration is the fact that globalization has opened up possibilities for investigating in greater depth the rationales behind everyday border-making by understanding borders as institutions, processes and symbols that not only reflect but condition societal change (see Popescu, 2012). Going beyond exclusively state-centred and territorial paradigms, the present state of debate emphasizes that borders are not given, they emerge through socio-political and cultural border-making or 'bordering' that takes place within society (Newman, 2006). Material borders, for example, do not emerge exclusively as a product of wars, agreement or high politics but are also made and maintained by cultural, economic, political and social activities. Bordering encompasses formal as well as everyday forms of border construction and is accomplished with the help of ideology, discursive and performative practices, and different forms of agency. Furthermore, everyday 'bordering and ordering' practices create and recreate social-cultural boundaries that are spatial in nature at the same time that they can also open up new spaces that reflect intersections, encounters and new affinities which emerge as a part of social life.

What this in fact indicates, is that globalization has challenged traditional disciplin-ary assumptions and certainties regarding borders. Talking about globalization has encouraged us to take seriously the interrelatedness of all previous thinking about studying and interpreting borders. In the contemporary practice of border studies, literature and art tell us as much about borders, borderlands and border crossings as do ethnographic or historical investigations. It is precisely the disruptive force of globalization – whether real or imagined – that drives home the main argument of border studies: that borders are in a constant process of confirmation, contestation, transformation and re-confirmation.

CONSEQUENCES OF GLOBALIZATION: BORDER STUDIES AND THE BORDERING PARADIGM

To begin with, globalization must be understood to be a non-teleological but transformative process that involves continuity and change. Moreover, in order to properly elaborate the interaction between globalization and the study of borders, a focus on borders as mediators, regulators and representations of processes associated with globalization is needed. In other words, in the context of globalization, border studies, in very broad terms, investigate processes of socio-spatial accommodation and adaptation to change that reflect non-finalizable makings and re-makings of borders. Consequently, the central border studies paradigm with regard to globalization is that of bordering, or the more fundamental process of creating socio-spatial distinctions at various scales by multiple actors (Van Houtum and Van Naerssen, 2002; Scott, 2012). Bordering is the everyday construction of borders, among others through ideology, cultural mediation, discourses, political institutions, attitudes and agency. The bordering paradigm thus brings diverse forms of social, cultural and economic globalization into a single frame of analysis. Moreover, the concept of bordering suggests that borders are not only semi-permanent, formal institutions but are also non-finalizable processes.

In terms of approaches and methodologies, globalization and its consequences are a major reason why border studies have largely abandoned descriptive, functional and positivistic approaches and instead developed highly critical perspectives, questioning, for example, the logics of border-making statecraft and focusing attention to the negative impacts of borders on individuals. Globalization has also challenged border scholars to more directly engage with bordering dilemmas and the ethical consequences of borders (Hing, 2010). Many of these dilemmas pertain to the contradictions between state territoriality and sovereignty and the heightened significance of mobility, the global reach of financial flows, information and ideas as well as environmental challenges and threats. At the same time, limited ability and willingness to come to terms with global interdependencies has also elicited a process of vigorous re-bordering, fencing and securitization of national spaces (see Rosière and Jones, 2012).

The bordering paradigm does not ignore the structuring power of state borders and formal border regimes. Although interdependence and processes of globalization have complicated the picture, the continuous (re)construction of borders based on forms of social-political organization and processes of nation-building remains a central problem in border studies. One criticism of contemporary border studies is its propensity to privilege agency at the expense of the wider social contexts and the structuring effect of state borders. Liam O'Dowd (2010) has been particularly blunt in his claims that recent post-national theorizing indicates a lack of historical reflexivity and careless 'epochal thinking'. On the contrary, states, state borders and their impacts are very much in evidence. As Paasi argues (2012: 2307) understanding borders remains inherently an issue of understanding how states function and thus: 'how borders can be exploited to both mobilize and fix territory, security, identities, emotions and memories, and various forms of national socialization'. Furthermore, according to Paasi 'this conceptualization of borders suggests that, while it is continually vital to examine how borders and bordering practices come about, it is also critical to reflect on the political rationalities and state-based ideologies embedded in these practices'. Looking at borders also forces us to take national and local experiences – and historical processes in general – seriously. State borders have a 'time print' – they symbolically reflect historical memories and can, in the sense of Megoran (2012) 'rematerialize' within changing national political trajectories. Similarly, the Finnish geographer Jussi Laine (2016: 467) writes that:

> The global primacy of state borders endures, but they are now commonly understood as multifaceted social institutions rather than as solely formal political markers of sovereignty. Borders are products of a social and political negotiation of space: they frame social and political action; help condition how societies and individuals shape their strategies and identities; and are re- and deconstructed through various institutional and discursive practices at different levels and by different actors.

However, border-making also proceeds through cognitive processes of creating space that reflect individual self-identifications with certain territories, cultures and political systems.

Taking these observations into consideration, we can argue that bordering serves to satisfy at least two basic needs, that of protection from external and internal threat and that of defining the territoriality, integrity and identity of groups and individuals. In a more traditional sense borders help determine both internal and external identities of territories, especially the states recognized by the international community: their right to maintain different relations, to create unions and associations, and to be represented in different unions, that is, to be 'legal' political actors. At a more subtle level, bordering is about a politics of difference, played out in different contexts and by different actors. The inculcation of political and social borders as a part of everyday life takes place through processes of socialization, for example in the family, at school and through the media. Border narratives, for example, have always, consciously and sub-consciously, thrown up the notion of difference which exists on both sides of a given (or supposed) socio-cultural border (Andrén et al., 2017).

If we accept the above, the bordering paradigm is in many ways a counterpart to the real and imagined impacts of globalization. Bordering indicates how globalization is 'domesticated' and incorporated, bordered, so to speak, within existing practices, conventions, cognitive spaces and lived places, not always in harmonious ways. Henk van Houtum (2005) uses the term 'b/ordering' to refer to the interplay between the ordering (of chaos) and border-making. Physical borders are not there only by tradition, wars, agreements and high politics but also made and maintained by other cultural, economic, political and social activities which accommodate the impacts of globalization. At the same time, everyday bordering and ordering practices connive to create and recreate new social-cultural boundaries and divisions which are also spatial in nature. Everyday lived experiences are also impacted by globalization – this results in intersections, differentiations and, under the most favourable circumstances, new convivialities in which gender, age and ethnicity meet and mutually constitute each other.

Table 4.1 captures some of the complexity of the bordering paradigm and indicates how globalization has affected border studies conceptually. The impacts have been profound. However, rather than challenge or question the centrality of borders as organizing elements of social life and human territoriality, globalization has confirmed their salience, while, at the same time, revolutionizing the ways in which we understand and interpret borders.

Table 4.1   Globalization's conceptual influence on border studies as reflected in the bordering paradigm

---

1) from state-centred to socially embedded perspectives
2) from bounded territoriality to multilocal and relational spaces
3) from normative assumptions to the questioning of border-making rationales
4) from 'objective' political rationality to the articulation of ethical concerns (violence and discrimination at borders)

CONCEPTUAL ELABORATIONS OF BORDERING

In its basic understanding, the bordering paradigm is quite general and can be rather abstract in terms of guiding empirical research. It is therefore undergoing constant development and refinement in order to accommodate different socio-spatial contexts as well as respond to the consequences of globalizing forces. In this section, some elaborations of the bordering paradigm that have linked bordering to other socio-spatial ideas will be provided. Power, ontological issues, the representational and symbolic nature of borders and the social ramifications of political bordering practices are some of the areas that will be briefly dealt with here.

Globalization and 'Post-national' Borders

Globalization is often conceptualized in terms of the transformation of state borders within the international economy and in response to global flows, environmental issues (e.g. global climate change and health issues), human rights and international terrorism, and so on. These issues have opened up space for scholarly work on the 'production' of borders, more specifically processes of bordering, de-bordering and re-bordering. Furthermore, border studies interrogates the control of mobility within the world system. Globalization has thus provided impetus for thinking of bordering in terms of a 'post-national' perspective. This could signify a new form of territorial sovereignty based on shared political responsibilities between states and/or the emergence of new territorial and cultural identities (Laine, 2016). With respect to Europe, post-national bordering approaches investigate the different ways in which national and European elements co-exist in the construction of borders within and between different political cultures and how these elements continue to shape opinions and attitudes on borders in different European countries (Scott, 2015).

Another important strand of post-national theorization is that of the emergence of new political and economic units that partly incorporate but are also beyond the context of the nation-state. The development of multinational (and geographically contiguous) zones of economic and political cooperation, such as the case of cross-border cooperation and transnational regionalism (see below), are one expression of the global forces that are restructuring the world system of individual states (Church and Reid, 1996; Goodwin et al., 2012). The concept of post-national borders can to an extent also be applied to new forms of territorial sovereignty that reproduce 'stateness' without traditional forms of external recognition. Examples of this are politically contested areas of the former Soviet Union such as Abkhazia, South Ossetia and Transnistria (Bahcheli et al., 2004).

Conceptualizations of post-national borders in no way suggest a disappearance of states or the decline of state territoriality per se. They instead suggest the potential emergence of new borders, new border functions and/or new methods of territorial control that go beyond traditional notions of state territoriality. Post-national borders might thus follow either sub- or supranational logics of political interaction. Such borders are post-national because they create new political functions of integration and interaction across state borders. Understood in these terms, post-national borders might define polities that transcend the jurisdictional and conceptual limits of state-centred orientations, for example as a community of states, as networks of cities or cross-border regions.

Borders and Place

Border studies represent an area of investigation that tempers emphasis of space-time convergence as a globally operating structuring force. It confronts 'spaces of flows' and 'power geometries' with ontological questions and the significance of place. The work, for example, of major urban theorists such as Castells (1997), Harvey (1996) and Massey (1993) has emphasized the socially constructed nature of place which increasingly reflects globally networked economic and social relations. On this view, places are interpenetrated by relational flows but, at the same time, locality is the space of bounded cultural formations and potential resistance to mobile capital. As Michael P. Smith (2001) has argued, emphasizing the global can result in a misreading of the significance of locale and place: rather than merely representing containers where global flows and networks become evident, cities are sites of localization, community and drivers of social and political innovation. Cresswell (2004) has also warned against theoretical rigidity in conceptualizing place as something either defined from within or determined by external (global) forces – places are characterized by adaptation to change and continuity. Bordering thus establishes a conceptual nexus between everyday practices of differentiating social space, the instrumentality of place-making, for example, as economic projects, and the ontological need for a sense of rootedness in place (Keresztély et al., 2017).

Along similar lines, Malpas (2012) has criticized what he sees as human geography's neglect of the ideational connections between space and place. Malpas (2012: 228) is particularly critical of sweeping aside ontological questions in favour of political commitments that include the 'theorization of spatial rhetoric and of spatial imagining as this forms the core of spatial politics'. In his development of the concept of place, Cresswell (2004) has voiced similar concerns, drawing attention to unproductive antagonisms between Place as something essential to existence ('Being in the World') and place as a product of negotiating spatial relationships ('Social Construct') when in fact both mutually contribute to place construction. In terms of its social operation, this interpretation of bordering draws attention to relationships between globalization therefore, territoriality *and* cognition. All three processes interact in creating a sense of socio-spatial difference and reinforcing a sense of groups and individual identity in space. This perspective addresses the shortcomings of dichotomous rather than comple-mentary understandings of relational and territorial borders which have been enumer-ated by a number of scholars, including Martin and Secor (2014), MacLeod and Jones (2007), and Paasi (2012). These authors suggest that the process of bounding space maintains its more general everyday significance despite the networked (and glo-balized) nature of life.

Border Walls and Bordering through Teichopolitics

Border studies increasingly deal with questions of border security and more general securitization policies that have proliferated during the new Millennium (Hall, 2012; Rosas, 2012; Weber and Pickering, 2011). Bordering is therefore also conceptualized in very concrete, physical terms as part of more general logics of control, security and border management. Globalization has, for example, engendered a politics of border closure that has been conceptualized as 'teichopolitics' (Ballif and Rosière, 2009) – the politics of building walls and barriers for security and economic reasons. Rosière and Jones (2012) in fact argue that 'teichopolitics' is best understood as the antithesis of the borderless world of globalization. As they suggest, new barriers are being built at asymmetric borders in the world where different spaces of economic, cultural, or political privilege converge. While the study of border walls and their evolution is an important agenda of practical borders research – this will be elaborated in more detail below – there are other important conceptual aspects involved as well. Despite its highly technical and legalistic nature border management is also a socio-political and cultural process: the basis for securitized differentiation is in fact generally a question of identity as well. Fabrizio Eva (2012) considers this form of physical bordering part of a process of 'self-caging' that is driven by interactions between iconographies of pyramidal (state) power and individual subjectivities regarding security and identity. Furthermore, Eva's notion of 'self-caging' reflects how discourses of threat and insecurity are internalized at the level of everyday life, creating more demand for physical, political bordering.

The Concept of the 'Border Multiple'

Others have developed the bordering paradigm by emphasizing the inherently multifarious nature of borders (Andersen et al., 2012). As 'multiple' processes, borders are not autonomous phenomena with their own unique ontology above and beyond their legal and institutional identities. Borders come to exist socially in the practices of different actors and the contested meanings attributed to them. Security borders, for example, are constructed by the performance of internal regulatory practices which challenge and constrain mobility across borders. The same borders, however, can have a very different significance, practically and symbolically, to other actors, for example as economic strategies or places of worship. The heteronomy of border practices, symbolisms, border-related identities and imaginations means that borders are both abstract and at the same time very concrete. The 'border multiple' is composed of Janus-faced, contested and contradictory narratives at different levels of practice, be it in the realm of memory and as imagined borders, in the realm of the political discourse and geopolitics or in practices enacting borders in the functional realm of administration.

In specific border zones, the geographic state border itself becomes embedded in everyday life and in the meanings attached to the local, as well as national, cultural environment, traditions, social habits and emotions. While it can be easy for individuals to cross the actual border, the border largely defines the spatial understanding of the local context. People make sense of their border-related social world in highly contextual and specific ways. The construction of meanings of borders can range from a desired barrier against a demonized 'other' and, as a means of exclusion, to its conception as an institution in potential need of reform but essential to economic survival. Border narratives should be read through their historicity and relationality. Understood in 'multiple' terms, bordering practices and social divisions affect one another, are constantly changing and can include as well as exclude. The 'border' and the divisions stemming from it are fluid, contextual and spatially manifested in the community and its relations with the state.

The Concept of Mobile Borders

Balibar's (1998) famous 'borders are everywhere' (BAE) proclamation has been criticized as a glib generality inspired by globalization and 'post-national' thinking (O'Dowd, 2010). However, with the bordering paradigm, BAE takes on a very concrete significance. As Paasi and Prokkola (2008) argue, borders are not 'located' merely in border areas but are everywhere in societies, for example in various forms of 'banal flagging' of the national in everyday life. Emotional bordering is loaded in national flag days and other national iconographies and practices – and this is also the 'location' of borders. Active 'borderwork' (Rumford, 2008) may deconstruct established and existing forms and codes of national socialization in some locations.

Moreover, the idea that borders exist everywhere where bordering takes places has given rise to the concept of mobile borders. It is more or less accepted that traditional dividing lines between the domestic and the international and between what is 'inside' and 'outside' of specific socio-spatial realms have been blurred as is strikingly depicted in the Moebius strip analogy of Bigo and Walker (2007). This has given way to understandings of borders embedded in new spatialities that challenge dichotomies typical to the territorial world of nation-states. Mobile borders are thus part of a security complex that goes beyond the level of physical walls and fences (Popescu, 2015). It includes a highly flexible system of delocalized and generalized border control mechanisms that differentiates between people their specific mobilities. Contemporary mobile borders can be created, shifted, and deconstructed by a range of actors (Amilhat Szary and Giraut, 2015). Luiza Bialasiewicz (2012) has documented that the mobility of political bordering mechanism also comprises 'off-shoring and out-sourcing', particularly in the case of the European Union where extraterritorial security perimeters have been set up in neighbouring regions far from the EU's physical external borders (Casas-Cortes et al., 2016).

Borderscapes

According to Jessop (2012: 74):

> [a]n imaginary is a semiotic ensemble (without tightly defined boundaries) that frames individual subjects' lived experience of an inordinately complex world and/or guides collective calculation about that world. There are many such imaginaries and they are involved in complex and tangled relations at different sites and scales of action.

Following the logic that geographical imaginations matter (see Gregory, 1994; Howie and Lewis, 2014), the bordering paradigm has appropriated the heuristic of imaginaries as a means of approaching complex socio-spatial processes (Brambilla, 2014; Bürkner, 2017; Vaughan Williams, 2012). Moreover, borders scholars such as Brambilla (2015a), Dell'Agnese and Amilhat Szary (2015), Grundy-Warr and Rajaram (2007) and Laine and Tervonen (2015), have suggested that the concept of bordering can be expanded to encompass 'borderscapes' – social/political panoramas that emerge around border contexts and that connect the realm of high politics with that of communities and individuals who are affected by and negotiate borders. As borderscapes, borders in fact cannot be reduced to instruments of terms of inclusion/exclusion as conveyed by biopolitical interpretations or metaphors such as 'Fortress Europe'. Instead, borders must be expanded to include what is happening in terms of everyday life at borders as reflected, for example in the agency of migrants. As borderscapes, borders emerge as fields in which processes of traversing and crossing meet those of reinforcement and blocking and in which borders are produced by social institutions and migration as a social force.

According to Brambilla (2015b: 111), borderscapes express 'the (geo)political and epistemic multidimensionality of the border, enabling a productive understanding of the processual, de-territorialized and dispersed nature of borders and their ensuing regimes in the era of globalization and transnational flows'. Much in accord with the 'border multiple' concept described above, the use of borderscapes as an epistemological tool involves connecting border experiences and border-making practices. Moreover, borderscapes involve an inclusive perspective on the political in which everyday practices, representations and border imaginaries as well as formal political processes are closely linked. The borderscapes concept also breaks down sharp divisions between territorial and relational understandings of borders. Moreover, the term borderscape expresses the representation of borders as individual as well as collective practices of border-making that express political subjectivities in particular ways. These bordering perspectives come together, among other ways, in the present geopolitical climate where, in stark contrast to the 1990s when discourses of 'de-bordering' Europe enjoyed substantial currency, Europe's (also worldwide) borders appear to have become formidable barriers symbolizing civilizational difference between East and West.

SELECTED RESEARCH PERSPECTIVES

The impact of globalization on research agendas in border studies is also evidenced by the increase in international comparative research and networks that have created a vibrant border studies community. The border studies community, which is now clearly global in scope, is represented by the *Journal of Border Studies*, the Association of Borderlands Studies (ABS) and informal groups such as the *Border Regions in Transition* network *(BRIT)*, which by 2018 has held sixteen international conferences, the last of these located in Nigeria and Benin. In addition, large projects have received funding in order to pursue investigation on the global roles of borders such as the aptly named Borders in Globalization (BIG) project, financed by the Canadian Social Sciences and Humanities Research Council. This project, inaugurated in 2014 seeks to 'build excellence in the knowledge and understanding of borders ... create new policy and foster knowledge transfer in order to address such globalization forces as security, trade and migration flows, and also to understand the forces of technology, self-determination and regionalization that are affecting borders and borderlands in regions around the world'.[1]

In addition, the EU-funded EUBORDERSCAPES project (full title: Bordering, Political Landscapes and Social Arenas: Potentials and Challenges of Evolving Border Concepts in a post-Cold War World), scrutinized conceptual change in understandings of borders in relation to transformations associated with globalization. Recognizing the close interrelationships between social change and paradigm shifts, the EUBORDERSCAPES project analyzed evolving concepts of borders in terms of their wider societal significance, critically interrogating 'objective' categories of state territoriality. This approach involved comparing and contrasting the ways in which different and often contested conceptualizations of state borders (in terms of their political, social, cultural and symbolic significance) resonate in concrete contexts at the level of everyday life.[2]

In this section, globalization will be related to border studies via discussion that indicates considerable thematic, theoretical and empirical diversity. We will specifically focus on: (1) borders, transnational spaces and cross-border cooperation, (2) the issue

---

[1]   The website address is: www.biglobalization.org (accessed 30 September 2017).

[2]   EUBORDERSCAPES, which ended in 2016, was funded by the EU's 7th Framework Programme for Research and Technological Development (Contract 290775). The website is accessible at www.euborderscapes.eu.

of borders as social and political resources and (3) key thematic areas that involve migration, mobility, securitization, state territoriality.

Transnational Spaces: Cross-border Cooperation

Globalization has led to the increasing importance of cross-border regions (Scott, 2017; Zimmerbauer, 2012). Cross-border cooperation between states has been the subject of interdisciplinary and comparative study for almost three decades. This research has been driven by at least one general core concern: i.e. transformations of nation-states and their consequences for economic, political, social and cultural life. Originally, research focused on urban and regional forms of 'subsovereign paradiplomacy' (Duchacek, 1986), and by the early 1990s the study of rapid cross-border urbanization on the US–Mexican border had also raised the question of local transnational planning and development responses (Herzog, 1990). This work has been subsequently developed in international comparisons of cross-border and transnational strategic alliances between cities, regions and other subnational actors (Amen et al., 2011; Goodwin et al., 2012; Setzer, 2015).

Partly spurred on by globalization – and in the guise of the European Union – the focus of research shifted during the 1990s from empirical research on transnational urban networks and their cooperation mechanisms to the study of local and regional forms of policy relevant cross-border interaction. Within this context, theories of state space rescaling and neoliberal governance have considerable influence (Brenner, 2004; Gualini, 2003). With specific regard to borders, a particular European characteristic of this research focus has been a more contextually sensitive understanding of the socio-political nature of cross-border cooperation. More directly, cross-border cooperation is defined by political projects carried out by private, state and, to an extent, third sector actors with the express goal of extracting benefit from joint initiatives in various economic, social, environmental and political fields (Perkmann, 2007; Svensson, 2014). Through new forms of political and economic interaction – both institutional and informal – it has been suggested that greater cost-effectiveness in public investment can be achieved, economic complementarities exploited, the scope for strategic planning widened and environmental problems more directly and effectively addressed. For these reasons, cross-border cooperation (CBC) is seen to promote the wider goals of European Cohesion.

Building upon the conceptual foundations of subnational paradiplomacy, border studies, particularly in the European case, developed during the 1990s and early 2000s a specific focus on cross-border policy integration as a form of 'multilevel governance' (Perkmann, 1999; Lepik, 2009). This focus remains important in terms of CBC policy within the EU. However, if the former approach positioned CBC within a context of globalization and transnational networks, the European perspective has been largely influenced by formal, structural understandings of transnational governance (see Blatter, 1997, 2004). For example, in order to overcome traditional forms of intergovernmentalism, institutionalization at the local and regional levels was seen as a necessary element for successful CBC (Scott, 2000). Prospects for transboundary regionalization have been thus defined by the outcomes of a gradual and complex process of institutional innovation and capacity-building at national, state and local levels. At the same time, the emergence of new planning forms across borders were prophesized in terms of regional dialogue. Dialogue, together with adequate strategies with which to reconcile and co-ordinate diverse interests, were seen to offer considerable promise for developing transboundary alliances between cities and their regions (van Geenhuizen et al., 1996; Leibenath et al., 2008).

Borders as Resources

Although state borders are often discussed as 'necessary evils' or formidable institutions of state control and sovereignty, border studies has for quite some time pursued the idea that, given the impacts and consequences of globalization, state borders also represent considerable resources, not just in economic terms but politically and culturally. The discussions briefly presented above regarding cross-border cooperation provides clear evidence of the importance of resource-oriented arguments in border studies. The economic potential of the border, expressed in wage and price differentials or economic complementarities is not a new story. Observers of cross-border trade on the US–Mexican border, for example, have developed a large body of work analyzing border effects in terms of economic integration and growth (Barajas et al., 2014; Hanson, 1996, 2001). However, globalization has also provided incentives for exploiting state borders as resources for economic, political and cultural cooperation. As discussed above, cross-border cooperation is premised on the idea that borders can represent 'bridges' between communities and regions in the pursuit of solutions to common challenges. Christophe Sohn (2014) has demonstrated that cross-border metropolitan regions have been able to use borders to achieve political recognition, exploit cost differentials, take advantage of cost-effective divisions of labour and promote themselves as international, global and multicultural places (at the same time, Sohn also mentions that the exploitation of border potentials can also exacerbate social inequalities, as in the case in US–Mexican metropolitan areas such as El Paso-Juárez). Eker and Van Houtum (2013) develop the ideas of borders as resources in a more cultural sense, as a place of common history, landscape formation, discovery and cooperation. In Europe's 'underused' border spaces, such as the Dutch-German borderland, the lack of controls and physical barriers is an invitation to jointly map, conceptualize and design common landscapes that invite exploration.

Mobility, Borders and Security

Mobility is about the ability to transcend borders whereas forced or self-imposed immobility strengthens the dividing power of borders. Mobility is thus a freedom, but it is not a freedom shared by all; in the neoliberal economy the hypermobility of the economically privileged can be contrasted with the marginalization of those with limited access to mobility networks. At another level, mobility is understood as conditioned by larger economic forces and the geopolitics that support their functioning. Vogt (2013: 764–765), for example, positions human mobility 'at the intersection between local and global economies' and 'embedded within logics of global capitalism and state militarism'. On this view, mobility is inherently subject to manipulation, control and punitive regulation. It is within this context that the positive characteristics of mobility has been questioned and subject to a highly politicized interpretation. On the other hand, as Doreen Massey (1993) has argued, mobility, such as that originally promoted by the EU, is a central element of a progressive sense of being that is at once robust due to its openness to the outside world and ability to connect with other cultures and other areas of the world.

Borders are at the core of discussions on security and the control of mobility at the national, European and global level: in terms of received understandings of security, borders represent an interface between domestic concerns and wider interstate and intercultural contexts. At the most basic level, borders serve to protect national societies from external threats while maintaining conditions for their economic sustainability. Beyond this, however, the functions and social significance of borders not only reflect the means in which security risks and challenges can be articulated and acted upon but also ethical questions of considerable importance. In tandem with the securitizing discourses that stigmatize mobility and migration, political pressure has increased for more formidable and militarized borders in order to defend national cultures, even at the risk of reducing cross-border mobility for everyday citizens and curtailing democratic rights (Jones, 2016). Border studies research has taken up ethical debate by questioning the Realpolitik of selective and restrictive mobility and interrogating the dilemma of bordering liberal societies (Jones and Johnson, 2016; Elden, 2009). Despite the European Union's visions of open borders, Europe's external boundaries are in many ways markers of inequality, exclusion and, as such, symbols of unfairness.

The European Union engages, for example, in humanitarian aid and conflict resolution in order to promote conditions for more equitable development on a global scale. But at the same time it has markedly restricted the possibility of asylum while invoking police powers and state violence in order to prevent, at a very high human cost, irregular entry into its territory. Through visa regimes and border politics the EU defines the ground rules of access for different groups depending on origin, citizenship, material situation and socio-professional background (Mau et al., 2012). In highly critical readings of the increasingly selective nature of EU external borders, the EU's practices have been compared with legalized apartheid: with the 'law of birth' determining an individual's degree of mobility across, and even occasionally within, state borders (Van Houtum, 2010). Furthermore, the violence of liberal states is not limited to their own territorial borders but is often extended to areas far beyond, such as in the case of the EU's security perimeter in the Southern Mediterranean (Casas-Cortes et al., 2016).

In de- and re-bordering processes, borders are mobile and territorially displaced. Border controls are, in principle, being carried out by anyone anywhere – by loyal inhabitants who inform police if they suspect the presence of 'illegals'. They also involve all-encompassing surveillance technologies and the compliance of private businesses, public agencies and social services (Popescu, 2015). Hence, borders as institutions are also ideal candidates for analyses of biopolitics and biopower, understood in their minimalist form: the control of population (Demetriou, 2013; Topak, 2014). The control of mobility takes place in rather mundane and unspectacular ways, such as through the use of biometrics and body scans at borders and airports. At the same time, borders can be seen as sophisticated techniques for the exercise of biopower and biopolitics and for the creation – discursively and physically – of spaces that are set outside everyday social contexts. These Agambenean (2003) exceptional spaces, such as the Guantánamo detention camp and refugee retention centres not only control mobility; they serve to externalize perceived societal threats and neutralize political resistance to security policies. In similar terms, limboscapes (detention centres), as portrayed by Ferrer-Gallardo and Abet-Más (2016), characterize the Moroccan–Spanish borderland at the exclave city of Ceuta, an iconic result of the EU's bordering practices.

THE EXAMPLE OF HUNGARIAN IDENTITY POLITICS AND REFUGEE 'CRISIS'

The vignette that follows encapsulates several of the conceptual and empirical links between globalization and border studies that have been elaborated in this chapter. Hungarian border politics in the period between 2015 and 2017 visibly express the nexus between globalization, mobility, national identity and borders. While 'humanitarian' issues loom large in border studies, equally complex ethical questions are raised by culturalist perspectives that emphasize strong borders as a means to protect the coherence and sustainability of national societies. As Vertovec (2011) indicates, cultural arguments are powerful instruments with which to marginalize accommodating political positions regarding asylum and migration. And this despite the fact that understandings of national culture are often based on highly questionable notions of homogeneity and singularity. Consequently, a focus on Hungarian border politics indicates ways in which powerful self-referential narratives of authenticity, national belonging and autonomy are influencing the securitization of mobility.

Radicalized national-conservative understandings of national identity have become a volatile element in Hungarian political discourse due to their non-negotiable and foundational nature. Moreover, this form of identity politics arguably contains 'revanchist' elements that promote a vigorous re-assertion of cultural chauvinism and political xenophobia. Such revanchism reflects a 'cognitive liberation' (Kallis, 2013) of nationalist and right-wing ideologies that feed cultural nationalism and directly challenge globalist assumptions that privilege mobility and cosmopolitanism. The electoral victory of Hungary's national conservative government in 2010 was a watershed in the emergence of contemporary culturalist politics and an apparent vindication of foundationalist understandings of nation. It was also facilitated by a deep sense of disillusionment with liberalism and EU membership (Pataki, 2013).

In this context, Hungarian Prime Minister Viktor Orbán's skilled use of the 'illiberal' idea provides a concrete example of how revanchism is put into political operation through an extreme self-referential view of Europe. Hungary's contestation of European Union attempts to find a community-wide approach to deal with the needs of refugees and migrants has received much attention.[3] With the installation and reinforcement of barbed wire fences and internment camps along its border with Serbia and Croatia, Hungary's mobility control practices have since 2015 achieved a new intensity. In legitimizing border closures and with a dismissive approach to Europe's refugee crisis,

---

[3]   'Hungary's zero refugee strategy', http://www.politico.eu/article/hungary-zero-refugee-strategy-viktor-orban-europe-migration-crisis/ (accessed 15 February 2017).

the present Hungarian government under Viktor Orbán has stylized itself as a guardian of Europe's historical legacy and Christian culture.[4] It has warned constantly of the dangers of 'unnatural migration' and the emergence of parallel (Islamic) societies that will threaten Europe's welfare, security and identity.[5] This discourse is supported by constant negative Hungarian media coverage of Europe's refugee crisis and conspiracy theories that suggest an 'externally' driven exploitation of Europe's open societies. Pointedly, in mobilizing support for fences, Hungary's prime minister suggests that 'illiberal' values are needed in order to protect national societies and guard against naive notions of openness and tolerance (Zalan, 2016).

Hungary, as dominated by the present national conservative (FIDESZ) government, is a key example of how border politics mobilize globalization as a threat to national identity and security and hence as a political resource. In this case, the national-conservative government has positioned Hungary as a major player in its quest to promote traditional values and as a defender of national sovereignty and identity.[6] Hungary is portrayed as not a follower, but an innovator and maker of Europe, according to notions of 'national Europe' against the 'political correctness' of Brussels bureaucrats and other major politicians (Szarka, 2017). Anti-globalist and anti-EU impulses are furthermore reflected in a repudiation of several major achievements of modernity and liberal society – achievements that can be associated with secular humanism, tolerance and mobility.

The relation to mobility is clear: the closure of borders and the installation of 'transit zones' cum internment camps along the southern borders is an expression of hostility to the change and encounters that mobility engenders: these are seen to dilute the authenticity of national culture and ethnic purity. With its border politics, the Hungarian government seeks, furthermore, to reduce the mobility and the visibility of refugees. Commenting on his government's 'zero refugees strategy', Hungarian Prime Minister Viktor Orbán argued in 2015 that:

> [T]he basics are that each nation is defined by its borders. Borders must be respected. And borders must be defended by the state. … And if you are a member of the European Union, especially the Schengen Area, you have an obligation to defend your national border, which is the European border, to stop them. Everybody who would like to cross the border in an illegal way: stop them and defend the border to defend your community and to defend Europe.

Hungary's open animosity towards the welcome culture idea expressed in 2015 by German Chancellor Angela Merkel is of course shared by many political groupings within the EU. The basis for the antagonism is the fear of socio-ethnic and religious

---

[4]   'Migration crisis: Hungary PM says Europe in grip of madness', https://www.theguardian.com/world/2015/sep/03/migration-crisis-hungary-pm-victor-orban-europe-response-madness (accessed 31 February 2017).

[5]   "'You're DESTROYING Europe" Hungary PM predicts "parallel Muslim society" due to migration', http://www.express.co.uk/news/world/715040/Hungary-referendum-Viktor-Orban-parallel-Muslim-societies-EU-migration-refugees (accessed 21 March 2017).

[6]   'Protecting Hungary's "national identity"? – Orbán's constitutional amendments and EU law', http://www.migszol.com/blog/protecting-hungarys-national-identity-orbans-constitutional-amendments-and-eu-law (accessed 31 March 2017).

tensions as well as increased social costs of caring for refugees. On the other hand, crime, terrorism and insecurity are openly associated not only with refugees but also illegal migrants (Fekete, 2016). The Hungarian government's securitization of mobility culminated in the creation during 2016 of four 'transit zones' on the border with Serbia and Croatia in order to 'fix', as it were, asylum-seekers in time and space and make them invisible to mainstream society. Since 2016, applications for asylum can only be processed at the border camps, anyone apprehended crossing Hungary's borders at other points will be sent immediately back to Serbia. The immobilization of refugee flows is both a deterrent as well as a clear expression of the government's resolve to reject any EU-level quotas or regulations that impinge upon national sovereignty. This policy is seen by critics as tantamount to imprisonment and a blatant violation of human rights.[7]

Popular Euroscepticism is not strong in Hungary but there is strong underlying agreement with the government's refusal to accept refugees and asylum-seekers. The October 2016 referendum against EU-wide resettlement quotas was not voted on by a majority of the population but its results did reflect overwhelming support. As such, while the revanchist discourses of the Hungarian government are for the most part a product of ideology and national conservative party politics, they resonate with everyday fears of the 'other' and a sense of imminent threat to national identity. Groups protesting Orbán's hardline nationalism and treatment of refugees have become more visible and vocal but are still marginalized within the domestic political landscape.

The forceful nature of the Hungarian government's arguments against refugee and migrant mobility is based on a deft strategy of invoking the inviolability of national borders and exaggerating threats to national cohesion, identity and sovereignty. As a result, the Hungarian government's politics of contestation could have the effect of making xenophobia less objectionable, particularly within the context of securitized understandings of mobility and migration. Kallis (2013) has warned that both right-wing contagion as well as weakening mainstream resolve to combat anti-immigrant sentiment are hindering a more open European debate about accommodating mobility, migration and asylum. This contagion effect is visible in the case of Hungary's border politics which have achieved general acceptance in Central Europe and elsewhere.

CONCLUSIONS AND A FUTURE RESEARCH AGENDA

Through the investigation of borders, we realize that there can be no hegemonic dominance of any specific social theory, whether critical or not, in the understanding of space and its social significance. And whereas space is abstract and absolute, we now understand that it is borders that make space intelligible, for example as everyday social places. Borders not only have different meanings for different actors but are also manifestations of power relations in society at different scales. By the same token, the present state of border studies indicates that processes associated with globalization

---

[7]   'Hungary's transit zones are actually prisons where even pregnant women are handcuffed', http://hungarianspectrum.org/2017/06/14/hungarys-transit-zones-are-actually-prisons-where-even-pregnant-women-are-handcuffed/ (accessed 11 June 2018).

have deeply changed the power of borders, modifying the dialectical relation between their fixed institutional nature and constantly changing, fluid processes of bordering within and between societies.

The progress made by border studies in terms of interdisciplinary and complex understandings of borders and their significance is unquestionable. Here, globalization has played a major role. The question is: whereto now? What might be possible future research agendas that maintain the innovative and integrating momentum as well as social relevance of border studies? One good place to develop research is in the ongoing ethical debate regarding the 'why' and 'how' of borders. An important target of these deliberations could be 'bordering dilemmas' within the context of globalization – a situation in which exclusion is often seen as a result of creating or maintaining inclusive and open societies (Elden, 2009; Jones and Johnson, 2016). For example, despite the European Union's visions of open borders, Europe's boundaries are in many ways markers of inequality, exclusion and, as such, symbols of unfairness. Nevertheless, at least in the 'Western' case, there exist few feasible alternatives to liberal notions of an 'inclusive' but self-defined and thus bounded community as a necessary precondition of democracy (Batt, 2002). It can be therefore argued that without a sense of closure and boundedness, the development and nurture of community and place identities is virtually impossible.

How then, can borders be 'de-securitized'? Bauder (2014) proposes that instead of regulating mobility with borders and arbitrary politics of inclusion and exclusion, a constructive alternative would be to link mobility (and migration) to the possibility of acquiring certain rights, including domicile-based citizenship, thus avoiding a future of uncertainty, statelessness and increasing social tensions. While such scenarios seem distant from political reality, at least at this point in time, desecuritization would be an important first step in opening up notions of community, belonging and citizenship to include an ever larger cross-section of humanity. It is also here where the concept of borderscapes can provide constructive impetus. Thinking in terms of borderscapes has direct ethical implications, adding to a rich social sciences and humanities engagement with borders that takes inspiration from the realm of philosophy. A potential way forward is offered, for example, by the political philosophy of Hannah Arendt, the reappraisal of which is indicative of the ethical issues involved in the securitization of mobility. Arendt (1968) has warned of world alienation and 'losing a sense of being in the world' and with it, identity. Identity is disclosed in the public sphere, the exclusion from which results in a loss of identification with the political system. Alienation also gives rise to the deterioration of the public sphere itself (d'Entreves, 1994). Making persons visible or invisible in the public realm is about bordering, about creating distinction. This idea resonates with Arendt's (1968) notion of the politics of appearance or the making evident of positions, interests and actors that represent them. Borren (2008) has suggested that Arendt's political philosophy can be adapted to criticize European policies that disenfranchise non-citizens through exposing them (as threats) and/or obscuring their claims, problems and motivations. Conversely, thinking of borders as borderscapes that reveal human conditions might stimulate positive agency and a politics of visibility could signify an expression of social acceptance and integration.

REFERENCES

Agamben, G. (2003), *Stato di eccezione (Homo sacer II.1)*, Torino: Bollati Boringhieri.

Amen, M., N.J. Toly, P.L. McCarney and K. Segbers (eds) (2011), *Cities and Global Governance: New Sites for International Relations*, Farnham: Ashgate Publishing.

Amilhat Szary, A.-L. and F. Giraut (2015), *Borderities and the Politics of Contemporary Mobile Borders*, London: Palgrave Macmillan.

Andersen, D.J., M. Klatt and M. Sandberg (eds) (2012), *The Border Multiple: The Practicing of Borders Between Public Policy and Everyday Life in a Re-Scaling Europe*, Farnham: Ashgate Publishing.

Andrén, M., T. Lindkvist, I. Söhrman and K. Vajta (eds) (2017), *Cultural Borders of Europe: Narratives, Concepts and Practices in the Present and the Past*, New York: Berghahn.

Arendt, H. (1968), *The Human Condition*, Chicago: University of Chicago Press.

Bahcheli, T., B. Bartmann and H. Srebrnik (eds) (2004), *De Facto States. The Quest for Sovereignty*, London and New York: Routledge.

Balibar, E. (1998), 'The Borders of Europe', in: P. Cheah and B. Robbins (eds), *Cosmopolitics: Thinking and Feeling Beyond the Nation*, Minneapolis, MN: University of Minnesota Press, pp. 216–229.

Ballif, F. and S. Rosière (2009), 'Le défi des teichopolitiques. Analyser la fermeture contemporaine des territoires', *L'Espace Géographique*, **3**: 193–206.

Barajas, I., N. Sisto, E. Gaytán, J. Cantú and B. López (2014), 'Trade Flows Between the United States and Mexico: NAFTA and the Border Region', *Articulo*, **10/2014**, https://articulo.revues.org/2567 (accessed 11 June 2018).

Batt, J. (2002), 'Introduction: Region, State and Identity in Central and Eastern Europe', in: J. Batt and K. Wolczuk (eds), *Region, State and Identity in Central and Eastern Europe*, London: Frank Cass and Co. pp. 1–14.

Bauder, H. (2014), 'Possibilities of Open Borders and No Border', *Social Justice*, **39**(4): 76–96.

Bialasiewicz, L. (2012), 'Off-shoring and Out-sourcing the Borders of Europe: Libya and EU Border Work in the Mediterranean', *Geopolitics*, **17**(4): 843–866.

Bigo, D. and R.J.B. Walker (2007), 'Political Sociology and the Problem of the International', *Millennium: Journal of International Studies*, **35**(3): 725–739.

Blatter, J. (1997), 'Explaining Crossborder Cooperation: a Border-Focused and Border-External Approach', *Journal of Borderlands Studies*, **XII**(1/2), 151–175.

Blatter, J. (2004), 'From "Spaces of Places" to "Spaces of Flows"? Territorial and Functional Governance in Cross-Border Regions in Europe and North America', *International Journal of Urban and Regional Research*, **28**(3): 530–548.

Borren, M. (2008), 'Towards an Arendtian Politics of In/visibility. On Stateless Refugees and Un-documented Aliens, Ethical Perspectives', *Journal of European Ethics Network*, **15**(2): 213–237.

Brambilla, C. (2014), 'Shifting Italy/Libya Borderscapes at the Interface of EU/Africa Borderland: a "Genealogical" Outlook from the Colonial Era to Post-colonial Scenarios', *ACME: An International E-Journal for Critical Geographies*, **13**(2): 220–245.

Brambilla, C. (2015a), 'Exploring the Critical Potential of the Borderscapes Concept', *Geopolitics*, **20**(1): 14–34.

Brambilla, C. (2015b), 'Navigating the Euro/African and Migration Nexus through the Borderscapes Lens: Insights from the LampedusaInFestival', in: C. Brambilla, J. Laine, J.W. Scott, G. Bocchi (eds), *Borderscaping: Imaginations and Practices of Border Making*, Farnham: Ashgate Publishing, pp. 111–121.

Brambilla, C., J. Laine, J.W. Scott and G. Bocchi (eds) (2015), *Borderscaping: Imaginations and Practices of Border-Making*, Farnham: Ashgate Publishing.

Brenner, N. (2004), 'Urban Governance and the Production of New State Spaces in Western Europe, 1960–2000', *Review of International Political Economy*, **11**, 447–488.

Bürkner, H.J. (2017), 'Bordering, Borderscapes, Imaginaries: From Constructivist To Post-Structural Perspectives', in: E. Opiłowska, Z. Kurcz and J. Roose (eds), *Advances in European Borderlands Studies*, Baden-Baden: Nomos, pp. 85–107.

Casas-Cortes, M., S. Cobarrubias and J. Pickles (2016), 'Good Neighbours Make Good Fences: Seahorse Operations, Border Externalization and Extra-Territoriality', *European Urban and Regional Studies*, **23**(3): 231–251.

Castells, M. (1997), *The Power of Identity*, Oxford and Malden, MA: Blackwell Publishers.

Ceglowski, J. (2000), 'Has Globalization Created a Borderless World?', in: Patrick O' Meara, Howard D. Mehlinger and M. Krain (eds), *Globalization and the Challenges of a New Century*, Indianapolis: Indiana University Press, pp. 101–109.

Church, A. and P. Reid (1996), 'Urban Power, International Networks and Competition: The Example of Cross-Border Co-operation', *Urban Studies*, **33**(8): 1279–1318.

Cochrane, A. and K. Pain (2000), 'A Globalizing Society?' in D. Held (ed.), *A Globalizing World? Culture, Economic, Politics*, London and New York: Routledge, pp. 5–45.

Cresswell, T. (2004), *Place. An Introduction*, London: Blackwell.

Dell'Agnese, E. and A.-L. Amilhat Szary (2015), 'Introduction. Borderscapes: From Border Landscapes to Border Aesthetics', *Geopolitics*, **20**(4): 4–13.

Demetriou, O. (2013), *Capricious Borders: Minority, Population, and Counter-Conduct between Greece and Turkey*, Oxford and New York: Berghahn.

d'Entreves, M.P. (1994), *The Political Philosophy of Hannah Arendt*, London: Routledge.

Donnan, H. and T. Wilson (2012), *A Companion to Border Studies*, Malden and Oxford: Wiley-Blackwell.

Duchacek, I. (1986), 'International Competence of Subnational Governments: Borderlands and Beyond', in: O.J. Martínez (ed.), *Across Boundaries. Transborder Interaction in Comparative Perspective*, El Paso: Texas Western Press, pp. 11–28.

Eker, M. and H. van Houtum (2013), *Grensland. Borderland. Atlas, Essays and Design. History and Future of the Border Landscape*, Eindhoven: Blauwdruk.

Elden, S. (2009), *Terror and Territory. The Spatial Extent of Sovereignty*, Minneapolis, MN: University of Minnesota Press.

Eva, F. (2012), 'Caging, Self-Caging, Materiality, Pyramids and Memes as Better Tools for Geopolitical Analysis: An Epistemological Anarchist Approach', *Human Geography*, **5**(3), https://hugeog.com/eva-53/ (accessed 11 June 2018).

Fekete, L. (2016), 'Hungary: Power, Punishment and the 'Christian-National Idea'', *Race and Class*, **57**(4): 39–53.

Ferrer-Gallardo, X. and A. Albet-Más (2016), 'EU-limboscapes: Ceuta and the Proliferation of Migrant Detention Spaces across the European Union', *European Urban and Regional Studies*, **23**(3): 527–530.

Goodwin, M., M. Jones and R. Jones (2012), *Rescaling the State: Devolution and the Geographies of Economic Governance*, Manchester: Manchester University Press.

Gregory, D. (1994), *Geographical Imaginations*, Oxford: Blackwell.

Grundy-Warr, C. and P.K. Rajaram (eds) (2007), *Borderscapes. Hidden Geographies and Politics at Territory's Edge*, Minneapolis and London: University of Minnesota Press.

Gualini, E. (2003), 'Cross-Border Governance: Inventing Regions in a Trans-national Multi-level Polity', *DISP*, **152**: 43–52.

Hall, A. (2012), *Border Watch: Cultures of Immigration, Detention and Control*, London: Pluto.

Hanson, G. (1996), 'Economic Integration, Intra-Industry Trade, and Frontier Regions', *European Economic Review*, **40**(3–5): 941–949.

Hanson, G. (2001), 'U.S.–Mexico Integration and Regional Economies: Evidence from Border-City Pairs', *Journal of Urban Economics*, **50**(2): 259–287.

Harvey, D. (1996), *Justice, Nature and the Geography of Difference*, Oxford: Blackwell.

Held, D. (ed.) (2000), *A Globalizing World? Culture, Economics, Politics*, London and New York: Routledge.

Herzog, L.A. (1990), *Where North meets South. Cities, Space and Politics on the US–Mexico Border*, Austin: University of Texas.

Hing, B. (2010), *Ethical Borders: NAFTA, Globalization, and Mexican Migration*, Philadelphia: Temple University Press.

Howie, B. and N. Lewis (2014), 'Geographical Imaginaries: Articulating the Values of Geography', *New Zealand Geographer*, **70**(2): 131–139.

Jessop, B. (2012), 'Social Imaginaries, Structuration, Learning, and Collibration: Their Role and Limitations in Governing Complexity', *Zarządzanie Publiczne*, **19**: 71–83.

Jones, R. (2016), *Violent Borders. Refugees and the Right to Move*, London: Verso.

Jones, R. and C. Johnson (2016), 'Border Militarization and the Re-articulation of Sovereignty', *Transactions of Institute of British Geographers*, **41**(2): 187–200.

Kallis, A. (2013), 'Breaking the Taboos and "Mainstreaming" the Extreme: The Debates on Restricting Islamic Symbols in Europe', in: B. Mral, R. Wodak and M. Khosravinik (eds), *Right-wing Populism in Europe: Politics and Discourse*, London: Bloomsbury Academic, pp. 55–70.

Keresztély, K., J.W. Scott and T. Virág (2017), 'Roma Communities, Urban Development and Social Bordering in the Inner City of Budapest', *Ethnic and Racial Studies*, **40**(7): 1077–1095.

Laine, J. (2016), 'European Civic Neighbourhood: Towards a Bottom-Up Agenda Across Borders', *Tijdschrift voor Economische en Sociale Geografie*, September 2016, DOI: 10.1111/tesg.12211.

Laine, J. and M. Tervonen (2015), 'Remaking the Border: Post-Soviet Borderscapes in the Finnish Media', in: C. Brambilla, J. Laine, J.W. Scott and G. Bocchi (eds), *Borderscaping: Imaginations and Practices of Border Making*, Farnham: Ashgate Publishing, pp. 65–76.

Leibenath, M., E. Korcelli-Olejniczak and R. Knippschild (eds) (2008), *Cross-Border Governance and Sustainable Spatial Development*, Berlin: Springer.

Lepik, K. (2009), 'Euroregions as Mechanisms for Strengthening of Cross-Border Cooperation in the Baltic Sea Region', *TRAMES*, **13**(3): 265–284.

MacLeod, G. and M. Jones (2007), 'Territorial, Scalar, Networked, Connected: In What Sense a "Regional World"?', *Regional Studies*, **41**(9): 1177–1191.

Malpas, J. (2012), 'Putting Space in Place: Philosophical Topography and Relational Geography', *Environment and Planning D: Society in Space*, **30**(2): 226–242.

Martin, L. and A.J. Secor (2014), 'Towards a Post-Mathematical Topology', *Progress in Human Geography*, **38**(3): 420–438.

Massey, D. (1993), 'Power-geometry and a Progressive Sense of Place', in: J. Bird, B. Curtis, T. Putnam, G. Robertson and L. Tickner (eds), *Mapping the Futures: Local Cultures, Global Change*, London: Routledge, pp. 59–69.

Mau, S., H. Brabandt, L. Laube and C. Roos (2012), *Liberal States and the Freedom of Movement. Selective Borders, Unequal Mobility*, Basingstoke: Palgrave Macmillan.

Megoran, N. (2012), 'Rethinking the Study of International Boundaries: A Biography of the Kyrgyzstan-Uzbekistan Boundary', *Annals of the Association of American Geographers*, **102**(2), 464–481.

Newman, D. (2006), 'The Lines that Continue to Separate us: Borders in our "Borderless" World', *Progress in Human Geography*, **30**(2): 143–161.

Newman, D. (2011), 'Contemporary Research Agendas in Border Studies: An Overview', in: D. Wastl-Walter (ed.), *The Ashgate Research Companion to Border Studies*, Farnham: Ashgate Publishing, pp. 33–47.

O'Dowd, L. (2010), 'From a "Borderless World" to a "World of Borders": Bringing History Back In', *Environment and Planning D: Society and Space*, **28**: 1031–1050.

Ohmae, K. (1995), *The End of the Nation State: The Rise of Regional Economies*, New York: The Free Press.

Orbán, V. (2016), Speech, http://www.kormany.hu/en/the-prime-minister/the-prime-minister-s-speeches/speech-by-viktor-orban-at-the-round-table-of-the-bratislava-global-security-forum (accessed 10 February 2017).

Paasi, A. (2012), 'Commentary. Border Studies Reanimated. Going beyond the Territorial/Relational Divide', *Environment and Planning A*, **44**(10): 2303–2309.

Paasi, A. and E.-K. Prokkola (2008), 'Territorial Dynamics, Cross-Border Work and Everyday Life in the Finnish–Swedish Border Area', *Space and Polity*, **12**(1): 13–29.

Pataki, F. (2013), *Hosszú menetelés (A Fidesz-jelenség)*, Budapest: Noran Libro.

Perkmann, M. (1999), 'Building Governance Institutions across European Borders', *Regional Studies*, **33**(7): 657–666.

Perkmann, M. (2007), 'Policy Entrepreneurship and Multilevel Governance: a Comparative Study of European Cross-Border Regions', *Environment and Planning C*, **25**(6): 861–879.

Popescu, G. (2012), *Bordering and Ordering the Twenty-first Century: Understanding Borders*, London: Rowman and Littlefield.

Popescu, G. (2015), 'Controlling Mobility: Embodying Borders', in: A.L. Amilhat Szary and F. Giraut (eds), *Borderities and the Politics of Contemporary Mobile Borders*, London: Palgrave Macmillan, pp. 100–115.

Raffestin, C. (1992), 'Autour de la fonction sociale de la frontière', *Espaces et sociétés*, **70/71**: 157–164.

Rosas, G. (2012), *Barrio Libre: Criminalizing States and Delinquent Refusals of the New Frontier*, London, Durham, NC: Duke University Press.

Rosière, S. and R. Jones (2012), 'Teichopolitics: Re-considering Globalisation through the Role of Walls and Fences', *Geopolitics*, **17**(1): 217–234.

Rumford, C. (2006), 'Theorizing Borders', *European Journal of Social Theory*, **9**: 155–170.

Rumford, C. (2008), 'Introduction: Citizens and Borderwork in Europe', *Space and Polity*, **12**(1): 1–12.

Scholte, J. (1997), 'Global Capitalism and the State', *International Affairs*, **73**(3): 427–452.

Scott, J.W. (2000), 'Transboundary Co-operation on Germany's Borders: Strategic Regionalism through Multilevel Governance', *Journal of Borderlands Studies*, **XV**(1), 143–167.

Scott, J.W. (2012), 'European Politics of Borders, Border Symbolism and Cross-Border Cooperation', in: T. Wilson, and H. Donnan (eds), *A Companion to Border Studies*, Hoboken: Wiley-Blackwell, pp. 83–99.

Scott, J.W. (2015), 'Bordering, Border Politics and Cross-Border Cooperation in Europe', in: F. Celata and R. Coletti (eds), *Neighbourhood Policy and the Construction of the External European Borders*, Cham: Springer, pp. 27–44.

Scott, J.W. (2017), 'Cross-Border, Transnational, and Interregional Cooperation', in: D. Richardson (ed.), *The International Encyclopaedia of Geography. People, the Earth, Environment and Technology*, Washington, DC: AAG-Wiley, DOI: 10.1002/9781118786352.

Setzer, J. (2015), 'Testing the Boundaries of Subnational Diplomacy: The International Climate Action of Local and Regional Governments', *Transnational Environmental Law*, **4**(2): 319–337.

Smith, M.P. (2001), *Transnational Urbanism*, London: Blackwell.

Sohn, C. (2014), 'The Border as a Resource in the Global Urban Space: A Contribution to the Cross-Border Metropolis Hypothesis', *International Journal of Urban and Regional Research*, **38**: 1697–1711.

Svensson, S. (2014), 'Forget the Policy Gap: Why Local Governments Really Decide to Take Part in Cross-Border Cooperation Initiatives in Europe', *Eurasian Geography and Economics*, **54**(4): 409–422.

Szarka, S. (2017), 'Ellen kell állni. A Brüsszeli bürokráták nem a nemzeti érdekeket képviselik', *Magyar Demokrata*, XXI/11 (16 March 2017): 10–12.

Topak, Ö.E. (2014), 'The Biopolitical Border in Practice: Surveillance and Death at the Greece–Turkey Borderzones', *Environment and Planning D: Society and Space*, **32**(5): 815–833.

van Geenhuizen, M., B. van der Knaap and P. Nijkamp (1996), 'Transborder European Networking: Shifts in Corporate Strategy?' *European Planning Studies*, **4**(6): 671–682.

van Houtum, H. (2005), 'The Geopolitics of Borders and Boundaries', *Geopolitics*, **10**(4): 672–679.

van Houtum, H. (2010), 'Human Blacklisting: The Global Apartheid of the EU's External Border Regime', *Environment and Planning D Society and Space*, **28**(6): 957–976.

van Houtum, H. and T. van Naerssen (2002), 'Bordering, Ordering, and Othering', *Journal of Economic and Social Geography*, **93**(2): 125–136.

Vaughan Williams, N. (2012), *Border Politics. The Limits of Sovereign Power*, Edinburgh: Edinburgh University Press.

Vertovec, S. (2011), 'The Cultural Politics of Nation and Migration', *Annual Review of Anthropology*, **40**: 241–256.

Vogt, W.A. (2013), 'Crossing Mexico: Structural violence and the Commodification of Undocumented Central American Migrants', *American Ethnologist*, **40**: 764–780.

Weber, L. and S. Pickering (2011), *Globalization and Borders, Death at the Global Frontier*, London: Palgrave.

Zalan, E. (2016), 'Hungary Is Too Small for Viktor Orban', *Foreign Policy* (October, 2016), http://foreign policy.com/2016/10/01/hungary-is-too-small-for-viktor-orban/ (accessed 26 June 2018).

Zimmerbauer, K. (2012), 'Conceptualizing Borders in Cross-Border Regions: Case Studies of the Barents and Ireland–Wales Supranational Region', *Journal of Borderlands Studies*, **26**(2): 211–229.

PART II

GLOBALIZED GEOGRAPHICAL PERSPECTIVES

5. World-systems analysis

Kees Terlouw

INTRODUCTION: THE ORIGINS OF WORLD-SYSTEMS ANALYSIS

World-systems analysis was formulated as an alternative to studying social change as the modernization process in nation-states. In the post Second World War world this modernization paradigm dominated. Inspired by American academics such as Walt Rostow (1960) the relation between the universal stages of modernization and the developmental path of individual societies dominated the social sciences. Differences between the general modernization process and the individual idiosyncrasies of national development were studied and linked to policies. For instance, the late industrialization of southern European states was linked to state policies to promote industrialization partially funded by the American Marshall Plan.

After the youth revolts in 1968, students increasingly questioned this modernization paradigm which dominated academic and political debates. The revival of the Marxist tradition undermined the assumption that everybody profited from modernization. This resulted in a growing interest in inequalities within society based on class struggle. Some shifted from studying exploitation within societies to studying the role of unequal exchanges between countries. The differences between developing and developed countries was no longer explained by the internal obstacles to development within societies, but through 'the development of underdevelopment' of Latin American, African and Asian states by Western states (Frank, 1989). The growing interest in comparisons, inequalities and international relations did not yet undermine the premise of separate national societies. The world-systems approach developed by Immanuel Wallerstein challenged the core assumption of national development within national borders. In his late twenties he studied social change in African states during the process of decolonization. He was especially interested in state formation and policies towards national integration. He tried to compare this with similar processes in other states which gained independence at other times and other places. Comparisons with Latin American states in the late eighteenth century and East European states in the beginning of the twentieth century did not seem very relevant for African states with their arbitrary colonial borders cutting through many social relations. Making such comparisons across space and time would not only be complicated but also problematic.

> Did it not call for some simplifying thrust? It seemed to me it did. It was as this point that I abandoned the idea altogether of taking either the sovereign state or that vaguer concept, the national society, as the unit of analysis. I decided that neither one was a social system and that one could only speak of social change in social systems. The only social system in this

scheme was the world-system. This was of course enormously simplifying. I had one type of unit rather than units within units. (Wallerstein, 1974: 7)

Wallerstein needed a unit of analysis big enough to embody all causes of structural social changes. External influences should only have a superficial influence (Wallerstein, 1974: 3–11, 1979: 4, 220). One problem is that there are many different kinds of human behaviour with spatial boundaries that do not necessarily overlap. Cultural, political, and economic processes can all operate in different only partially overlapping areas. Wallerstein chooses economic exchanges to delimit his social system, because these relations have the biggest influence on the other spheres of human activity (Wallerstein and Hopkins, 1977: 114). This results in what Wallerstein himself calls a 'basically materialist bias' of the world-system theory (Wallerstein and Mosely, 1978: 284).

SOCIAL SYSTEMS: MINI-SYSTEMS, WORLD-EMPIRES AND WORLD-ECONOMIES

The present global world-system is the most important, but certainly not the only, social system in world history. Wallerstein identifies two other kinds of social systems: mini-systems and world-empires. These social systems differ in the extent to which economic, political, and cultural boundaries coincide. Mini-systems are very small social systems, with a single culture and polity. They exhibit hardly any division of labour. This means that economic, political, and cultural boundaries are identical. These systems are comparable to what others call tribal societies. They have virtually disappeared. Only some isolated indigenous tribes might nowadays be classified as mini-systems (Wallerstein, 1979: 155). The two other social systems Wallerstein distinguishes are world-systems. World-systems are much larger than mini-systems, but only the modern world-system covers the entire globe. World-systems have more complex economies in which an elite profits from the production of the masses, which are separated through a spatial division of labour between rich urban and poor rural areas. These spatial divisions are institutionalized through cultural differences between areas. The main difference between world-empires and world-economies is their political structure. A world-empire is characterized by a central administration, whereas in a world-economy numerous states oppose each other. These systems achieve integration in different ways. A world-empire is united through the political control over its economic relations. A world-economy lacks central political control and is integrated through economic exchanges. World-empires are therefore generally more stable than world-economies. The modes of production are also different. Whereas the mode of production of a world-empire is characterized by political redistribution, the mode of production of a world-economy is based on economic exchanges regulated by market forces (Wallerstein, 1974: 348–349, 1979: 5, 256).

Many world-economies had existed since the agricultural and urban revolutions thousands of years ago. The wealth generated by these world-economies commonly attracted militarily conquest through which they were transformed into a world-empire. These empires like Egypt, Greece, Persia, Rome and China used their wealth to build imposing buildings which makes world-empires and their ancient civilizations much more visible than world-economies in world history. History is written by the victors. The European world-economy which emerged during the long sixteenth century was an exception. It was not conquered and transformed into a world-empire and over time subjugated all other mini-systems, world-economies and world-empires in the world.

THE DEVELOPMENT OF THE EUROPEAN WORLD-ECONOMY

According to Wallerstein (1974) the European world-economy emerged out of the crises of feudalism in the 'long sixteenth century' covering roughly the period from 1450 to 1640. The general stagnation at the end of the Middle Ages commonly associated with the Black Death and linked to a crisis in seigniorial revenue stimulated the search for new sources of wealth overseas. This resulted in the emergence of a European world-economy covering most of Europe and the American colonies.

The commercializing of the economy in the core areas was linked to the increasing coercion in the periphery. The increased wealth in European cities was based on the colonization of the Americas with a slave based plantation economy and a growing grain trade with the Baltic based on the 'second serfdom' of landless labourers on large estates. The profits generated by the cheap imports of gold, silver and sugar from the Americas and the grain and timber from Eastern Europe enabled entrepreneurs in core states to concentrate on highly profitable industries. The economic development in peripheral states suffered from their specialization in agricultural commodities with low profit margins. The relations with Northwest Europe also hindered the emergence of an independent class of entrepreneurs and weakened the peripheral states (Wallerstein, 1974). Although some historians question the importance of the international trade links in this period (Brenner, 1977), Wallerstein points out that – unlike the historians studying this period – his goal is not to describe that period, but to analyse the roots of the emerging capitalist world-economy which only later came to full fruition (Wallerstein, 1983).

In this period the rulers of the House of Habsburg tried to transform the emerging European world-economy into a world-empire based on their territories in Spain, Italy, Germany, the Balkans, the Low Countries and their colonies in the Americas. The House of Habsburg failed to do this, while others like the Ottoman sultans and the French kings had imperial ambitions of their own (Wallerstein, 1974: 182), or opposed it like the English and the Dutch. This struggle over control of the European world-economy ended in a stalemate that created an international political system of sovereign states. This was symbolized by the Treaty of Westphalia concluded in 1648 which ended decades of wars in Europe and further institutionalized the division of the European world-economy in a multitude of sovereign territorial states. This institutionalized rivalry between states gave the European world-system its dynamic thrust. It gave the economy the necessary relative autonomy and flexibility to develop. Growing economic integration and political fragmentation went hand-in-hand. Rivalry with other states forced states to concede to the European entrepreneurs the freedom to develop trade relations and accumulate profits. Individual states could not control the large-scale trade network and the mobility of capital in the European world-economy. This was a symbiotic relationship. States needed the financial support of the entrepreneurs to be successful in their competition with other states, while entrepreneurs in the core states profited from their political and military domination of peripheral areas. This continuing competition and the increasing tax base due to economic development, caused an ongoing build-up of their political and military strength (Wallerstein, 1974; Chase-Dunn, 1989).

After this period in which the modern world-system was created, it stagnated economically. International economic relations and production declined. The economies of the different countries became more self-sufficient in this period. Although the strength of the economic ties declined in this period, the basic structure of the European world-economy remained intact and it was not succeeded by a new world-system. It even strengthened its political structure in this period between roughly 1600 and 1750. The many wars in this period strengthened the core states and further institutionalized the interstate system of competing sovereign states. External conflicts forced European states to strengthen their bureaucracies and military. Internally, the territorial states made important progress in integrating their territories into national markets (Wallerstein, 1980).

This second phase of consolidation created the conditions for a phase of renewed expansion at the end of the eighteenth century. The military power of the European core states created new peripheries for the world-system in Africa and Asia. Only in this period did a significant gap in economic development and wealth develop between the European core and the rest of the world. This spatial expansion went hand-in-hand with the further intensification of the world-system. The economic specialization and social polarization increased to such a degree that many characterized it as an industrial revolution. Although agriculture no longer dominated the economy and the role of wage labourers increased, there was no fundamental break in the functioning of the modern world-system. The membership of the core, semi-periphery and periphery changed in this period, but the working of the market based capitalist world-economy remained the same (Wallerstein, 1980).

After this phase of expansion and intensification the modern world-system became more unstable while it became more integrated. After its further economic and political integration and spatial expansion in the previous phases, it became during the nineteenth century also more culturally integrated in the wake of the French Revolution and the subsequent spread of a liberal ideology which further stimulated the capitalist character of this world-economy. In this period quite similar brands of centrist liberalism became to dominate the state politics and social sciences creating a 'geoculture' (Wallerstein, 2011). After its glorious nineteenth century, the world-system entered into a more challenging period at the beginning of the twentieth century. Economic stagnations and world wars characterized the first half of the twentieth century. A renewed period of economic growth and a changing spatial division of labour characterizes the second half of the twentieth century. The European core states lost their dominant position in the world-system to new core states. The centre of the world-system shifted further westward. The USA and later Japan overtook West Europe. The more recent globalization further intensified integration and spreads semi-peripheral development towards some parts of the former periphery, especially in Asia. The modern world-system has become more dynamic and unstable and is according to Wallerstein entering in a phase of structural crisis which will end in the next decades and will be replaced by a new type of world-system or fragment into different world-systems (Wallerstein, 1979). This increased instability and volatility of the modern world-system makes it more difficult to clearly differentiate between the last different phases. After initially dividing the development of the modern world-system in four phases, Wallerstein now further divides the last two phases in several overlapping phases (Wallerstein, 1974, 2011). Figure 5.1 depicts how Wallerstein conceptualizes the emergence and development of the present world-system.

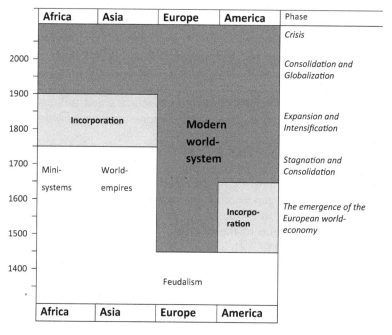

Figure 5.1   The emergence and development of the current world-system

STATES AND CITIES

Although starting as a critique of state centred developmentalist thinking of modernization theories, states still dominate in Wallerstein's (1974, 1980, 1989, 2011) voluminous analyses of our changing world-system. Some see this as a fundamental problem in the conceptualization of space in the world-systems approach. According to Neil Brenner the use of competing states as the defining characteristic of the modern world-system hinders the world-systems approach to look beyond this grid of nationally organized territories. 'The possibility that the process of capitalist development might unhinge itself from this entrenched national-global scalar couplet to privilege other subnational or supranational sociospatial configuration is thereby excluded by definitial fiat' (Brenner, 2004: 52).

Many stress the growing role of cities in globalization. Some, like Peter Taylor (2000, 2013) have further developed world-systems analysis through studying the changing role of urban networks. His studies, for instance, show that the dominant or hegemonic position of Dutch, British and American core states in the sixteenth, nineteenth and twentieth century, was preceded by the emergence of new cities. The growth of these cities preceding the rise of these states indicates that the new developments, on which a new phase in the world-system is based, are initiated in cities (Taylor et al., 2010). Others stress that cities are not only important as hubs in economic networks, but that cities and their administrations are becoming new important political actors in the world. Cities and their mayors become more important in the current era of globalization in which especially the role of states in economic regulation is undermined (Barber, 2013; Acuto, 2013). This growing role of cities and urban networks can form the basis of a new world-system in which cities succeed states as the key political institutions (Khanna, 2011).

FROM THE OUTSIDE TO THE INSIDE: THE INCORPORATION OF EXTERNAL ARENAS INTO THE MODERN WORLD-SYSTEM

The modern world-system has incorporated all other world-systems and mini-systems. The character of the relations between these areas changed fundamentally upon incorporation. External arenas became incorporated in the modern world-system as a new periphery which offered new opportunities for exploitation by the entrepreneurs in the core states. This transformed these new peripheries to the needs of the capitalist world-economy.

The way goods are produced changes fundamentally upon incorporation into the capitalist world-economy. Normally there was some trade between an external arena and the modern world-system, but the goods exchanged were not specifically produced for this trade. It was a trade of surpluses which were left over after internal demands were satisfied. It was a luxury trade of goods which did not become part of the commodity chains in the capitalist world-economy. External arenas traded also with many different other parties outside the modern world-system. All this changed upon incorporation. The trade between the new peripheral areas increased in volume, became less erratic and focused on specific core states. The goods exported to the core were now specifically produced for this trade, like on the plantations in the Americas, or on the large grain estates in the Baltic. This transformed the local social structure and the economy in the peripheries so that they were able to satisfy the requirements of the profit-seeking entrepreneurs in core states. This was enforced by the powerful core states. Indigenous production structures linked to a predominantly self-sufficient production were largely destroyed and replaced by mostly large-scale foreign domin-ated production in the form of plantations, mining companies and later labour intensive industries. Incorporation also transformed the local political structures. When the political structures in these new peripheries were too weak to protect trade, core states created their new colonial states in these areas. This happened to most mini-systems, such as in nineteenth century Africa. World-empires like the Ottoman and Chinese empires were usually defeated militarily and subsequently transformed into territorial states which were internally strong enough to enable the new trade from which the core states profited, but which were not so strong externally that they could resist domination by the European core states (Wallerstein, 1974; Terlouw, 1992: 65–78).

The capitalist world-economy is based on this exploitation of the periphery by the core. Based on low wages the economy in peripheries is based on the production of a few low-priced products. These cheap imports are beneficial for the core and enables the development of a much more diversified economy based on high wages and innovation. These differences between core and periphery were initially quite small, but have intensified in the last phases of the modern world-system. The types of goods exchanged between core and periphery vary between periods. The first, mostly agricultural division of labour in the modern world-system was based on importing commodities from the periphery. Grain and timber from the Baltic and plantation crops and silver from the Americas were exchanged with equipment and consumer goods for the local comprador elites. Later, the international division of labour shifted to the unequal exchange between cheap raw materials from the periphery with the expensive industrial products made in the core. Nowadays, many industrial products come from the periphery, and the core profits from its dominance of the high-tech service sector. The core and periphery in the modern world-system are not defined by their production of specific products, but by their unequal relation through which the core profits from the products produced by cheap labour in the periphery (Wallerstein, 1979).

Not only the products exchanged between core and periphery shift over time, also the position of specific areas and states in the world-system can change. Old peripheries can improve their position and become part of the semi-periphery which is in between the core and periphery. The semi-periphery also profits from trade with the periphery, but is still subject to exploitation by the core. Its economy is based on a combination of peripheral and core like products. Almost all areas which belonged to the periphery in the first phase of the modern world-system have now become semi-peripheral or even core states.

Figure 5.2 sketches the mobility in the world-system that emerges from many scattered remarks by Wallerstein on the position of states in different years (Terlouw, 2002). These maps show that only Portugal and most of Spain have always belonged to the semi-periphery. Usually, the older semi-peripheries have achieved core position by 1900, like Germany and the United States, or by 1980, like Sweden and Northern Italy. Most older peripheries have improved their position in the world-system. Only some Latin American states, like Peru, Colombia and Surinam, have always belonged to the periphery.

The existence of a semi-periphery not only helps to better classify individual states than the simple core–periphery dichotomy or describe the development of the modern world-system. The existence of a semi-periphery is also an important structural characteristic of the modern world-system. The semi-periphery depolarizes the relation between core and periphery. The exploited will always be divided and unable to unite and overthrow the system, because the strongest among them – the semi-periphery – profit from the exploitation of the periphery (Wallerstein, 1974: 348–350; Wallerstein and Hopkins, 1977).

The possibility to join the core also appeases the semi-periphery. As Figure 5.2 shows, the semi-periphery is the most dynamic part of the world-system. Its political and economic power is clearly subordinate to the core, but unlike the periphery, it has

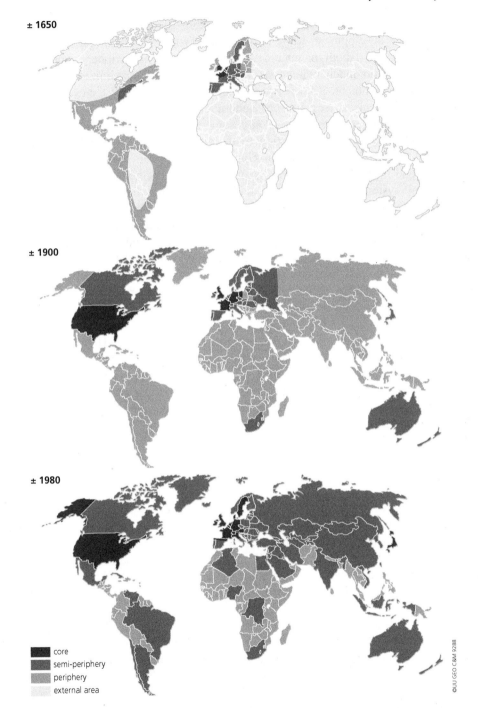

± 1650

± 1900

± 1980

core
semi-periphery
periphery
external area

©UU GEO C&M 9288

*Figure 5.2   The developing modern world-system: expansion and mobility between
external arena, periphery, semi-periphery and core*

some resources to resist exploitation. In many core states institutional sclerosis and congestion stifle development, while in the periphery the absence of good government, services and skilled labour hamper development. The semi-periphery tends to have enough, but not too much regulation. The semi-periphery maximizes the need and necessity for development. Semi-peripheral development is part of the cyclical renewal of the world-system. Crises in the Kondratieff like economic long waves give opportunities to the semi-periphery. Only some semi-peripheries can transform this temporary advantage into a promotion to the core (Wallerstein and Hopkins, 1977; Wallerstein, 1979). This happened in continental European states after the Second World War and in many states in the Asian Pacific Rim in the last decades.

Periods of system wide economic stagnation interrupt the development of new industries in the core, and stimulate their redeployment. Semi-peripheral areas attract new development, while they are unburdened by the negative side effects of previous development and do not suffer from the negative spiral of underdevelopment. Labour costs are the most visible cause of this global re-division of labour, but other, more hidden, production costs, like transportation costs and environmental protection measures, are also important. This combination of lax regulation and strong developmental pressures makes the semi-periphery vulnerable to social, cultural, political and ecological conflicts.

Semi-peripheral development focuses on semi-peripheral states. Their possibility for semi-peripheral development enabled capitalism to develop. However, there is not a uniform semi-peripheral developmental path (Arrighi, 1985). Each semi-peripheral state has its own developmental path. Economic factors are important for some, while other semi-peripheral developments are more based on political factors (Terlouw, 1992).

RIVALRY AND HEGEMONY

The modern world-system not only has cycles in its economy, but also in how its politics functions. The rivalry between core states, which characterizes and strengthened the modern world-system, sometimes escalated into world wars. These end with a wide-ranging peace agreement (1648 Westphalia, 1815 Vienna, 1945 Yalta), which heralds a new period of relative peace under the hegemony of a leading member of the winning coalition not directly affected by warfare. This was the case for the Dutch United Provinces in the mid-seventeenth, the United Kingdom in the mid-nineteenth, and the United States in the mid-twentieth century. Despite the differences between these three states, they all dominated the world-system in a similar way. However, rivalry and hegemony are not completely separate phases. 'Hegemony therefore is not a state of being, but rather one end of a scale which describes the rivalry relations of great powers to each other' (Wallerstein, 1984: 39). Hegemony exists when a single core state is superior to all others in the world-system. No other state, or coalition of states, is able to threaten the superiority of the hegemonic state. But it is not omnipotent. The international system of states is still functioning in the world-system.

The other states are still strong enough to prevent the hegemonic state from transforming the modern world-system into a world-empire with a central authority (Wallerstein, 1984: 37–46).

Hegemonic states have the most advanced production process. As a result, their products were even competitive on the home markets of their rivals. The hegemonic state always dominates world trade. It even has an important role in the trade between other core states. The hegemonic state is also superior in the financial sector. It has the highest rate of return on capital, it lends money to others, and it is a net exporter of capital. Furthermore, a hegemonic state also dominates world politics. This is based on its dominant position in the world-economy and manifests itself in the military field as well. The dominant, though not necessarily unrivalled, military strength of the hegemonic state, especially in sea power, is instrumental in maintaining its dominant position (Wallerstein and Hopkins, 1977: 121, 130, Wallerstein, 1980: 38–39, 1984: 37–46).

The existence of a hegemonic state affects the way in which the world-system functions. Because of its economic superiority, a hegemonic state profits from the unrestrained functioning of the world market. A hegemonic state rarely intervenes directly in the functioning of the world market. The hegemonic state only cracks down on other states when they try to compensate for the economic weakness of their bourgeoisie through state intervention. A period of hegemony is therefore a period of relatively free trade. During a period of hegemony, semi-peripheral states have consequently little opportunity to improve their position in the world-system (Wallerstein and Hopkins, 1977: 131, Wallerstein, 1980: 38, 61, 65, 269, 1984: 37–46). But hegemony is not a stable condition of the world-system. Within the hegemonic state the production costs in the hegemonic state rise because the workers negotiate increasingly better deals with the bourgeoisie in the hegemonic state, who appease the workers in order to capitalize on their hegemonic advantage over other states. After several decades the other core states also succeed in undermining the dominant position of the hegemonic state. The liberal ideology of the hegemonic state allows the new technologies developed in the hegemonic state to spread to the other states. The other states are therefore able to catch up in this period of declining hegemony like in our present era of globalization. This is then followed by a new long period of rivalry between core states, culminating again in a world war and the emergence of a new hegemonic power (Wallerstein, 1980: 211, 241, 1979: 99, 116, 1984: 37–46).

TRENDS AND THE CRISIS OF THE MODERN WORLD-SYSTEM

Besides these repeating economic and political cycles, there are also trends which transform the modern world-system. These changes are not reversible like hegemony or economic stagnation. The rivalry between states generates not only a political cycle of alternating rivalry and hegemony, but also a trend of a growing role of the state in society as each state tries to improve its position in relation to the other states by strengthening its machinery. This trend undermines the profitability of the modern world-system for the bourgeoisie. Also the rising wages and welfare redistributions in the core cut into the privileged position of the bourgeoisie (Wallerstein, 1979: 499).

The bourgeoisie tries to compensate for this trend of diminishing privilege by expanding the sphere of action of the modern world-system. Traditionally, incorporating new peripheries increased the possibilities of capital accumulation. Besides this spatial expansion, the capitalist world-economy also expands its social boundaries. All kinds of social relations become more and more commercialized. Relations that were previously regulated by moral principles are now 'freed' from this constraint. Social relations become further subject to market forces. Not only subsistence farming in the periphery is commercialized, but also activities in the core households are increasingly commercialized ranging from ready-made meals to commercial hospice services.

These trends of bureaucratization, diminishing privileges for the bourgeoisie, geographic expansion, and commercialization from the cradle to the grave cannot continue forever. The world-system can hardly expand beyond the globe and at a certain moment all social activities are commercialized. Also the growth of the power of the state and the redistribution of surplus towards the proletariat deplete the resolve of the bourgeoisie to defend the present world-system against the growing resistance in the form of all different kinds of anti-systemic movements. This is a mixed bag which includes terrorist, liberation, LGBT and climate movements. This will according to Wallerstein (1979) bring the modern world-system into a structural crisis. It will be replaced by another kind of world-system.

World-systems analysis has since its formulation in the early 1970s provided an alternative to the nation-states centred modernization perspective. World-systems analysis gives a powerful analysis of how the worldwide inequalities in our world were formed and perpetuated. It places current developments like globalization into a wider historical perspective. This account of the developing modern world-system provided new viewpoints to many traditional debates in the social sciences, like the crisis of feudalism, the character of capitalism, the role of industrial revolution and the consequences of (de-)colonization. World-systems analysis has been criticized as an overbearing systems-based explanation, which does not do justice to all the peculiarities of social development, which many specialists are eager to point out. World-systems analysis does not aim to give such detailed analyses. Its goal is to point out the importance of worldwide interconnection for social change over the last centuries. As such it provided an important contribution to the debates related to globalization.

REFERENCES

Acuto, M. (2013), *Global cities, governance and diplomacy: The urban link*, London: Routledge.
Arrighi, G. (1985), *Semiperipheral development*, Beverly Hills: Sage.
Barber, B.R. (2013), *If mayors ruled the world: dysfunctional nations, rising cities*, Newhaven: Yale University Press.
Brenner, N. (2004), *New state spaces: urban governance and the rescaling of statehood*, Oxford: University Press.
Brenner, R. (1977), 'The origins of capitalist development: a critique of neo Smithian Marxism', *New Left Review*, I(104), 25–92.
Chase-Dunn, C. (1989), *Global formation: structures of the world economy*, New York: Blackwell.
Frank, A.G. (1989), 'The development of underdevelopment', *Monthly Review*, **41**, 37–51.
Khanna, P. (2011), *How to run the world: charting a course to the next renaissance*, New York: Random House.

Rostow, W. (1960), *The stages of economic growth: a non-communist manifesto*, Cambridge: Cambridge University Press.

Taylor, P.J. (2000), 'World cities and territorial states under conditions of contemporary globalization', *Political Geography*, **19**(1), 5–32.

Taylor, P.J. (2013), *Extraordinary cities: Millennia of moral syndromes, world-systems and city/state relations*, Cheltenham, UK and Northampton, MA, USA: Edward Elgar Publishing.

Taylor, P.J., A. Firth, M. Hoyler and D. Smith (2010), 'Explosive city growth in the modern world-system', *Urban Geography*, **31**, 865–884.

Terlouw, K. (1992), 'The regional geography of the world-system: external arena, periphery, semiperiphery, core', Utrecht: NGS.

Terlouw, K. (2002), 'The semiperipheral space in the world-system', *Review*, **25**, 1–22.

Wallerstein, I. (1974), *The modern world system: capitalist agriculture and the origins of the European world economy in the sixteenth century*, New York: Academic Press.

Wallerstein, I. (1979), *The capitalist world economy*, Cambridge: Cambridge University Press.

Wallerstein, I. (1980), *The modern world system II: mercantilism and the consolidation of the European world economy 1600–1750*, New York: Academic Press.

Wallerstein, I. (1983), 'European economic development: a comment on O'Brien', *The Economic History Review*, **36**, 580–583.

Wallerstein, I. (1984), *The politics of the world economy: the states, the movements and the civilizations*, Cambridge: Cambridge University Press.

Wallerstein, I. (1989), *The modern world system III: the second era of great expansion of the capitalist world economy, 1730–1840s*, New York: Academic Press.

Wallerstein, I. (2011), *The modern world-system IV: centrist liberalism triumphant, 1789–1914*, New York: University of California Press.

Wallerstein, I. and T.K. Hopkins (1977), 'Patterns of development of the modern world system: research proposal', *Review*, **1**, 111–145.

Wallerstein, I. and K.P. Mosely (1978), 'Precapitalist social structures', *Annual Review of Sociology*, **4**, 259–290.

6. Globalization and sustainable development
Joyeeta Gupta

INTRODUCTION

In writing this chapter, I kept remembering the old limerick:

> *There was a young lady of Niger,*
> *Who went for a ride on a tiger,*
> *They returned from the ride,*
> *With the lady inside,*
> *and a smile on the face of the tiger.*

The global adoption of the Sustainable Development Goals (SDGs) (UNGA, 2015) is akin to an attempt to tame the processes of globalization. The SDGs are the young lady of Niger and the question is whether she is in a position to avert the fate that awaits her. I find this limerick also appropriate because the lady of Niger represents the gender and developing country perspective, although the approximative localization (obviously there are no tigers to ride in Niger) exposes the very colonial origin of this famous limerick.

Globalization, environment and development are closely linked. Amongst the many theories there are two contradictory hypotheses (Thai et al., 2007). The first argues that globalization promotes convergence and homogenization of discourses and common policymaking structures and instruments. It sees globalization as unleashing a process of discovering shared global values and knowledge that are adopted, for example, through the SDGs and co-interpreted and co-implemented at multiple levels of governance for the good of the world (cf. the role of transformationalists in Ezcurra and Rodriguez-Pose, 2013).

The other hypothesis is that globalization promotes reckless economic growth and little government interference which both violates the environment and social values. It sees globalization as an uncontrollable ad-hoc phenomenon that cannot be made subject to global rules in an anarchic, borderless world (Ohmae, 1990). Even if it is, the processes of co-interpretation and co-implementation will be captured by those in power thus calling for a re-territorialization, a counter-narrative and a counter-movement to re-establish context and identity as key to the new storyline (Burgh-Woodman, 2014).

This chapter examines the fields of international development and environmental studies which are gradually merging since socio-ecological systems are increasingly conceptualized as intertwined. Both fields are mediated by technologies and infrastructures and hence, this chapter also deals with these issues. It builds on previous papers (Gupta, 2014, 2016) and first discusses the ecological, technological and social dimensions of the sustainable development challenge, then the institutional infrastructure that attempts to deal with this challenge, before assessing the role the SDGs could potentially play in resolving this challenge.

THE DIMENSIONS OF THE CHALLENGE

Globalization and Environmental Dimensions: Acknowledging Limits

Globalization debates in the Anthropocene (Crutzen, 2006) recognize the impact of the accelerated use of natural resources and sinks (the carrying capacity of the air, water, and soil which provide so-called ecosystem services, Steffen et al., 2004; Millennium Ecosystem Assessment, 2005). When discussing the (potential) crossing of planetary ecological boundaries (Rockstrom et al., 2009), they tend to ignore social floors (Raworth, 2012). Such impacts occur through, inter alia, the medium of advanced technologies and infrastructures often exacerbating multi-dimensional inequalities and injustices.

In the Anthropocene, we are facing the limits of resources (land, water, minerals, biodiversity) and sinks. This is especially so in a world of a growing demand of an ever more greedy and rising global population. These limits lead to heightened struggles over these 'shrinking' resources and sinks. Land and fresh water are limited, implying less on a per capita basis as time moves on. Some abiotic resources (e.g. strategic metals and minerals like iron ore, zinc, bauxite, copper, aluminum, zinc, lead, copper and nickel and minerals like nitrogen and phosphorous) are increasingly in short supply and can only be further extracted at huge economic, ecological and social costs (Edixhoven et al., 2014; PBL, 2011). Our biodiversity, in addition, is declining (Steffen et al., 2004) and fish supplies dwindling. Our ecosystems are increasingly unable to provide the stable environment (e.g. a stable climate) that we have relied upon in the past. To prevent the destruction of our planetary ecosystem we need to reduce our pollution. For example, to keep the average rise of the temperatures below 1.5 to 2°C as required by the Paris Agreement (2015) we have to keep more than 80 per cent of our fossil reserves unused (Carbon Tracker, 2012).

Globalization and Technological Dimensions: Are Limits Desirable?

Globalization is both empowered by, and empowers, technological evolution and revolution. We are now moving into the sixth technological wave with biomimicry, robotization, nanotechnology, renewable energy, green chemistry, and radical resource productivity (Silva and Di Serio, 2016). On the positive side, technologies enhance the efficiency of production, distribution and consumption processes and related infrastructures and increase global wealth. However, on the socio-economic side, technologies may replace labour and hence jobs and income. Technologies may also facilitate the concentration of wealth as well as tax avoidance/evasion and the dark web. On the ecological side, technologies enable the large-scale use of resources. This often leads to the externalization of pollution and allows for the commodification and privatization of the free goods and services of nature such as seeds, water, and even electromagnetic waves (Paige, 1997). On the political side, the use of technologies and algorithms are increasingly manipulating economic investments in capital markets, political elections and are even playing an increasingly important part in modern warfare. As artificial intelligence and robotization are becoming more important, the world economy is moving into dangerous territory (Harari, 2017; Hanley, 2017 citing Elon Musk). Fundamental questions regarding the ethics of new technologies and their impact on labour, financial transfers, and privacy and in particular the 'meaningful human control' over artificial intelligence are being raised.

Globalization and Socio-economic Dimensions: Respecting Rights, Reducing Inequality?

While development issues focused on how to enable the South and the poor to develop, prescriptions on how societies develop have changed almost every decade. These prescriptions have moved from being Keynesian in nature to neoliberal approaches which advocate a small state and more freedom for the private sector. Throughout this period, many developing countries demanded a New International Economic Order (Schrijver, 2008) and the Right to Development (UNGA, 1986) as they felt that the international order was skewed against the interests of the periphery – the South. While developing countries have become integrated in the global order, this has reduced overall absolute poverty (Bergh and Nilsson, 2014) and generally enhanced gender equality (Chen et al., 2013), but it may have also contributed to more inequality within these countries (Ezcurra and Rodriguez-Pose, 2013; Milanovic, 2016). In the meanwhile, the neoliberal capitalist order has led to both the over-use of resources and sinks, and the concentration of power and wealth in the hands of the rich in the North and the South, thereby frequently neglecting the basic socio-economic rights of other people.

There are evidently hard ecological limits to what we can use and how much we can pollute. Technologies are being developed in areas that challenge democratic processes, the expectation of jobs and human rights and, hence, call for a discussion regarding the governance of technologies. In the Anthropocene, it is no longer possible for the South to follow in the footsteps of the North; the South has to take a completely different short cut to sustainable development. There is a need for a completely different socio-economic, environmental and technological order.

GOVERNANCE DIMENSIONS

Governing Socio-economic, Technological and Environmental Challenges

The formal infrastructure of governing globalization and sustainable development has developed along many paths and only recently converged within the Sustainable Development Goals. The mainstream development strategy post Second World War was the establishment of the Bretton Woods Institutions (to promote monetary cooperation, exchange rate stability and economic development) and free trade agreements (to promote the flow of goods and services to enable production to take place where it was cheapest). These regimes aimed at creating a global system of cooperation based on extracting resources and creating infrastructure, producing goods and services and trading them worldwide. Trade and development brought prosperity but came at the cost of direct environmental externalities (Werksman, 1993; White, 1996) as well as indirect socio-environmental externalities through the impact of the developing country debt crises on the environment (see for example Miller, 1991; George, 1992). Increasingly, these regimes enabled the commodification of nature and the use of market mechanisms to protect the environment (e.g. payment for ecosystem services, the Clean Development Mechanism under the Climate Change Agreement). Human rights regimes developed in parallel, though most were never ratified by the USA, and focused primarily on political rights rather than social-economic rights (Gavison, 2003).

Development cooperation regimes coordinated through the Development Assistance Committee (DAC) of the OECD tried to come up with decadal prescriptions of how development could be promoted in the developing world and this was financed by aid strategies (Gupta and Thompson, 2010). Between 1985 and 2005, OECD countries paid about 55 billion US dollars annually, and since 2005 it is about 100 billion US dollars. The Paris Agreement argues for another 100 billion US dollars annually for combating climate change. While arguably these resources should enable greater cooperation and wealth in the South, closer examination in recent years reveals that for every dollar going to the South in terms of foreign direct investment (44¢), repatriation (34¢), aid (10¢), portfolio equity (6¢), charities (3¢), and other flows (3¢); 2 dollars return to the developed world through illicit financial flows (93¢), loans to developed countries (59¢), profits taken out by foreign companies (42¢) and interest payments on foreign debt (14¢) (Griffiths, 2014). The illicit flows are highest from the emerging economies and they all appear to be flowing into the developed world.

The field of environmental governance emerged in 1972 with the first global conference on the human environment and led to the establishment of the United Nations Environment Programme (UNEP), now UN Environment. As issues became important, treaties were negotiated. However, the negotiation of the climate treaty was not entrusted to UNEP in 1990 but to the General Assembly as it was seen as much too important. Many other UN agencies and bodies such as the FAO, UN Water, and UN Energy govern resource use – and UN Environment tries to coordinate the activities of these other bodies in relation to the environment. Increasingly such governance is based on insights from science organized in epistemic communities such as the Inter-governmental Panel on Climate Change.

Technology and infrastructure is an area where there has been much less multilateral governance. Technology developments are protected de jure by intellectual property and contract laws, de facto by the nature of technology which has erased privacy and autonomously replaced other ways of doing things. Specific technologies and related infrastructure may lead to path dependency and lock-in (e.g. into the fossil fuel economy). As these technologies are typically controlled by powerful actors including a few global firms, they also have leverage over political agendas. For example, the Internet uses an Internet Protocol and the World Wide Web and enables the transfer of knowledge, data and finance. It has empowered individuals to communicate, but at the same time it has concentrated power in the hands of a few commercial, state and military actors that control the system (AIV, 2014). It enables new business opportunities such as *Uber* and *Airbnb* which provide cheaper services to people while often flagrantly ignoring existing laws on transport and hotels. David Clark (1992), an Internet pioneer, has reportedly said: 'We reject presidents, kings and voting. We believe in rough consensus and running code'. Although the UN has tried to govern multinationals and the Internet, powerful countries and actors have ensured that this does not go beyond soft voluntary codes of conduct (e.g. the Equator Principles[1] adopted by Banks, the OECD Guidelines for Multinational Enterprises, the Global Sullivan Rules of Social Responsibility, the UN Global Compact) (cf. Sauvant, 2015). Technology companies 'wield their power relationally with other state and non-state actors' such that it is 'impossible to disentangle interests' and states and companies have hijacked each other's agendas (Mikler, 2013: 7).

Changing Patterns of Global Stratification

The brief history of the governance process mentioned above is taking place against three changing patterns of global stratification. First, we see an increasing concentration of resources: In 2016, eight men had more wealth than the bottom half of the world; while the top 1 per cent earned more than a million dollars per year and together have more than the remaining 99 per cent, and this difference is growing annually. Ten corporations have revenues greater than the government revenues of 180 countries combined (Oxfam, 2017). The top implicitly (see Leonardo Di Caprio's film on how the richest men buy off members of the US senate on climate change) and explicitly (by running for President) pull the strings. Many in the top are evading/avoiding taxes (see Panama Papers, Paradise Papers) which reduces the resources for actually governing. Furthermore, 'The globalization of world politics is said to necessitate global governance ... but increasingly global companies are key players in the drafting of international agreements' (Mikler, 2013: 11). They increasingly 'attempt to pre-empt public regulation' (Fuchs, 2013: 85).

Second, the rise of the middle classes in the South and the shrinking middle classes in the North (Sumner, 2010) are linked by the offshoring of employment made possible by ICT. On the one hand, the size of the middle class can determine tax revenues for financing public and/or merit goods, on the other hand it also has a greater environmental footprint. However, growing under- and unemployment worldwide is creating greater distrust of government institutions.

Third, some argue that technological emancipation has implied the onset of a post capitalist era (Mason, 2015). However, there remains a generation of haves and have-nots in the technological era. Furthermore, everyone who uses the Internet of Things is increasingly losing their privacy as not only the substance of their ideas and communications are recorded into posterity, but also the frequency with which they contact others, use transport, visit places and so on are all recorded in big data. And 'power is at its most effective when least observable' (Lukes, 2005: 64).

---

[1]   See http://www.equator-principles.com/index.shtml.

Inferences

Global governance of trade and investment up till 2008 has been predominantly laissez-faire and under the aegis of neoliberalism. It has supported the 'oligopolistic' nature of technology governance which has both enabled global wealth development while at the same time resulted in environmental tipping points and increased inequality. Against this background there are fragmented attempts at issue-by-issue environmental governance, which is often hampered by ratification politics. Oil and gas dependence has ensured that the US did not take on legally binding quantitative obligations under the climate change regime since 1992. Governance has been mainly reactive and diffuse while together the concept of 'free trade' and autonomous globalization has contributed jointly to crossing planetary boundaries. At the same time, basic needs have been frequently neglected and, within many countries, inequality has increased. There is a widespread agreement among scholars that this is no longer sustainable, and the Right to Development of developing countries as recognized in the UN General Assembly Declaration of 1986 has morphed into a Right to Promote Sustainable Development in the Climate Convention (Gupta and Arts, 2017).

THE SUSTAINABLE DEVELOPMENT GOALS AND GLOBALIZATION

Unleashing Globalization: the Tiger

The autonomous processes of globalization through trade, Internet and other technologies, investment and migration have had many impacts. It has increased global wealth, but has come at the cost of destroying biodiversity and depleting ecosystem services; the financialization and patenting of natural resources; and the mainstreaming of market instruments leading to a de jure and/or de facto transfer of resources from one party to another (e.g. through land and water 'grabbing' (Zoomers, 2010)); expanded the use of transfer pricing, tax evasion and avoidance reducing resources for the state; while 'normalizing' corrupt practices. Furthermore, as financial power becomes concentrated in a handful of individuals and companies, countervailing powers are emerging. Local communities are fighting against the extractive industries worldwide from fossil fuel fields in Ecuador to the production of natural gas in the Netherlands. Movements against the financial system (e.g. Occupy) and at city level are trying to counter such dominance (Acuto, 2013). At the national level the emerging economies are creating their own bank (i.e. the New Development Bank) to counter the dominance of the World Bank. At the global level, scholars and policymakers are trying to govern these processes possibly within the context of a transformationalist approach to globalization.

The SDGs: the Lady of Niger

A relatively new global governance storyline is governing by goals as stated in Agenda 2030 (Kanie et al., 2017; Young, 2017; Young et al., 2017). This follows the

Millennium Development Goals and individual treaty regimes. However, such goal setting was also established by the UN Charter (1945) which aimed at promoting international peace and security, and cooperating to address economic, social, cultural and/or humanitarian problems while recognizing human rights. Nevertheless, Agenda 2030 is a watershed moment in human history both procedurally and substantively.

The SDG negotiations were procedurally innovative: first, they were based on the largest global online and offline surveys to understand what people worldwide wanted (Gellers, 2016). Second, they enhanced ownership, trust and participatory effectiveness by allocating 30 negotiating seats distributed region-wise which had to be shared by 70 countries, often with vastly different interests, forcing them to come to a combined position to the negotiations (Chasek and Wagner, 2016). The SDGs are substantively innovative aiming at promoting coherence in multi-level and multi-party action in relation to five P's – people (i.e. social issues), prosperity (i.e. economic issues comprising much more than profit), planet (i.e. ecological issues), peace (addressing dealing with conflicts), and partnerships (i.e. the role of governance, emphasizing co-creation rather than hierarchies). They emphasize that all goals are interrelated, integrated and indivisible whose implementation will be monitored by the High Level Political Forum, the Global Sustainable Development Reports and the multiple partnerships (Boas et al., 2016).

In operational terms, the SDG document specifies 17 Goals, 169 targets and hundreds of indicators. The 17 Goals aim at ending poverty and hunger, promoting healthy lives, education, gender equality, access and management of water, access to sustainable energy, economic growth and employment, resilient infrastructure, industrialization and innovation, reducing inequality, sustainable settlements, sustainable consumption and production patterns, action to combat climate change, sustainably manage oceans and terrestrial ecosystems, promote peaceful societies, and strengthen the partnership for implementation. The Goals themselves are broad and open-ended; the targets and instruments are narrower and in line with results-based management approaches. The Goals lay out a programme of multi-level action that aim to counter the negative externalities of globalization on the environment and social aspects including human wellbeing.

One can see the SDGs as incoherent, or the logical outcome of a democratic process; as top-down or as reflecting evolving history and bottom-up processes in international negotiations; as caricatures that are Twitterable or as the need to 'keep it simple, stupid' (KISS) learning from the experiences of the mostly ignored voluminous and exhaustive Agenda 21 of 1992; as too vague or as open to contextual adaptation; as a paper tiger or a game changer; as focused on average trends and 'normalizing' dominant neoliberal values or as being revolutionary in aiming at addressing the extreme problems and universalizing local values and human and community rights; and as endorsing a business-as-usual approach or being truly transformative in potential.

I believe that the SDGs open up a huge space for transformative and even revolutionary thinking, but know that that will not happen if neoliberals capture the space and post-development thinkers and critical scholars exclude themselves. Although the SDGs are revolutionary in implicitly framing planetary boundaries and social floors while also aiming at reducing inequalities between and within states; they do not go far enough in terms of planetary floors and social boundaries. They do not tackle head on the privatization and hegemonization of global resources. They do not tackle the challenges ahead in terms of sharing our resources and sinks – sharing our Earth. They do not call for a precautionary approach to technology governance. Their assumption of partnership is too simplistic – and they do not acknowledge that each country will have to take its own route to sustainable development.

The bottom line is that the SDGs even if they do not go far enough, are in themselves quite dramatic and aim at being a first step to control globalization and limit its negative impacts globally by putting obligations on states but also on non-state actors.

Inferences

Agenda 2030 includes Goals no one could object to. But it represents the outcome of a battle between neoliberal capitalist perspectives, those who feel that it is high time that human rights and socio-economic rights are respected, and those who argue that ecological issues need to be addressed now rather than later. The problem is that addressing the goals requires trillions of dollars and such resources are only available in the hands of the neoliberal capitalists who would be unwilling to use these resources for global re-distribution (Voituriez et al., 2017). This battle is not only a financial one but very much multifaceted. It is also a discursive one in that some contest 'the epistemic and normative basis for sustainable development' (Haas and Stevens, 2017: 157) and others argue that the North–South issues, though ironed out in the consensus, remain unresolved (Bernstein, 2017). This is also seen in the contestation with respect to the choice of words within the document: for example 'growth' is mentioned 16 times, while 'inclusive' is mentioned 41 times, and 'sustainable development' 85 times (Gupta and Vegelin, 2016). It can also be seen as a controversy over measurability with choices being made for simplistic measurable targets that avoid the big issues (Pintér et al., 2017). Related to this is an instrumental challenge as the proposed measures and instruments may counter the very goals they aim to achieve. One can also point at a challenge with respect to agency and the focus on partnerships where some are modestly enthusiastic about the role of philanthropists (Andresen and Iguchi, 2017) and the private sector (Yamada, 2017), while others are sceptical. Finally, there is the issue of governability where the need for resources itself leads to a call for collaboration with the private sector and other actors given the nature of the small state and thus implicitly provides a preference for hybrid governance of what I believe are essentially public goods that are unlikely to lead to the kinds of profits that the private sector needs if they are implemented in the spirit of the goals.

While the SDGs could be a first step in the process towards institutionalizing transformation globally, possibly the SDGs 'cannot be expected to generate major structural reforms ... that will significantly reduce the fragmentation of the global governance system' (Underdal and Kim, 2017: 254).

CONCLUSION

I would argue that the development community and the environment community should rejoice at their success in having mainstreamed their agendas in the SDGs. True, the SDGs represent a compromise between different discourses – the neoliberal capitalist agenda and the demand for de-growth or contestation of the notion of development itself within alternative world visions (e.g. Ubuntu and Buen Vivir and those which focus on happiness instead of GDP). However, for the first time reducing multi-dimensional inequality and meeting social floors while living within planetary boundaries and the need to deal with the different goals in an integrated, interrelated and indivisible manner has been recognized universally.

In my view, the SDGs represent the transformationalist view on globalization. They are the result of the mainstreaming of the counter-movement to the massive destruction of the Earth's environment and the blatant disregard of human wellbeing. But there is a very big risk of ideological capture where the SDGs are interpreted and implemented by the dominant neoliberal frame using its huge financial power to sway scholarship, NGOs and governments to allow for the commodification of both public and merit goods, leading to an ever-growing inequality.

Let me return to the Lady of Niger. The SDGs are the Lady of Niger, who is riding on the wave of globalization. The SDGs are ostensibly beautiful – who could object to these Goals? They are leading the way – research institutions, UN agencies, global business have all rhetorically committed to the SDGs. The lady has the reins! She could lead to the universalization of just values and the homogenization of narratives including upscaling local indigenous values and science to a global level (cf. Pieterse, 1994; Sassen, 1998). But the Achilles' heel of the SDGs is that it requires trillions of dollars for its implementation.

The tiger is globalization – it represents the inexorable forces of the capital market, global trade, investment, the Internet of Things; it is probably being manipulated by a chip under the skin of the tiger inserted by the ultra-rich – the eight richest men in collusion with the top ten companies and the top 1 per cent to serve their ends. The notion of partnership so strategically inserted in the SDG document, ostensibly to please those who oppose hierarchy may, de facto, lead to control by the rich partners – after all, he who pays the piper calls the tune. The rich may interpret and implement the SDGs using the neoliberal frame to privatize and commodify resources and ecosystem services, and achieve the goals with a minimal interpretation of human rights while ignoring the calls for redistribution and 'normalizing' neoliberal values to the exclusion of other value systems. The question is whether those who critique the SDGs as representing an arrogant intentional interventionist approach to development and exclude themselves from the discussions are also complicit in allowing the Lady of Niger to get eaten up during the ride into the future?

REFERENCES

Acuto, M. (2013), *Global Cities, Governance and Diplomacy: The Urban Link*, London: Routledge.

AIV (2014), 'The Internet: A Global Free Space with Limited State Control', *Advisory Council on International Affairs*, No. 92.

Andresen, S. and M. Iguchi (2017), 'Lessons from the Health Related Millennium Development Goals', in: Kanie, M. and F. Biermann (eds), *Governing through Goals: Sustainable Development Goals as Governance Innovation*, Cambridge, MA: MIT Press, pp. 165–186.

Bergh, A. and T. Nilsson (2014), 'Is Globalization Reducing Absolute Poverty?', *World Development*, **62**, 42–61.

Bernstein, S. (2017), 'The United Nations and the Governance of Sustainable Development Goals', in: Kanie, M. and F. Biermann (eds), *Governing through Goals: Sustainable Development Goals as Governance Innovation*, Cambridge, MA: MIT Press, pp. 213–240.

Boas, I., F. Biermann and N. Kanie (2016), 'Cross-sectoral Strategies in Global Sustainability Governance: Towards a Nexus Approach', *International Environmental Agreements: Politics, Law and Economics*, **16**(3), 449–464.

Burgh-Woodman, H.C. de (2014), 'Homogeneity, "Glocalism" or Somewhere in Between? A Literary Interpretation of Identity in the Era of Globalization', *European Journal of Marketing*, **48**(1/2), 288–313.

Carbon Tracker (2012), 'Unburnable Carbon: Are the World's Financial Markets Carrying a Carbon Bubble?' http://www.carbontracker.org/wp-content/uploads/downloads/2011/07/Unburnable-Carbon-Full-rev2.pdf (accessed 11 June 2018).

Chasek, P.S. and L.M. Wagner (2016), 'Breaking the Mold: A New Type of Multilateral Sustainable Development Negotiation', *International Environmental Agreements: Politics, Law and Economics*, **16**(3), 397–413.

Chen, Z., Y. Ge, H. Lai and Cc. Wan (2013), 'Globalization and Gender Wage Inequality in China', *World Development*, **44**, 256–266.

Clark, D. (1992), 'A Cloudy Crystal Ball – Visions of the Future', in: Davies, M., C. Clark and D. Legare (eds), *Proceedings of the Twenty-Fourth Internet Engineering Task Force*, Cambridge, MA: MIT Press, pp. 539–545.

Crutzen, P.J. (2006), 'The "Anthropocene"', in: Eckksrt, T. and T. Krafft (eds), *Earth System Science in the Anthropocene*, Heidelberg: Springer, pp. 13–18.

Edixhoven, J.D., J. Gupta, and H.H.G. Savenije (2014), 'Recent Revisions of Phosphate Rock Reserves and Resources: A Critique', *Earth System Dynamics*, **5**(2), 491.

Ezcurra, R. and A.S. Rodriguez-Pose (2013), 'Does Economic Globalization affect Regional Inequality? A Cross-country Analysis', *World Development*, **52**, 92–103.

Fuchs, D. (2013), 'Theorizing the Power of Global Companies', in: Mikler, J. (ed.), *The Handbook of Global Companies*, London: Wiley Blackwell, pp. 77–95.

Gavison, E. (2003), 'The Relationships between Civil and Political and Social and Economic Rights', in: Coicaud, J.-M. et al. (eds), *Globalization of Human Rights*, New York: UNU Press, pp. 23–55.

Gellers, J.C. (2016), 'Crowdsourcing Global Governance: Sustainable Development Goals, Civil Society, and the Pursuit of Democratic Legitimacy', *International Environmental Agreements: Politics, Law and Economics*, **16**(3), 415–432.

George, S. (1992), *The Debt Boomerang: How Third World Debt Harms Us All*, London: Pluto Books.

Griffiths, J. (2014), 'The State of Finance for Developing Countries, 2014: An Assessment of the Scale of all Sources of Finance Available to Developing Countries, European Network on Debt and Development' http://www.eurodad.org/files/pdf/1546315-the-state-of-finance-for-developing-countries-2014-1450105202.pdf (accessed 11 June 2018).

Gupta, J. (2014). *Sharing our Earth*, Inaugural address as Professor of Environment and Development in the Global South, University of Amsterdam, 5 June 2014. http://www.oratiereeks.nl/upload/pdf/PDF-3450weboratie_Gupta.pdf (accessed 31 August 2018).

Gupta, J. (2016), 'Toward Sharing Our Ecospace', in: Nicholson, S. and S. Jinnah (eds), *New Earth Politics*, Cambridge, MA: MIT Press, pp. 271–292.

Gupta, J. and K. Arts (2018), 'Achieving the 1.5°C Objective: Just Implementation Through a Right to (Sustainable) Development', *International Environmental Agreements: Politics, Law and Economics*, **18**(1), 11–28.

Gupta, J. and N. van de Grijp (eds) (2010), *Mainstreaming Climate Change in Development Cooperation: Theory, Practice and Implications for the European Union*, Cambridge: Cambridge University Press.

Gupta, J. and M. Thompson (2010), 'Development and Development Cooperation Theory', in: Gupta, J. and N. van de Grijp (eds), *Mainstreaming Climate Change in Development Cooperation: Theory, Practice and Implications for the European Union*, Cambridge: Cambridge University Press, pp. 33–66.

Gupta, J. and C. Vegelin (2016), 'Sustainable Development Goals and Inclusive Development', *International Environmental Agreements: Politics, Law and Economics*, **16**(3), 433–448.

Haas, P.M. and C. Stevens (2017), 'Ideas, Beliefs, and Policy Linkages: Lessons from Food, Water, and Energy Policies', in: Kanie, M. and F. Biermann (eds), *Governing through Goals: Sustainable Development Goals as Governance Innovation*, Cambridge, MA: MIT Press, pp. 137–164.

Hanley, S. (2017), 'Musk & 116 Others Tell UN To Ban Weapons Controlled By Artificial Intelligence', *The Guardian*, 21 August.

Harari, Y.N. (2017), *Homo Deus: A Brief history of Tomorrow*, London: Random House.

Kanie, N., S. Bernstein, F. Biermann and P.M. Haas (2017), 'Introduction: Global Governance Through Goal Setting', in: Kanie, M. and F. Biermann (eds), *Governing through Goals: Sustainable Development Goals as Governance Innovation*, Cambridge, MA: MIT Press, pp. 1–28.

Lee, B.F., F. Preston, R. Bailey, G. Lahn and B. Kooroshy (2012), *Resources Future: A Chatham House Report*, London: The Royal Institute of International Affairs. http://www.chathamhouse.org/sites/default/files/public/Research/Energy,%20Environment%20and%20Development/1212r_resourcesfutures.pdf (accessed 26 June 2018).

Lukes, S. (2005), *Power: A Radical View*, London: Palgrave.

Mason, P. (2015), *Post-Capitalism: A Guide to our Future*, London: Penguin Books.

Mikler, J. (2013), 'Global Companies as Actors in Global Policy and Governance', in: Mikler, J. (ed.), *The Handbook of Global Companies*, London: Wiley Blackwell, pp. 1–16.

Milanovic, B. (2016), *Global Inequality; A New Approach for the Age of Globalization*. Cambridge, MA: The Belknap Press of Harvard University Press.

Millennium Ecosystem Assessment (2005), *Ecosystem Services and Human Well-being: Policy-responses, Millennium Ecosystem Assessment*, Vol. 3, Washington: Island Press, pp. 489–523.

Miller, M. (1991), *Debt and the Environment: Converging Crises*, New York: United Nations Publications.

Ohmae, K. (1990), *The Borderless World: Power and Strategy in the Interlinked Economy*, London: Collins.

Oxfam (2017), *An Economy for the 99%*, Oxfam Briefing Paper: https://www.oxfam.org/en/research/economy-99 (accessed 11 June 2018).

Paige, E. (1997), 'Electromagnetic Spectrum: Key to Success in Future Conflicts', *Defense Issues*, **11**(83): http://cryptome.org/jya/emskey.htm (accessed 20 June 2018).

PBL (2011), 'Scarcity in a Sea of Plenty', Netherlands Environmental Assessment Agency.

Pieterse, J.N. (1994), 'Globalisation as Hybridisation', *International Sociology*, **9**(2),161–184.

Pintér, L., M. Kok and D. Almassy (2017), 'Measuring Progress in Achieving the Sustainable Development Goals', in: Kanie, M. and F. Biermann (eds), *Governing through Goals: Sustainable Development Goals as Governance Innovation*, Cambridge, MA: MIT Press, pp. 99–134.

Raworth, K. (2012), 'A Safe and Just Space for Humanity. Can We Live Within the Doughnut?', Oxfam Discussion Papers, February 2012.

Rockström, J., W. Steffen, K. Noone, et al. (2009), 'A Safe Operating Space for Humanity', *Nature*, **461**(7263), 472–475.

Sassen, S. (1998), *Globalization and its Discontents: Essays on the New Mobility of People and Money*, New York: New Press.

Sauvant, K.P. (2015), 'The Negotiations of the United Nations Code of Conduct on Transnational Corporations: Experience and Lessons Learned', *The Journal of World Investment & Trade*, **16**, 11–87.

Schrijver, N. (2008), *The Evolution of Sustainable Development in International Law: Inception, Meaning and Status*, Pocketbooks of the Hague Academy of International Law. Leiden/Boston: Martinus Nijhoff.

Silva, G. and L.C. Di Serio (2016), 'The Sixth Wave of Innovation: Are We Ready?', http://dx.doi.org/10.1016/j.rai.2016.03.005 (accessed 12 June 2018).

Steffen, W., R.A. Sanderson, P.D. Tyson, et al. (2004), *Global Change and the Earth System: A Planet Under Pressure*, Berlin: Springer.

Sumner, A. (2010), 'Global Poverty and the New Bottom Billion: What if Three-Quarters of the World's Poor Live in Middle-income Countries?', *IDS Working Papers*, **2010**(349), 1–43.

Thai, K.V., D. Rahm and J.D. Coggburn (2007), 'Globalization and the Environment: An Introduction', in: Thai, K.V., D. Rahm and J.D. Coggburn (eds), *Handbook of Globalisation and the Environment*, London: Taylor and Francis, pp. 1–18.

UN Charter (1945), Charter of the United Nations (San Francisco) 26 June 1945, and amended on 17 December 1963, 20 December 1965 and 20 December 1971, ICJ Acts and Documents No. 4.

UN Paris Agreement (2015), Paris Agreement to the Climate Convention: http://unfccc.int/files/home/application/pdf/paris_agreement.pdf (accessed 12 June 2018).

Underdal, A. and R.E. Kim (2017), 'The Sustainable Development Goals and Multilateral Agreements', in: Kanie, M. and F. Biermann (eds), *Governing through Goals: Sustainable Development Goals as Governance Innovation*, Cambridge, MA: MIT Press, pp. 241–258.

UNGA (1986), 'Declaration on the Right to Development', Resolution A/RES/41/128, 4 December 1986.

UNGA (2015), *Transforming Our World: The 2030 Agenda for Sustainable Development*, Draft resolution referred to the United Nations summit for the adoption of the post-2015 development agenda by the General Assembly at its sixty-ninth session. UN Doc. A/70/L.1 of 18 September.

Voiturez, T., K. Morita, T. Giordano, N. Bakkour and N. Shimuzu (2017), 'Financing the 2030 Agenda for Sustainable Development', in: Kanie, M. and F. Biermann (eds), *Governing through Goals: Sustainable Development Goals as Governance Innovation*, Cambridge, MA: MIT Press, pp. 259–274.

Werksman, J.D. (1993), 'Greening Bretton Woods', in: Sands, P. (ed.), *Greening International Law*, London: Earthscan, pp. 65–84.

White, N.D. (1996), *The Law of International Organisations*, Manchester: Manchester University Press.

Yamada, T. (2017), 'Corporate Water Stewardship: Lessons for Goal Based Hybrid Governance', in: Kanie, M. and F. Biermann (eds), *Governing through Goals: Sustainable Development Goals as Governance Innovation*, Cambridge, MA: MIT Press, pp. 187–210.

Young, O. (2017), 'Goal Setting as a Strategy for Earth System Governance', in: Kanie, M. and F. Biermann (eds), *Governing through Goals: Sustainable Development Goals as Governance Innovation*, Cambridge, MA: MIT Press, pp. 31–52.

Young, O., A. Underdal, N. Kanie and R.E. Kim (2017), 'Goal Setting in the Anthropocene: The Ultimate Challenge of Planetary Stewardship', in: Kanie, M. and F. Biermann (eds), *Governing through Goals: Sustainable Development Goals as Governance Innovation*, Cambridge, MA: MIT Press, pp. 53–74.

Zoomers, A. (2010), 'Globalisation and the Foreignisation of Space: Seven Processes Driving the Current Global Land Grab', *The Journal of Peasant Studies*, **37**(2), 429–447.

7. An economic-geographic perspective on globalization

Robert C. Kloosterman and Pieter Terhorst

INTRODUCTION

Given its nearly all-embracing, multifaceted character, it is not surprising that globalization has been looked at from many academic disciplinary angles. This plurality of approaches still holds, albeit to a somewhat lesser extent, for the economic dimensions or aspects of processes of globalization. Historians, sociologists, urbanists, planners, political scientists, and, evidently economists and (economic) geographers have cut their teeth on studying a plethora of economic aspects of globalization.

In this chapter, we explore more in detail what the specific contribution of economic geographers to the debates on the economic dimension of globalization consists of. We present, in general terms, some key characteristics of economic geographical thinking on globalization. Thus we focus on positioning the economic-geography perspective. Contra the usual imperialist claim by many mainstream economists that only they are equipped to study the economy (cf. Chang, 2014), we contend that geography has a distinct and valuable contribution to make to debates on economic globalization. This distinct contribution is deeply rooted in a particular understanding of social reality which is anchored in a rich ontology which stresses the multidimensionality (and hence relative uniqueness) of places and actors, and acknowledges the role of a wide variety of social structures (from capitalism to ethnic solidarity) in determining the drivers and the outcomes of concrete, historically and geographically specific, economic processes. Mainstream economics, by contrast, conceptualizes places and actors in a much narrower way (often reduced to one or just a few quantitative indicators) and typically ignores the salience of structures. The ontological viewpoint of economic geography is very much intertwined with a pluralistic epistemological position allowing drawing theoretical notions from a wide set of (sub)disciplines and an openness towards methodologies (which encompass quantitative as well as qualitative methods). Or to quote Allen Scott (2004: 485) economic geography is characterized by 'practices of research that are grounded, open, polycentric, focused on rich empirical description, and deeply conscious of the contingency of things'.

We first sketch the key characteristics of the recent phase of 'new' or 'hyperglobalization' (Baldwin, 2016). After that we present the key characteristics of an explicit economic-geographic perspective on these processes of globalization. Given the lack of a hard disciplinary core and the inherent plurality of the (sub)discipline of economic geography and the by now large scope of studies on globalization, we are very much aware of our partiality and bias in singling out this approach. Still, we contend that these conceptual building blocks are crucial to an economic-geographic understanding of globalization. More specifically we emphasize how and why globalization is simultaneously a medium and an outcome of national and regional developments. In the concluding section, we briefly highlight the differences between the approaches of globalization in economic geography and those in (spatial) economics.

THE ECONOMIC CONTOURS OF HYPER-GLOBALIZATION

From 1980 onwards to the outbreak of the credit crisis in 2008, the pace of economic globalization evidently accelerated (WTO, 2015; Milanovic, 2016; Mason, 2016; Baldwin, 2016). Cross-border flows of goods, services, and capital spearheaded by Transnational Corporations (TNCs) and large financial institutions increased significantly. Chains of interdependence became more complex and more dispersed thereby connecting ever more parts of the globe. The enabling conditions of this acceleration in globalization have been, though analytically distinct, closely intertwined. Rapidly declining communication costs (thanks to the new information and communication technologies – ICT), advancements in transport (both the adoption of standardized containers on a worldwide scale and a massive increase in jet-propelled air transport) as well as a broad wave of trade liberalization (including the opening up of China and the fall of communism in the Soviet bloc) lowering the barriers to cross-border flows have gone hand-in-hand to create a space in which global flows are facilitated and fostered. Economic globalization in itself is nothing new. This recent phase, however, differs in several respects from earlier phases.

First, it has been more favourable to cross-border flows of capital, goods, and services, than of workers. The global mobility of labour has been highly selective. On the one hand, we see high-skilled and mostly highly paid workers such as bankers, engineers, footballers, architects, students and academics easily crossing borders. On the other, we can observe a population of (both documented and undocumented) low-skilled migrants who are overly presented in jobs involving routine tasks and often can be seen as part of the precariat (OECD, 2017). For the vast majority of workers, however, mobility remains limited to the national labour market – unless they travel as tourists. The ideology of global 'free markets', then, primarily applies to markets of capital, goods, and services, and much less to labour.

Second, the spatial scope, the extent, and the intensity of linkages between different locations seem to be substantially larger than ever before (Dicken, 2015; Baldwin, 2016). Moreover, the nature of the relationship between what were once seen as core and peripheral countries has also changed (Scott, 2012). The fast rise of notably East-Asian countries as locations of high-value added production has created a whole new global division of labour.

Third, the current phase of globalization is characterized by what Baldwin (2016) calls the 'second unbundling': the carving up of production processes in several phases which each can take place in separate locations. Driven by cost arbitrage – regarding both the costs of labour and of environmental regulations – production processes have increasingly been split up in many separate parts and located in different locations often spanning different continents (Coe and Yeung, 2015). This spatio-organizational fragmentation on a global scale has been enabled by the twin effects of technological and institutional change. The first refers to a low-tech innovation (the introduction of a standard-size steel box – or container – and the subsequent extension of the appropriate infrastructure) (Levinson, 2006), and a cascade of high-tech innovations in storing, manipulating and communicating large chunks of data or information over large distances (thanks to the ICT). The latter has been a crucial driver of a nearly global wave of liberalization which has opened many parts of the world to trade since the 1980s which, in its turn, has been key in enabling the acceleration of globalization (Frieden, 2006; Baldwin, 2016).

Fourth, this second unbundling comprises much more than the spatial separation of the various phases of the goods-handling components of value chains. Increasingly it also involves the offshoring and outsourcing of service activities (Eichengreen and Gupta, 2011, 2013; Kloosterman et al., 2015; Kleibert, 2015; Lambregts et al., 2017).

More generally, one could say that the recent acceleration and extension of globalization is deeply intertwined with a series of structural changes in the production system amounting to what is actually a new phase in capitalism (Scott, 2008, 2012; Mason, 2016). In advanced economies, the production system has been moving away from Fordist mass-production methods based on churning out large numbers of standardized goods. Instead activities where the main input consists of specialized knowledge – high-tech, high-finance, high-concept and high-craft – now form the mainstay of advanced economies in the West and, increasingly also elsewhere, particularly in East Asia. This overarching structure impacts on cross-border flows and, hence, also on the global division of labour. A whole array of labels have been used to describe this new phase in capitalism including: post-Fordism, flexible specialization, digital economy and cognitive-cultural capitalism. Whatever term is used, they share the emphasis on the importance of knowledge and the widespread use of digital technology in the production system in combination with increasing cross-border flows and concomitant integration of global linkages. How then, has economic geography tried to make sense of this new phase of hyper-globalization and cognitive-cultural capitalism?

AN ECONOMIC-GEOGRAPHIC GAZE

Economic geography is a rather recent academic (sub)discipline. It emerged in the immediate post-Second World War period (Scott, 2000: 485). Initially, its approach was very descriptive and, quite similar to history, focusing on case studies and eschewing generalizations and lacking in theoretical ambition. During its more positivist, quantitative phase, from the late 1950s to the early 1970s, economic geography was very much influenced by regional economics. With the shift to sophisticated quantitative methods, economic geography also became more theoretically ambitious trying to uncover more general relationships. With some exaggeration, one can say that these approaches were very much part of modernist projects where, it appeared, governments could stimulate the national and the regional economy through investments in calculable ways. Location theories and regional input-output models were crucial instruments in paving the way for these Keynesian endeavours (Scott, 2000).

In the 1970s, these approaches lost much of their appeal. Keynesian economic policies were unable to combat persistent unemployment in general and more in particular in declining regions because of deindustrialization. In addition, there was a growing unease with the rather narrow positivist approach which ignored social, cultural and political factors. Instead, geographers became much more sensitive to the economic, cultural and power structures which lay behind the more direct world of figures. Like other social sciences, economic geography became more critical. It looks beyond economics and increasingly started to draw insights from economic sociology and (comparative) political science. This epistemological shift coincided with an ideological turn as well as Marxian, feminist, and postmodern theories started to inform the research questions and analytical frameworks (Scott, 2000). Economic geography thus has become much more pluralistic, thereby displaying according to Allen Scott (2004: 481) an 'extreme intellectual hybridity'.

More concretely, economic geographers have abandoned the formal models of regional science for a more discursive approach and the discipline has become more eclectic and empirically oriented. Neoclassical location theories have been displaced by a variety of heterodox theories and approaches such as Myrdal's theory of circular cumulative causation, (neo-)Marxism (particularly David Harvey's life-long project to promote historical-geographical materialism), the French regulation school of political economy with its emphasis on the role of (national) institutions and the ensuing varieties of capitalism, Schumpeterian and evolutionary economics, institutional economics, and economic sociology with its emphasis on various forms of embeddedness (from Max Weber to Karl Polanyi and Mark Granovetter; see Peck, 2005 and 2013 for 'economic sociologies in space' and 'Polanyian economic geographies' respectively). Economic, political, and social institutions, which tend to be ignored by economists, have been playing a key role in understanding how landscapes of production and consumption evolve – on a local scale but also as part of a globalizing economy (Peck, 2005).

A fundamental shift in the epistemological orientation, inevitably, also brought about a change in the underlying ontological assumptions and in the preferred methodologies. The ontological reorientation can be seen in the way space has come to be viewed. For economists space is exogenous and can be reduced to distance and the measurable characteristics of the physical environment (cf. Krugman, 1991). Economic geographers, however, tend to conceptualize space in a much more complex way. For them, space and economy are simultaneously a medium and outcome of each other. Consequently, neither geography nor economy are fixed; each is continually shaping the other through dialectical processes of co-constitution (Soja, 1980). This also implies that in order to be able to analyse the historically and geographically specific character of spatial economic developments such as the impact of globalization, one has to take into account the richness of real places and their concrete historical trajectories (Martin, 1999: 77). A much richer ontology, then, has come into view to understand spatial economic dynamics.

To open up this rich social reality and achieve 'a fine-grained substantive appreciation of diversity' (Scott, 2004), a wide variety of methods have been deployed by economic geographers. Extensive case studies, often comprising both quantitative (e.g. data on employment, trade, GDP) and qualitative methods (e.g. historical description, interviews, ethnographic observation), are arguably the archetypical methodological tool of economic geographers.

Opting for more discursive approaches and rich case studies in economic geography instead of using formal mathematical economic models has not led to an overall abandonment of aiming at formulating more generalized statements. Notably David Harvey's (1982) widely influential theories on the spatial fix (or better temporal-spatial fix) and the Global Production Network approaches (Gereffi, 2014; Coe and Yeung, 2015) are examples of more abstract and generally applicable theorizing. Andrew Sayer (1984 and 1985), Doreen Massey (1984) and Bob Jessop (2001) have grounded the relationship between abstract theory and concrete, historical spatial developments using insights from critical realism, which makes a distinction between abstraction (when a one-sided or partial aspect of an object is isolated in thought) and generalization, necessary (internal) and contingent (external) relations, and levels of abstraction. At a lower level of abstraction, necessary (structural), theoretical relations interact with and are modified by contingent relations. With this ontological (and epistemological) point of departure, it comes as no surprise that the methodology of 'intensive research' by means of case studies is widely used by many economic geographers.

There is, however, a trade-off between the richness of the ontology and the potential scope for generalizations. A narrow ontology, for instance, reducing places to a small set of quantifiable indicators, allows for more sweeping generalizations compared to richer, place descriptions. In the latter case, one can aim at middle-range theories – generalizations circumscribed by place- and time-specific contexts. We will illustrate this approach by looking at how economic geographers have dealt with the relationship between globalization and, respectively, the institutional set-up of the nation state and the role of regions and their cross-border linkages.

STRUCTURES AND PLACES: GRASPING ECONOMIC GLOBALIZATION

Globalization and the Nation State

The sensitivity to the historical and geographical specifics of spatial economic development and their embeddedness in institutional structures is reflected in the popularity of the French regulation school among economic geographers. Adherents of this approach go beyond Marx's autonomous laws of accumulation and instead argue that a high rate of economic growth is only possible if there is a right fit between a 'regime of accumulation' and a 'mode of regulation' (Boyer, 2004). This mode consists of specific institutional forms of capital–labour relations, inter-firm competition, monetary and financial regulation, specific state policies regarding managing aggregate demand, welfare, taxation, and uneven spatial development, and a specific international configuration. The Fordist regime of accumulation, for instance, was, hence, deeply intertwined with Keynesian policies, strong trade unions, a particular social stratification and gendered division of labour in conjunction with international relations dominated by the United States (the Pax Americana after the Second World War (Frieden, 2006; Scott, 2008, 2012; Baldwin, 2016)).

These national institutional frameworks tend to be reproduced over longer periods of time – they display, in other words, strong path-dependent characteristics even in an era of hyper-globalization. Political-economic institutions dealing with state-business relations, capital–labour relations, inter-firm relations, the financial system, corporate governance, and vocational training and education are interlocking and constituting a coherent and complementary complex set of rules, regulations, practices and attitudes (Hall and Soskice, 2001; Whitley, 1999; Kloosterman, 2010b). American shareholders' capitalism, for instance, in which managers have to 'score' in the short term, fits well with flexible labour markets but conflicts with firm-specific labour training programs (cf. Amable, 2003). The strength of the complementarity of these institutions determines to a large extent the distinctiveness of a particular national form of capitalism (Amable, 2003; see for an incisive criticism of the variety of capitalism Crouch, 2005). In addition, each specific national institutional configuration provides a particular competitive advantage, thereby fostering a certain form of economic specialization – for instance in making high-quality cars in Germany. Such a specialization, in turn, reinforces the specific national institutional configuration. The different national institutional frameworks are, hence, key elements in the emergence and reproduction of the global division of labour.

Whereas economists such as Krugman have tried to formulate a universal theory of economic globalization driven by declining transport- and communication costs and trade liberalization, economic geographers inspired by the French regulation school do not deny the importance of declining transport- and communication costs and trade liberalization, but explain the acceleration in (economic) globalization since the early 1980s by the breakdown of Fordism. During the Fordist era the global movements of financial capital were very limited as currencies were tied to the dollar which in turn was tied to gold according to the Bretton Woods agreement of 1944 (Mason, 2016). Nationally specific institutional frameworks were the outcomes of grand coalitions between labour and capital. In the post-war period, rising wages went hand-in-hand with high profit rates. As a result, firms made most of their investments in their home country and were not yet stimulated to globalize. When at the end of Fordism, during the crisis-ridden 1970s and early 1980s, profit rates fell, firms felt forced to increasingly invest abroad.

The shift from Fordism to post-Fordism and from standardized mass production to flexible specialization has also been characterized by the abandonment of Keynesianism in favour of neoliberal policies pushing back the role of the state in economic affairs. This entailed a horizontal transfer of political authority from the state to private actors (flexibilization of labour markets, privatization of state firms and public services), and the formation of quangos and public–private partnerships.

In the view of many economic geographers, this has gone hand-in-hand with a vertical rescaling of both the economy and state (Swyngedouw, 2000; Brenner, 2004; Jessop, 2002). The main building blocks of the world economy are no longer national economies and national states, but urban and regional economies (Scott, 2012). Consequently, the key dialectical relationship in the current phase of globalization is the one between (in)formal institutions of urban and regional governance on the one hand, and global economic processes and their key actors (such as financial institutions, multi-national companies, workers, international strategic alliances, supra-national political organizations, and NGOs) on the other. Upward and downward rescaling is, however, not a uniform process: there are various forms of 'glocalization'. Some countries have become more globalized than others, and some regions have remained largely institutionally complementary to national institutions, whereas others have become much more loosely coupled (Crouch and Voelzkow, 2010; Schröder and Voelzkow, 2015; Terhorst, 2006; Erkuş-Öztürk and Terhorst, 2012).

Globalization, Regions and Linkages

A downward rescaling of economy and state meant that a resurgence of (urban) regions within a globalizing world (Storper, 1995; Scott, 1996, 2008, 2012) has prompted the emergence of a 'new economic regionalism' which comprises a number of approaches, each aiming to explain the role of regions in a global economy from various angles: flexible specialization and industrial districts (Amin, 2000; Becattini et al., 2003; Sabel, 1994), new industrial spaces (Scott, 1988, 1998), the regional world (Storper, 1997), innovative milieus (Camagni, 1991), regional innovation systems (Braczyk et al., 1998), and learning regions (Cooke and Morgan, 1998; Cooke, 2002). Notwithstanding their subtle and, in some cases, rather significant differences, they share the idea that the competitive advantage of regions in the global economy is based on a number of characteristics.

First, these approaches typically depart from the view that economies encompass amalgams of various modes of economic coordination: markets, hierarchies (large firms), associations, communities, networks, and the state (for more see Rogers Hollingsworth and Boyer, 1997). Hierarchies and the national state were the dominant modes of economic coordination in the Fordist era. In the current post-Fordist era, however, we witness a growing importance of intricate mixtures of markets, networks, associations, communities generating and maintaining non-privatized resources, and state institutions on regional or local scale. A spatial clustering of similar firms may stimulate competition (Porter, 2000), but spatial clusters can only be successful in the global economy if markets are combined with above-mentioned other modes of economic coordination.

Second, the shift to a post-Fordist or cognitive-cultural production system has fundamentally altered the (regional) landscapes of production. More specifically the economies of large cities are now dominated by a plethora of combinations of high-tech, high-finance, high-concept and high-craft activities (cf. Scott, 2012; Folmer and Kloosterman, 2017). Most, if not all, of these combinations are heavily dependent on the use of digital information and communication technologies. A myriad of producer services, ICT development, bio-tech, high-end consumer services, cultural and creative industries are now the dominant drivers of advanced urban and regional economies. Many of these firms tend to be disintegrated and, hence, dependent on outsourcing and on external suppliers. Vertical disintegration, in its turn, has stimulated spatial clustering as this reduces transaction costs between firms, particularly in case of irregular, unpredictable or complex transactions. These transaction costs are not so much related to pecuniary externalities or traded interdependencies as to non-pecuniary externalities or untraded interdependencies which occur outside the market realm (Storper, 1997; Scott, 2008, 2012).

Third, whereas the dominant form of competition in Fordist times was based on prices, nowadays competition based on quality (technological, environmental, esthetical) is essential in many markets. Firms in leading sectors tend to be more focused on innovation than on cost reduction through mass production. Innovation, it should be said, comprises much more than radical technological innovations. New producer or consumer-services concepts or a stylistic innovation in music or design may also provide a (temporary) competitive advantage (Kloosterman, 2010a).

Fourth, the increasing prominence of clustering is also based on the presence of pools of specialized workers whose skills match the skills profile demanded by the local firms. Local clusters then become 'hot spots' of particular fields of knowledge reducing the costs of matching between the supply and the demand for specialized labour. It is not just matching, but also learning and sharing that are fostered by spatial proximity. Because specialized knowledge is increasingly codified into information which can be easily disseminated digitally, the competitive position of regions has become increasingly based on the local circulation of tacit knowledge and a local buzz which both have so far escaped codification (Storper and Venables, 2004; Kloosterman, 2008, 2010a). Such knowledge can only be learned by doing and picked up by 'being there' (Gertler, 1995), in face-to-face contacts fostered by social and cultural proximity between actors who share a certain vocabulary and a set of practices and attitudes (Grabher, 2004). Physical proximity, then, still matters in an age of globalization and seemingly frictionless communication.

Fifth, innovation and learning, crucial to compete in a global arena, are collective processes that are supported by coalitions of private and public actors. The state no longer plays a dominant role in it but is at best a primus inter pares. Due to the increasing regionalization of the economy and the ongoing devolution of state power to the regional and urban level, innovation and learning are increasingly sustained by regional alliances of businesses, trade unions, universities and research institutes, vocational training colleges, economic development agencies, and local governments.

Regional clusters are, hence, essential building blocks of the global economy; clusters should be seen as 'Marshallian nodes in global networks' (Amin and Thrift, 1992) where 'local buzz interacts with global pipelines' (Bathelt et al., 2004). Studies have shown that extra-regional, that is, national and increasingly global, untraded interdependencies are also important for the competitiveness of firms (Clark and Tracey, 2004; Simmie, 2004; Sturgeon, 2003; Wolfe and Gertler, 2004). There is also a growing awareness that tacit knowledge may circulate among 'micro-communities of knowledge' and 'communities of practice' despite being globally dispersed (Gertler, 2003). Proximity, it appears, does not only arise in the form of being physically close, but may also emerge through shared social, cultural and professional orientations (Boschma, 2005). These other forms of proximity can help to develop and sustain relations of trust and shared cognitions that allow for the sharing of tacit knowledge. Geographical proximity, then, may be less important for some groups of professionals, than forms of relational proximity.

These different forms of proximity may in practice strongly overlap and reinforce each other – especially if one looks at it from a more dynamic perspective. Anna Saxenian (2006) has shown that Silicon Valley has attracted many immigrant entrepreneurs from Taiwan, India, Israel and China. These entrepreneurs formed or joined local networks according to their country of origin, which were very helpful in starting their businesses, but were able to insert themselves in other networks based on physical and professional proximity. Many of these entrepreneurs returned to their home country and started new firms there (often with help of the state), thus contributing to the diffusion of ICT-clusters on a more global scale. In many cases, they maintained their contacts with others elsewhere, contacts that were initially formed through physical proximity, but afterwards are maintained thanks to relational proximity.

CONCLUSION

Economic geography has developed its own, rather pluralistic, interpretation of processes of economic globalization. These various approaches tend to share a rich view of places, regions and actors which is also sensitive to their particular historical trajectories of development thus typically avoiding straightforward accounts of homogenization and convergence through globalization. Economic geographers, in addition, often position specific cases within broader contexts where larger, not directly visible structures such as capitalism, institutional frameworks and socio-cultural practices are highly relevant in shaping the local outcomes of globalization. This layered ontology necessitates an eclectic epistemology, open to borrowing from different theoretical approaches and from other disciplines. It implies a wide variety of methodologies, often within the framework of case-study approaches.

These philosophical points of departure are quite different from those in mainstream economics. As a result, there is hardly any dialogue between economic geographers and geographical economists (for an exception see Garretsen and Martin, 2010). The assumptions underlying many methods used by economic geographers (including interviews, observations, historical descriptions) are apparently too exotic for mainstream economists to allow a meaningful exchange. Moreover, mainstream economics for years has been anything but outward looking and willing to engage with other disciplines (Chang, 2014). More recently, however, and partly triggered by the dismal performance of economists vis-à-vis the credit crisis of 2008, we see stirrings of more open attitudes within the discipline (Earle et al., 2016; Raworth, 2017). Maybe this will foster more meaningful exchanges between economic geographers and economists in discussing globalization.

In the face of the lack of recognition among economists, we appraise a rich harvest (and many examples are to be found in this *Handbook*) of case studies on globalization by geographers. They show, among other things, the diversity of responses to processes of globalization as these are filtered and shaped by national and local institutional frameworks as well as by conscious actors interacting on different spatial scales. These analyses also make clear that globalization and its impact is anything but a given, exogenous force. This observation is corroborated by the more recent trade statistics of the WTO (2015) which show not just a serious dip after the credit crisis but also a more modest recovery of the ratio of trade and services in total GDP. We can also point at the current rise of economic nationalism propagated by right-wing populists (and others) in many countries which conceivably will change, if not reverse, the course of economic globalization. Economic geographers will probably be among the first to look deeper into such a U-turn as their discipline tends to be 'notably responsive to changes in external economic condition' (Scott, 2000: 484).

REFERENCES

Amable, B. (2003), *The Diversity of Modern Capitalism*, Oxford: Oxford University Press.

Amin, A. (2000), 'Industrial districts', in: Eric Shepard and Trevor J. Barnes (eds) *A Companion to Economic Geography*, Oxford: Blackwell: 149–168.

Amin, A. and N. Thrift (1992), 'Neo-Marshallian nodes in global networks', *International Journal of Urban and Regional Research*, **16**(4), 571–587.

Baldwin, R.E. (2016), *The Great Convergence: Information Technology and the New Globalization*, Cambridge, MA: The Belknap Press of Harvard University Press.

Bathelt, H., A. Malmberg and P. Maskell, (2004), 'Clusters and knowledge: local buzz, global pipelines and the process of knowledge creation', *Progress in Human Geography*, **28**(1), 93–109.

Becattini, G., M. Bellandi, G. Dei Ottati and F. Sforzi (2003), *From Industrial Districts to Local Development: an Itinerary of Research*, Cheltenham, UK and Northampton, MA, USA: Edward Elgar Publishing.

Boschma, R. (2005), 'Proximity and innovation: a critical assessment', *Regional Studies*, **39**(1), 61–74.

Boyer, R. (2004), *Théorie de la Régulation. Les Fondamentaux*, Paris: La Découverte.

Braczyk, H.J., P. Cooke and M. Heidenreich (eds) (1998), *Regional Innovation Systems*, London: UCL Press.

Brenner, N. (2004), *New State Spaces. Urban Governance and the Rescaling of Statehood*, Oxford/New York: Oxford University Press.

Camagni, R. (1991), *Innovation Networks. Spatial Perspectives*, London/New York: GREMI/Belhaven Press.

Chang, H.J. (2014), *Economics: The User's Guide*, London: Penguin.

Clark, G.L. and P. Tracey (2004), *Global Competitiveness and Innovation. An Agent-Centred Perspective*, Basingstoke: Palgrave Macmillan.

Coe, N. and H. Yeung (2015), *Global Production Networks: Theorizing Economic Development in an Interconnected World*, Oxford: Oxford University Press.

Cooke, P. (2002), *Knowledge Economies. Clusters, Learning and Cooperative Advantage*, London/New York: Routledge.

Cooke, P. and K. Morgan (1998), *The Associational Economy. Firms, Regions, and Innovation*, Oxford: Oxford University Press.

Crouch, C. (2005), *Capitalist Diversity and Change: Recombinant Governance and Institutional Entrepreneurs*, Oxford: Oxford University Press.

Crouch, C. and H. Voelzkow (2010), *Innovation in Local Economies: Germany in Comparative Context*, Oxford: Oxford University Press.

Dicken, P. (2015), *Global Shift, Mapping the Changing Contours of the World Economy* (7th edn), Los Angeles/London/New Delhi/Singapore/Washington, DC: Sage.

Earle, J., C. Moran and Z. Ward-Perkins (2016), *The Econocracy; The Perils of Leaving to the Experts*, Manchester: Manchester University Press.

Eichengreen, B. and P. Gupta (2011), *The Service Sector as India's Road to Economic Growth* (No. w16757), Washington, DC: National Bureau of Economic Research.

Eichengreen, B. and P. Gupta (2013), 'The two waves of service-sector growth', *Oxford Economic Papers*, **65**(1), 96–123.

Erkuş-Öztürk, H. and P. Terhorst (2012), 'Two-micro-models of tourism capitalism and the (re)scaling of state-business relations', *Tourism Geographies*, **14**(3), 494–523.

Folmer, E. and R.C. Kloosterman (2017), 'Emerging intra-urban geographies of the cognitive-cultural economy: evidence from residential neighbourhoods in Dutch cities', *Environment and Planning A*, **49**(4), 801–818.

Frieden, J. (2006), *Global Capitalism, Its Fall and Rise in the Twentieth Century*, New York/London: W.W. Norton and Company.

Garretsen, H. and R. Martin (2010), 'Rethinking (new) economic geography models: taking geography and history more seriously', *Spatial Economic Analysis*, **5**(2), 127–160.

Gereffi, G. (2014), 'Global value chains in a post-Washington Consensus world', *Review of International Political Economy*, **21**(1), 9–37.

Gertler, M.S. (1995), '"Being there": proximity, organization and culture in the development and adoption of advanced manufacturing technologies', *Economic Geography*, **71**(1), 1–26.

Gertler, M.S. (2003), 'The spatial life of things: the real world of practice within the global firm', in: J. Peck and H. Wai-chung Yeung (eds) *Remaking the Global Economy: Economic–Geographical Perspectives*. London: Sage: 101–113.

Grabher, G. (2004), 'Learning in projects, remembering in networks? Communality, sociality, and connectivity in project ecologies', *European Urban and Regional Studies*, **11**(2), 103–123.

Hall, P.A. and D. Soskice (eds) (2001), *Varieties of Capitalism: The Institutional Foundations of Comparative Advantage*, Oxford: Oxford University Press.

Harvey, D. (1982), *The Limits to Capital*, Oxford: Blackwell.

Jessop, B. (2001), 'Capitalism, the regulation approach and Marxism', in: A. Brown, S. Fleetwood and J.M. Roberts (eds) *Critical Realism and Marxism*, London/New York: Routledge: 88–115.

Jessop, B. (2002), *The Future of the Capitalist State*, Cambridge: Polity Press.

Kleibert, J.M. (2015), *Expanding Global Production Networks: The Emergence, Evolution and the Developmental Impact of the Offshore Service Sector in the Philippines*, Amsterdam: Dissertation Universiteit van Amsterdam.

Kloosterman, R.C. (2008), 'Walls and bridges: knowledge spillover between "superdutch" architectural firms', *Journal of Economic Geography*, **8**(4): 545–563.

Kloosterman, R.C. (2010a), 'Building a career: labour practices and cluster reproduction in Dutch architectural design', *Regional Studies*, **44**(7), 859–871.

Kloosterman, R.C. (2010b), 'This is not America: embedding the cognitive-cultural urban economy', *Geografiska Annaler: Series B, Human Geography*, **92**(2), 131–143.

Kloosterman, R.C., N. Beerepoot and B. Lambregts (2015), 'Service-sector driven economic development from a historical perspective', in: B. Lambregts, N. Beerepoot and Robert C. Kloosterman (eds) *The Local Impact of Globalization in South and Southeast Asia: Offshore Business Processes in Services Industries*, Oxford/New York: Routledge: 17–28.

Krugman, P. (1991), *Geography and Trade*, Leuven: Leuven University Press.

Lambregts, B., N. Beerepoot and J. Kleibert (2017). 'Globalisation and services-driven economic growth: an introduction', in: N. Beerepoot, B. Lambregts and J. Kleibert (eds) *Globalisation and Services-driven Economic Growth*, London/New York: Routledge: 1–18.

Levinson, M. (2006), *The Box: How the Shipping Container Made the World Smaller and the World Economy Bigger*, Princeton: Princeton University Press.

Martin, R. (1999), 'Critical survey. The new geographical turn in economics: some critical reflections', *Cambridge Journal of Economics*, **23**(1), 65–91.

Mason, P. (2016), *Postcapitalism; A Guide to Our Future*, London: Penguin Books.

Massey, D. (1984), *Spatial Divisions of Labour. Social Structures and the Geography of Production*, Basingstoke: Macmillan.

Milanovic, B. (2016), *Global Inequality; A New Approach for the Age of Globalization*, Cambridge, MA: The Belknap Press of Harvard University Press.

OECD (2017), *International Migration Outlook 2017*, Paris: OECD Publishing.

Peck, J. (2005), 'Economic sociologies in space', *Economic Geography*, **81**(2), 129–175.

Peck, J. (2013), 'For Polanyian economic geographies', *Environment and Planning A*, **45**(7), 1545–1568.

Porter, M.E. (2000), 'Locations, clusters, and company strategy', in: G.L. Clark, M.P. Feldman and M.S. Gertler (eds) *The Oxford Handbook of Economic Geography*, Oxford: Oxford University Press: 253–274.

Raworth, K. (2017), *Doughnut Economics; Seven Ways to Think Like a 21st-Century Economist*, London: Random House Books.

Rogers Hollingsworth, J. and R. Boyer (1997), 'Coordination of economic actors and social systems of production', in: J. Rogers Holingsworth and R. Boyer (eds) *Contemporary Capitalism. The Embeddedness of Institutions*, Cambridge: Cambridge University Press: 1–47.

Sabel, C. (1994), 'Flexible specialisation and the re-emergence of regional economies', in: A. Amin (ed.) *Post-Fordism: a Reader*, Oxford: Blackwell: 101–156.

Saxenian, A.L. (2006), *The New Argonauts. Regional Advantage in a Global Economy*, Cambridge, MA: Harvard University Press.

Sayer, A. (1984), *Method in Social Science. A Realist Approach*, London: Hutchinson.

Sayer, A. (1985), 'The difference that space makes', in: D. Gregory and J. Urry (eds) *Social Relations and Spatial Structures*, Basingstoke: Macmillan: 49–66.

Schröder, M. and H. Voelzkow (2016), 'Varieties of regulation: how to combine sectoral, regional and national levels', *Regional Studies*, **50**(1), 7–19.

Scott, A.J. (1988), *New Industrial Spaces. Flexible Production Organization and Regional Development in North America and Western Europe*, London: Pion.

Scott, A.J. (1996), 'Regional motors of the global economy', *Futures*, **28**(5), 391–412.

Scott, A.J. (1998), *Regions and the World Economy.: the Coming Shape of Global Production, Competition, and Political Order*, Oxford: Oxford University Press.

Scott, A.J. (2000), 'Economic geography: the great half-century', *Cambridge Journal of Economics*, **24**(4), 483–504.

Scott, A.J. (2004), 'A perspective of economic geography'. *Journal of Economic Geography*, **4**(5), 479–499.

Scott, A.J. (2008), *Social Economy of the Metropolis: Cognitive-Cultural Capitalism and the Global Resurgence of Cities*, Oxford: Oxford University Press.

Scott, A.J. (2012), *A World in Emergence, Cities and Regions in the 21st Century*, Cheltenham, UK and Northampton, MA, USA: Edward Elgar Publishing.

Simmie, J. (2004), 'Innovation and clustering in the globalised international economy', *Urban Studies*, **41**(5/6), 1095–1112.

Soja, E. (1980), 'The socio-spatial dialectic', *Annals of the Association of American Geographers*, **70**(2), 207–225.

Storper, M. (1995), 'The resurgence of regional economies, ten years later', *European Urban and Regional Studies*, **2**(3), 191–221.

Storper, M. (1997), *The Regional World: Territorial Development in a Global Economy*, New York/London: Guilford Press.

Storper, M. and A.J. Venables (2004), 'Buzz: face-to-face contact and the urban economy', *Journal of Economic Geography*, **4**(4), 351–370.

Sturgeon, T.J. (2003), 'What really goes-on in Silicon Valley? Spatial clustering and dispersal in modular production networks', *Journal of Economic Geography*, **3**(2), 199–225.

Swyngedouw, E. (2000), 'Elite power, global forces and the political economy of "glocal" development', in: G.L. Clark, M.P. Feldman and M.S. Gertler (eds) *The Oxford Handbook of Economic Geography*, Oxford: Oxford University Press: 541–558.

Terhorst, P. (2006), 'The scaling of the Dutch vegetables-under-glass cluster: sweet peppers, tomatoes and cucumbers', *Tijdschrift voor Economische en Sociale Geografie*, **97**(4), 434–442.

Whitley, R. (1999), *Divergent Capitalisms: The Social Structuring and Change of Business Systems*, Oxford: Oxford University Press.

Wolfe, D.A. and M.S. Gertler (2004), 'Clusters from the inside and out: local dynamics and global linkages', *Urban Studies*, **41**(5/6), 1071–1093.

WTO (2015), *International Trade Statistics 2015*. https://www.wto.org/english/res_e/statis_e/its2015_e/its15_toc_e.htm (accessed 2 January 2018).

8. Globalization in cultural and media geographies
Soyoon Choo and Elizabeth Currid-Halkett

> Then came the film and burst this prison-world asunder by the dynamite of a tenth of a second ... With the close-up space expands, with slow motion, movement is extended ... Evidently, a different nature opens itself to the camera than opens to the naked eye – if only because an unconsciously penetrated space is substituted for a space consciously explored. (From Walter Benjamin's *Illuminations* (1936[1969]: 236))

In his essay, 'The Work of Art in the Age of Mechanical Reproduction' (1936[1969]), philosopher and cultural critic Walter Benjamin wrote about the 'revolutionary demands' of art and cultural production that comes with technological development and socio-economic transformations. Using film and photography as the prime examples, Benjamin expounded that even if an artistic product loses its originality and authenticity in the process of reproduction – its 'aura' – the viewer is able to have aesthetic interpretations of the reproduced image. Art and cultural products are utilized and perceived beyond 'art for art's sake' appreciation; the value is now created and placed on the experience it brings to the everyday life, the message it delivers, and the imagination it evokes. Geographer David Harvey states the aforementioned Benjamin's quote as the reason why the development of cultural production and marketing on a global scale has itself been a primary agent in time-space compression (Harvey, 1989: 349) – the effect of making distant places grow 'closer'.

Fast forward to 2017: the global scope and intense interconnectedness of culture and media have now become an ordinary aspect of daily life. Simply put, culture is everywhere (Hannerz, 1996) – and the 'next new' technologies and platforms that enable the 'new viral' cultures and cultural products appear within shorter periods. From scholars to businesses to everyday users, it is becoming harder to keep track of these changes and updates while at the same time they *have to* constantly pay keen attention to shifting tastes and trends, or else it is a matter of months or even weeks to become 'outdated'. What does this entail? Is culture, taste, and the impact of culture itself becoming even more ephemeral than ever? Does place even matter anymore in the ever-growing virtualization of space? What communities and values are being formed in this twenty-first century cultural globalization?

In the twentieth century, the diffusion of culture and media were criticized as imperialistic; globalization depicted as Westernization, Americanization, and McDonaldization (Ritzer, 2015) contained a dire outlook on local cultures, values, and ways of living. A more 'geographically sensitive' approach to globalization (Gregory et al., 2009: 310) highlighted new forms of uneven development involving both deterritorialization and reterritorialization; this brought 'glocalization' under attention as a way of exploring 'reciprocal local–global relations' that avoids end-state, end-of-geography ideas about global flattening (Friedman, 2006; Gregory et al., 2009; Swyngedouw, 2004) or homogenization. The cultural and media globalization in the twenty-first century provide a dramatic shift in this phenomenon. While glocalization has been commonly understood as business practices and strategies (Gregory et al., 2009) on the localized adaptation of a global product (for instance, kimchi and bulgogi burgers at Korean McDonald's and lattes with regional flavours in Japanese Starbucks), it is also beyond that: it is the 'dialectical local–global relations' through which local regions mediate and change global processes even as they are remade and rescaled themselves (Swyngedouw, 2004). Further, it indicates the local(ized) acceptance of the social values created through such cultural exchange and consumption. What the local and non-local consumers – as a 'community of sentiment' (Appadurai, 1990) – make of these values, and image the places and products they are from, in turn affect the production and distribution process.

This chapter will focus on the more contemporary cultural globalization (2000s and onwards), where there is 'no historical equivalent' of the global reach and volume of cultural traffic (Held et al., 1999) and the globalization of both cultural and media geographies reinforce the existence of each other at an ever-intimate level. In particular, it will look into the concept of globalization in these two geographic subfields in terms of the social aspects of cultural production, consumption, dissemination, and value creation. By going over the historical trajectories leading up to the twenty-first century cultural globalization and its spatiotemporal and organizational characteristics, it examines the symbiotic relationship between culture, media, and 'global cities' (Sassen, 1991) – both well-known and emerging ones – and how altogether they play a crucial role in the 'imaging' and 'imagining' of place as well as community.

THE CONCEPTUALIZATION OF GLOBALIZATION IN CULTURAL AND MEDIA GEOGRAPHY

The conventional definition of cultural geography is the geographic study of cultural products, practices, and the norms and meanings elucidated in/by their distributional patterns across spaces and places. However, with growing diversity on the idea of culture and challenging notions on technology's relationship between space and place (Sassen, 1997; Castells, 2004, 2009; Currid-Halkett, 2010), the breadth of the field in the twenty-first century is much wider in scope to be merely confined as a single branch of human geography (Atkinson, 2005). As dealing with important questions such as identity formation and the construction of citizenship and belonging have become central in cultural geography, the field has for some become more of a 'style of thought' seeking and linking the ideas and imaginations with the changing material world (Anderson, 2003; Atkinson, 2005). Specifically, the social relations and dynamics are critical to acknowledge within the (need for a) renewed understanding on cultural geography, as it illuminates the shifting and multifaceted dimensions of cultural production and dissemination in the twenty-first century.

While the field of cultural geography has been broadened, and received attention as a lively, interdisciplinary field under the larger umbrella of human geography, media geography has had a more difficult 'identity search'. The 'elusive' nature of where media and communication studies lie on – encompassing fields such as linguistics, philosophy, sociology, and rhetoric – has prohibited the 'geographers' response' to have an articulated paradigm up until the late 1990s (Adams, 2016). However, the dramatic evolution of media – in form (i.e. media organizations, operations, traditional and emerging/alternative platforms) as well as content (i.e. images, messages, and the discourses that shape or are shaped by them) – has led to a flourishing of media studies by geographers at the turn of the twenty-first century. Critical questions on representation, power relations, and social meanings became inevitable in the study of the media and its accompanying technological advancements (Sassen, 1997; Banet-Weiser, 2012; Jenkins et al., 2015).

The phenomenon of globalization is a central component and driver in the aforementioned developments in cultural and media geography, and has become the conduit by which these two fields are intimately and intricately connected. The production, consumption, and dissemination processes in both fields cannot be limited to traditional contexts and perceptions. For instance, the understanding that cultural homogenization is the sole result of the globalizing society neglects the multilevel, layered processes that occur by individuals/groups in a certain place or country. Further, film and television are no longer the main bridges or only mediums to the 'world'. As each field does not operate separately in a linear manner, the increasing importance of spatiality in media production and representation (Tinic, 2008) is key to enhancing place effects in cultural geography, and vice versa. Likewise, globalization itself should also be understood as an evolving concept with diversified causal dynamics, communication networks, and flows between place, people, and values via cultural and media representations.

GLOBALIZATION AND ITS CONTEXT: MEDIA AND CULTURE

Globalization in media and cultural geography is conceptualized under the following contexts: the cultural turn in human geography informed by postmodernity (Murray, 2006; Jackson, 1989; Johnston and Sidaway, 2004), and the globalization of culture itself. Several major topics are crucial in understanding the conditions of this rather grand topic. First, the socio-economic, cultural, political, and epistemological shifts from modernity to postmodernity (Harvey, 1989) and subsequent changes in the capitalist society and the way economies and organizations function and as skills illuminate the cultural discourse and value systems that are interplayed in cultural production and consumption (Sennett, 2008). Second, the development in communication technologies both created the network society and further developed the global network society, the latter of which propelled globalization at an unprecedented speed and with profound implications in the spatial distribution of human activities and social structures (Sassen, 1991; Castells, 2004, 2009). Third, the social context and dynamics of cultural production and consumption has arisen as critical features especially in the contemporary era, in terms of how art/cultural products are produced and distributed, and what aesthetic and market values are generated in the process (Becker, 1982; Caves, 2002). Lastly, mass mediation in the cultural geography framework plays a key role in the cultivation of place-branding (Molotch, 2002), and the construction of imagined selves, communities, and worlds that compel the transformation of everyday discourse (Anderson, 1983; Appadurai, 1996). The physical centrality of these processes has manifested in cities, which have become the critical production and consumption zones as an embodiment of materialized intangible values (Zukin, 1991; Glaeser et al., 2001; Clark, 2011).

Then how is globalization conceptualized in the twenty-first century in these contexts? As concisely laid out by Murray, the earlier global cultural interactions before the Second World War in the form of European imperialism, 'global empires', and the establishment of nation-states marked the legacy and connection to the contemporary era (Murray, 2006). The shift from industrial to post-industrial in the 1970s was followed by the 'cultural turn in social sciences' in the 1980s (Murray, 2006), where culture was emphasized in bringing social change and thus the cultural turn in human geography is highlighted. Further, the notion of postmodernity – decentralization, mass media and popular culture, global economy, consumerism and so on – culminated in all aspects of industry, productivity, and artistic endeavours. Culture as in values and practices 'enter the picture' as the glue to the specialized yet fragmented postmodern business cultures and community sentiments (Sennett, 2008).

The 'age of migration' in the 1990s brought an explosion of international population flows – from the professionally skilled migrants, to unskilled labourers, and to refugees – which characterized the globalized spatial patterns in terms of cultural hybridity and diaspora populations (Castles and Miller, 1993). At the turn of the twenty-first century leading up to today, the networked economy, adaptation to local ('glocalized') conditions/contexts, and so forth reached the pinnacle on the cultural globalization, or globalization of culture. Transnational businesses in culture, communication technologies and rapid media exchanges, and especially, consumption culture and brands (both place and products) as markers of identity (Held et al., 1999; Murray, 2006) hallmarked as features of globalization.

Such sociological, anthropological, and geographical theoretical inquiries are now informed with real-life empirical case studies by many contemporary scholars – some of which will be introduced later in this chapter. Research studies in this realm often take the form of comparative studies, area/regional studies, mixed-methods, and innovative attempts at aggregating data/phenomena. Despite the diversity in topics, methods, and scope of research, they all lead to common findings: (1) the global reach of the spatiotemporal dimensions, or connections, between cultural production/consumption, media gatekeepers, and creation of place-specific tastes and values; and (2) the networked organizational dimensions that set the stage for that global reach. In the following sections, we will discuss each in turn.

SPATIOTEMPORAL DIMENSIONS

A simplistic yet convenient way of describing the causal dynamics of globalization would be the 'flattening' of the world and nation-state boundaries, which enabled extensive networks and global linkages in production, marketing, and distribution of cultural values and products (Friedman, 2006). To unpack this a bit more, Castells observes that industrialism is subsumed by informationalism, which became the new paradigm that is based on the augmentation of human capacity in information processing and communication technologies (Castells, 2004). The diversified and digitized platforms in mass mediation (i.e. from televisions and radios to computer software, telecommunications and so on) has created a network society where instantaneity and real-time feedback in cultural production on events, media images, and so forth became innovative sources for economic productivity, creativity, and political power-making (Harvey, 1989; Castells, 2001). Such developments amplified the age of globalizing migration (Castles and Miller, 1993) and global cultural flows: 'migratory audiences' (Appadurai, 1996) of high mobility – migrants, tourists, diasporas, transnational businesses/partnerships and so on – become 'the link between globalization and the modern [world]' (Appadurai, 1996: 4), where the imagination has become a collective, social fact and a community of sentiment (Appadurai, 1990) is created via collective experiences through media. However, the proliferation of media platforms, communication technologies, and digital/virtual spaces and experiences for cultural production/consumption does not indicate that place is now meaningless (Meyrowitz, 1985). In fact, the time–space compression (Harvey, 1989) that impacts the intensity and propensity of (global) interconnectedness shows how much place still matters. It is rather a 'converse thesis' that exemplifies locational advantages and spatially different qualities (Harvey, 1989: 293), and further new spatial structures consisted of networks that connect places through information and communication flows (Castells, 2004).

According to Castells, in a network society, the network(s) itself is a set of interconnected nodes without a centre; within nodes, information is absorbed and processed, and it depends on each network's ability to communicate that determines the success or failure of whatever goal or mission the network(s) possesses (Castells, 2004). When it comes to globalization in cultural and media geographies, global cities (or world cities) are places specific in themselves and also shaped by external relations (Massey, 1984) as nodes in networks (Hannerz, 1996). This process is elaborated well in Lorenzen and Mudambi's study on Bollywood and software clusters each in emerging global cities Mumbai and Bangalore in India (further explained in the next section). The global linkages are the social networks with local and non-local actors as the nodes, and according to their findings, the intangible soft infrastructures were more significant in these nodes in creating successful clusters of production and further diffusion of products (Lorenzen and Mudambi, 2012). In other words, the social aspects and discourses are key in illuminating the spatio-temporality in cultural and media geography's context of globalization.

As the opening quote from Walter Benjamin elucidates, the 'unconsciously penetrated space' (1936[1969]: 236) is 'postmodernism's most patently transparent feature' (Harvey, 1989: 63), the deep, intimate embeddedness of culture and the meanings behind the cultural forms/products in everyday life. In other words, by blurring the distinction between what is constituted as high and low culture, culture under postmodernism is characterized by its full immersion in everyday practices. Harvey links such culture with consumerism in the late capitalist society (Harvey, 1989), where '[postmodernism] raids and parodies past art' (Brooker, 1992: 3). However, even accounting for this reproduction and consumption process, the meaning and value of authenticity is maintained through a branded cultural space, and as a relationship between consumers and branders (Banet-Weiser, 2012), the latter including the cultural producers and media gatekeepers. Branded places, and especially branded cities, as global hubs of culture provide/project intangible exchange values in lieu of tangible use values, and thus cities become both a physical, real place and an abstract space (Greenberg, 2008). The goal or outcome from such place-branding or place-imaging is not the homogenizing effect that was under much criticism in the twentieth century. Rather, it is about the individualized creation of values, culture, and identity, which in turn can build up into a collective level as part of formulating Appadurai's notion of 'community of sentiment' in multiple physical places/virtual spaces at the same time (which does not, especially today, necessarily 'start off' in the US/Western countries). The identity of place that is born from this place-imaging and what ever-increasingly proactive consumers and audiences make of it, are critical in this cultural globalization.

The main drivers of such spatiotemporal dimensions in globalization can be identified as the following, each leading up to and/or reinforcing the next. First, as mentioned earlier, are media and communicative technologies. With the empowerment of consumers/audiences, such technologies are now emphasizing the interactive, participatory flows of information. Cultural producers are now in need to seek, try out, and familiarize themselves in new technologies; in fact, it is often the cultural producers (of cultural products, events and so on) who are the pioneers in using the 'next big' communicative/media platforms. For instance, the commonly used social media outlets like *Facebook*, *Twitter*, and *Instagram* are becoming 'old school' – both cultural start-ups and existing institutions are investing in mobile application-based platforms such as *Snapchat*, *Vine*, and *Reddit* to reach out effectively to consumers/users in 'even more' real time. Furthermore, the organic 'word of mouth' marketing – which Serazio calls 'regimes of dialogue' – is also incorporated as a 'program of flexibility and authenticity' (Serazio, 2013: 106) in a brand's socialization, which embodies the enhanced powers of flexibility and mobility in time–space compression (Harvey, 1989).

Second, as an outcome of the first driver, are the symbolic communications that deliver media images and the conveyed ideas, messages, and emotions. Such visual and social representations place emphasis on the brand recognition and associated images/ideological value rather than on the actual product, for instance better quality, socially conscious/transparent production process, and self-empowerment and entrepreneurism (Harvey, 1989; Banet-Weiser, 2012; Currid-Halkett, 2017). Also, industries shape, reshape and influence the symbolic images of regions, thus producing an 'authentic' reputation (Scott, 2001; Rantisi, 2004) that glues the association between place-based advantages and cultural products/industries.

This brings us to the intimately linked third driver; the place factors that appeal to cultural production, consumption (Molotch, 2002; Currid and Williams, 2010), and 'remote' and 'actual' experiences that can serve as a global catalyst. The place of origin where a product is made, an event has happened, and the symbolic communication is circulated – the nodes – becomes a powerful influential in viewing particular industries and subsequent aesthetic/market values; in short, 'places become products themselves' (Currid and Williams, 2010: 438).

Finally, yet importantly, is the imaginary of community as collective, social facts (Appadurai, 1990; Hannerz, 1996; Currid-Halkett, 2017) because of cultural flows and place-imaging. The more instantaneously cultural exchanges and consumption occur, such communities that valorize a certain quality/image/nature of the culture or cultural product can be formed across national boundaries and, in other words, become communities that are not based on localities but on cultural consumption.

Globalization in the twenty-first century and the new capitalism of flexible accumulation influenced a change in business culture and practice (Lorenzen and Mudambi, 2012; Sennett, 2008). As the consumers and audiences of the changing media ecosystem are not merely passive recipients, cultural producers, marketers, and media producers pour much resources and keen investment in understanding that system to the fullest to encourage an even more fluid and participatory-networked society (Serazio, 2013). Having the 'right ingredients' as the drivers are critical for creating an intense global interconnectedness and subsequent impact, where 'space of flows' and 'timeless time' are *the* cultures of the networked society (Castells, 2004). In the following sections, we will consider case studies in the intersection of globalization, culture and media.

CASE STUDIES: GLOBAL FLOWS IN CULTURE, MEDIA, AND PLACE-IMAGING

Globalization as a propeller in cultural production, value creation, and place-imaging has been studied in various cities and regions. The following section will briefly look into the main findings of three case studies that – within their unique contexts in some aspects – reinforce how cultural production, consumption and the values created through gatekeepers that transcend borders also lead to transnational place-based communities. The first set of studies delve into the socio-spatial correlations in cultural and media geography. First, Currid and Williams (2010) compare the social context of cultural consumption, the 'mechanism of buzz' and social milieu in cultural production in place branding, which sets the stage further for Currid-Halkett and Ravid's 2012 study on world-city cultural hubs ('star markets'). Second, Lorenzen and Mudambi (2012) combine theory of clusters that stresses importance of local geography and international business literature that stresses the importance of the global context, comparing the Bollywood entertainment and software cluster's connectivity and the types of global linkages each has. Lastly, Otmazgin (2016) looks into the regionalization and region-making via Hallyu ('Korean Wave') and how translational dissemination of popular culture happens across Asian societies, as well as dissemination of communities of lifestyles and concepts.

STARS, BUZZ, AND CULTURAL-LOCATIONAL VALUES: CASE OF WORLD CITY CULTURAL HUBS

While the spatial analysis in Currid and Williams' 2010 study is in a US context comparing Los Angeles and New York, the non-random spatial clustering of local cultural events and industries provide a compelling case on how 'art worlds' work (Becker, 1982) and can be extended beyond localities (as exemplified in the next two case studies). Based on the fundamental notion that cultural consumption is social, and buzz motivates such consumption and creates market(able) value (Caves, 2002), the authors aggregated the geographical form of the social milieu by using an original dataset of Getty images of buzzworthy events in Los Angeles and New York. By conducting a just-in-time *in situ* analysis (Currid and Williams, 2010: 433), the authors were able to explore localized socio-spatial dynamics of cultural consumption/ production, as well as the spatial analysis of Los Angeles and New York of what makes them the 'hot spots' for arts and cultural events.

There are two important findings from this study that amplify the spatiotemporal dimensions of, and symbiotic relationship between cultural and media geography. The first one is that social consumption of art and culture is not spatially random, and that places become products themselves (Currid and Williams, 2010: 438). At first glance, it may seem obvious in a common-sensical way that Los Angeles and New York host and generate the most buzzworthy cultural events that viewers/consumers all around the world 'tune in'. However, it is important to note that it is also certain neighbourhoods within those cities that have the iconic infrastructure (i.e. the Kodak Theater in Hollywood) and other 'urban boosters' that enable the 'place branding' (Molotch, 2002) and subsequent 'image' of the event as a powerful influencer (Currid and Williams, 2010: 429). The second finding concerns media as an industry 'in its own right' (Currid and Williams, 2010: 443) with market-driven incentives, in which together with the social milieu it creates the flow of information and value dissemination of certain cultural products. The authors explain that the presence of cultural products at particular spaces brought 'unanticipated' media clustering and 'unintentional influence' on place development and identity.

Such roles of the social milieu and media gatekeepers create the 'mechanisms of fascination' (Thrift, 2008) that partially explain how and why the 'hyperagglomeration' of cultural hubs – or 'star markets' – in certain world cities happen (Currid-Halkett and Ravid, 2012). Similar to the 2010 case study, Currid-Halkett and Ravid employed a quantifying method on the connectivity of human capital across these world cities/star markets. Again, using Getty photographic data, this analysis looked at the 'geo-behaviour' of individual stars in the film industry, and how certain cities are media-driven and/or industry-prestige-driven based on the individuals' geographical movements. While the previous 2010 study considered the clustering of cultural events and media of particular neighbourhoods in Los Angeles and New York, and thereby the clustering of social/cultural consumption and valorization towards each location, this study looks into how these star markets have a 'cooperative oligopoly' relationship amongst themselves. Thus, Los Angeles, New York, and London reinforce their world-city status as a cultural hub by sharing the (elite) human capital and informal milieus central to the buzz of the place itself and the cultural product – either event and/or presence of the individual star. The authors also found out how there is a 'second tier of cities', or 'niche star markets', that are linked to such world cities due to their specialization. The most prominent example would be Park City, Utah, where the largest independent film festival, Sundance Film Festival, takes place. Although 'seasonal' in terms of media attention and star presence, the 'Sundance effect' contributes to the place-branding of Park City.

These two sets of studies provide a macro picture of how place effects and cultural production/consumption go hand-in-hand, and the role of media gatekeepers that reinforce the social aspects behind 'validated' consumption. In turn, such 'cumulative reputational effects' (Currid-Halkett and Ravid, 2012) in the globalization era created a certain global hierarchy in the cultural as well as media industries. As noted in the historical trajectories of globalization in cultural and media geography, the influencers of this global hierarchy have been largely in the US/Western cities, exemplified in the aforementioned cases. Partly in critique of the dominant narrative, and partly to recognize how global cultural influencers are created in a non-Western context, much scholarly attention is being placed – and needed – in studying those arenas today. Next, we will turn to the case of Bollywood, perhaps the most salient example of non-Western contemporary globalized media and cultural clustering.

CLUSTERS AND PERSONAL RELATIONSHIPS: CASE OF BOLLYWOOD

Currid-Halkett and Ravid mention in their 2012 study of star markets, that the case of Bollywood is 'counterfactual' to the process of becoming a cultural hub. From their aggregated Getty image database, Bollywood events had minimal exposure (thus 'low' on the global buzz barometer) and found that the industry had very limited to no engagement with other film capitals or cultural milieu around the world. Nevertheless, Bollywood has been the only 'late-moving cluster' in the film industry that 'caught up' with Hollywood (Lorenzen, 2009), and Lorenzen and Mudambi's 2012 case study sheds light on how local geography matters in creating global linkages necessary for cultural dissemination across borders. While Bollywood does not capture Western markets as much as Hollywood, Bollywood indeed has a global reach and attracts more viewers and fans across the world than Hollywood (Currid-Halkett, 2010).

The case study compares the Bollywood film industry in Mumbai and the information and communication technology (ICT) cluster in Bangalore, India. From this comparison, the importance and significance of social and human capital cultural industries (in this case, Bollywood) are reaffirmed: the players involved are 'intrinsically motivated' for 'art's sake' (Caves, 2002). In the previous section of this chapter, this case study was introduced as a good example of Castells' notion of the network society: the (global) linkage is seen as a social network itself with non-local actors as the nodes. Lorenzen and Mudambi explain that the network structure and communication technologies of the local actors matter in the growth and facilitation of an industry cluster; in their case comparisons of Bollywood and Bangalore's ICT cluster, having a decentralized network structure in a micro/local context was pivotal in their success in creating global linkages.

What is notable in the case of Bollywood is the decentralization of personal relationships in its network structure. From the first global linkages–personal relationships with the European film industry in the 1920s and 1930s, the Bollywood industry shifted its priorities to local markets since the country's independence in 1947. These localized linkages played a crucial role in the 2000s and onwards, as it became the direct bridge between the industry and diaspora population outside of India, particularly in the UK and North America. In other words, Bollywood's global success today is a result of the intense, global connections through the decentralized personal relationships to the Indian diaspora.

Compared to Bangalore's ICT sector's decentralized networks focused on 'in-depth catch-up' by creating pipelines where knowledge spills over to domestic firms, the Bollywood industry's flows of knowledge and capital were dependent on the engagement of the diasporas, making it a more social production and consumption process. The authors found three types of flows that were facilitated through such relationships: information on global trends and styles, mobility of global talents, and 'value capture activities' consisting of capital. This enabled the active engagement and shared human and market capital between the Indian diasporas and Bollywood followers abroad and the industry in Mumbai.

Thus, this case contributed in the relatively under-researched role of personal relationships on creating value and connectivity with the global economy, without the need to be 'validated' at a broader scale so to speak by widespread media gatekeepers or by other world city cultural hubs or their social milieu. The Bollywood industry was essential in creating and bridging creative talent for the industry and ultimately a community that contributes to its value creation and further export potential.

IMAGINING EAST ASIA VIA POPULAR CULTURE: CASE OF HALLYU, THE 'KOREAN WAVE'

Hallyu, translated as the 'Korean Wave', pretty much encompasses all things cultural that are originated from South Korea; while it is mostly associated with music (K-Pop), media entertainment (K-Dramas) and the individual stars in those industries, Hallyu also includes fashion and beauty products, cuisine, and other commodities. Like the case of Bollywood, the Hallyu scene also exemplifies how cultural consumption creates a community. However, as Otmazgin examines in his recent article, Hallyu serves as a shared language for transnational communication and enhances a 'sense of we-ness' that also leads to 'region-making', 'regionalization of taste', and identity formation amongst Hallyu consumers/fans.

From the late 2000s and onwards, Hallyu has been successful in its dissemination through the internet (YouTube, in particular) and smartphone technologies; consequently, Hallyu consumers tend to be young urban residents with a relatively elevated standard of living (Otmazgin, 2016). According to Otmazgin, the images, ideas, and emotions conveyed in the form of popular culture and cultural products were critical in creating a 'new sort of community' that shares the 'consequential perceptions, lifestyles, and thoughts [the cultural products] offer' (Otmazgin, 2016). The dissemination of Hallyu further led to transnational collaborations amongst the key actors in East Asia. In turn, such 'we-ness' permeated through popular culture influences the way a region – in this case, East Asia – is perceived.

While the Korean government has made noteworthy investments in the culture industry in the 1990s, it was not until the 2000s when such export capacity and 'surprising national achievement' of culture was acknowledged (Shim, 2006). Further, the networks and mechanisms of distribution and consumption of popular culture in

East Asia has not been considered much as a scholarly inquiry in the process of region-making (Otmazgin, 2016). Case in point, Hallyu/popular culture illuminates the 'bottom-up logic of regionalization', where the creation and success of transnational creative platforms, industries, and cultural consumption from the 'ground level' led governments to realize the impact of 'soft power'. Culture, thus, is recognized as an economic sector in its own right and becomes rigorously promoted as government initiatives. The K-Pop industry alone, for instance, has received tremendous financial support from the Korean government for export power and subsequent socio-cultural value production (Shim, 2006).

The mobility of human capital (producers, consumers, and social milieu), meanings and ideas through the media (Hannerz, 1996) were key to Hallyu's presence in East Asia, which is now extended to Southeast Asia and also a smaller but growing fan base in non-Asian countries as well. Such phenomenon will continue to provide a different or contextualized examination of cultural globalization in future research.

ORGANIZATIONAL DIMENSIONS

The above case studies strongly suggest that the infrastructure of globalization is based on the (trans)national connection between culture, places and people (Hannerz, 1996). With the flattening of cultural boundaries, how is the concept of globalization institutionalized? Globalization of culture and cultural/media geographies can start from the bottom-up, with governments realizing the impact of 'soft power' as initiated on the grassroots level and later implementing it as government initiative/policy (e.g. Hallyu); or involve government support when national interests are at stake (e.g. Hollywood). This combination of top-down and bottom-up logic make the case that it is not just having marketing and distribution networks (as well as media networks) in strategic locations, but the interpersonal relationships with the product/industry that can fuel the global diffusion. This was exemplified in the aforementioned cases of Hallyu and the demand from diaspora communities and successful global linkages in the Bollywood entertainment industry, and the social consumption of art and culture in Los Angeles/New York that had led to media clustering and place-branding. As the development of new communicative technologies and the taste/trends in culture happen and shift at an unprecedented pace, these circumstances stipulate an institutional adaptation (Beck et al., 1994; Rantisi, 2004). Cultural producers need to be mindful of the unpredictability from both macro- and micro-level – the global and local demands – in the twenty-first-century context of globalization.

Governance structures in a network society and in businesses/industries are therefore decentralized organizations, but with power relationships which depend on the ability of the 'programmers and switchers' as articulated by Castells in his extensive study on the network society. These power positions are not power elites per se, but are more subtle, complex, negotiated systems of social actors (Castells, 2004). It could be the media, one of the main actors participating in the 'urban growth machine' that bolster and maintain the 'predisposition' for growth (Logan and Molotch, 2007) or reputation, the gatekeepers within the media geography, and the collective base of consumers that could catalyze a global process of valorization (Currid and Williams, 2010). Local input of the programmers and switchers are just as, or perhaps now even more, important to cultural production on a global scale, which includes the place-based advantages, resources, and symbolic images in creating intrinsic value.

However, the 'who and where' that determine the brand recognition and the media gatekeepers that propel the branded culture and space is being shaken by a growing generation of new media platforms and the change of dynamics it brings to the production–consumption–dissemination process. One example would be the notion of stars, or celebrities, and how they are created to have their value (or the value of their work) validated by the media and cultural consumers globally. Being, or becoming, a celebrity in any field is not solely dependent on one's talent and skills; it is a social construction and careful, deliberate collaboration amongst an industry of publicists, networking strategies, and 'being with the right person' and/or 'at the right place' at the 'right time' (Currid-Halkett, 2010). This chapter previously mentioned that the flattening of cultural boundaries meant that the globalization process can start from the bottom up. In addition to that, there is a 'flattening of hierarchies' with the changing media ecosystems and integrated marketing communication (IMC) strategies of cultural industries, where 'horizontally integrated networks' of evolved buzz agencies and crowdsourced marketing enable brands – or celebrities – to be represented in more flexible capacities (Serazio, 2013).

Globalization of cultural and media geography in this sense can be encapsulated in what Jenkins (2006) calls 'convergence culture'. As Serazio quotes Jenkins in explaining this flexible and engagement-oriented process of culture and branding, it is 'both a top-down corporate-driven process and a bottom-up consumer-driven process', where more 'active, migratory, more socially connected, noisy and public' consumers profoundly impact the cultural production process, by embracing disinterested spaces and participatory flows (Serazio, 2013: 25). Traditional gatekeepers become irrelevant (Diaz-Hurtado, 2017), as cultural producers turn to online/digital platforms such as YouTube, Vimeo, and Netflix in a more independent fashion and 'make their marks'. One example would be Issa Rae, who first started her show 'The Misadventures of Awkward Black Girl' on YouTube in 2011. The popularity of the show led to a successful Kickstarter campaign at the finish of the season, and now Rae is a Golden Globe nominated actress and producer of her own *HBO* series (Diaz-Hurtado, 2017). It will be an exciting agenda for future research in cultural and media geography to explore the globalization and consumption/valorization process in these increasing scenarios. How much do physical place effects contribute to these online, new media platforms for cultural production, and to what extent do traditional gatekeepers and mediators still play a role?

CONCLUSION

We believe that despite the wave of new media platforms and faster pace of change in taste and technologies, the impact *of* place effects and impact *on* place imaging through culture and cultural consumption will continue. Where the technologies are developed, which traditional media gatekeepers will capture the value of an online-based show, and where the cultural producers and consumers travel to still matter greatly. Hence, the speed, intensity, and sharing between culture and communities will quicken. How to effectively capture and aggregate changing media ecosystems and the cultural globalization within will become a pressing issue in the field of geography.

Another interesting turning point in the global era and for world cities/cultural hubs is the trend to 'shy away' from conventional globalization and a tendency towards artisanal production and localism (i.e. Made in Brooklyn, farmers' markets). The backlash against globalization, the rise of information and demand for a more transparent and authentic production-to-consumption process, and a market economy with postmodern values, which prize ethical production over monetary profit, are key forces that explain this phenomenon, perhaps most vividly observed in cities (Currid-Halkett, 2017). Paradoxically, it can be cautiously determined that this backlash against globalization is becoming a global phenomenon itself – affecting cultural production, consumption, tourism policies/strategies (more authenticity than modernism), and so forth. What this means geographically, and both positive and negative implications it may have for places, people, and societal values in a global sense remains to be seen.

At any rate, cultural and media geography – and perhaps even the process and spectacle of globalization itself – is at an interesting crossroads. This is a time when critical reflections on cultural globalization and urban dynamics are being made, and thus perhaps what may seem like retaliation might become a catalyst for new narratives, processes, and development towards a community of sentiment that is not dispersed, divided, or selective, and is at once place-specific and ubiquitous in its production, consumption and messaging.

REFERENCES

Adams, P.C. (2016), 'Geographies of media and communication I: Metaphysics of encounter', *Progress in Human Geography*, 30913251662825. doi:10.1177/0309132516628254.

Anderson, B.R.O. (1983), *Imagined Communities: Reflections on the Origin and Spread of Nationalism*, London: Verso Editions/NLB.

Anderson, K. (2003), *Handbook of Cultural Geography*, London; Thousand Oaks, CA: Sage.

Appadurai, A. (1990), 'Topographies of the self: Praise and emotion in Hindu India', in: C.A. Lutz and L. Abu-Lughod (eds), *Language and the Politics of Emotion* (pp. 92–112). Cambridge: Cambridge University Press.

Appadurai, A. (1996), *Modernity at Large: Cultural Dimensions of Globalization*, Minneapolis, MN: University of Minnesota Press.

Atkinson, D. (2005), *Cultural Geography: A Critical Dictionary of Key Concepts*, London; New York: New York: I.B. Tauris; In the United States distributed by Palgrave Macmillan.

Banet-Weiser, S. (2012), *Authentic™: The Politics of Ambivalence in a Brand Culture* (1st edn), New York: NYU Press.

Beck, U., A. Giddens and S. Lash (1994), *Reflexive Modernization: Politics, Tradition and Aesthetics in the Modern Social Order*, Stanford, CA: Stanford University Press.

Becker, H.S. (1982), *Art Worlds*, Berkeley, CA: University of California Press.

Benjamin, W. (1936[1969]), 'The work of art in the age of mechanical reproduction', in: Arendt, H. and Zohn, H. (1969, eds), *Illuminations* (pp. 217–252), New York: Schocken Books.

Brooker, P. (1992), *Modernism/Postmodernism* (Longman Critical Readers), Abingdon; New York: Routledge.

Castells, M. (2001), *The Internet Galaxy: Reflections on the Internet, Business, and Society*, Oxford; New York: Oxford University Press.

Castells, M. (2004), *The Network Society: A Cross-Cultural Perspective*, Cheltenham, UK and Northampton, MA, USA: Edward Elgar Publishing.

Castells, M. (2009), *Communication Power*, Oxford; New York: Oxford University Press.

Castles, S., and M.J. Miller (1993), *The Age of Migration: International Population Movements in the Modern World*, New York: Guilford Press.

Caves, R. (2002), *Creative Industries: Contracts between Art and Commerce*, Cambridge, MA; London: Harvard University Press.

Clark, T.N. (2011), *The City as an Entertainment Machine*, Lanham, MD: Lexington Books.

Currid, E. and S. Williams (2010), 'The geography of buzz: Art, culture and the social milieu in Los Angeles and New York', *Journal of Economic Geography*, **10**(3), 423–451.

Currid-Halkett, E. (2010), *Starstruck: The Business of Celebrity* (1st edn), New York: Faber and Faber.

Currid-Halkett, E. (2017). *The Sum of Small Things: A Theory of the Aspirational Class*, Princeton, NJ: Princeton University Press.

Currid-Halkett, E. and G. Ravid (2012), '"Stars" and the connectivity of cultural industry world cities: An empirical social network analysis of human capital mobility and its implications for economic development', *Environment and Planning A*, **44**(11), 2646–2663.

Diaz-Hurtado, J. (2017), 'Why storytellers of color ignore usual gatekeepers, take a chance on the *National Public Radio*'. http://www.npr.org/sections/codeswitch/2017/02/11/514487439/why-storytellers-of-color-ignore-usual-gatekeepers-take-a-chance-on-the-internet (accessed 11 February 2017).

Friedman, T.L. (2006), *The World is Flat: A Brief History of the Twenty-First Century* (1st update and expanded edn), New York: Farrar, Straus and Giroux.

Glaeser, E.L., J. Kolko and A. Saiz (2001), 'Consumer city', *Journal of Economic Geography*, **1**(1), 27–50.

Greenberg, M. (2008), *Branding New York: How a City in Crisis was Sold to the World*, New York: Routledge.

Gregory, D., R. Johnston, G. Pratt, M. Watts, S. Whatmore and Credo Reference (2009), *The Dictionary of Human Geography* (5th edn), Malden, MA: Blackwell.

Hannerz, U. (1996), *Transnational Connections: Culture, People, Places*, London and New York: Routledge.

Harvey, D. (1989), *The Condition of Postmodernity: An Enquiry into the Origins of Cultural Change*, Oxford; Cambridge, MA: Blackwell.

Held, D. (1999), 'Globalization, culture and the fate of nations'. In: A. McGrew, D. Goldblatt, D. Held and J. Perraton (eds), *Global Transformations: Politics, Economics and Culture* (pp. 327–375), Stanford, CA: Stanford University Press.

Jackson, P. (1989), *Maps of Meaning: An Introduction to Cultural Geography* (1st edn), London; Boston: Unwin Hyman.

Jenkins, H. (2006), *Convergence culture: Where Old and New Media Collide*, New York: New York University Press.

Jenkins, H., M. Itō, and D. Boyd (2015), *Participatory Culture in a Networked Era: A Conversation on Youth, Learning, Commerce, and Politics*, Malden, MA; Cambridge: Polity Press.

Johnston, R. and J.D. Sidaway (2004), 'The trans-Atlantic connection: "Anglo-American" geography reconsidered', *Geojournal*, **59**(1), 15–22.

Logan, J. R. and H.L. Molotch (2007), *Urban Fortunes: The Political Economy of Place* (20th anniversary edn), Berkeley, CA: University of California Press.

Lorenzen, M. (2009), 'Creativity in context: content, cost, chance, and collection in the organization of the film industry', in: P. Jeffcut and A. Pratt (eds), *Creativity and Innovation in the Cultural Economy* (pp. 93–118). London: Routledge.

Lorenzen, M. and R. Mudambi (2012), 'Clusters, connectivity and catch-up: Bollywood and Bangalore in the global economy', *Journal of Economic Geography*, **13**(3), 501–534.

Massey, D. (1984), *Spatial Divisions of Labour: Social Structures and the Geography of Production*, New York: Methuen.

Meyrowitz, J. (1985), *No Sense of Place: The Impact of Electronic Media on Social Behaviour*, New York: Oxford University Press.

Molotch, H. (2002), 'Place in product', *International Journal of Urban and Regional Research*, **26**(4), 665–688.

Murray, W. (2006), 'Globalizing cultural geographies', in: W. Murray, *Geographies of Globalization* (pp. 219–260), New York: Routledge.

Otmazgin, N. (2016), 'A new cultural geography of East Asia: Imagining a "region" through popular culture', *The Asia-Pacific Journal*, **14**(7), No. 5.

Rantisi, N.M. (2004), 'The ascendance of New York fashion', *International Journal of Urban and Regional Research*, **28**(1), 86–106.

Ritzer, G. (2015), *The McDonaldization of Society* (8th edn), Los Angeles: Sage.

Sassen, S. (1991), *The Global City: New York, London, Tokyo*, Princeton, NJ: Princeton University Press.
Sassen, S. (1997), 'Electronic space and power', *Journal of Urban Technology*, **4**(1), 1–17.
Scott, A.J. (2001), 'Capitalism, cities, and the production of symbolic forms', *Transactions of the Institute of British Geographers*, **26**(1), 11–23.
Sennett, R. (2008), *The Culture of the New Capitalism*, New Haven: Yale University Press.
Serazio, M. (2013), *Your Ad Here: The Cool Sell of Guerrilla Marketing*, New York: New York University Press.
Shim, D. (2006), 'Hybridity and the rise of Korean popular culture in Asia', *Media, Culture & Society*, **28**(1), 25–44.
Swyngedouw, E. (2004), 'Globalisation or "glocalisation"? Networks, territories and re-scaling', *Cambridge Review of International Affairs*, **17**(1): 25–48.
Thrift, N. (2008), 'The material practices of glamour', *Journal of Cultural Economy*, **1**(1), 9–23.
Tinic, S. (2008), 'Dossier: Media space in perspective: Mediated spaces: Cultural geography and the globalization of production', *The Velvet Light Trap – A Critical Journal of Film and Television*, 74–75.
Zukin, S. (1991), *Landscapes of Power: From Detroit to Disney World*, Berkeley, CA: University of California Press.

9. Political geographies of globalization

Sami Moisio, Juho Luukkonen and Andrew E.G. Jonas

INTRODUCTION

During the past decades, the word globalization has become the mantra through which politicians, scholars, policy commentators, the media and even members of the public seek to understand and explain the ways the contemporary world works. This chapter seeks to elaborate upon a phenomenon which can be called political geographies of globalization (see also Kelly, 1999; Sparke, 2006). By this we refer to the different political discourses and related imaginaries, policy practices and regimes of governance through which globalization can be understood as being constantly produced in and through political geographical formations.

Rather than conceptualizing the global as a discrete scale or level of analysis we highlight specific policies and processes which are globalizing in nature. In particular, we comprehend globalization both as an actually existing process which links places – cities, regions and so on, institutions (especially the state) and people (notably workers) – and creates interdependencies between them, and as a politically loaded rhetorical device used to rationalize and legitimate myriad political decisions and policy practices. In doing so, we stress the need to examine the ways in which globalization can be understood as being constituted through the 'national' and the 'local', and how the state – or its absence at least in respect of policies that support workers rather than capital – has been crucial in enhancing the emergence of these as relational spaces of globalization.

Given the huge amount of literature on the political geographies of globalization, we present a decidedly partial mapping of the existing literature. In the ensuing sections, we single out three interlinked and partly overlapping issues through which the political geographies of globalization can be mapped out: (1) the spatial formations of globalization and the state; (2) the 'globalizing' role and 'globalized' nature of public policy; and (3) the globalizing regimes and policies of labour. These are areas of interest which sometimes come together in the attempts of governments, international organizations, business associations, consultant companies and the like to govern and come to terms with the purportedly messy sphere of the global. Often, however, these function as separate spheres of political intervention and, as such, are represented as discrete spaces of state intervention in the academic literature on globalization. For example, work on labour geographies tends to examine national spaces of labour regulation quite separately from local spaces of labour control. Nonetheless, towards the end of our chapter we consider ways in which these processes increasingly converge within particular cities and regions, producing new political geographies of globalization.

POLITICAL GEOGRAPHICAL FORMATIONS OF GLOBALIZATION AND THE TERRITORIAL STATE

A great variety of spatial formations has been discussed under the rubric of globalization to depict the political economic changes that began to take shape in the 1990s – in particular as a consequence of rapid development of information and communications technologies, pervasive financialization and the increasing mobility of at least some segments of capital along with ideas and commodities. It has been typical to argue that the economy is a driver of globalization to which the state is forced to merely 'react'. Rather than repeating this economic determinism we elaborate three themes which disclose how the territorial form of the state is being reimagined and represented in new ways, and how the state can be understood as a key driver of globalization.

First, since the 1990s, spatial formations such as networks and what may be called exceptional economic spaces have been associated with political geographies of globalization. Globalization, in this context, is understood as being constituted by relational spaces that challenge the national territorial hierarchies that purportedly characterized the Fordist–Keynesian production of welfare statehood. Perhaps the most discussed theory of the networked nature of globalization was developed by Manuel Castells in the 1990s. Accordingly, the digital revolution and concomitant information age and 'new economy' is built on networks which link states, institutions and individuals in different tapestries. But rather than producing a flat world, differentiation between those who are and those who are not part of the networks of informational capitalism characterizes the contemporary condition. What Castells (1996) calls network society consists of new network enterprises, states, places, regions, and subjects, and is both integrating and fragmenting. Even if the network society diffuses throughout the entire world, it does not include all people: it is rather characterized by a growing gap between winners and losers. In order to be located on the winners' side of the network society, states and cities are forced to tailor new kinds of spatial strategies and build a kind of spatial exception (Ong, 2006) within their fabric.

The concept of spatial exception resonates with the argument by Saskia Sassen on the structuration of the global inside the national. The national can therefore be understood as a pivotal site for the global, and the global gets constituted inside the national through particular spatial exceptions. The state can for instance participate in the making of protected jurisdictions for firms operating globally (Sassen, 2013: 37). The list of these spatial exceptions or transversal spaces – territories or jurisdictions within nation state territories – is extensive and would include territorial formations such as financial centres, special economic zones, innovation centers and offshore production zones which are often exempted from usual national regulations covering corporate taxation, labour standards, and suchlike. This argument is theoretically important as it implies that the national territory of the state can be effectively denationalized in the attempts of the state to tap into or enhance the processes of globalization. This may also lead to a particular empowering of some agencies of the state as they become more important given their operational importance in connecting the state to the processes of global economy (Sassen, 2010) or global governance. Correspondingly, other agents – especially workers – are disempowered.

Second, the political geographies of globalization are being disclosed in literature on the rise of global cities, global city-regions and city-regionalism. Large urban agglomerations in the Global North in particular are often understood as sites of globalization. These cities are understood through their functions and substances and the degree of integration into the purportedly 'global' economy. What Richard Florida (2017) calls super-star cities are sites of investment or venture capital, foreign exchange, as well as sites of high order business services related to marketing, accounting and law, innovation activities developed by talented people, cultural industries and urban amenities. These cities are nests of headquarters of multinational and national corporations, leading international organizations, media organizations, and professional and trade organizations. This kind of spatial concentration of economic and governance activities, as well as the related regional clustering of other activities around these cities, has led to a powerful political geographical articulation of city-regions. Allen Scott (2001: 813, 817), for instance, writes about 'an extended archipelago or mosaic of large city-regions' which 'increasingly function as the spatial foundations of the new world system that has been taking shape since the end of the 1970s'.

Mapping out these spatial formations discloses political geographies of capitalist globalization and the associated re-organization of the spaces of finance, production and governance. It has therefore been argued that the increasingly transnational form of capitalism and the associated technological developments have from the 1990s onwards generated a spatial structure of the world economy that is characterized by a rapidly growing economic and political role of city-regions that may also cross state borders (Jonas and Moisio, 2018). The literature on global city-regions generally deals with major urban agglomerations as key spaces of capital accumulation and within which human capital and productivity increasingly concentrate (see e.g. Scott et al., 2001: 11–12). The rise of such global city-regions is, however, fraught with all sorts of distributional and political tensions not just within them but also across the wider state territory in which they are situated (Jonas, 2013).

It has become commonplace not only for urbanists, consultant companies, globalization boosters and businesses firms (which also construct and finance urban infrastructures) but also for nation-state policymakers and politicians to argue that the strategic role of cities and city-regions has grown fundamentally during the past two or three decades. In such a geopolitical imaginary, the growth of major cities and city regions is conceived of as an inevitable global phenomenon orchestrated by market forces, one which proceeds beyond politics, political regulation and the territorial state. Policy analyst Parag Khanna's (2017) *Connectography* is one among the many attempts to tell a story about the contemporary de-territorializing global processes and the related ways in which the future is being shaped less by states than by connectivities of urban hubs and flows.

Particular notions of urbanization as an apolitical global trend has become one of the key constituents of the geopolitical discourses of globalization. We believe that these geopolitical discourses on the inevitability of large-scale urbanization and 'connectography' play a significant constitutive role of contemporary capitalist expansion which is fundamentally grounded in the circulation of capital through urban built environments, technologies and infrastructures, urban consumption and the related urban needs and desires.

Finally, rather than understanding that the state has been hollowed out by globalization, the political geographies of globalization point to the qualitative transformation of the state from the 1980s onwards. Accordingly, the contemporary conjuncture is marked by back and forth movement of the processes of state de-territorialization and re-territorialization and the relativization of scale. In such a perspective, the state is not a static object but rather its contemporary geography and re-worked scalar organization both results from and is a constitutive element of the capitalist process of restructuring (Brenner, 2004). In other words, the national state has a crucial role as an agent, site and medium of global restructuring that is often labelled as globalization. Similarly, the seemingly relational, mobile and de-territorializing capitalist globalization inescapably hinges upon the production of relatively fixed and immobile socio-spatial configurations (Brenner, 2004: 33).

What has been said above reminds us that the state is a constantly changing social organization, a process rather than a thing. As Bob Jessop (2016: 10) suggests, the state and its constituent powers can be understood as made up of 'an institutionally and discursively mediated condensation of a changing balance of forces that seek to influence the forms, purposes, and content of polity, politics and policy'. The spatial re-working of the territorial state thus brings together different actors who operate with different sociocultural resources. Since the 1980s, this re-working has also increasingly involved what Leslie Sklair (2001) calls the transnational capitalist class which arguably operates through distinctively 'global' or 'post-national' priorities, interests and loyalties, and whose actions are often predicated upon a particular geopolitical imaginary of a flat, borderless and networked world. Indeed, the transnationalization of the state apparatus has been associated with the experienced tensions in many state contexts between the drive to equalize capital investment across space on the one hand, and the pressure to differentiate such investment in order to exploit place- and territory-specific conditions for accumulation, on the other (Brenner, 2004).

GLOBALIZING AND GLOBALIZED PUBLIC POLICIES

Political geographies of globalization are produced in public policy discourses and practices. Urban policy, for instance, is not simply a blueprint or outcome of political processes developed in response to globalization. Rather the policy should be understood as playing a constitutive role in the production of globalization. To put it concretely, the competition between cities through marketing and branding strategies, for instance, is not just a response to globalization but it also constructs a space of globalization in the form of marketing and branding practices and strategies that circulate from one city to another across the globe (e.g. McCann, 2011). From our perspective, the globalization of policies involves two significant dimensions, that is, the globalization of policymaking processes and the globalization of the policy substances.

First, the globalization of policymaking processes denotes the reassembling of governance structures as new actors and activities become involved in solving policy problems which spread across jurisdictional and administrative boundaries. In political geography terms, the shifts in the geographies of governance involve the emergence of new territorial and relational modes of governance and alliances of power as well as the rescaling of state powers and responsibilities 'vertically' and 'horizontally' to various trans- and supranational institutions, regional authorities and non-state actors. One manifestation of the new geographies of the internationalized governance structures is the emergence of global policy networks, a sort of 'boundary crossing web of influences that shape political and policy decisions' (Prince, 2012: 189).

Global policy networks affect significantly policymaking practices by engaging local, situated policies to broader international contexts. Among the government actors such as supra-national, national and regional public authorities, global policy networks grasp together non-government actors such as policy consultants and advisors, lobbyists, academics and other experts to international policy communities which guide and shape policy practices in national or local political contexts (Prince, 2012; Larner and Laurie, 2010). Within these networks, policy ideas developed elsewhere pass and become localized in new situations and geographical contexts.

The global policy networks are part of the extension of the pre-existing spaces of governance to multiple 'elsewheres' simultaneously, as well as the emergence of entirely new ones. Spaces of governance refer both to physical locations and social spaces which gather policymakers around particular policy issues. In these sites, policymakers produce, circulate and modify policy ideas and seek solutions to shared policy problems at stake (Bachmann, 2015; Temenos and McCann, 2013). While administrative territories still remain central components in defining the composition and scope of these spaces, the emergence of global policy networks and involvement of new actors have extended these spaces beyond their territorial boundaries. However, this has not only meant the exposure of national or regional policies under the influence of international political organizations such as the EU, World Bank or OECD but also the opening of the possibilities for policymakers to engage and articulate policy practices and goals through wider political rationalities and more powerful political actors (cf. Cox, 1998).

Second, the globalization of policy substances does not refer only to the contingent transferring or copying of individual policy ideas between places but also relates to the emergence of shared transnational or global policy cultures (cf. Toly, 2011). This involves the spreading and adoption of similar kinds of policy goals, measures and discourses.

In critical political geographies, the globalization of policy substances is often associated with the emergence and entrenchment of neoliberalism as 'a globally hegemonic regime of governance and as a system of governmentality' (Sparke, 2006: 364) (see e.g. Brenner and Theodore, 2002; Peck and Theodore, 2010; Prince, 2012). Neoliberal governance has become globally spread and a widely adopted 'standardized policy framework' (Sparke, 2006) advocating market-oriented policy strategies such as deregulation, pro-risk attitudes, incorporation of private sector actors, reduction of public expenses, increasing the efficiency of public sector, self-evaluation and business friendly decision making (cf. Sparke, 2013). A key tenet of the neoliberal policymaking is the rise of a sort of audit culture which refers to the internationalization and institutionalization of the principles and techniques of calculation, accountancy and financial management as an essential part of policy practices and governance processes (Shore and Wright, 2015: 24). This governing by numbers has become a normalized part of the day-to-day routines of public institutions and policy organizations across the world. The European Union, for instance, has been eager to apply calculative practices of measurement and ranking for evaluating the policy performances around the EU territory (e.g. Luukkonen and Moisio, 2016). These standardized evaluation methods have become popular because they are easily inserted to new places and they produce harmonized and simplified information of the often complex and messy context-specific situations (Prince, 2016).

The adoption of the audit culture in policy practices involves two kinds of spatial implications. First, audit techniques contribute to the formation of 'topological connections' between particular places and policies. In geographical policy mobility literature these connections refer to processes of presencing and proximity (Robinson, 2013: 10) or virtual relations between places and policies – distinct from the physical or material connections of the policy networks – emerging from the standardized and harmonized practices of measurement and comparison (Prince, 2016). A concrete example of the topological relations are the global city-rankings which have become a routine part of the international urban policies. These rankings bring seemingly distant and differing cities and the associated urban policies together into international league tables and make them amenable to comparisons on a global level. The global city-rankings not only drive cities to compete with each other over standings in the league tables but also improve the capacity of the international consultancies, the European Commission, the OECD and other powerful centres of calculation which often compile the listings, to regulate and govern cities at distance (cf. Kitchin et al., 2015).

Second, different auditing techniques play a central role in the so-called 'territory work' (Moisio and Luukkonen, 2017) – that is, in the reformulation and emergence of territories as objects of policy and governance. The rising popularity of the audit culture illustrates how policies are understood as 'political technologies' which remove political problems from the realm of politics and recast them in the neutral language of science (Dreyfus and Rabinow, 1982: 196). However, although calculative practices of measurement and ranking are often deemed as value-free tools for policy management, several authors have pointed out how they are essentially connected with particular globally hegemonic political imaginaries such as 'the knowledge-based economy' (e.g. Luukkonen and Moisio, 2016) and 'city-regionalism' (e.g. McCann, 2008). These imaginaries form mental frameworks which delimit policymakers' perceptions of territories as policy objects. They guide policymakers to prefer particular modes of geographical knowledge and choose particular indicators and indices through which territories are examined and converted as manageable and auditable economic and/or political entities.

GLOBALIZING POLICIES AND REGIMES OF LABOUR

Approaching globalization as a set of geographically mediated political processes requires that due consideration is paid to how class relations and tensions in capitalism are mediated through a variety of state and non-state policies and spatial practices. From the political vantage-point of capital, some of these policies and practices are designed to suspend the conflict between capital and labour at the point of production. For example, place-based labour control practices – or local labour control regimes – often emerge within and around particular localities in order to enable capitalist firms to integrate labour into the local production system and forestall worker resistance (Jonas, 1996). Such practices tend to be found in industrial localities and regions where workers' institutions and political identities have been profoundly shaped by social interactions that extend from the workplace into the living place. From the perspective of workers, however, it is dangerous to assume that the outcomes of local political struggles around globalization in such places best serve their objective class interests. Instead, local factions of capital and organized labour might enter into political alliances designed to promote further rounds of economic growth rather than to redistribute income and resources. In such situations, state interventions, such as tax concessions for multinational corporations or the negotiation of harsher labour laws, lend a territorial rather than class political inflection to struggles around the globalization of capitalist production and economic development (Cox and Mair, 1988).

Following Gramsci, the rise to dominance of a particular regime of production (e.g. the Fordist system of mass production in the early twentieth century) is generally associated with a sustained period of cultural hegemony in which the ruling class adopts a range of cultural and political practices in order to control the workforce and accumulate capital (Gramsci, 1971). Neo-Gramscian regulationist approaches to the state (see e.g. Jessop, 2016) refer to such periods as regimes of accumulation, which in turn are supported and sustained by corresponding modes of societal regulation, the broad aims of which are to regulate and distribute the social product between capital and labour, respectively (see e.g. Aglietta, 1979[2005]). However, the emergence of a particular mode of social regulation and the degree to which it coalesces at the national scale are contingent political outcomes reflecting national as well as regional and local differences in the composition of class interests and political forces in civil society. Here the concept of political geographies of globalization is useful for investigating the social, economic and political conditions under which spatio-regulatory fixes are achieved not just nationally but also locally within different national territories (Peck and Tickell, 1994).

At a general level, under Fordism the national state played a strategic role in setting the regulatory context for collective bargaining agreements between organized labour and capital. National spatial regulation in turn underpinned a geopolitical context in which multinational firms could make decisions about how to allocate production between different places, regions and countries on the basis of certain expectations about productivity, wages and potential profits. This led to the development of a finely tuned international division of labour centred around the major industrial regions of North America, Europe, and parts of Asia. Nonetheless, factory owners and managers operating in particular regions and localities continued to experiment with local labour control regimes depending on how local labour requirements and skills could be coordinated with national production objectives (Jonas, 1996). In some cases, the state intervened to enhance the local supply of labour by, for example, encouraging the development of new towns in peripheral regions (Cox, 2016). Similarly, local political jurisdictions used variations in productivity, skills and labour costs as a basis for competing for jobs and a specific place in the wider territorial division of labour.

Following the closure of Fordist branch-plant industries in the 1970s and 1980s, subsidies by regional governments and, in Europe, the European Union (EU) compensated, at least in part, for shortfalls in income and national public investment. The net general effect was eventually to establish a worldwide production system which was nonetheless based on highly organized national political economies. At the same time, there remained in place finely grained local and regional political differences resulting from the effects of different spatial structures of production on local class, gender and race relations (Massey, 1984, 1994).

As uneven development intensified even further, places continued to jostle for a position within wider divisions of production, contributing to a politics of urban and regional development often accentuated by national differences in state structures and institutions. At the national level, for example, the contemporary political landscape of globalization is increasingly fought around immigration laws, producing a variegated landscape of political struggle around national labour regulation. At the local level, it is marked by rapidly diverging 'high road' and 'low road' local accumulation strategies (Gough, 2012).

Those who recognize capital's ongoing ability to innovate, experiment and compete tend to refer to selective cases of 'high road' local political responses where new labour organizations have emerged to extract limited concessions from capital. Whilst making the case that the United States (US) is an example of national political exceptionalism, Cox (2016: 226) nonetheless points to the recent rise of new initiatives within the organized labour movement, which by-and-large have been focused around major urban centres:

> These are the initiatives of the various urban-centered organizations around the country that emerged in the 1990s, funded by labor and foundations, with an agenda focused on labor and development issues. Some of them contained 'new economy' in their titles as in the Los Angeles Alliance for a New Economy (LAANE) and the Connecticut Center for a New Economy (CCNE), and very loosely one might say that, to a degree, they were responding to problems thrown up by a period of rapid economic transformation: in particular the emergence of a bipolar income distribution, a related growth of low-paying service jobs, the erosion of the minimum wage and a continued decline of union membership. (Cox, 2016: 226)

However, local initiatives to challenge growing wage disparities, such as campaigns for a living wage, are becoming far more widespread and, in terms of their organizational geography, have much more in common with each other than is implied by Cox. Indeed, Peck (2001) points to a more dominant 'low road' tendency in respect of labour regulation, namely, the gradual dismantling of national scale forms of welfare provision and their replacement by local-scale arrangements structured around principles of workfare and contingent labour. The accelerated emergence of such experimental city-based 'workfare states' in the first two decades of the twentieth century is symptomatic of the rise of an arguably more insidious and decidedly global form of cultural hegemony based around the free market, competition and a relaxation of national labour regulation, a.k.a. neoliberalism.

Nevertheless, the difference that geography makes to the globalization of labour regulation continues to matter. For example, the pressure on workers, communities and civic organizations in different places to experiment and innovate rather than resist in order to survive and compete appears to lend credence to claims by politicians, consultants and academics that contemporary forms of globalization are being reconstructed around exceptional spaces (Ong, 2006) – that is, particular territories, such as new industrial spaces, global city-regions and so on within which the knowledge economy and the creative class tend to converge. However, it is less clear how the liberal economic policy rhetoric associated with the rise of this new urban economy can be reconciled with the emergence of local labour control regimes within major cities (especially former global industrial centres) around which the relationship between capital and labour is being (re)constructed in quite new and subtle, yet no less exploitative, ways (Hastings and Mackinnon, 2017). And if labour agency has not completely disappeared, and all sorts of new social movements are indeed coalescing around issues that affect labour's reproduction, it seems that class tensions have far from disappeared from the political landscape of globalization; instead, such tensions are best revealed in the new territorial political formations which are to be found in new economic spaces and city-regions.

So if recent years have arguably witnessed the unraveling of so-called 'organized capitalism', the fraught balance of power between capital and labour continues to produce divergent political geographies of globalization. For example, the globalization of production is accentuating differences between the Global North and Global South in how labour is socially integrated into the production process. In countries, such as the Philippines, which are heavily dependent upon foreign direct investment by US multinational firms, the development of local labour control regimes depends upon the activities of a variety of non-state political actors ranging from national labour organizations and private recruitment agencies to local village and community leaders (Kelly, 2001). In the US, by way of comparison, capital continues to recruit internationally for cheap labour, and immigrant workers have developed their own transnational support networks in order to compensate for the absence of social support from the US state. Miraftab (2016) documents how migrant meatpacking workers in the US Midwest today draw upon social relations and resources in their communities of origin – in Mexico and western Africa – in order to facilitate their integration into the local production system. If the new economic geographies of capitalism depend upon the construction of new labour control regimes, these in turn are premised upon a relational politics of social reproduction which increasingly connects places in the Global North to those in the Global South.

Finally, in the post-Fordist regulatory vacuum created by state withdrawal from social welfare provision one might have expected capital to step in and develop strategies to compensate, at least in part, for the social predicament of workers. There is an assumption in much of the literature on the new urban economy that, in lieu of the state, capital develops its own systems of collection action in order to address various negative externalities arising from spatial agglomeration and rapid urban growth (Storper, 2013). Whilst this might arguably be the case for employers operating in certain successful or high-profile global city-regions, such as Silicon Valley, there is also evidence from other places of a lack of corresponding levels of city-regional collaboration and emerging territorial-distributional struggles around issues of labour regulation. An example was recently reported in the *Chicago Sun Times* (see Garcia, 2017) in which suburban jurisdictions in the Chicago metropolitan area have resisted efforts by Cook County officials to introduce a minimum wage to match that of the City of Chicago, which plans to raise its minimum wage from US$11 per hour in the summer of 2017 to US$13 per hour by 2019. Local businesses in some Chicago suburbs have lobbied local politicians to use local powers of 'home rule' to opt out of a county-wide ordinance designed to raise the minimum wage. This brief vignette suggests that new regimes of labour regulation and control continue to be shaped by nationally and locally specific political geographies of globalization.

CONCLUDING REMARKS

In this chapter, we have sought to highlight a set of overlapping objects, subjects and processes that have to do with political geographies of globalization. We have first argued that the contemporary condition may be understood as being structured partly around 'hub and flow imaginaries' and related 'exceptional' spatial formations that are its constituents. In our reading, the contemporary processes associated with hubs and flows and the set of other 'relational spaces' do not signal a de-territorial geo-economic condition in which the territorial state is hollowed out. Rather, these imaginaries reveal the back and forth nature of the contemporary state territoriality. At the present conjuncture, the state constantly seeks to re-territorialize, nationalize and 'fix in place' the relational spaces of contemporary capitalism. But at the same time governments seek to de-territorialize and internationalize the state through increasingly spatially selective strategies in order to be connected to all sorts of 'global networks' of money, talent, innovations and ideas. Much empirical work is still needed in order to disclose the ways in which states are spatially transformed within such a process (Moisio, 2018).

Second, we have highlighted that the globalization of policies is a process which not only connects policy actors and regimes across space and time but also contributes to the emergence of new spaces of governance and to the formation of new governable spaces. Thus, while in the geographical literature on globalization focus has usually been in either the globalization of policy processes or policy substances, they ought to be understood as inseparable dimensions which both need to be taken into account when scrutinizing the effects of globalization on the daily practices of policymakers.

Third, we argue that globalization can be examined as a set of geographically mediated political processes involving various state and non-state policies and spatial practices designed to manage class relations and tensions in capitalism. Here we can identify both globalizing and localizing political tendencies at work in respect of the social control and regulation of labour. On the one hand the extension of capitalist production on a global scale and the corresponding shrinkage of the welfare functions of the national state has forced workers to develop international networks to support their social reproduction as a class. On the other hand, capitalist firms increasingly rely upon local experimentation in order to adjust to the changing social requirements of production. At the same time, widening of wage disparities within the workforce has engendered a variety of local and regional political initiatives on behalf of labour, which nonetheless continue to be resisted by employers and their political allies. Despite the much-vaunted benefits of collaboration and collective action that inform contemporary discussions of globalization and the new urban economy, social divisions and conflict remain the order of the day in many places.

Finally, globalization can take different progressive and regressive forms, and it can be qualitatively directed to different dimensions of social life, as suggested by our efforts to focus on the state, public policy and labour, respectively, rather than transnational capital, which often is assumed to be the prime driver and beneficiary of globalization. Political geographies of globalization are therefore historically contingent, bound to various and often competing social forces, and inescapably open to contestation.

REFERENCES

Aglietta, M. (1979 [Reprinted 2005, trans. David Fernbach]), *A Theory of Capitalist Regulation: The US Experience*, London: Verso.

Bachmann, V. (2015), 'Global Europa, ESPON and the EU's regulated spaces of interaction', *Journal of European Integration*, **37**, 685–703.

Brenner, N. (2004), *New State Spaces*, Oxford: Oxford University Press.

Brenner, N. and N. Theodore (2002), 'Cities and the geographies of "actually existing neoliberalism"', *Antipode*, **34**, 349–379.

Castells, M. (1996), *The Rise of the Network Society*, Oxford: Blackwell.

Cox, K.R. (1998), 'Spaces of dependence, spaces of engagement and the politics of scale, or: looking for local politics', *Political Geography*, **17**, 1–23.

Cox, K.R. (2016), *The Politics of Urban and Regional Development and the American Exception*, Syracuse, NY: Syracuse University Press.

Cox, K.R. and A. Mair (1988), 'Locality and community in the politics of local economic development', *Annals of the Association of American Geographers*, **78**, 307–325.

Dreyfus, H.L. and P. Rabinow (1982), *Michel Foucault: Beyond Structuralism and Hermeneutics*, Chicago: University of Chicago Press.

Florida, R. (2017), 'Why America's richest cities keep getting richer?', *The Atlantic Cities*, https://www.theatlantic.com/business/archive/2017/04/richard-florida-winner-take-all-new-urban-crisis/522630/ (accessed 12 April 2017).

Garcia, M. (2017), 'Cook County hiked the minimum wage, but maybe not really for you', *Chicago Sun Times*, http://chicago.suntimes.com/opinion/cook-county-hiked-the-minimum-wage-but-maybe-not-really-for-you/amp/ (accessed 17 March 2017).

Gough, J. (2012), 'Capital accumulation in space, capital-labour relations, and political strategy', in: Andrew E.G. Jonas and Andrew Wood (eds), *Territory, the State and Urban Politics*, Farnham: Ashgate Publishing, pp. 89–112.

Gramsci, A. (1971), *Selections from the Prison Notebooks*, London: Lawrence and Wishart.

Hastings, T. and D. Mackinnon (2017), 'Re-embedding agency at the workplace scale: workers and labour control in Glasgow call centres', *Environment and Planning A*, **49**(1), 104–120.

Jessop, B. (2016), *The State: Past, Present, Future*, Cambridge: Polity.

Jonas, A.E.G. (1996), 'Local labour control regimes: uneven development and the social regulation of production', *Regional Studies*, **30**, 323–338.

Jonas, A.E.G. (2013), 'City regionalism as a "contingent geopolitics of capitalism"', *Geopolitics*, **18**, 284–298.

Jonas, A.E.G. and S. Moisio (2018), 'City regionalism as geopolitical processes: a new framework for analysis', *Progress in Human Geography*, **42**, 350–370.

Kelly, P.F. (1999), 'The geographies and politics of globalization', *Progress in Human Geography*, **23**, 379–400.

Kelly, P.F. (2001), 'The political economy of labor control in the Philippines', *Economic Geography*, **77**, 1–22.

Khanna, P. (2017), *Connectography: Mapping the Future of Global Civilization*, New York: Random House.

Kitchin, R., T.P. Lauriault and G. McArdle (2015), 'Knowing and governing cities through urban indicators, city benchmarking and real-time dashboards', *Regional Studies, Regional Science*, **2**, 6–28.

Larner, W. and N. Laurie (2010), 'Travelling technocrats, embodied knowledges: globalising privatisation in telecoms and water', *Geoforum*, **41**, 218–226.

Luukkonen, J. and S. Moisio (2016), 'On the socio-technical practices of the European Union territory', *Environment and Planning A*, **48**, 1452–1472.

Massey, D. (1984), *Spatial Divisions of Labor: Social Structures and the Geography of Production*, New York: Methuen.

Massey, D. (1994), *Space, Place and Gender*, Cambridge: Polity Press.

McCann, E. (2008), 'Expertise, truth, and urban policy mobilities: global circuits of knowledge in the development of Vancouver, Canada's "four pillar" drug strategy', *Environment and Planning A*, **40**, 885–904.

McCann, E. (2011), 'Urban policy mobilities and global circuits of knowledge: toward a research agenda', *Annals of the Association of American Geographers*, **101**, 107–130.

Miraftab, F. (2016), *Global Heartland: Displaced Labor, Transnational Lives, and Local Placemaking*, Bloomington: Indiana University Press.

Moisio, S. (2018), 'Towards geopolitical analysis of geoeconomic processes', *Geopolitics*, **23**, 22–29.

Moisio, S. and J. Luukkonen (2017), 'Notes on spatial transformation in post-Cold War Europe and the territory work of the European Union', in: Peeter Vihalemm, Anu Masso and Signe Opermann (eds), *The Routledge International Handbook of European Social Transformation*, London: Routledge, pp. 224–238.

Ong, A. (2006), *Neoliberalism as exception. Mutations in citizenship and sovereignty*, Durham: Duke University Press.

Peck, J. (2001), *Workfare States*, New York: Guilford Press.

Peck, J. and N. Theodore (2010), 'Mobilizing policy: models, methods, and mutations', *Geoforum*, **41**, 169–174.

Peck, J. and A. Tickell (1994), 'Searching for a new institutional fix: the *after*-Fordist crisis and global–local disorder', in: Ash Amin (ed.), *Post-Fordism: A Reader*, Oxford: Blackwell, pp. 280–316.

Prince, R. (2012), 'Policy transfer, consultants and the geographies of governance', *Progress in Human Geography*, **36**, 188–203.

Prince, R. (2016), 'The spaces in between: mobile policy and the topographies and topologies of the technocracy', *Environment and Planning D: Society and Space*, **34**, 420–437.

Robinson, J. (2013), '"Arriving at" the urban/urban policy: traces of elsewhere in making city futures', in: Ola Söderström, Didier Ruedin, Shalini Randeria, Gianni D'Amato and Francesco Panese (eds), *Critical Mobilities*, London: Routledge, pp. 1–28.

Sassen, S. (2010), 'The global inside the national. A research agenda for sociology', *Sociopedia.isa*. http://www.saskiasassen.com/pdfs/publications/the-global-inside-the-national.pdf (accessed 12 June 2018).

Sassen, S. (2013), 'When territory deborders territoriality', *Territory, Politics, Governance*, **1**, 21–45.

Scott, A.J. (2001), 'Globalization and the rise of city-regions', *European Planning Studies*, **9**, 813–826.

Scott, A.J., J.A. Agnew, E. Soja and M. Storper (2001), 'Global city-regions', in: Allen J. Scott (ed.), *Global City-Regions: Trends, Theory, Policy*, Oxford: Oxford University Press, pp. 11–32.

Shore, C. and S. Wright (2015), 'Governing by numbers: audit culture, rankings and the new world order', *Social Anthropology/Anthropologie Sociale*, **23**, 22–28.

Sklair, L. (2001), *The Transnational Capitalist Class*, Oxford: Blackwell.

Sparke, M. (2006), 'Political geography: political geographies of globalization (2) – governance', *Progress in Human Geography*, **30**, 357–372.

Sparke, M. (2013), *Introducing Globalization: Ties, Tensions, and Uneven Integration*, Chichester: Blackwell Publishing.

Storper, M. (2013), *Keys to the City*, Princeton, NJ: Princeton University Press.

Temenos, C. and E. McCann (2013), 'Geographies of policy mobilities', *Geography Compass*, **7**, 344–357.

Toly, N. (2011), 'Cities, the environment and global governance: a political ecological perspective', in: Mark Amen, Noah Toly, Patricia McCarney and Klaus Segbers (eds), *Cities and Global Governance: New Sites for International Relations*, Farnham: Ashgate Publishing, pp. 137–150.

PART III

GEOGRAPHIES OF FLOWS

10. The geographies of mobility and migration in a globalizing world

Anne-Laure Amilhat Szary

INTRODUCTION

In the social sciences, by using words operating as our first tools of understanding, we examine the structures of and changes in the world we build. Two of these analytical categories, 'mobility' and 'migration', do not have the same historical status, although both are quite recent classifications for people's movements at different scales of space and time.

Mobility is a quality (namely, of being in motion), and the term has been in use more and more since the recent phase of globalization, at the turn of the twentieth and twenty-first centuries, and that process of globalization imposed a faster pace on all moves – not only of human beings but also of ideas, capital, and goods.

Migration, by contrast, is a subtler operator that looks very neutral but is in fact more political than it seems to be. A quarter of a century ago, we used to refer to emigration or immigration, thinking that the international movement of human beings followed a straight trajectory from point A to point B. The loss of a prefix has accompanied a vast change to the status of moving and the design of increasingly differentiated mobility regimes. Not all humans are welcome to cross borders. For some, there might be no horizon out of the condition of 'being a migrant': Because their migratory status entails them being considered criminals, they are denied the possibility of legal integration as citizens into the political community of their place of arrival. It now appears that, for some people, there could be no end to the original movement of crossing an international boundary: Migration could become a defining feature for their whole life. More generally, mobility and migration do not escape policy making, and we are seeing an ever more complex scheme of multilevel management of these spatial practices. Nevertheless, personal agency remains extremely influential in this domain, and mobility–migration patterns are constantly being renegotiated. In that sense, they certainly constitute a very interesting marker with which to analyse globalization from below.

HOW IS GLOBALIZATION CONCEPTUALIZED IN MOBILITY AND MIGRATION STUDIES?

Society has traditionally been described with *fixist* lenses within a scheme that we only very recently became aware of and have characterized as a 'territorial epistemology' (Lapid, 2001). This way of thinking also implies 'nationalist methodologies': Social science analysis has traditionally been conducted at the scale of states (i.e. national histories, regional geographies as components of national frameworks and so on) without the bias even being noticeable. Considering this legacy, the 'mobility turn' (Sheller and Urry, 2006) of the beginning of the twenty-first century may appear to be 'new' because for many, 'being mobile' has become 'a way of life' (Urry, 2002: 256). It has also been reconfiguring the methods and epistemology of all social sciences. 'Thinking on the move' suggests the renewal of traditional conceptual categories that used to be based on a fixed understanding of society. The formulation of recent 'mobile epistemologies' (Ricketts et al., 2008) could well allow us to do away with the long-criticized 'territorial trap' (Agnew, 1994) in all fields of social knowledge. How did this 'turn' come into existence?

Movements of all kinds accelerated at the end of the twentieth century, which propelled what came to be known as 'globalization' in English or 'worldization' in the Latin languages: fluxes of goods, people, capital and ideas came to the fore. For the first decade or so, many analysts believed their liberalization from state containers was to become the rule (Moore, 2003; Ohmae, 1995). Closer insights have revealed that this momentum probably only corresponded to a renewed phase of both a quantitative and a qualitative acceleration of world exchanges and not to a totally new globalization (Gruzinski, 2004). Reading our past through these lenses, we have since learnt to emphasize links, exchanges and correlations and to build more connected histories.

This emphasis on fluxes rather than fixed containers could have been conceptualized before. For decades, famous precedents were left almost untouched after they had been written, such as Simmel's digression on the stranger in his huge seminal work, *Soziologie* (Simmel, [1908]1997) – a mere seven pages as part of the ninth chapter, which was devoted to 'space and the spatial order of society'. For the first time, the question of the role of mobile people within a spatially fixed social group was discussed. Simmel answered this question in a dual way by suggesting that mobility can contribute to the group's uniformization and to its division. Interestingly, he then proceeded to differentiate between individual expressions of mobility (contrasting figures of positively and negatively appreciated types, such as the tramp and the explorer) and between mobile communities, such as nomadic collectives. It is thought-provoking for us today to realize that one of the founders of sociology may have conceptualized the sedentary condition that Westerners consider as widespread as being not quite universal.

Nevertheless, it took almost a century before social scientists were able to focus on our mobile condition again in a non-reductive manner, by showing that movement could not be reduced to a characteristic of non-Western, less developed populations that are unable to access proper livelihoods in a stable manner. This 'mobility turn' corresponds to an incredible increase in movements of all kinds. According to commonly aggregated data, the intensity of people's travels has grown to over 1 billion legal international arrivals each year, compared to 25 million in 1950 (IATA figures cited by Sheller and Urry, 2006: 207), while international financial transactions have come to represent 100 times the sum of all national incomes (in the 1990s, this ratio was only 15.3). Renewed infrastructure, and especially the increasing use of commercial planes, as well as the spread of high-technology transmissions permitting the development of the worldwide web, conditioned this recent phase of globalization.

In this picture, a very special type of displacement has evolved according to specific patterns: migration, that is, people's mobility across an international border. Some data are available on immigration to Western countries from the end of the nineteenth century, namely official arrival figures and departures from the big ports to the New World. However, these figures are not only heterogeneous but also very incomplete. At the turn of the twentieth century, political interest was directed at European emigrants; within Europe, other kinds of statistics were being consolidated regarding the presence of foreigners on national soil, and they focused less on migration (Renard, 2014, 2018). Between the two world wars, it was up to the International Labour Office (ILO) to collect the figures that would become those of 'international migration' and were negotiated between the ILO, the United States and the main colonial empires (Stricker, 2017). During the first decades of precise international monitoring (i.e. post-Second World War), these figures have remained quite stable (cf. the United Nations annual reports): From the 1950s to the early 2010s, 3 per cent of the world's population changed country – a figure that grew in absolute terms as the planet's population more than tripled but has remained almost stable in terms of percentage (growing to 3.2 per cent in 2010 and dropping slightly since). This picture has been challenged by the recent increase in the number of refugees, that is, people forced to change countries in order to flee a crisis: this figure has risen from 2 million worldwide in 1950 to 31 million in 2010 and almost 60 million in 2017 (UN, 2017). These numbers have to be added to the 200–250 million 'official' migrants who are registered as such and therefore have a regular legal status. This increase is linked to the end of the Cold War and the renewal of geopolitical threats, notably the shift from bilateral conflicts to civil wars and asymmetrical struggles that have a severe impact on civilians.

A study of the various ways in which people move around the planet may be undertaken at all scales. However, migration studies, initially defined as the move from country A to country B, long constituted a domain of its own, separate from mobility studies, which dedicated itself to a better understanding of more mundane human movements, such as work- or tourism-based. One significant impact of globalization and the way it is being studied is that it is viewed as straddling both domains. Thus, it bridges the divide between the two, which had previously been seen as a scientific dead-end (Zelinsky, 1983). It does this especially through the notion of scale: long-scale journeys and everyday mobilities have proved to be related, as migrants are generally keen to spend more time commuting within the cities where they settle (Dureau et al., 2000).

THE SPATIOTEMPORAL DIMENSIONS OF MOBILITY AND MIGRATION

Mobility and migration are defined by the movement of people through space and, implicitly, also through time. As the stress was gradually more and more being put on motorized movement, speed rapidly became an implicit and almost the unique criterion of mobility. As this was happening, social science research disregarded all the corporeal qualities of moving bodies and their kinesthetic dimensions of extent, direction and weight, only to reincorporate them again very recently in works on walking or dancing and more generally in studies of the performing dimension of movement (Schiller and Rubidge, 2014). Thus, there is great potential to renew the analysis of the spatio-temporal components of human movement.

Mobility and migration concern both individuals and communities, and they have to be analysed by connecting personal practices and collective patterns. Thus, the livelihoods of many traditional communities are based on mobility, and it has often been noted that 'leaving is a means of staying' (Cortès, 2000). Traditionally, some members of a village from the Alps or the Andes mountain regions would leave to work in the plains during particular times of the year that corresponded to quieter moments on the agricultural agenda in order to return with useful resources for their communities. Recently, these patterns have expanded to include international migration, and members of such communities regularly move abroad, sometimes to another continent, to help secure the living of their kin in their places of origin. This shows how mobile behaviour can develop in a circular manner and in a time horizon that allows for regular returns. The recent surge in the securitization of borders has made these movements more difficult and paradoxically drawn people who did not want to remain in the immigration countries to settle there more permanently (Rus and Rus, 2014).

At the local scale, the time–space component of mobility is a key indicator of social inequality. The capacity of moving quickly and being connected to the greatest possible number of nodes in a given territory is a privilege. It is linked especially to the ownership of a means of transportation, for example, having a car enables suburban women to feel emancipated (cf. Law, 1999) and having a passport and a visa to board a plane means having the status to move (Salter, 2003). However, this mobile capital may not be enough to achieve adequate accessibility to all desired places, in particular because of the overload of infrastructures on which the means of transportation rely: Whether new or old, cheap or expensive, all types of cars are stuck in traffic in the same way, unless a special transit regime, possibly with paying tax discs or tolls, is organized to select flows on a financial basis and thus increases accessibility for some while decreasing it for others (cf. the growing restrictions on car access in West European cities).

A good segment of contemporary research in this domain thus pertains to mobility policies, that is, ways and means of orienting individual choices in this field, with two kinds of objectives: reducing inequality regarding access to places and the opportunities they offer (notably access to jobs, but also housing, education and so on) and improving global sustainability (by reducing the environmental impact of movement, calculating the amounts of energy needed to transport goods and people and to sustain the mobilities of ideas and capitals; consider, for example, the electricity needed to maintain cables and servers for Internet exchanges). More specifically, we must remember that transportation now accounts for 20.5 per cent of the world's greenhouse gas emissions, without taking into account the energy needed to build cars, aeroplanes, ships, trains and so on. These figures have been rising globally since 1960, but the peak (22 per cent) seems to have been reached in 2000 (cf. World Bank, n.d.).

Mobility is indeed linked to standards of living. If the ability to move is as old as mankind, it has developed in direct correlation with the development of transportation technologies. The first phase of globalization, seen from that perspective, was linked to the advent of navigating techniques that made the regular crossing of oceans possible (sixteenth to eighteenth centuries). The second phase was directly linked to the invention of steam power in ships and, somewhat later, trains, which made the commercialization of agricultural goods possible at a global scale as foodstuffs could be transported across the globe in less than one week (nineteenth century). The third phase was characterized by the massive use of cars, trucks and, at a later stage, aeroplanes (first two-thirds of the twentieth century), whereas the fourth stage, which started in the 1980s and in which we still find ourselves, relies on microelectronic conductivity that, together with the satellite triangulation, made the Internet possible.

As the level of access to consumer goods continues to increase, especially in emerging economies, the conditions of mobility are also changing fast. Annual sales on the global car market increased from 60 million units in 1980 to 102 million in 2009 and are projected to reach 205 million in 2020, 65 per cent of which will be in Asia: This rise amounted to over 1 billion vehicles in the world in 2010, a figure expected to double by 2030–2035 (Rodrigue et al., 2017; and reports such as Statista.com, 2018). Between 1955 and 2005, the rate of car growth was three times faster than that of demographic growth. The fact that 80 per cent of migration flows are observed between southern countries also suggests that, in the future, mobility and migration studies should broaden their perspectives by focusing outside the Western regions, too. This globalization of case studies could well lead to a reformulation of the existing theoretical frameworks that have been designed to match Western patterns of mobility and migration.

Through the combined enhancement of transportation conditions and the promotion of free movement (notably of goods and, subsequently, of services, as promoted first by the General Agreement on Tariffs and Trade and then by the World Trade Agreements), the facilitation of movement has not been accompanied by fewer regulatory obstacles on people's cross-border mobility. At the start of the twenty-first century, a growing concern for global security has mitigated the movement of people at an international level. In Europe, the Schengen Agreement signed in 1985 (and supplemented by a convention implemented in 1995) and its recent questioning or suspension by many European Union member states illustrates both the power of the idea of a universalization of the freedom of people's movement and the strong resistance to its implementation. This drawback finds its roots in an increasing resistance among rather broad swathes of the population against migration in general and certain groups in particular, which illustrates the tension between processes of globalization and socially marked responses on both the local and the national level.

Thus, the main drivers in this particular case of globalization spanning mobility and migration are the numerous restrictions on people's mobility and the tension between a global extension of material and immaterial circulations[1] and the long-heralded

---

[1] The relative stability of migration patterns (3 per cent of the total world population) is mitigated by other flows conditioning it. The acceleration of the movement of goods can be perceived through the explosion of container traffic: from 100 million containers in 1992 to 650 million in 2015. And although the commercial exchanges have increased 260-fold, from US$60 billion in 1949 to over US$16 trillion in 2016, they participate in an annual world

freedom of movement. The restrictions can be seen in two ways. In a context of unequal access to a resource, the ability to move on an everyday basis is firmly linked to socio-economic conditions, whereas at an international level, migration control through the enforcement of increased border securitization calls into question the generalization of a right to worldwide mobility. Even if at one point all of globalization was described as a 'mobility regime' (Shamir, 2005), this notion is now being specified within mobility and migration studies: The definition of mobility or migration regimes is becoming essential to understanding the individualization of access to places and of socio-spatial practices. This notion, taken from the vocabulary of international relations, mixes policy- and people-based approaches. In terms of migrations and borders, it insists on the multiple iterations of different state-centred sets of regulations combined with the expression of counter-power by mobile individuals in general and migrants in particular (Koslowski, 2011). The idea of *borderities* (Amilhat Szary and Giraut, 2015) expresses this nuance when the movement of people comes up against the complex political apparatus of border crossing and the multiple spatialities that emerge from individualized social practices at international boundaries.

ORGANIZATIONAL DIMENSIONS

The agency of mobility and migration is highly fragmented. Migrants, human traffickers and others in the migration chain are crucial in the process, although literature often reduces analysis to policy implications because all kinds of public actors are involved in public action aimed at organizing/facilitating or/and restraining the movement of people at the different scales they deploy themselves. The spatial practices that involve movement are not necessarily collective action. Personal positions are also very important to explain how we move according to our own interests and affects. Mobility is indeed at the crossroads of rational interest and subjective choices. The decision to move to another country typically involves making a calculus regarding the chances for higher standards of living (e.g. in terms of income, conditions of work or a cleaner or safer environment etc.) but also depends on the contingencies of an individual's life (some move for love or to be closer to family members).

Mobility and migration are not organized at the same levels of government. While mobilities at the national scale may rely on states, more generally, everyday mobilities are linked to regional or local governments. What is at stake, then, is these local governments' capacity for interlinkage; given that urban displacements might cross many municipal boundaries, the capacity for intervention in the field of mobility policies implies coordinated governance at the metropolitan level. One cannot work to reduce car access to central cities without transforming major roads or railways in the region and without taking the risk of exacerbating major congestion points at the intersection between areas covered by public transportation and ones that are not.

Concerning migration, states' interventions are decisive in the design of the conditions for leaving and entering a national territory. They materialize in the possibility

---

wealth production of US$77,328 trillion, to be added to the US$5 trillion traded *every day* on financial markets (US$1,825 trillion annually), cf. Rodrigue et al., 2017.

to hold (or not) the official documentation required to proceed, for example, a passport, together with a visa, where needed, to gain access to specific countries. This legal condition of international mobility is the sovereign responsibility of citizens' states, unless these competencies have been transferred to a continental organization. The latter is the case for the European Union with the Schengen visa, but also with the free-circulation agreements signed with countries such as Switzerland. This could become the case in Africa as well, where a growing number of supporters are calling for the effective implementation of the ECOWAS Free Movement of Persons Protocols and the ECOWAS Common Approach on Migration. For the first period of implementation of a legal framework for cross-border mobility (1850s to 1990s), the issue was to gain access to the right to leave a country (either because of the difficulty to access identity papers or because of the political restrictions on travel that had been imposed by certain governments, especially during the Cold War). Today, the right to gain official access to a territory has become a serious problem, with major restrictions placed on access at the borders of countries in the Western world.

However, the state can delegate this sovereign capacity and even externalize it to private or external stakeholders (MIGREUROP and Clochard, 2012). The richer countries that have restricted access to their territories have done so with three complementary moves: First, by building partnerships with transportation companies that now operate passport and visa control on behalf of governments. Second, by interfering with migration services in the countries of origin (e.g. the compulsory presence of a network of EU immigration liaison officers in West African countries since 2004). Third, by increasing negotiations with third (so-called *transit*) countries to ensure that migration flows can be stopped during people's journeys. Since 2005, Libya and Turkey have played this role vis-à-vis Europe, which saw a major increase after 2011 as the countries had to manage growing numbers of undocumented travellers on their own territory. This development has allowed destination countries, which do abide by international protective texts such as the 1951 Geneva Refugee Convention, to avoid the need to respect this very legislation.

A whole industry of migration and border crossing has taken root in a context that has become ever more complex (Hernández-León, 2013; Gammeltoft-Hansen and Nyberg Sørensen, 2013). This diverse sector is growing rapidly in order to accommodate demands ranging from paying for help with paperwork (small companies that facilitate the procurement of tourist visas, law services to accelerate access to a working permit and so forth) to a whole ecosystem of accompaniment and smuggling. The latter is certainly organized by international criminal networks. However, it also relies on the local resources of poor neighbourhoods crossed by state borders where, traditionally, one can find both a smuggler and a border patrol agent in the same family, given that these two jobs are the main possible sources of income in remote areas (cf. Diaz, 2015; Kraudzun, 2017; and personal fieldwork interview on the US/Mexico border as well as on the Alpine border).

Given this framework of collective actions, one should not minimize the agency of the people on the move, that is, their capacity to choose routes and interlocutors, their negotiating power with both official and criminal agents, as well as their political stamina, which leads them to build active partnerships, notably with NGOs and activist networks who foster their empowerment (De Genova, 2017). It becomes clear that migration forcibly questions notions of human rights and their links with the dimensions of contemporary political inclusion, notably in democratic regimes. Issues like the right to vote granted to citizens abroad or/and citizens inside the country challenge exclusionary definitions of belonging and destabilize the idea of borders as the site of exclusion for strangers from a national community. For the migrants, the choice between voting in their country of origin or destination is increasingly acute, and the growing political power of diasporas is slowly – but profoundly – transforming the modernist tautological self-evidence of a stable state defined by fixed borders over a territory from which sovereignty arises. The step beyond is the claim of 'active citizenship' on behalf of migrants without a legal status who work their way through social inclusion via numerous political initiatives (Schwenken and Ruß-Sattar, 2014; Swerts, 2017).

A CASE STUDY: STILL POINTS ON ANIMATED MAPS

Because mobility and migration are challenging more established forms of space–time analysis, it has become very hard to represent them. And the mere use of animated data does not suffice in producing a dynamic vision of the phenomenon. Without a proper explanation of the processes of map making or an appropriate examination of the data used for their geo-localization, a map can easily be deployed for a very ideological purpose. For example, the Finnish website *Lucify*, which shows animated comet-like symbols converging towards the north-western parts of Europe, conveys the impression of a locust invasion by way of migration flows (https://www.lucify.com/the-flow-towards-europe/). Why? Because the idea that mobility and migration obey a simple spatial pattern dominates this representation and erases the complexity of the journeys, the tos-and-fros and workarounds, as well as the standstills and the moments when all moves are suspended. They silence the standstills on the migrants' journeys that, in some cases, last such a long time that they become much more than linking points between two life phases.

Even when we are aware that all maps are inevitably ideologically biased representations of social realities, we realize that this kind of map, while pretending to render the time component of mobility, silences it. The duration of migration journeys and the complexity of the migration routes cannot be rendered by a simplified interpretation of the spatialities and temporalities of mobile lives. To account for this complexity, there are many options, one of which is to make explicit the subjective dimension of these practices: This is the path that has been followed in the participatory workshops in Grenoble in 2013 called 'Crossing Maps' (Mekdjian and Amilhat Szary, 2015). The subtitle of the *Crossing Maps* paper was: 'spaces of never-ending arrival', which alludes to the fact that many of those who undertake migration journeys do not consider themselves as settled once they have set foot in their destination country. The first reason is that the place may not be what they initially wanted to reach when they undertook the voyage, but they stop there because of legal reasons. For example, they are immobilized because they are forced to stay in the country, notably under the Dublin rule in contemporary Europe, a European regulation that stipulates that a refugee may only officially claim asylum in the first EU member state that he or she enters, or there might be personal motivations, such as exhaustion, a lack of monetary resources and so on. The second reason is that it often takes a long time before they can acquire, if not official citizenship, at least a practical understanding of belonging to their new place of residence, especially since they are far from welcome.

This is one of the ways of drawing our attention to the growing numbers of people who find themselves 'lost in migration'. However, these workshops and such partici-patory endeavours are only partially able to unveil an emerging issue in mobility studies, namely the importance of stillness within displacement (Cresswell, 2010). Let's start by recalling how, when sitting in a car or even standing in a packed bus in peak traffic, an individual is in certain ways motionless and thus able to undertake activities usually linked to immobile environments (sleep, read, watch the news on a smart-phone). At the scale of mundane everyday movements, this alternation of movement and immobility only breaks up the course of the day. At the scale of international migrations, a similar phenomenon takes on a radically more poignant meaning as the duration of the periods of stillness and motion is growing.

To illustrate this point, let us look at the detention facilities that European countries and their neighbouring partners have built over the past 10 years to lock in migrants because of their status vis-à-vis the current restrictions on people's movement at the borders of the Schengen Area and within the area it seals off. Since 2008, an EU rule (CE/115/2008 Art. 15), often named the 'return directive', both proclaims a wish to bring all detention to 'the shortest possible term' and restricts the maximum period of custody to 18 months (Arbogast, 2016). These camps are intended for people who are either waiting for their admission request to be processed or awaiting expulsion after they were caught on one of the EU national territories without appropriate paperwork. National legislations differ in an important way, and one can represent the modalities of the time of migratory routes towards Europe (see map, Arbogast, 2016: 16). This kind of cartographic representation can be complemented by mapping the actual localization of all the facilities that are used for this encampment, some on European soil, others in neighbouring countries, notably in states that do not comply with the Geneva convention on asylum and are therefore not bound by international law in this respect. By no means should these states be considered safe. As a growing number of so-called 'hotspots', that is, mobile control points introduced on the migratory routes, are organized to protect and prevent migrants from proceeding further on their routes, the duration of their journeys is growing dramatically. People can now be on the road for years and eventually neither arrive where they intended to go nor go back to where they came from because of the ongoing political crisis they have fled (war or authoritarian governments). Economic factors might also prevent them from reimbursing the debts incurred to undertake the journeys in the first place. The causes may also be symbolic, since it is very difficult to confront conservative social environments after all that migrants have endured and learnt on their long journeys.

CONCLUSION AND RESEARCH AGENDAS

As this short overview of what is currently at stake in mobility and migration studies reveals, taking into account the mobile component of our existence radically modifies our analytical tools of interpretation of spatiality. Notions like 'home' and 'belonging' become 'networks' and 'pluri-localization'. It is essential to understand that these notions play with both space and time in a complex variation of all kinds of scales. Moreover, the movement of people on the surface of the earth appears to be cumulative: People who have already had mobile practices over the course of their existence are more likely to move than those whose past has been more stable. This trend can be observed at the scale of an individual lifetime and also with a transgenerational dimension, meaning that mobility can be transmitted as a family legacy. This phenomenon has been conceptualized as 'mobility capital', a term whose definition is in line with Bourdieu's concept of social capital. Thus, sociologist Vincent Kaufmann speaks of 'motility' as our personal capacity to use the resources of a mobile world (Kaufmann et al., 2017; Kaufmann, 2017).

Following these attempts to link the multiple dimensions of mobility and migration in time and space, recent research efforts have also addressed the interactions between human flows and the other mains flows of globalization, that is, the movement of things and data. Some interesting projects have shifted, for example, from the analysis of cultural and political diasporas in places to the expansion of e-diasporas (cf. the first @-atlas of the kind, edited by Dana Diminescu, http://e-diasporas.fr), featuring the links of migrant communities online, as well as their practices of online relations, such as doing homework with children left in their country of origin through instant messaging services. Another topical question is the evolution of security issues and their impact on the movement of both goods and peoples.

Last but not least, another set of innovative approaches addresses mobility at the scale of the moving bodies: The recent results of dance and performance studies that analyse the quality of movement and qualify its kinetic component appear as resources to renew mobility and migration studies. The manner in which we move, slowly or quickly, heavily or lightly, regularly or irregularly, and so forth, considerably influences the definition of motion and its relation to spatiality in ways that geographers have not yet fully explored. Since its first formulation by Zelinsky (1971), not only has the hypothesis of mobility transition been confirmed, but mobility studies have moved from an analysis of transportation to one of movement. This has been done in an iteration of focus from man to technique, and from technique back to man, with recently cleared ground to study the embodiment of movement, walking and so forth in greater detail. This could be another way of including stillness in mobility studies and the fact of being stranded or stuck, that is, 'lost in migration', in critical border studies (Bauder, 2011; Brambilla, 2014).

All of the material produced by the analysis of movement at all scales encourages us to avoid causal categories to analyse migration. In policy-making terms, it has proved almost impossible and very dangerous to distinguish economic from political exile, and it is no easier to define climate-change refugees. They do not exist as such: People leave their homes when their livelihoods are destroyed, be it for economic or political reasons. The day they leave their country, they do not blame climate change but poverty (Baldwin and Bettini, 2017). This confirms the opening lines of this chapter: Mobility and migration are analytical categories that are far from neutral. The words we use to describe movement often contribute to fixing it, and it is not easy to escape the possible counter-uses of all the categories we produce, whether in the academic literature or in the political spheres. Mobility and migration are part of these, and when working on mobile realities, we should probably be even more careful to explain why we refer to the terms we use and how they contribute to stabilizing social categories. The notion of 'illegal' migrant and the criminalization of human flows around the world was unthinkable 20 years ago, when an 'alien' called to mind the image of a fictional figure from outer space. This vocabulary has now invaded all kinds of narratives and contributed to naturalizing the worst possible travel conditions. If every time a boat capsized in the Mediterranean Sea, the news headlines in the media were about the number of human deaths rather than the figures of migrants, the framing of this issue of mobility and migration within the public debate would be quite different.

What is certain is that all the components of our worlds are in motion, whether they are human or non-human, alive or not (e.g. the movement of mobile phones). The notion of invasive plants or the circulation of germs also contributes to modifying our traditionally fixed territorial views (Dobson et al., 2013). What remains no less assured is that people and things are not equal towards mobility and migration and that the analysis of mobile worlds is a good way to start understanding globalized socio-spatial inequalities. Mobility and migration studies are therefore not only an established field of research (cf. the publication of handbooks such as Adey et al., 2017) but also one that can help critical methods and epistemologies in social sciences to come into view (Söderström et al., 2013).

REFERENCES

Adey, P., D. Bissell, K. Hannam, P. Merriman and M. Sheller (eds) (2017), *The Routledge Handbook of Mobilities*, London and New York: Routledge.

Agnew, J. (1994), 'The Territorial Trap: the Geographical Assumptions of International Relations Theory', *Review of International Political Economy*, **1**(1), 53–80.

Amilhat Szary, A.-L. and F. Giraut (eds) (2015), *Borderities: The Politics of Contemporary Mobile Borders*, Basingstoke: Palgrave Macmillan.

Arbogast, L. (ed.) (2016), *Migrant Detention in the European Union: A Thriving Business. Outsourcing and Privatisation of Migrant Detention*, Bruxelles: MIGREUROP/Rosa-Luxemburg-Stiftung. http://www.migreurop.org/IMG/pdf/detention-migrants-eu-fr.pdf (accessed 26 June 2018).

Baldwin, A. and G. Bettini (eds) (2017), *Life Adrift: Climate Change, Migration, Critique. Geopolitical Bodies, Material Worlds*, London; New York: Rowman & Littlefield International.

Bauder, H. (2011), 'Toward a Critical Geography of the Border: Engaging the Dialectic of Practice and Meaning', *Annals of the Association of American Geographers*, https://doi.org/10.1080/00045608.2011.577356.

Brambilla, C. (2014). 'Exploring the Critical Potential of the Borderscapes Concept', *Geopolitics*, 1–21. https://doi.org/10.1080/14650045.2014.884561.

Cortès, G. (2000), *Partir pour rester. Survie et mutation des sociétés paysannes andines (Bolivie)*, Paris: IRD.

Cresswell, T. (2010), 'Mobilities II: Still', *Progress in Human Geography*, **36**(5), 645–653.

De Genova, N. (ed.) (2017), *The Borders of Europe: Autonomy of Migration, Tactics of Bordering*, Durham: Duke University Press.

Diaz, G.T. (2015), *Border Contraband: A History of Smuggling across the Rio Grande*, Austin, TX: University of Texas Press.

Dobson, A., K. Barker and S.L. Taylor (eds) (2013), *Biosecurity: The Socio-Politics of Invasive Species and Infectious Diseases*, New York, London: Routledge.

Dureau, F., V. Dupont, E. Lelièvre, J.-P. Lévy and T. Lulle (eds) (2000), *Métropoles en mouvement. Une comparaison internationale*, Paris: Anthropos.

Gammeltoft-Hansen, T. and N. Nyberg Sørensen (eds) (2013), *The Migration Industry and the Commercialization of International Migration*, London/New York: Routledge.

Gruzinski, S. (2004), *Les quatre parties du monde: histoire d'une mondialisation*, Paris: Ed. de la Martinière.

Hernández-León, R. (2013), 'The Migration Industry: Brokering Mobility in the Mexico–U.S. Migratory System', *Program on International Migration*, février, 14.

Kaufmann, V. (2017), *Les paradoxes de la mobilité: bouger, s'enraciner*. 3ème édition, Lausanne: Presses polytechniques et universitaires romandes.

Kaufmann, V., Y. Dubois and E. Ravalet (2017), 'Measuring and Typifying Mobility Using Motility', *Applied Mobilities*, septembre, 1–16.

Koslowski, R. (2011), *Global Mobility Regimes*, Basingstoke: Palgrave Macmillan.

Kraudzun, T. (2017), 'Sovereignty as a Resource: Performing Securitised Borders in Tajikistan's Pamirs', *Geopolitics*, **22**(4), 837–862.

Lapid, Y. (2001), 'Introduction: Identities, Borders, Orders: Nudging International Relations Theory in a New Direction', in: Mattias Albert, David Jacobson and Yosef Lapid (eds), *Identities, Borders, Orders: Rethinking International Relations Theory*, Minneapolis, MN: University of Minnesota Press: 1–20.

Law, R. (1999), 'Beyond "Women and Transport": Towards New Geographies of Gender and Daily Mobility', *Progress in Human Geography*, **23**(4), 567–688.

Mekdjian, S. and Amilhat Szary, A.-L. (2015), 'Crossing Maps: Spaces Of Never-Ending Arrival', *Visions Carto*, 29 May 2015 http://visionscarto.net/crossing-maps (accessed 28 June 2018).

MIGREUROP and Olivier Clochard (eds) (2012), *Atlas des migrants en Europe. Géographie critique des politiques migratoires*, Paris: Armand Colin.

Moore, M. (2003), *A World Without Walls: Freedom, Development, Free Trade and Global Governance*, Cambridge: Cambridge University Press.

Ohmae, K. (1995), *The End of the Nation State: The Rise of Regional Economies*, New York/ London/ Toronto: Simon & Schuster.

Renard, L. (2014), 'The statistical construction of alterity: Governing national population by numbers in France and Germany (1860-1900)', in: Per Wisselgren, Peter Baehr and Kiyomitsu Yui (eds), Proceedings of the Research Committee on History of Sociology from the XIVth ISA World Congress of Sociology in Yokohama, Japan. Jul 2014, Yokohama: 225–240.

Renard, L. (2018), 'Mit den Augen der Statistiker. Deutsche Kategorisierungspraktiken von Migration im historischen Wandel', in: *Zeithistorische Forschungen*, 2018/3.

Ricketts, J., J. Evans and P. Jones (2008), 'Mobile Methodologies: Theory, Technology and Practice', *Geography Compass*, **2**(5), 1266–1285.

Rodrigue, J.-P., C. Comtois and B. Slack (2017), *The Geography of Transport Systems* (4th edn), London; New York: Routledge, Taylor & Francis Group.

Rus, D.L. and J. Rus (2014), 'Trapped Behind the Lines: The Impact of Undocumented Migration, Debt, and Recession on a Tsotsil Community of Chiapas, Mexico, 2002–2012', *Latin American Perspectives*, **41**(3), 154–177.

Salter, M.B. (2003), *Rights of Passage: The Passport in International Relations*, Boulder: CO: Lynne Rienner.

Schiller, G. and S. Rubidge (eds) (2014), *Choreographic Dwellings: Practising Place*, Basingstoke: Palgrave Macmillan.

Schwenken, H. and S. Ruß-Sattar (eds) (2014), *New Border and Citizenship Politics. Migration, Diasporas and Citizenship*, Basingstoke: Palgrave Macmillan.

Shamir, R. (2005), 'Without Borders? Notes on Globalization as a Mobility Regime', *Sociological Theory*, **23**(2), 197–217.

Sheller, M. and J. Urry (2006), 'The New Mobilities Paradigm', *Environment and Planning A*, **28**, 207–226.

Simmel, G. (1908[1997]), 'The Sociology of Space', in: D. Frisby and M. Featherstone (eds), *Simmel on Culture: Selected Writings*, 137–170. London: Sage.

Söderström, O., D. Ruedin, S. Randeria, G. D'Amato and F. Panese (eds) (2013), *Critical Mobilities*, *Urbanism*. Lausanne: EPFL Press.

Statista.com. (2018), 'Automotive Industry', Statista. 2018. https://www.statista.com/study/9644/automotive-industry-statista-dossier/ (accessed 26 June 2018).

Stricker, Y. (2017), 'Migration Statistics and the Making of an International Point of View in the Interwar Period', in: *History of Knowledge. Research, Resources, and Perspectives*, https://historyofknowledge.net/2017/10/05/migration-statistics-and-the-making-of-an-international-point-of-view-in-the-interwar-period/ (accessed 2 May 2018).

Swerts, T. (2017), 'Creating Space For Citizenship: The Liminal Politics of Undocumented Activism', *International Journal of Urban and Regional Research*, **41**(3), 379–395.

UN (2017), International Migration Report 2017, https://www.un.org/development/desa/publications/international-migration-report-2017.html (accessed 28 June 2018).

Urry, J. (2002), 'Mobility and Proximity', *Sociology*, **36**(2), 255–274.

World Bank (n.d.) Data repository, https://data.worldbank.org/indicator/EN.CO2.TRAN.ZS (accessed 12 June 2018).

Zelinsky, W. (1971), 'The Hypothesis of the Mobility Transition', *Geographical Review,* **61**(2), 219–249.

Zelinsky, W. (1983), 'The Impasse in Migration Theory: A Sketch Map for Potential Escapes', in: P.A. Morrison (ed.), *Population Movements: Their Forms and Functions in Urbanization and Development*, 19–46. Liege, Belgium: Ordina Editions.

11. Geographies of citizenship and identity in a globalizing world

Tatiana Fogelman

INTRODUCTION

After decades spent under much of the academic radar, citizenship has in the past thirty years re-emerged as one of the core concepts and areas of contemporary scholarship in social science. Attracting interest across disciplinary boundaries, citizenship studies has in fact established itself as a relevant, attractive and fertile sub-field of academic inquiry, complete with an identically named and respected journal *Citizenship Studies*. Addressing manifold aspects of membership in a political community from legal status with its attendant system of rights and obligations to the issues of identity and belonging, it seems that there is 'no other notion more central to politics' (Skhlar, 1991: 1 in Bosniak, 2000: 450) than citizenship. Additionally, citizenship embodies the 'highest normative value' (Bosniak, 2000: 451) as social and moral goods to be striven for across the political spectrum. As such, citizenship is both a foundational and capacious concept, making it well poised to an illustrious career. But I suggest it is especially the territorial, and more broadly speaking the spatial relations of citizenship that geographic perspectives have centred on, that have made citizenship such a fertile multidisciplinary research field. After all, it is the spatiality of socio-political life that has been most profoundly challenged and reshaped through the processes of the late twentieth century globalization.

This chapter then traces recent transformations of citizenship, starting with the impacts of global migrations on citizenship. Following upon the critiques of early assessments of citizenship as denationalized through migration, the chapter highlights geographic advances in understanding citizenship as a multi- and inter-scalar political and social relation, while the last section discusses the implications of the integrationist turn in state-migrant population relations for contemporary citizenship formations (Marston and Mitchell, 1999), including their embeddedness in neoliberal rationality (Brown, 2015, 2003).

GLOBALIZATION, MIGRATION AND CITIZENSHIP

The current era of globalization has been often referred to as an 'age of migration' (Castles and Miller, 1998). The world is said to be witnessing an accelerating flow of increasingly ethnically, linguistically and religiously diversified transnational migrants, who cross ever-longer distances to and from an ever-increasing number of countries, engage in progressively more complex migratory trajectories and maintain strong active involvement with and in their countries of origin. Before submitting unquestioningly to this portrayal, seized upon also by recent anti-immigrant populism in the Global North, it warrants pointing out that many of its claims have not been yet investigated empirically at length. The perception of increased diversity of immigrants in terms of their countries of origin and religious, ethnic, linguistic and racial background has, for example, been recently shown to reflect specifically European, rather than global experience (Czaika and de Haas, 2014). Likewise, it is easy to forget that cross-border migration is actually a marginal social practice: migrants continue to represent an exceedingly minor proportion of the world population, hovering around 3 per cent. Yet, even with such qualifications noted, it is hard to deny the impacts that contemporary migrations have had on destination societies. After all, even slight percentage-wise changes in migrant population (rising globally from 2.9 per cent in 1990 to 3.4 per cent in 2017 (United Nations)) translate into tens of thousands of migrants arriving in relatively short periods of time in an actually shrinking number of destinations (Czaika and de Haas, 2014). In many locations, they are joining minority communities of (descendants of) large post-Second World War migration waves of labour and post-colonial migrants and refugees, whose presence, modes of incorporation and loyalties have recently been contested anew in popular and public discourses.

Moreover, thanks to massive advances in cheap and speedy modes of communication and transportation technologies, migrants today are able to nurture substantial cross-border ties with their families and local communities in the countries they left behind to a qualitatively different extent and with more ease than earlier generations. These encompass not just the maintenance of social ties with family members and friends or cultural and religious connections, but for many extend to a variety of political engagements, especially during election periods, as well as economic and financial investment in local communities through remittances and participation in home associations, widely popular for example in Latin America. In short, everyday life unfolds today for many, if not most migrants – and in fact in many ways also for non-migrants – in at least some way in transnational social spaces or fields (Faist, 2000) rather than just in a specific location in the country of settlement.

And finally, countries of origin as well as destination countries have for their own part become ever keener to capitalize on their emigrants' transnationalism (Basch et al., 1992), seeking to reap 'globalization dividends' through the diasporic, or extra-territorial citizenship. Accordingly, emigration states (Gamlen, 2008; Coutin, 2007) have intensified their deployment of a diversified portfolio of diaspora strategies (Ho et al., 2015), ranging from specific citizenship regulations and rights extensions to emigres – for example the right to elect their own representation in the home country legislative bodies as in the case of 13 countries (Collyer, 2014) – to discourses that seek to reinforce their emotional attachments to their original home (Cohen, 2011). Not all diasporic citizens are created the same, for not all can be thought of equally strongly in terms of now dominant human capital (Larner, 2007). Affluent overseas citizens are often times offered privileged treatment for purposes of investment or resettlement back in the country of original citizenship (Cohen, 2016). Other countries differentiate between those who had naturalized and those who had (ethnically) belonged for generations, granting the former weaker status than the latter ones, despite them both being holders of the same passport (Palop-García and Pedroza, 2016).

It is precisely the net emigration countries that have been behind the steady expansion of dual citizenship (Leitner and Ehrkamp, 2006). Emigration countries' 'scramble for citizens' (Cook-Martin, 2013) is by no means limited to Europe, where the dominance of single formal citizenship model started breaking down after the 1997 Council of Europe's European Convention on Nationality started encouraging more liberal policies (Ehrkamp and Leitner, 2003). Just within the last decade for example half of African states likewise approved constitutional changes to legally incorporate, albeit to a varied extent, dual nationality provisions into their citizenship regimes (Pailey, 2017).

Such transformations in cross-border politics, especially when seen in conjunction with broader globalized processes of capital accumulation, are seen as a massive challenge to citizenship because they challenge the tight territorial overlap between the nation and the political community of state (Castles and Davidson, 2000). That the early analyses of these changes conflated citizenship with specifically national citizenship – or citizenship of the nation-state – is understandable in that this is indeed what we came to understand under citizenship by the middle of the past century. By then the nation has namely come to denote the prime political community that citizenship addresses itself to, thanks to modern nationalism's successful promotion of nation-state as the dominant state form (Goswami, 2002). Citizenship, which has always been about some form of territorialized belonging (Isin, 2007), had become tied to the territory of the nation-state (Mountz, 2009).

The nation-state institutionalized citizenship through a variety of monopolizations. The monopolization of the legitimate means of movement across space for example gave us the iconic modern passport (Torpey, 2000). The passport did not document nation-state population as much as it helped co-constitute it. Moreover, as a codifying and solidifying technology it helped create its holder's national (state) identity (Torpey, 2000). This aspect of the technology of passports was clearly on the minds of EU officials when they insisted on a uniform, burgundy-and-gold look of the passports of its member states (the so-called EU passport) in order to foster joint European identity amongst its holders (Wiener, 1997). The modern state's quest for such a 'unique and unambiguous identification' of its nationals-cum-state citizens has served not just the exclusion of unwanted subjects but equally also a more effective 'embrace' of those belonging, whether for purposes of taxation, military conscription or welfare (Torpey, 2000: 16). This process has then been crucial to the expansion of liberal democratic citizenship to include not just the civic and political but also significant social rights under the modern welfare state of the twentieth century (Marshall, 1950), thus safeguarding robustly the relationship between the nation-state and its national citizens.

SCALE AND CITIZENSHIP

Even if a neat overlap between the nation, state and territory is 'a "past" that has never been, and will never be' (Derrida, 1982: 21 in Balibar, 1990: 329), what coupling there ever was, has arguably become looser over the past half a century. A post-Second World War rise of denizenship (Hammar, 1985), namely a condition where migrant residents have access to most (if not all) social, civic and economic, albeit not political rights, that formal citizens have, has uncoupled rights from formal membership and – most troublingly for populist commentators – possibly also identity aspects associated with formal membership. In her book tellingly titled *Limits of Citizenship*, Yasemin Soysal (1994) interpreted this ability of denizens to enjoy quasi-citizenship status based on their lawful residency as an outcome of an intensification of human rights discourse that has put liberal democracies under increasing pressure to extend membership rights to resident aliens. For Soysal, this amounted – both descriptively and normatively – to a shift to a post-national membership where 'universal personhood' trumps national belonging as a source of rights (Soysal, 1994: 136).

Soysal's and others' claims about the process of denationalization of citizenship in favour of post-national, global or cosmopolitan forms, echoing the claims of the demise of the state under globalization, have been duly critiqued. Over two decades later, formal national citizenship continues to matter in a number of ways. In the first place, the difference between denizens having de jure access to social and civic rights and their de facto ability to get those acknowledged and implemented persists. Such a gap between formal and substantive citizenship affects also those with formal citizenship, especially ethnic, racial and religious minorities and indigenous people (e.g. Christensen, 2013). In a recent study of ethnic Vietnamese Cambodians in possession of a variety of legal citizenship-pertinent documents Laurie Parsons and Sabina Lawreniuk (2018) show how even those with a full set of legal documents share the same restrictions on their everyday life in regard to work, schooling and housing opportunities as those without, resulting in a condition of de facto or 'liminal statelessness'. As they highlight, much citizenship relies in practice on bureaucrats' 'aesthetic judgments of ethnicity' (Parsons and Lawreniuk, 2018: 6). Whether outright intentional or deeply subconscious, such judgments come from reading and categorizing bodies of claimants as either normal or dissident ones.

But it has likewise become obvious that the national scale remains crucial in citizenship politics. The nation-state has retained its position as the ultimate arbiter that can, and has recently been taking away denizens' rights, often justified through discourses of threats to national security (Ben-Porat and Ghanem, 2017). Not even formal citizenship status is immune from this power, as attested to by the practice of stripping some Islamist subjects of their citizenship status in the UK and France in the post-9/11 world and global 'war on terror' (Fargues, 2017). And it is the national scale that is the baseline even in the most pronounced, if not the only actually existing case of supranational citizenship, the EU citizenship, which is directly derivative of the national one (Aradau et al., 2010). Finally, the importance of 'the national' is clear also from migrants' perspectives. They often 'consider formal citizenship a prerequisite for personal security and protection under the law and equal access to social and political rights' even if they are aware that formal status does not guarantee actual equality and lack of discrimination in practice (Leitner and Ehrkamp, 2006: 1629). And while many right claims might be couched in terms of international human rights, they are oriented most often to the national governments where migrants reside (Koopmans and Statham, 1999).

Assessed from today's vantage point, the critical debate spurred by a variety of denationalization claims has thoroughly transformed the landscape of citizenship studies by making explicit the multi-scalar nature of citizenship. At the same time the attention to scales other than national has become rather unavoidable, not simply because citizenship can, and should always be understood as multi-scalar, but because citizenship has been significantly rescaled. The question, and in fact much of citizenship research since Soysal's claims, has become about the extent and consequences of such rescaling. Rescaling of citizenship has been of course underpinned by the broader rescaling of statehood. Political geographers have in many ways focused more on the sub-national rather than supra-national aspects of rescaling, researching especially how cities have become the locus for much contemporary governance (Antonsich and Matejskova, 2015). This is due to the decline of more redistributive policies that spurred in turn more entrepreneurial modes of urban governance. The result of those – besides the increasing competition as well as socio-spatial segregation in most cases – has been a stronger political rather than just economic clout of major cities, with implications for citizenship politics.

Such strong leverage of major cities forms the context of the other major spatial reorientation of citizenship scholarship towards the urban scale, the historical origins of (at least Western) citizenship. The second contextual aspect of this reorientation rests on the fact that despite some diversification of national geographies of settlement, large cities and metropolitan areas continue unabatedly to provide a home for the majority of immigrant populations and their offspring. It is not uncommon for the population of major cities in, for example, Northwestern Europe to be for example 30 to 40 per cent foreign-born. This means that the 'crisis of territoriality' (Benhabib, 2005), expressed in the lacking match between a subject's citizenship status and rights on the one hand and her territorial mooring in citizenship, is most acutely experienced here. But this is also why such cities have tended to be at the forefront of extending social rights to non-citizen migrant residents, further enabled by the increasing implementation responsibilities as well as decision-making powers of cities. In order to reinforce this tendency some political theorists of citizenship have argued normatively for the formalization of the existing urban citizenship, promoting residence or *ius domicile* as the most meaningful as well as just basis for citizenship (Bauböck, 2003). Those grounded more in critical social theory have found further inspiration for their advocacy of urban citizenship in Henri Lefebvre's notion of right to the city as a collective right that stresses more radically primacy of (residence-based) usage over the ownership of city (e.g. Purcell, 2003), inspiration that has exerted strong impact on urban grassroots movements.

The focus on cities as 'key sites of struggles of citizenship' (Bell and Binnie, 2004: 1808) and thus spaces of hope for redressing inequities of current citizenship regimes stems in the first place from them being spaces of actual everyday life. Cities have also come to be seen as having been more pragmatic than the national governments (e.g. Barber, 2014). In fact, in much critical citizenship and migration scholarship the urban has in fact emerged as the scale that is somehow inherently progressive in comparison to that of the (nation-)state, conceived as always already 'a site of oppression' – presumably because of its historical complicity in the rise to dominance of mono-cultural nationhood (Antonsich and Matejskova, 2015). This has been this way despite the evidence that municipal neoliberal policies focused on revenue-generation have encroached on rights of many migrant (as well as non-migrant) residents just as much as immigration or integration legislation enacted at the national scale. Urban discursive embrace of citizen diversity likewise often remains not only highly selective but also declaratory. And finally, many localities have been assertive promoters of harsh anti-immigrant policies, whether passing ordinances punishing landlords across the US for renting out to undocumented migrants (Walker and Leitner, 2011) or using police to carry out immigration status checks, prerogative of the national scale authorities, during routine policing (Coleman, 2007).

Additionally, some of the same critiques that can and have been raised against the denationalization thesis also apply to the urbanization of citizenship. As other sub-national forms of citizenship, the urban one remains most 'often subordinated to the demands of national citizenship as a matter of domestic law' (Bosniak, 2000: 457). Thus it has been unable to immunize especially those most vulnerable subjects, namely irregular migrants (Varsanyi, 2006) from excesses of state power, as highlighted for example by a nation-wide, Trump administration-ordered federal raid against the increasingly vocal coalition of Sanctuary cities in the US in the autumn of 2017. Yet the main point in critiquing the current privileging of the urban is not to somehow tip the scales, analytically speaking, back to the nationally scaled citizenship. This would just reinforce the most dominant tendency in citizenship research, namely its opposi-tional, binary-driven conceptualizations (Mhurchu, 2014). Rather, it is to stress how citizenship is constantly created through 'processes operating at different scales [that] interact to create conditions, opportunities and constraints that political agents confront and use' (Staeheli, 1999: 66).

RELATIONALITY OF CITIZENSHIP

That citizenship is such an inter- rather than just multi-scalar construction becomes clear further when considered beyond institutional and state-, or rather sovereignty-centered perspective and more from a subject-centered one. Sociocultural geographers just like many sociologists focus namely on citizenship as a social relation and 'a social practice that individuals engage in beyond the state, through organizations of civil society and civic actions' (Ehrkamp and Leitner, 2003; Benhabib, 1999). Especially the communal and identity aspects of citizenship – in other words issues of belonging[1] – get substantiated, negotiated and challenged through the social relations, including those of the everyday life. As spaces where multitudinous and disparate subjects encounter one another with strongest immediacy, urban neighbourhoods work espe-cially strongly as 'performative spaces implicated in citizenship formation' (Phillips, 2015: 71). Recent anti-immigrant protests in neighbourhoods of southern Tel Aviv have, for example, become stages of defensive citizenship through which their residents, Middle Eastern origin Jews, contest the wider citizenship regime, one in

---

[1] Over the last fifteen years belonging has replaced identity, a concept that saw its highpoint in the 1990s, as the concept of choice here (e.g. Antonsich, 2010). Capturing the dynamic nature, as well as the emotional charge of subject's social location, it then namely replaces identity's essentializing and reificatory tendencies (Brubaker and Cooper, 2000), sharp boundary-drawing, and reliance on static categorization with its homogenizing implications with more robust relationality, situatedness, and commonness (rather than sameness) (Pfaff-Czarnecka, 2011).

which they are culturally, politically and socio-economically marginalized, while simultaneously having to cope with deteriorating local social infrastructure, under pressure from migrants (Cohen, 2015). This example points to the deep relationality of not just differently scaled processes and events but also to the interconnectedness of sociocultural, political and economic relations 'between social groups and structures of power that mediate the standing of individuals in the polity' (Staeheli, 1994: 850).

By focusing on relations between different sites, locations and scales of citizenship such 'relational view[s] of citizenship' (Kurtz and Hankins, 2007: 3) enable an integration of legal-political focus of more state-centric perspectives with those subject-centric perspectives that prioritize the everyday and the sociocultural aspects of citizenship. Relationality thus offers potential for surpassing such a dualistic frame of citizenship studies. Lynn Staeheli and colleagues have further proposed that we approach citizenship as ordinary because ordinariness merges 'conceptually legal structures, normative orders, and the practices and experiences of individuals, social groups and communities' (Staeheli et al., 2012: 630; see also Neveu, 2015). Citizenship, in other words, is only graspable through how it is actually animated in life, through individual subjects, communities, organization and state-making actors.

What is furthermore brought here to the forefront is how citizenship is thoroughly imbued with the emotional and the affective (Ho, 2009). Not only do individuals experience citizenship through a gamut of often contradictory feelings. State, non-state and other actors likewise make citizenship the site of emotional investment, even engaging at times in emotional blackmail tactics, attempting to produce citizen or denizen fear or guilt in order to enforce citizens' compliance. The centrality of the emotional and affective to citizenship becomes blatant when considering the viscerality of public debates that flares up regularly.

INTEGRATIONISM, NEOLIBERALIZATION AND CITIZENSHIP

In the past decades strong affective charge has in fact become more or less a consistent feature of the new focus of public discourse and policy on integration and social cohesion. The reasons behind the turn towards integrationism, most pronounced in Northwestern welfare countries with significant non-Western migrant population (Goodman, 2010), are complex, including huge concerns about the hollowing out of citizenship. Citizenship had become seen, especially by libertarian and conservative critics, as having become reduced to just a status with attached rights that state owes to its bearer, rather than carrying with it obligations, and especially, deeper meaning. Proportionally tiny but highly visible segments of elite migrants, able to accumulate several passports – including by purchasing them through special provision for investors – helped popularize the idea that citizenship has become thoroughly instrumentalized. While such flexible citizenship (Ong, 1999) might indeed aptly describe practices of those transnationally highly mobile, migrants more generally have fallen under the suspicion of not being emotionally invested in their adopted country citizenship.

As part of the integrationist agenda, an increasing number of destination countries have in the past years then introduced citizenship ceremonies in order to imbue citizenship with more meaning for migrants (Byrne, 2014; Badenhoop, 2017). In countries where ceremonies are voluntary, like Norway, they seem to have a strong emotional value for the minority of the naturalized who choose to participate, making them feel like they now 'belong to this country' (Hagelund and Reegard, 2011). Yet even so, new (especially) non-white citizens, drew the line between becoming Norwegian citizens and becoming Norwegians. The latter, they explained, was reserved for the white migrants. The potential of belonging stops at the threshold between the Norwegian state and society on one hand and Norwegianness on the other.

The rise of integrationism in 'post-migration societies' (Fourutan et al., 2014) could be interpreted as a re-nationalization of citizenship (Joppke, 2010). It is certainly so in the sense that (nation-)states seek more agency to shape again the parameters of belonging (Favell, 2001). Besides introducing citizenship ceremonies, states have namely become assertive about migrants' language acquisition, labour market incorporation, knowledge about state history and functioning of liberal democracy. Integration has also refocused on the cohering of society as something unfolding at the nation-state scale. This is unsurprising, given the term's rootedness in the Durkheimian functionalist sociology that presumes 'the idea of the "nation-state" as the principle organizing unit of society' (Favell, 2003: 13).

But there continues to be a debate about the more specific meaning of re-nationalization. Some point to strong reverence and demand for cultural sameness in integration discourse and policies, arguing that it makes integration equivalent to assimilationism (Schinkel and van Houdt, 2010). Here re-nationalization amounts to 'culturalization of citizenship' (Schinkel, 2011; Duyvendak et al., 2016), or more precisely its mono-culturalization. Such analyses of integrationist nativism tend to originate from particular national contexts, especially Dutch and Danish (e.g. Olwig and Pærregaard, 2011; Fernández and Jensen, 2017). But for most scholars writing from elsewhere, integrationism has effectuated more of a civic turn, understood as culturally neutral, or, at worst, operating with a 'thin' notion of culture as the shared desirability of the rule of law, individual autonomy and gender equality (e.g. Joppke, 2007). In Germany for example the turn towards integration has in fact opened up the state-defined parameters of Germanness to those previously firmly excluded from it despite their multi-decade residence and even birth in the German territory. Together with heavily liberalized citizenship law, integration has thus territorialized Germanness and made it less ethno-cultural. Such opening up has, however, been enmeshed with the neoliberalization of citizenship (Matejskova, 2013), that in Germany as in most other countries, underpins integrationist agenda.

Neoliberalization has been probably the most general and pervasive tendency in the transformation of citizenship over the past few decades, prioritizing narrowly economistic understanding of the relationship between an individual and the state and in fact individual subjects themselves (Brown, 2016). States have been thus enabling an ever smoother movement across territorial borders for the elite subjects while raising barriers for the low-educated, poor and unskilled (Sparke, 2006). Accompanying the shift from welfare to workfare societies, neoliberalized domestic policies of settlement, integration and naturalization have likewise by now increasingly stratified migrants according to their economic worth, granting most favourable conditions to those affluent enough to invest and speeding up the access to permanent residence and formal membership to those with steady, well-paid employment and penalizing those without.

Those on welfare have in many countries been in fact hit with coercive measures in the name of integration, such as a threat to cuts in benefits unless they participate in language courses or vocational training.

But 'native' citizens have likewise been affected. Long-term unemployed citizens have been hit most severely. It is them who constitute the antithesis of the contemporary citizenship ideal of a self-governing subject who is not only never economically dependent on the state but also one who manages her life like an entrepreneurial project, with the view of maximizing her economic potential (Davidson, 2008). Such ideal of super-citizens who are only ever an 'asset' is often made explicit by state officials during citizenship ceremonies for the new citizens (Badenhoop, 2017). Those very much unlike super-citizens are not only devalued in terms of their standing in the polity but also in terms of trimmed social rights, such as family unification, made increasingly conditional and in need of earning.

The contractualization and responsibilization have become hallmarks of citizenship in later capitalism (van Houdt et al., 2011; Suvarierol and Kirk, 2015). And while these processes effectuate individualization of citizenship, they are not incompatible with the communitarian thrust of late capitalist governance (Rose, 2000). In the Dutch context for example '"Community" is ... selectively seen as mobilized and present (when immigrant integration is concerned) or as latently present and still in need of mobilization (when indigenous [citizens] are concerned)', amounting to neoliberal communitarian transformation of citizenship (Schinkel and van Houdt, 2010: 696). Most importantly, these kinds of changed norms of citizenship have become increasingly internalized by individuals, whether it is long-term unemployed migrants in Germany (Matejskova, 2013), or Spanish citizens discounting culturally and linguistically more familiar migrants in favour of those with high education and work contributions (Cook-Martin, 2013).

Neoliberal conditionality of citizenship is often thought of as a unique break with the history of modern citizenship. More historical studies have, however, pointed out that much of the development of social citizenship in the twentieth century was conditional on one's contributions to the state, whether through war-effort, work or parenthood (Turner, 2001; Cowen, 2005). Even in the state-socialist context, social rights were earned through work, which was as much an obligation as a right. The true distinction of the contemporary era might be in the fact that citizenship is increasingly conditional on one's net economic contributions, while at the same time potentials for such contributions have been simultaneously eroded for many citizens by the precariousness and casualization of employment (Turner, 2016, 2001).

CONCLUSION

A decade ago in their editorial to a special issue on geographies of citizenship, Hilda Kurtz and Katherinse Hankins described citizenship as a 'fundamentally geographic concept' (Kurtz and Hankins, 2007). Rescaling, re- and extra-territorialization and multi-sitedness of citizenship outlined in this chapter lay bare the complexity as well as the contested nature of spatiality that has given rise to contemporary 'citizenship constellations' (Bauböck, 2010). What the bustling scholarship on actually lived and existing citizenship regimes has likewise repeatedly contested is the idea of citizenship as full membership, no matter how useful and inspirational the concept of citizenship might be politically. As a social scientific concept citizenship then appears as a gradient category (Cohen, 2009), always working through mechanisms of differential inclusion (Mezzadra and Neilson, 2013). Or rather, taking the processual nature of citizenship seriously it can be approached as an emergent (cf. Hepworth, 2014) socio-political relation between the subject and the community, a relation that is both in its practice and construction multi-sited and inter-scalar.

Two aspects of this relation will most likely continue to stand out, as the most recent directions suggest. First is the issue of community, a crucial political fault line of citizenship as exemplified by regularly intense discussions of who belongs, when, how, to what extent, and according to whom (Staeheli, 2008). Theoretically, community also remains crucial, with some wondering whether in fact citizenship can do without or go beyond community (e.g. Balibar, 2012; Stephens and Squire, 2012). And second is the nature of the relationship between the social and the political (see also Painter and Philo, 1995). The focus on the embeddedness of citizenship in the everyday life of sociality-centric approaches has become challenged by the prioritization of more politically saturated 'acts of citizenship' (Isin and Nielsen, 2013) and insurrection as the active mode of citizenship (Balibar, 2015). Here citizenship is thought of in the first place as 'a right to be political, a right to constitute oneself as an agent to govern and be governed' (Isin, 2002: 1). One could already object that the eventful and momentous insurgency of citizenship must inevitably develop through engagements with social relations of everyday belonging, however pre-political those might seem. Still, the rich conceptual field surrounding these, as well as other aspects of citizenship, together with new empirical developments will certainly ensure ongoing relevance of citizenship studies in the near future.

REFERENCES

Antonsich, M. (2010), 'Searching for Belonging – an Analytical Framework', *Geography Compass*, **4**(6): 644–659.

Antonsich, M and T. Matejskova (2015), 'Immigration Societies and the Question of "the National"', *Ethnicities*, **15**(4): 495–508.

Aradau, C., J. Huysmans and V. Squire (2010), 'Acts of European Citizenship: a Political Sociology of Mobility', *Journal of Common Market Studies*, **48**(4): 945–965.

Badenhoop, E. (2017), 'Calling for the Super Citizen: Citizenship Ceremonies in the UK and Germany as Techniques of Subject-Formation', *Migration Studies*, **5**(3): 409–427.

Balibar, É. (1990), 'The Nation Form: History and Ideology', *Review (Fernand Braudel Center)*, **13**(3): 329–361.

Balibar, É. (2012), 'The "Impossible" Community of the Citizens: Past and Present Problems', *Environment and Planning D. Society and Space*, (30): 437–449.

Balibar, É. (2015), *Citizenship*, Cambridge: Polity Press.

Barber, B.R. (2014), *If Mayors Ruled the World*, New Haven, CT: Yale University Press.

Basch, L., C. Blanc-Szanton and N. Glick Schiller (1992), *Towards a Transnational Perspective on Migration: Race, Class, Ethnicity, and Nationalism Reconsidered*, New York: The New York Academy of Sciences.

Bauböck, R. (2003), 'Reinventing Urban Citizenship', *Citizenship Studies*, **7**(2): 139–160.

Bauböck, R. (2010), 'Studying Citizenship Constellations', *Journal of Ethnic and Migration Studies*, **36**(5): 847–859.

Bell, D. and J. Binnie (2004), 'Authenticating Queer Space: Citizenship, Urbanism and Governance', *Urban Studies*, **41**(9): 1807–1820.

Ben-Porat, G. and A. Ghanem (2017), 'Introduction: Securitization and Shrinking of Citizenship' *Citizenship Studies*, **21**(8): 861–871.

Benhabib, S. (1999), 'Citizens, Residents, and Aliens in a Changing World: Political Membership in the Global Era', *Social Research*, **66**(3): 709–744.

Benhabib, S. (2005), 'Borders, Boundaries, and Citizenship', *Political Science & Politics*, **38**(4): 673–677.

Bosniak, L. (2000), 'Citizenship Denationalized', *Indiana Journal of Global Legal Studies*, **7**(2): 447–509.

Brown, W. (2003), 'Neo-Liberalism and the End of Liberal Democracy', *Theory and Event*, **7**(1): 1–43.

Brown, W. (2015), *Undoing the Demos*, Cambridge, MA: MIT Press.

Brown, W. (2016), 'Sacrificial Citizenship: Neoliberalism, Human Capital, and Austerity Politics', *Constellations*, **23**(1): 3–14.

Brubaker, R. and F. Cooper (2000), 'Beyond "Identity"', *Theory and Society*, **29**(1): 1–47.

Byrne, B. (2014), *Making Citizens*, London: Palgrave Macmillan UK.

Castles, S. and A. Davidson (2000), *Citizenship and Migration: Globalization and the Politics of Belonging*, London: Psychology Press.

Castles, S. and M.J. Miller (1998), *The Age of Migration: International Population Movements in the Modern World* (2nd edn). New York City: Guilford Press.

Christensen, J. (2013), '"Our Home, Our Way of Life": Spiritual Homelessness and the Sociocultural Dimensions of Indigenous Homelessness in the Northwest Territories (NWT), Canada', *Social & Cultural Geography*, **14**(7): 804–828.

Cohen, E.F. (2009), *Semi-Citizenship in Democratic Politics*, Cambridge: Cambridge University Press.

Cohen, N. (2011), 'Rights Beyond Borders: Everyday Politics of Citizenship in the Israeli Diaspora', *Journal of Ethnic and Migration Studies*, **37**(7): 1137–1153.

Cohen, N. (2015), 'Southern Discomfort: Defensive Urban Citizenship in Tel Aviv', in: *Governing Through Diversity: Migration Societies in Post-multiculturalist Era*, T. Matejskova and M. Antonsich (eds), London: Palgrave Macmillan: pp. 241–271.

Cohen, N. (2016), 'A Web of Repatriation: the Changing Politics of Israel's Diaspora Strategy', *Population, Space and Place*, **22**(3): 288–300.

Coleman, M. (2007), 'Immigration Geopolitics Beyond the Mexico–US Border', *Antipode*, **39**(1): 54–76.

Collyer, M. (2014), 'Inside Out? Directly Elected "Special Representation" of Emigrants in National Legislatures and the Role of Popular Sovereignty', *Political Geography*, **41**(2014): 64–73.

Cook-Martin, D. (2013), *The Scramble for Citizens*, Stanford: Stanford University Press.

Coutin, S.B. (2007), *Nation of Emigrants: Shifting Boundaries of Citizenship in El Salvador and the United States*, Ithaca, NY: Cornell University Press.

Cowen, D. (2005), 'Welfare Warriors: Towards a Genealogy of the Soldier Citizen in Canada', *Antipode*, **37**(4): 654–678.

Czaika, M. and H. de Haas (2014), 'The Globalization of Migration: Has the World Become More Migratory?', *International Migration Review*, **48**(2): 283–323.

Davidson, E. (2008), 'Marketing the Self: the Politics of Aspiration Among Middle-Class Silicon Valley Youth', *Environment and Planning A*, **40**(12): 2814–2830.

Duyvendak, J.W., P. Geschiere and E. Tonkens (eds) (2016), *The Culturalization of Citizenship*, London: Springer.

Ehrkamp, P. and H. Leitner (2003), 'Beyond National Citizenship: Turkish Immigrants and the (Re)-Construction of Citizenship in Germany', *Urban Geography*, **24**(2): 127–146.

Faist, T. (2000), 'Transnationalization in International Migration: Implications for the Study of Citizenship and Culture', *Ethnic and Racial Studies*, **23**(2): 189–222.

Fargues, É. (2017), 'The Revival of Citizenship Deprivation in France and the UK as an Instance of Citizenship Renationalisation', *Citizenship Studies*, **21**(8): 984–998.

Favell, A. (2001), 'Integration Policy and Integration Research in Europe: A Review and Critique', in: *Citizenship Today: Global Perspectives and Practices*, T.A. Aleinikoff and D. Klusmeyer (eds), Washington, DC: Brookings Institute, pp. 349–399.

Favell, A. (2003), 'Integration Nations: The Nation-State and Research on Immigrants in Western Europe', *Comparative Social Research*, **22**: 13–42.

Fernández, C. and K.K. Jensen (2017), 'The Civic Integrationist Turn in Danish and Swedish School Politics', *Comparative Migration Studies*, **5**(1): 5.

Fourutan, N., C. Canan, S. Arnold, B. Schwarze, S. Beigang, and D. Kalkum (2014), 'Deutschland Postmigrantisch I', Berlin: Humboldt-Universität Zu Berlin, Berliner Institut Für Empirische Integrations-Und Migrationsforschung (BIM) Forschungsprojekt.

Gamlen, A. (2008), 'The Emigration State and the Modern Geopolitical Imagination', *Political Geography*, **27**(8): 840–856.

Goodman, S.W. (2010), 'Integration Requirements for Integration's Sake? Identifying, Categorising and Comparing Civic Integration Policies', *Journal of Ethnic and Migration Studies*, **36**(5): 753–772.

Goswami, M. (2002), 'Rethinking the Modular Nation Form: Toward a Sociohistorical Conception of Nationalism', *Comparative Studies in Society and History*, **44**(4): 770–799.

Hagelund, A. and K. Reegard (2011), '"Changing Teams": A Participant Perspective on Citizenship Ceremonies', *Citizenship Studies*, **15**(6–7): 735–748.

Hammar, T. (1985), 'Dual Citizenship and Political Integration', *International Migration Review*, **19**(3): 438–450.

Hepworth, K. (2014), 'Encounters with the Clandestino and the Nomad: The Emplaced and Embodied Constitution of Non-Citizenship', *Citizenship Studies*, **18**(1): 1–14.

Ho, E.L.E. (2009), 'Constituting Citizenship Through the Emotions: Singaporean Transmigrants in London', *Annals of the Association of American Geographers*, **99**(4): 788–804.

Ho, E.L.E., M. Hickey and B.S.A. Yeoh (2015), 'Special Issue Introduction: New Research Directions and Critical Perspectives on Diaspora Strategies', *Geoforum*, **59**: 153–158.

Isin, E.F. (2002), *Being Political: Genealogies of Citizenship*, Minneapolis: University of Minnesota Press.

Isin, E.F. (2007), 'Who Is the New Citizen? Towards a Genealogy', *Citizenship Studies*, **1**(1): 115–132.

Isin, E.F. and G.M. Nielsen (2013), *Acts of Citizenship*, London: Zed Books.

Joppke, C. (2007), 'Transformation of Immigrant Integration – Civic Integration and Antidiscrimination in the Netherlands, France, and Germany', *World Politics*, **59**(2): 243–273.

Joppke, C. (2010), 'The Inevitable Lightening of Citizenship', *Archives Europeennes De Sociologie*, **51**(1): 9–32.

Koopmans, R. and P. Statham (1999), 'Challenging the Liberal Nation-State? Postnationalism, Multi-culturalism, and the Collective Claims Making of Migrants and Ethnic Minorities in Britain and Germany', *American Journal of Sociology*, **105**(3): 652–696.

Kurtz, H. and K. Hankins (2007), 'Guest Editorial: Geographies of Citizenship', *Space and Polity*, **9**(1): 1–8.

Larner, W. (2007), 'Expatriate Experts and Globalising Governmentalities: The New Zealand Diaspora Strategy', *Transactions of the Institute of British Geographers*, **32**(3): 331–345.

Leitner, H. and P. Ehrkamp (2006), 'Transnationalism and Migrants/Imaginings of Citizenship', *Environment and Planning A*, **38**(9): 1615–1632.

Marshall, T.H. (1950), *Citizenship and Social Class*, Cambridge: Cambridge University Press.

Marston, S.A. and K. Mitchell (1999), 'Citizens and the State: Citizenship Formations in Space and Time', in: *Women, Citizenship and Difference (Postcolonial Encounters)*, C. Barnett and M. Low (eds), London: Zed Books, pp. 93–113.

Matejskova, T. (2013), '"But One Needs to Work!": Neoliberal Citizenship, Work-Based Immigrant Integration, and Post-Socialist Subjectivities in Berlin-Marzahn', *Antipode*, **45**(5): 984–1004.

Mezzadra, S. and B. Neilson (2013), *Border as Method, or, the Multiplication of Labor*, Durham, NC: Duke University Press.

Mhurchu, A. Ni. (2014), *Ambiguous Citizenship in an Age of Global Migration*, Edinburgh: Edinburgh University Press.

Mountz, A. (2009), 'Citizenship', in: *Key Concepts in Political Geography*, C. Gallagher, C. Dahlman, M. Gilmartin, A. Mountz and P. Shirlow (eds), London: Sage, pp. 288–298.

Neveu, C. (2015), 'Of Ordinariness and Citizenship Processes', *Citizenship Studies*, **19**(2): 141–154.

Olwig, K.F. and K. Pærregaard (2011), 'Introduction: "Strangers" in the Nation', in: K.F. Olwig and K. Pærregaard (eds), *The Question of Integration*, Newcastle upon Tyne: Cambridge Scholars Publishing, pp. 1–29.

Ong, A. (1999), *Flexible Citizenship: the Cultural Logics of Transnationality*, Durham, NC: Duke University Press.

Pailey, R.N. (2017), 'Between Rootedness and Rootlessness: How Sedentarist and Nomadic Metaphysics Simultaneously Challenge and Reinforce (Dual) Citizenship Claims for Liberia', *Migration Studies*, online ahead of press.

Painter, J. and C. Philo (1995), 'Spaces of Citizenship: an Introduction', *Political Geography*, **14**(2): 107–120.

Palop-García, P. and L. Pedroza (2016), 'Beyond Convergence: Unveiling Variations of External Franchise in Latin America and the Caribbean From 1950 to 2015', *Journal of Ethnic and Migration Studies*, **81**(22): 1–20.

Parsons, L. and S. Lawreniuk (2018), 'Seeing Like the Stateless: Documentation and the Mobilities of Liminal Citizenship in Cambodia', *Political Geography*, **62**(January): 1–11.

Pfaff-Czarnecka, J. (2011), 'From "identity" to "Belonging" in Social Research. Plurality, Social Boundaries, and the Politics of the Self', *Working Papers in Development Sociology and Social Anthropology*, **368**, University Bielefeld.

Phillips, D. (2015), 'Claiming Spaces: British Muslim Negotiations of Urban Citizenship in an Era of New Migration', *Transactions of the Institute of British Geographers*, **40**(1): 62–74.

Purcell, M. (2003), 'Citizenship and the Right to the Global City: Reimagining the Capitalist World Order', *International Journal of Urban and Regional Research*, **27**(3): 564–590.

Rose, N. (2000), 'Community, Citizenship, and the Third Way', *American Behavioral Scientist*, **43**(9): 1395–1411.

Schinkel, W. (2011), 'The Nationalization of Desire: Transnational Marriage in Dutch Culturist Integration Discourse', *Focaal*, **2011**(59): 99–106.

Schinkel, W. and F. van Houdt (2010), 'The Double Helix of Cultural Assimilationism and Neo-Liberalism: Citizenship in Contemporary Governmentality', *The British Journal of Sociology*, **61**(4): 696–714.

Soysal, Y.N. (1994), *Limits of Citizenship: Migrants and Postnational Membership in Europe*, Chicago, IL: University of Chicago Press.

Sparke, M.B. (2006), 'A Neoliberal Nexus: Economy, Security and the Biopolitics of Citizenship on the Border', *Political Geography*, **25**(2): 151–180.

Staeheli, L.A. (1994), 'Empowering Political Struggle: Spaces and Scales of Resistance', *Political Geography*, **13**(5): 387.

Staeheli, L.A. (1999), 'Globalization and the Scales of Citizenship', *Geography Research Forum*, **19**: 60–77.

Staeheli, L.A. (2008) 'Citizenship and the Problem of Community', *Political Geography*, **27**(1): 5–21.

Staeheli, L.A, P. Ehrkamp, H. Leitner and C.R. Nagel (2012), 'Dreaming the Ordinary', *Progress in Human Geography*, **36**(5): 628–644.

Stephens, A.C. and V. Squire (2012), 'Politics Through a Web: Citizenship and Community Unbound', *Environment and Planning D. Society and Space*, **30**(3): 551–567.

Suvarierol, S. and K.K. Kirk (2015), 'Dutch Civic Integration Courses as Neoliberal Citizenship Rituals', *Citizenship Studies*, **19**(3–4): 248–266.

Torpey, J.C. (2000), *The Invention of the Passport: Surveillance, Citizenship and the State*, New York: Cambridge University Press.

Turner, B.S. (2001), 'The Erosion of Citizenship', *British Journal of Sociology*, **52**(2), 189–209.

Turner, B.S. (2016), 'We Are All Denizens Now: on the Erosion of Citizenship', *Citizenship Studies*, **20**(6): 679–692.

UN (2017), 'United Nations, Population Division, Department of Economic and Social Affairs (UN DESA), *International migrant stock: the 2017 revision*, available from http://www.un.org/en/development/desa/population/migration/data/estimates2/estimates17.shtml

van Houdt, F., S. Suvarierol and W. Schinkel (2011), 'Neoliberal Communitarian Citizenship: Current Trends Towards "Earned Citizenship" in the United Kingdom, France and the Netherlands', *International Sociology*, **26**(3): 408–432.

Varsanyi, M.W. (2006), 'Interrogating "Urban Citizenship" Vis-à-Vis Undocumented Migration', *Citizenship Studies*, **10**(2): 229–249.

Walker, K.E. and H. Leitner (2011), 'The Variegated Landscape of Local Immigration Policies in the United States', *Urban Geography*, **32**(2): 156–178.

Wiener, A. (1997), 'Making Sense of the New Geography of Citizenship', *Theory and Society*, **26**(4): 529–560.

12. Migration, families and households in globalizing Asia

Brenda S.A. Yeoh, Shirlena Huang and Theodora Lam

INTRODUCTION

Particularly since the 1970s, the increasingly pervasive forces of globalization expressed in 'increased levels of education, proliferation of international media, improved transport systems and the internationalization of business and labour markets' (Hugo, 2005: 94), have moved in tandem with the amplified scale, scope and complexity of international migration. Geographical scholarship has been particularly productive in examining the intertwined relationship between the economic, cultural, social and spatial processes constitutive of globalization on the one hand, and people's increasing engagement in a wide range of differentiated mobilities within and beyond the nation-state – including permanent, temporary, circular and return migrations – on the other hand (Bailey and Yeoh, 2014). It has become amply clear that global processes have not only led to a multiplication of cross-border transactions in the economic sphere, but also shaped the social sphere, including human relationships of intimacy. In this light, rethinking globalization as an integral rather than separate process from the social order is an important task, as 'the impact of globalization on gender ideologies, work–family relationships, conceptualizations of children, youth, and the elderly have been virtually absent in mainstream approaches' (Trask, 2010: 182).

With the quickening pace and intensity of people on the move, social worlds in an increasingly interconnected globe experience considerable flux, and in the process, new opportunities for the 'melting, blending and re-solidifying' of old and new subjectivities are produced (Jarvis et al., 2009: 111). Such processes may be highly contradictory, and can be 'potentially both emancipating (enabling) and confining (disabling and elitist)' (Jarvis et al., 2009: 111) in its effects on social institutions and social relations. An important arena where these paradoxical processes are particularly salient is the sphere of the family which, in whatever configuration, 'remains a strategic arrangement that meets certain social and economic needs' and where decisions about work, care, movement, and identity are negotiated, contested, and resolved (Trask, 2010: 55). Families and households absorb, resist, process and act on opportunities or threats caused by major structural changes, thus acting as the crucial link between macro and micro factors of socio-economic and political change such as those implicated in the processes of transnational migration. In other words, 'migration' and the 'family' in the age of globalization have mutually constitutive effects because migration affects the family in ways that can be destabilizing or affirming; at the same time, family norms, relations and dynamics also exert manifold influences on specific migration decisions and outcomes (Asis, 2000; Yeoh et al., 2002).

This chapter begins by sketching the development of conceptual tools that have been helpful in understanding the links between globalization, transnational migration, families and households in the context of Asia. To illustrate the significance of these relationships, the chapter develops two themes that have particular traction in Asian societies: first, how families and households in sending communities are reproducing themselves in the light of a feminization of migration in Southeast Asia and accompanying gender politics; and second, how families and households are being transformed by the intergenerational provisioning of care in view of rapid ageing in many East Asian countries. A brief conclusion points to the need for putting families and households on the research agenda in globalizing Asia.

CONCEPTUALIZING FAMILIES AND HOUSEHOLDS IN A TIME OF ASIAN MIGRATIONS

In the context of Asia, the rising tide in studies of transnational migration – involving the mobility flows of talent, labour, students, marriage partners, caregivers, care recipients, retirees, lifestyle-seekers and more – has contributed significantly to the opening up of the Asian 'family' as a material and ideological construct to academic scrutiny. This is partly motivated by the increasing levels of intra-Asian migrations, given the rise of new global migration magnets in 'the Persian Gulf regions (particularly since the 1970s), the "Asian Tiger" economies (principally Singapore and Korea) as well as Japan' (Czaika and de Haas, 2014: 294). As the prevailing labour migration regimes in Asia are underpinned by rotating-door principles of enforced transience, the overwhelming majority of migrants – particularly those seeking low-skilled, low-waged work – are admitted into host nation-states on the basis of short-term, time-bound contracts, with little or no possibility of family reunification or permanent settlement at destination. The quickening of migration flows in the region is hence governed by temporary migration schemes, which in turn require a sustained, if not permanent, transborder household division of labour for family/household formation and maintenance.

In the process, the increasing attention to temporary migration and its implications for 'doing family' has paved the way for a more critical understanding of how social reproduction is underpinned by gender identities and relations. By giving attention to power geometries and household politics, the burgeoning body of work helps to challenge the assumption that household decisions regarding migration are guided by 'principles of consensus and altruism' to take on the notion that they may 'equally be informed by hierarchies of power along gender and generational lines' (Mahler and Pessar, 2006: 33). This is a significant step forward given the taken-for-granted nature of the Asian family as 'the primary unit of regulation and the vehicle of state power' (Ong, 1999: 71). In many Asian countries, state constructions of the 'family' are often based on a 'nostalgic vision of femininity' where decision-making is expected to be hierarchical (read 'patriarchal'), and where individual desires are usurped by the 'greater good' of the family (Stivens, 1998: 17).

Accompanying the increasing attention on the 'family' and 'household' as sites and scales of analysis in the literature on globalization and migration, several conceptual approaches have gained traction since the turn of the new millennium. We discuss two complementary approaches – transnational families and global householding – that have particular resonance for studying families and households in globalizing Asia.

Transnational Families

Focusing on the continued significance of the 'family' in the face of distance, dispersal and translocality, Bryceson and Vuorela's (2002) pioneering work on transnationally split families in the European context, proposes that family members no longer depend on physical reunification in order to experience the intimacy of familial relations. In their words, family members 'that live some or most of the time separated from each other' continue to maintain 'a feeling of collective welfare and unity, namely "family-hood", even across national borders' (Bryceson and Vuorela, 2002: 3). Transnational familyhood is sustained through two complementary strategies: 'frontiering' or 'the ways and means transnational family members use to create familial space and network ties in terrain where affinal connections are relatively sparse'; and 'relativizing', referring to 'the ways individuals establish, maintain or curtail *relational* ties with specific family members' (Bryceson and Vuorela, 2002: 11–14). In other words, the sense of belonging to a family unit in a transnational migratory context can be a matter of choice and negotiation; members choose to maintain emotional and material attachments of varying degrees of intensity with certain kin while opting out of transnational relationships with others.

As reviewed elsewhere (see Yeoh et al., 2018), at least three interrelated strands can be discerned in the scholarship on Asian transnational families. First, transnational families draw on ideologically laden imaginaries to give coherence to notions of belonging despite the physical dispersal of their members. For example, despite the feminization of labour migration in Southeast Asia where women in less developed economies are fashioning themselves as international breadwinning migrants respond-ing to the growing gender-segmented demand for domestic and care workers in more developed economies, normative gender ideologies are often at work to simultaneously exalt them as heroes of foreign exchange while also casting them as in need of protection to preserve their sexual and moral purity for the sake of their families (Silvey, 2006). And while notions of good long-distance mothering have been reconsti-tuted to incorporate breadwinning, it also continues to retain expectations that migrant mothers demonstrate a strong sense of maternal responsibility (Graham et al., 2012). Second, transnational families are realized through lived experiences, where varying degrees of intimacy are negotiated across transnational spaces in the context of new communication technologies. The changing technologies, economic costs, and emo-tional pains and gains of 'staying connected' in order to 'do family' and perform care work across national borders are central concerns to understanding the inner workings of the transnational family (Asis et al., 2004). At the same time, scholars such as Madianou (2016: 185) caution us not to 'romanticize the role of communication technologies for "doing family" because, as with non-mediated practices, acts of mediated communication can have complex consequences, both positive and negative, depending on a number of factors, including the relationships themselves'. Third, families often assume transnational morphologies with the strategic intent of generating remittances for economic survival, or accumulating social and economic capital so as to maximize social mobility for the family. For low-income families, economic remittances generated by having one or more family members embark on labour migration are critical to the well-being of the family and 'are often at the centre of socio-economic mobility strategies' (Carling et al., 2012: 202). Among middle-class families in East Asia, transnationalizing the family by educating children abroad provides an important route to garner international social mobility and prestige (Waters, 2005; Baldassar and Wilding, 2014; Huang and Yeoh, 2011).

Global Householding

Another conceptual pathway to understanding family dynamics in the age of globalization and transnational migration was developed by Douglass (2007) in the context of East Asia. Concerned with the links between migration, household reproduction and the reproduction of societies with rapid pace and penetrating depth of globalization, the concept of 'global householding' describes how day-to-day 'householding' is predicated on the transnational migration of people and their transactions across borders in the context of flexibilized household life cycles. By focusing on the different elements including 'marriage, having and raising children, maintaining the household materially and socially, and caring for elders' required for household formation and sustenance over generations, 'the household as a social institution … is fully animated … through the interactions' and outcomes of each element with the others (Douglass, 2014: 313). In this formulation, the household should not be treated either as 'a harmonious and cooperative black box' or reduced to 'a disguised labor process and site of exploitation', but instead carefully unpacked as 'a complex social institution that includes affection and altruism as well as material dimensions' (Douglass, 2014: 313).

This is of increasing relevance in many parts of Asia as institutional support for household reproduction has either been minimal or considerably cut back as a result of economic restructuring in the region. In the region's more developed economies, neoliberal strategies have resulted in the privatization and commercialization of care work, and as a result women are expected to subsidize the economy with their caring work to sustain households. Where female citizens are unable to fill care deficits in middle-class households, these gaps have been plugged primarily by migrant women from less developed economies in the region. In turn, this has triggered the formation of gendered circuits of labour migration linked to the globalization of care work, or what Hochschild (2000: 132) calls 'global care chains', referring to 'a series of personal links between people across the globe based on the paid or unpaid work of caring'. In the process, families in less developed economies are compelled to assume transnational formations in order to send a female member – a wife, mother or daughter – abroad as the primary breadwinner, in response to highly gender-segmented markets for care and domestic work. While these global care chains transfer social capital from the poorer to the richer countries and receive economic capital in return, an important observation is that the economic value of the labour dwindles as it moves down the care chain (Parreñas, 2012), thus resulting in diminishing returns for each subsequent woman. Parreñas' (2012: 269) notion of the 'international division of reproductive labour' reminds us that by moving down the care chain as opposed to across the gender divide, a widening system of the gender substitution of care labour is put in place, based on exploitative practices of extracting cheap labour from migrant women. These migrant women who leave their families are increasingly incorporated as 'the global world's newest proletariat, into the global capitalist activities of the North, all under severely diminished citizenship regimes' (Dobrowolsky and Tastsoglou, 2006: 3).

CHANGING FAMILIES/HOUSEHOLDS AND THE POLITICS OF GENDER IN SOURCE COMMUNITIES

With the increasing incorporation of the developing economies of Southeast Asia into the global context in recent decades, transnational labour migration in the region is experiencing a so-called 'feminization'. Gender-segmented overseas markets for care and domestic work create a strong demand for female migrant labour, facilitated by the development of commercial migration channels under the control of migration brokers and other intermediaries (Khoo and Yeoh, 2017). This shift in favour of women migrants is an increasingly significant driver of contemporary social transformation of the family/household in sending communities, as seen in its impact on changing arrangements and relationships of care for large numbers of children. Often known as 'left-behind' children, many spent a large proportion of their childhood in the absence of a migrant father, mother, or both, and under the care of a 'single' parent or other surrogate caregivers.

In view of these developments, recent studies on migrating mothers and their left-behind children called attention to the gender politics implicit in a range of mutually constitutive interactions between migrants and the left behind. In areas such as rural Indonesia and Vietnam where feminized out-migration streams predominate, established gender ideologies may either be challenged by changing social practices where women assume breadwinner roles, or continue to regulate traditionally scripted roles for men and women but in new ways (Elmhirst, 2000; Resurreccion and van Khanh, 2007).

A major strand in current research highlights the enactment of family interaction and negotiation of family intimacy through distanced communication as the family takes on transnational dimensions (Parreñas, 2001; Asis et al., 2004). Studies focusing on migrant mothers have emphasized the resilience of gender ideals surrounding mother-hood even under the transmigration context. While mothering at a distance reconstitutes 'good mothering' to incorporate breadwinning, it also continues maternal responsibility of nurturing by employing regular (tele)communications to transmit transnational circuits of care and affection to their families and children left in source countries. Scholars observed that most migrant mothers actively worked to ensure a sense of connection across transnational spaces with their children through modern communication technologies (Asis, 2002; Graham et al., 2012). Sobritchea (2007: 183) characterized 'long-distant mothering' as an intensive form of emotional labour that involves activities of 'multiple burden and sacrifice', spending 'quality time' during brief home visits, and reaffirming the 'other influence and presence' through surrogate figures and regular communication with children. Migrant mothers often make consid-erable efforts to ensure that care work and responsibilities are transferred to other family members in their absence, although the available evidence seems to suggest that the 'intangibles' of the mothering identity are less yielding and not so easily reassigned. For example, Asis et al. (2004: 208) show that migrating mothers who depend on long-distance mothering tend to leave the children 'in an indeterminate state of being "neither here nor there" – of having a mother, yet not being able to enjoy her daily involvement in their lives'. At the same time, it is important to guard against undue focus on 'a discourse of maternal loss and absence' as such a discourse 'not only works through conventional, potentially conservative notions of the family, but it can and does quickly turn to blame' (Pratt, 2012: 79). As Parreñas (2005) notes, migrant mothers are often stigmatized for leaving their children behind, while fathers working overseas are not. Long-distance mothering, often performed for the sake of advancing the children's education opportunities as part of what 'sacrificial breadwinner mothers' do, thus has to contend with highly entrenched views of what constitutes a sanctioned gender division of labour in the household. Gender ideals, particularly those concerning motherhood, continue to remain resistant to change, even under conditions of long-term physical separation resulting from the region's temporary migration regime.

Current research also suggests that the care vacuum resulting from the absence of migrant mothers is often filled by female relatives such as grandmothers and aunts (Gamburd, 2000; Parreñas, 2005; ECMI/AOS-Manila et al., 2004; Save the Children, 2006). The continued pressure to conform to gender norms with respect to caring and nurturing practices explains men's resistance to, and sometimes complete abdication of, parenting responsibilities involving physical care in their wives' absence. These studies conclude that the 'delegation of the mother's nurturing and caring tasks to other women family members, and not the father, upholds normative gender behaviors in the domestic sphere and thereby keep the conventional gendered division of labor intact' (Hoang and Yeoh, 2011: 722). At the same time, there is also emerging evidence that gender practices of care in source communities are becoming more flexible. In place of the image of the delinquent left-behind man who is resistant to adjusting his family duties in the woman's absence, some Southeast Asian men strive to live up to masculine ideals of being both 'good fathers' and 'independent breadwinners' when their wives are working abroad. These men repackage fatherhood to include at least some care functions that signified parental love and authority while holding on to paid work (even if monetary returns are low) for a semblance of economic autonomy (Lam and Yeoh, 2018). For example, given the longstanding migration of Filipino women to become global care workers over the last three decades (Asis, 2005; Asis et al., 2004), role reversal of the traditional gender ideology that men are the 'pillars' while women are the 'lights' of the home is increasingly common in Filipino society. Pingol's (2001) pioneering study on Filipino migrant wives and househusbands found that male respondents project themselves as important providers of care, even if they perform care differently from their wives. Her study reveals that while Filipino men may 'experience sudden shifting of gears that heavily disorients them', the maintenance of their productive self keeps them going (Pingol, 2001: 220). Filipino men shared that their ability to perform the additional caring tasks well 'enhances their sense of self worth ... gives them pride not only as they view themselves but also as they are looked upon by the community' (Pingol, 2001: 221).

In the current phase of neoliberal capitalist globalization, the emerging scholarship on source communities in Southeast Asia has highlighted the plural and dynamic nature of gender subjectivities, as well as the intersectional character of gender identities and relations (Malam, 2008; Yea, 2013). On the one hand, mothering identities appear to be particularly 'sticky' despite the increasing normalization of migrant women who carry primary breadwinning roles. On the other hand, hegemonic masculinities, constructed on the basis of state-sanctioned hetero-patriarchal narratives of economic prowess, entrepreneurial mobility and macho-ness, are increasingly challenged by the rising educational attainment among Southeast Asian women and the feminization of Southeast Asian global labour. Transnational families are in this sense in continual flux as they become a key arena for the interplay of gender politics and normative gender scripts in the region.

CHANGING FAMILIES/HOUSEHOLDS AND THE PROVISION OF INTERGENERATIONAL CARE

In Asia, the notion of 'family' and familial obligations often encompasses extended family members, and usually includes the provision of both material and emotional care to and from different members of the family in varying permutations. For instance, an adult may be caring for one's parents and/or own children while the older generation may also be caring for their adult children and/or grandchildren. In the more developed economies in the region, and particularly in the major cities such as Hong Kong, Seoul, Singapore and Taipei, plummeting fertility rates to below replacement levels, increasing life expectancy and rapid ageing have amplified pressure on adult children to care for their aged parents for longer periods of time. To remain responsive to global changes, both working-class and middle-class families have to juggle between the increasing demands of paid work (including internationalized careers in some cases) and new household exigencies arising from delayed marriages and familial re-structuring. This twin problem of a heightened demand for eldercare, and the increasing challenges faced by adult children to meet that need is further exacerbated by the fact that modern families are now more dependent on dual incomes to sustain their household expenditures. With women's growing participation in the workforce in globalizing cities, families experience difficulty shouldering the triple burden of career building, child-raising, and eldercare, not least because the household division of labour by gender remains relatively rigidly drawn in many households.

In this context, households faced with a growing care deficit have turned to external sources of care labour to plug the gaps. Gender based strategies of care substitution have developed, drawing on the paid and unpaid labour of different groups of transnational subjects, mainly women from less developed parts of Asia. In particular, these global householding strategies are dependent on the care labour of two groups of transnational female subjects – transnational domestic workers and transnational marriage migrants working in the privatized sphere of the home. The gender and class contours of these transnational flows provide a useful lens to observe the more intimate dimensions of globalization, migration and social change.

Whether recruited as a live-in domestic worker or a commercially matched foreign wife whose 'lot' is to take care of domestic affairs, migrant women's diverse pathways to transnationality represent a continuum of solutions to plug the household reproductive gap under different, class-specific circumstances. As Lan (2008: 834, 2003) has argued, 'unpaid household labour and paid domestic work are not dichotomous categories but constitute' different facets of the same structural means of absorbing migrant female reproductive labour – either as waged work to fill a 'care deficit' in the home or the public sphere, or as kin labour, to fill a 'bride deficit' in the private arena. The practice of middle-class urban families in recruiting foreign domestic workers for householding purposes (e.g. eldercare) is thus not unrelated to, but somewhat analogous with, the way their working-class counterparts procure reproductive labour (for childbearing, childcare, eldercare and domestic work) by enrolling 'foreign brides' as family members, sometimes through the brokerage of commercial agencies. The two sets of householding strategies are arguably most distinguishable from each other along class lines. At one end of the continuum of care strategies, employing a foreign domestic worker as care substitute is only affordable if the citizen-woman that she is substituting for is earning a wage that puts her in at least the middle-income bracket. At the other end of the continuum, marrying a 'foreign bride' hailing from poorer nations to fulfil the social reproduction needs of the family is a more viable option for working-class citizen-men who are unable to attract better-educated citizen-women who are not amenable to 'marrying down' (Yeoh and Huang, 2014). Both means of global householding are consistent with the mode of social reproduction preferred in neoliberal capitalist societies where the cost of care is internalized within the household. As Douglass (2014) observed, 'while the household is not a mini-market economy or arm of the state, it must nonetheless reflexively interact with and adjust to market forces and state power'.

Beyond enrolling migrant women as 'domestic workers' and 'foreign wives' to fill care deficits in the family/household, care-driven migration and intergenerational relationships also intersect in other ways, as can be seen from the global householding strategies that more privileged families are able to afford. In Asian societies, cultural obligations often mean that adult family members who have migrated are expected to provide care from a distance for elderly parents left behind in the country of origin. In fact, elderly parents may construct such care when not given as a lack of attentiveness to their emotional care needs. Hence, even when actual physical care is not possible because of the physical separation, long-distance care is conveyed through care advice, the maintenance of regular communication about physical and emotional health issues, and financial remittances, even if not needed given the importance of such acts as symbolic representations of the emotional work of caring (Lam et al., 2002). In her study of skilled migrants from Sri Lanka residing in Australia, de Silva (2018) found that while the receiving of long-distance emotional care from their migrant children (e.g. through frequent communication, the sharing of everyday events) was very important to the ageing parents, this is more challenging for sons than daughters, with the former tending to demonstrate care through financial remittances and failing to understand the significance of emotional transnationalism. In yet another global householding permutation facilitated by the emergence of a 'retirement industry' as a national development policy in Malaysia, the Philippines and Indonesia, later-life migrants with financial wherewithal – usually citizens of developed countries within and beyond Asia including Japanese and western pensioners – are targeted as retirement migrants. Drawn by the lower cost of living, as well as the easy availability and affordability of live-in domestic workers in these countries, 'average pensioners in one country' are transformed into 'high-power[ed] consumers in another' through retirement migration (Toyota and Xiang, 2012: 716). This form of global householding allows families to take advantage of the uneven development and vast differentials in healthcare costs among countries in the region.

It should also be remembered that from the perspective of global householding, later-life migrants, especially in their retirement years, may also act as care providers, either to their grandchildren, or their even older parents (Toyota and Xiang, 2012: 711; Thang et al., 2011). Beyond providing physical childcare, transnational grandparenting (usually provided by grandmothers) plays a crucial role in the social reproduction of affluent migrant families in terms of language maintenance, religious development, cultural identification and so on of the next generation – both when grandparents fly (Lie, 2010) to take care of their grandchildren, or when young children from professional families are sent back to the home country to be taken care of by their grandparents (Da, 2003; Mujahid et al., 2011). While the phenomenon has not been given much attention in the literature, ethnic Chinese grandmothers are among the most documented global grandparents; their transnational grandmothering is tied not only to Chinese gender ideology, but also to the Confucian belief in the reciprocity of intergenerational care (Mujahid et al., 2011: 193–194; see also Da, 2003; Lie, 2010; Lee et al., 2015). The engagement of the elderly in transnational caregiving challenges first, the tendency in the literature 'to conceptualise the mobilities of care as one-way traffic' (Baldassar and Merla, 2014: 26), *à la* the notion of the care chain; and second, the view of the elderly as merely recipients of care. Instead, we see how transnational familial care is multidirectional, multigenerational and evolves over time as the care needs within households change (Baldassar and Wilding, 2014; Baldassar et al., 2007).

CONCLUSION

Across a diverse Asian landscape, what this brief survey of the impact of 'globalization' on the 'family-household' illustrates is the centrality of families and households as social formations to understanding geographies of globalization. At least two lines of inquiry deserve further attention.

First, we reiterate the need for the by now abundant scholarship on economic and cultural globalization to give greater weight to the dynamics of the household scale as a route into understanding the inner workings of 'intimate globalization' (Sunanta and Angeles, 2013). Studies of globalization thus far has privileged the level of the 'nation' or 'society' in general, or focus on organizations, institutions and to a lesser extent, individuals. As one of the most important material and ideological bases for the conduct of personal relations, the reproduction of values and the socialization of new generations, how the family-household is responding in the wake of globalization forces is deserving of more attention than it has received. This calls for a view of globalization not so much as an inevitable external force occurring hierarchically but as materially and discursively constructed out of complex interactions and power struggles between diverse social, political and economic actors occurring simultaneously across various geographical scales. Rather than privileging the global as a given, it is more important to understand globalization as 'rooted in material processes of qualitative change in the global system whereby social relations across space [and scales] are being integrated in more intensive and extensive ways' (Kelly, 1999: 395).

Second, and perhaps parallel to the work which has explored the possible end of the nation-state in the face of globalization, research should continue to ask whether the 'family-household', as a concept as well as a material reality, is being subverted or strengthened by globalization forces, and if so in what ways. Transnational families that are being stretched across boundary lines and geographically split up in different ways, are becoming more prevalent, as are households which are made more porous to members of different ethnicities and nationalities. Attention to the material and ideologically laden social relations (of gender and generation especially) that constitute and dissolve families and households in an era of quickened transnational migration and how they are being reconfigured pose significant research puzzles, not just for academic discourses but also state policies and legal regulation of who belongs to the family/household (Asis, 2000; Brickell and Yeoh, 2014). In bringing to the foreground different subjects and cross-border flows related to the family/household, new areas of discussion on the multiple ways in which the 'intimate' meets the 'global' unfold.

ACKNOWLEDGEMENTS

This work was supported by the Singapore Ministry of Education Academic Research Fund Tier 2 (Grant number MOE2015-T2-1-008, PI: Brenda Yeoh); and the Social Science and Humanities Research Council of Canada Partnership Grant (File No: 895–2012–1021, PI: Ito Peng).

REFERENCES

Asis, M.M.B. (2000), 'Imagining the future of migration and families in Asia', *Asian and Pacific Migration Journal*, **9**, 255–272.

Asis, M.M.B. (2002), 'From the life stories of Filipino women: personal and family agendas in migration', *Asian and Pacific Migration Journal*, **11**, 67–94.

Asis, M.M.B. (2005), 'Caring for the world: Filipino domestic workers gone global', in: Yeoh, B.S.A., Huang, S. and Rahman, N.A. (eds), *Asian Women as Transnational Domestic Workers,* Singapore: Marshall Cavendish, 21–53.

Asis, M.M.B., S. Huang and B.S.A. Yeoh (2004), 'When the light of the home is abroad: unskilled female migration and the Filipino family', *Singapore Journal of Tropical Geography*, **25**, 198–215.

Bailey, A.J. and B.S.A. Yeoh (2014), 'Migration, society and globalisation: introduction to Virtual Issue', *Transactions of the Institute of British Geographers*, **39**, 470–475.

Baldassar, L. and L. Merla (2014), 'Locating transnational care circulation in migration and family studies', in: Baldassar, L. and Merla, L. (eds), *Transnational Families, Migration and the Circulation of Care: Understanding Mobility and Absence in Family Life*, New York and London: Routledge, 25–58.

Baldassar, L. and R. Wilding (2014), 'Middle-class transnational caregiving: the circulation of care between family and extended kin networks in the global north', in: Baldassar, L. and Merla, L. (eds),

Transnational Families, Migration and the Circulation of Care: Understanding Mobility and Absence in Family Life, New York: Routledge, 253–251.

Baldassar, L., C. Baldock and R. Wilding (2007), *Families Caring Across Borders: Migration, Aging and Transnational Caregiving*, London: Palgrave Macmillan.

Brickell, K. and B.S.A. Yeoh (2014), 'Editorial: geographies of domestic life: "householding" in transition in East and Southeast Asia', *Geoforum*, **51**, 259–261.

Bryceson, D. and U. Vuorela (2002), 'Transnational families in the twenty-first century', in: Bryceson, D. and Vuorela, U. (eds), *The Transnational Family. New European Frontiers and Global Networks*, Oxford: Berg, 3–30.

Carling, J., C. Menjívar and L. Schmalzbauer (2012), 'Central themes in the study of transnational parenthood', *Journal of Ethnic and Migration Studies*, **38**, 191–217.

Czaika, M. and H. de Haas (2014), 'The globalization of migration: Has the world become more migratory?', *International Migration Review*, **48**, 283–323.

Da, W.W. (2003), 'Transnational grandparenting: child care arrangements among migrants from the People's Republic of China to Australia', *Journal of International Migration and Integration*, **4**, 79–103.

de Silva, M. (2018), 'Making the emotional connection: transnational eldercare circulation within Sri Lankan–Australian transnational families', *Gender Place and Culture*, **25**, 88–103.

Dobrowolsky, A. and E. Tastsoglou (2006), 'Crossing boundaries and making connections', in: Tastsoglou, E. and Dobrowolsky, A. (eds), *Women, Migration and Citizenship: Making Local, National and Transnational Connections*, Aldershot: Ashgate Publishing, 1–36.

Douglass, M. (2007), 'The globalization of householding and the social reproduction in Pacific Asia', *Philippine Studies*, **55**, 157–182.

Douglass, M. (2014), 'Afterword: global householding and social reproduction in Asia', *Geoforum*, **51**, 313–316.

ECMI/AOS-Manila (Episcopal Commission for the Pastoral Care of Migrants and Itinerant People/ Apostleship of the Sea-Manila), SMC (Scalabrini Migration Center) & OWWA (Overseas Workers Welfare Administration) (2004), *Hearts Apart: Migration in the Eyes of Filipino Children*. Quezon City: Scalabrini Migration Center.

Elmhirst, R. (2000), 'A Javanese diaspora? Gender and identity politics in Indonesia's transmigration resettlement program', *Women's Studies International Forum*, **23**, 487–500.

Gamburd, M. (2000), *The Kitchen Spoon's Handle: Transnationalism and Sri Lanka's Migrant Housemaids*, Ithaca and London: Cornell University Press.

Graham, E., L.P. Jordan, B.S.A. Yeoh, T. Lam, M.M.B. Asis and Su-kamdi (2012), 'Transnational families and the family nexus: perspectives of Indonesian and Filipino children left behind by migrant parent(s)', *Environment and Planning A*, **44**, 793–815.

Hoang, L.A. and B.S.A. Yeoh (2011), 'Breadwinning wives and "left-behind" husbands: men and masculinities in the Vietnamese transnational family', *Gender and Society*, **25**, 717–739.

Hochschild, A.R. (2000), 'Global care chains and emotional surplus value', in: Hutton, W. and Giddens, A. (eds), *On the Edge: Living with Global Capitalism*, London: Jonathan Cape, 130–146.

Huang, S. and B.S.A. Yeoh (2011), 'Navigating the terrains of transnational education: children of Chinese "study mothers" in Singapore', *Geoforum*, **42**, 394–403.

Hugo, G. (2005), 'The new international migration in Asia', *Asian Population Studies*, **1**, 93–120.

Jarvis, H., J. Cloke and P. Kantor (2009), *Cities and Gender*, Abingdon: Routledge.

Kelly, P. (1999), 'The geographies and politics of globalization', *Progress in Human Geography*, **23**, 379–400.

Khoo, C.Y. and B.S.A. Yeoh (2017), 'Responsible adults-in-the-making: intergenerational impact of parental migration on Indonesian young women's aspirational capacity', *Geoforum*, **85**, 280–289.

Lam, T. and B.S.A. Yeoh (2018), 'Migrant mothers, left-behind fathers: the negotiation of gender subjectivities in Indonesia and the Philippines', *Gender, Place & Culture*, **25**, 104–117.

Lam, T., B.S.A. Yeoh and L. Law (2002), 'Sustaining families transnationally: Chinese-Malaysians in Singapore', *Asian and Pacific Migration Journal*, **11**, 117–144.

Lan, P.C. (2003), 'Maid or madam? Filipina migrant workers and the continuity of domestic labor', *Gender & Society*, **17**, 187–208.

Lan, P.C. (2008), 'Migrant women's bodies as boundary markers: reproductive crisis and sexual control in the ethnic frontiers of Taiwan', *Signs: Journal of Women in Culture and Society*, **33**, 833–861.

Lee, Y.S., A. Chaudhuri and G. Yoo (2015), 'Caring from Afar: Asian H1B migrant workers and aging parents', *Journal of Cross-Cultural Gerontology*, **30**, 319–331.

Lie, M.L.S. (2010), 'Across the oceans: childcare and grandparenting in UK Chinese and Bangladeshi households', *Journal of Ethnic and Migration Studies*, **36**, 1425–1443.

Madianou, M. (2016), 'Ambient co-presence: transnational family practices in polymedia environments', *Global Networks*, **16**, 183–201.

Mahler, S.J. and P.R. Pessar (2006), 'Gender matters: ethnographers bring gender from the periphery toward the core of migration studies', *International Migration Review*, **40**, 27–63.

Malam, L. (2008), 'Bodies, beaches and bars: negotiating heterosexual masculinity in Southern Thailand's tourism industry', *Gender, Place & Culture*, **15**, 581–594.

Mujahid, G., A.H. Kim and G.C. Man (2011), 'Transnational intergenerational support: implications of aging in mainland China for the Chinese in Canada', in: Cao, H. and Poy, V. (eds), *The China Challenge: Sino–Canadian Relations in the 21st Century*, Ottawa: University of Ottawa Press, 183–204.

Ong, A. (1999), *Flexible Citizenship: The Cultural Logics of Transnationality*, Durham, NC: Duke University Press.

Parreñas, R.S. (2001), 'Mothering from a distance: emotions, gender, and intergenerational relations in Filipino transnational families', *Feminist Studies*, **27**, 361–390.

Parreñas, R.S. (2005), *Children of Global Migration: Transnational Families and Gendered Woes*, Stanford: Stanford University Press.

Parreñas, R.S. (2012), 'The reproductive labour of migrant workers', *Global Networks*, **12**, 269–275.

Pingol, A. (2001), *Remaking Masculinities: Identity, Power, and Gender Dynamics in Families with Migrant Wives and Househusbands*, Philippines: UP Center for Women's Studies.

Pratt, G. (2012), *Families Apart: Migrant Mothers and the Conflicts of Labor and Love*, Minneapolis and London: University of Minnesota Press.

Resurreccion, B. and H.T. van Khanh (2007), 'Able to come and go: reproducing gender in female rural–urban migration in the Red River Delta', *Population, Space and Place*, **13**, 211–224.

Save the Children (2006), *Left Behind, Left Out: The Impact on Children and Families of Mothers Migrating for Work Abroad*, Sri Lanka: Save the Children.

Silvey, R. (2006), Consuming the transnational family: Indonesian migrant domestic workers to Saudi Arabia. *Global Networks*, **6**, 23–40.

Sobritchea, C. (2007), 'Constructions of mothering: female Filipino overseas workers', in: Devasahayam, T. and Yeoh, B.S.A. (eds), *Working and Mothering in Asia: Images, Ideologies and Identities*, Singapore: NUS Press, 195–220.

Stivens, M. (1998), 'Theorising gender, power and modernity in affluent Asia', in: Sen, K. and Stivens, M. (eds), *Gender and Power in Affluent Asia*, London: Routledge, 1–34.

Sunanta, S. and L.C. Angeles (2013), 'From rural life to transnational wife: agrarian transition, gender mobility, and intimate globalization in transnational marriages in northeast Thailand', *Gender, Place & Culture*, **20**, 699–717.

Thang, L.L., K. Mehta, T. Usui and M. Tsuruwaka (2011), 'Being a good grandparent: roles and expectations in intergenerational relationships in Japan and Singapore', *Marriage and Family Review*, **47**, 548–570.

Toyota, M. and B. Xiang (2012), 'The emerging transnational "retirement industry" in Southeast Asia', *International Journal of Sociology and Social Policy*, **32**, 708–719.

Trask, B. (2010), *Globalization and Families: Accelerated Systemic Social Change*, New York: Springer.

Waters, J.L. (2005), 'Transnational family strategies and education in the contemporary Chinese diaspora', *Global Networks*, **5**, 359–377.

Yea, S. (2013), 'Masculinity under the Knife: Filipino men, trafficking and the black organ market in Manila, the Philippines', *Gender, Place & Culture*, **22**, 123–142.

Yeoh, B.S.A., E. Graham and P. Boyle (2002), 'Migrations and family relations in the Asia Pacific region', *Asian and Pacific Migration Journal*, **11**, 1–11.

Yeoh, B.S.A. and S. Huang (2014), 'Singapore's changing demography, the eldercare predicament and transnational "care" migration', *TRaNS: Trans-Regional and -National Studies of Southeast Asia*, **2**, 247–269.

Yeoh, B.S.A., S. Huang and T. Lam (2018), 'Transnational family dynamics in Asia', in: Triandafyllidou, A. (ed.), *Handbook on Migration and Globalisation*, Cheltenham, UK and Northampton, MA, USA: Edward Elgar Publishing, 413–430.

13. Labour geographies in a globalizing world[1]
Dennis Arnold

INTRODUCTION

Over the past several decades, the growing power and reach of global capital has exceeded the ability of nations and labour movements to regulate it, exacerbating inequality, accumulation of debt, precarious work, and broader social precarity. Numerous labour trends have been associated with neoliberal globalization, including a decline in attachment to employers, an increase in long-term unemployment, growth in perceived and real job insecurity, increasing nonstandard and contingent work, risk shifting from employers to employees, a lack of workplace safety, and an increase in work-based stress and harassment. The lack of public and private investment in skills and development is accompanied by a lack of access to schooling, where women and ethnic and racial minorities disproportionally bear the brunt of these disadvantages, particularly those in or from the South. Yet precarity is about not only the disappearance of stable jobs and other workplace-specific issues, but also the questions of housing, debt, welfare provision, and the availability of time for building effective personal relations.

These trends are not entirely new, and follow previous patterns. Capital is continually in search of spatial, technological, and product fixes, but with each phase of innovation the intensity of capital investment and productivity requirements increases (Harvey, 2003; Silver, 2003). Not only have global production networks stretched across national boundaries to cover greater geographic scope and new 'frontiers', but also lead times have become shorter to respond to oscillations in consumer demand. As the geographies of production continue to expand, the process is often reproduced with tighter margins for lower tier producers and employers, which tend to provide lower remuneration for workers. Greater flexibility is the mantra of the day, from labour-intensive garment factories, to retail and service sectors, to knowledge work in the academe. In short, precarious labour regimes, and the central role states have played in configuring and reproducing them, have reconfigured the geography of the reserve army of labour.

In this chapter, I will unpack these and other trends that highlight increasing disparities in power and wealth in the global economy. This will be done using three analytical lenses: the 'bordering' of precarious and migrant labour, labour geography, and labour regimes. The three varyingly utilize geographical tools including space,

[1] Portions of this chapter draw on Arnold, Dennis (2017) 'Labour Migration', in Richardson, D. (ed.) *International Encyclopedia of Geography: People, the Earth, Environment, and Technology*. London: Wiley Blackwell and Association of American Geographers.

place and scale, however, rather than restrict analysis to intra-Geography literature the article instead briefly investigates these three approaches from a wider disciplinary optic.

BORDERING PRECARIOUS AND MIGRANT LABOUR

Taken in broad terms, Schierup and Jørgensen (2016) drawing on Bourdieu's (1999) view of precarity as 'flexploitation', speak of precarity as a mode of keeping the 'reserve army of labour in labour', thereby maximizing both productive activity and placing downward pressure on wages. They assert that precarization of work develops in tandem with a precarization of citizenship as the dual contingencies of the (global) restructuring of economy and labour markets alongside a fracturing of frameworks of citizenship. Precarity is structured as a condition embodying imperatives of flexibility and multilocality, designating the centrality of mobility in workers' behaviour and struggles (Schierup and Jørgensen, 2016; Standing, 2011). Scholars and activists have long argued that working conditions migrants experience, as well as the labouring poor in the Global South addressed in the following section, are becoming increasingly common among formerly privileged white male workers in the advanced industrialized economies (Bernstein, 2007; Breman and Van der Linden, 2014). Migrants, whether rural–urban in China and Myanmar, or cross-border from Mexico to the USA or Africa to Europe, are increasingly at the centre of the contemporary labour process (Wills et al., 2010).

Studying migrant labour reveals a general shift of responsibility that follows the capitalist dream of an available labour force disconnected from the need for its reproduction. Not only has capital increasingly outsourced beyond national borders in the contemporary wave of globalization, so too has the social reproduction of labour been outsourced across (national, rural–urban) borders, generally to migrants' home communities and countries. The neoliberalization of social reproduction entails a reduction in state-provided social assistance for its citizens. Migrants are often not afforded any support for the social reproduction of their own families, whether in their home or host country. This is a clear example of re-regulation that places burdens on families and individuals, away from capital and states.

In addition to their appeal to capital, they also have major political advantages as a flexible labour supply, even those who cross borders legally and find themselves politically disenfranchised. Lacking citizenship, they are more likely to encounter restrictions on their access to employment, welfare, and the political process. In contemporary globalization, migrants are not only less likely to organize trade unions, they are the sine qua non of flexibility that many employers seek. Migrant workers are attractive to employers precisely because they are migrants (Wills et al., 2010).

This line of analysis leads to a simple equation: the less social support and the fewer rights migrants have, the more vulnerable and docile they are likely to be (Samers, 2010). While not a universal truth, this contention helps to understand the rights claims that migrant labour may or may not make of employers and states. This is most salient when workers fear being sacked and/or deported. Clearly, a precarious migrant labour force has been central to processes reworking the balance of power between capital and labour. With that in mind, consideration should be given to how it is that migrant labour has come to be viewed by many analysts as a key contributor to economic development, and how migrant subjectivities are understood in relation to global-scale transformations.

Labour migration has always been a contested field in state regulation of populations in capitalism. Capital's solicitation of labour mobility has always gone hand-in-hand with manifold attempts, particularly on the part of states, to filter, to curb, and even to block it (Mezzadra and Neilson, 2013). Geographical literature has been attuned to heterogeneous migrant, state, and capital practices through the continuous reworking and reshaping of both social and geographical borders and scales. Cohen (2006) provides a summary of the shift from open immigration policies in much of Western Europe and the United States during the post-war era, to the more restrictive policies from the 1970s when sharp restrictions were first imposed (while in the United States there has not been an absolute decline in legal immigration, there have been significant qualitative changes in the occupational and legal categories admitted – from immigrants to refugees and from agricultural and mass production workers to the professional, technical, and independent proprietor categories. Japan has not permitted significant numbers of labour migrants throughout the post-war era). Cohen attributes the shift to several factors, including:

- The oil crisis, which contributed to a wave of redundancies in energy-intensive industries.
- The rise of xenophobia, particularly among the working class fearing competition in the labour market.
- The organization of migrant workers who increasingly pushed for family reunification policies, contributing to many politicians and interest groups pressing for more restrictions.
- The rise in the cost of reproduction in terms of child care, language training, and education.
- Economic restructuring, or the general shift from Fordist mass production to flexible specialization and off shoring of production and services.

Speaking to contemporary state policy, Cohen contends that the modern state has sought to differentiate the various people under its sway by including some in the body politic and according them full civic and social rights while seeking to exclude others from entering this 'charmed circle' (Cohen, 2006). Currently, formal citizenship and visa regulations are the most prominent form of regulation at national governments' disposal. Typically, they are used to control access to and the duration of different kinds of work. This is known as 'differential exclusion', describing the incorporation of migrants into some areas of national society, especially the labour market, and their exclusion from others, such as social welfare and citizenship. While a good starting point, this binary concept is easily complicated when considering the different experiences of skilled and unskilled migrants, those 'legal' or not, and variation across national contexts. It is more useful to view migrant labour management as creating different degrees of precarity and vulnerability, as well as opportunities and empowerment.

Today, globalization has both deepened and extended these dynamics and altered the effects they have. Far from flattening the world and reducing the significance of borders, the contemporary social regime of capital has multiplied borders and the rights they differentially allocate across populations (Casas-Cortes et al., 2015). To understand how migrants traverse and reshape the multiple spaces of globalization, the study of borders and labour migration provides much insight. Borders are complex social institutions that can both connect and divide not only migrant labour, but also the times and spaces of global capitalism. This includes both political borders, whose relevance has altered yet certainly not gone away, and the multiplication of social borders that very often shape migrant workers' experiences. In migration studies, the border has typically been comprehended as a wall, both literally and figuratively, that excludes migrants from a national territory. The US–Mexico border is the prototype of the border-as-wall, an 'economic dam' that seeks to prevent migrant flows from 'contaminating' the US body politic, an approach and discourse that has been replicated elsewhere. Border securitization has not only failed, it has contributed to riskier efforts on the part of migrants to evade authorities, while also leading to a proliferation of human traffickers and others capitalizing on the criminalization of certain migrant labour populations – very often police and military officials mandated to 'protect' the border.

Borders have become a key site of articulation for capital, especially as a means for differentiating between labourers according to skill level, nationality, race, gender and class. Not only have borders multiplied, but labour has as well with the intensification of labour processes combined with the tendency of work to colonize life. Viewed in relation to one another, a range of state and economic actors use borders to create territories and spaces of regulatory inclusion and exclusion for labour (Mezzadra and Neilson, 2013). Indeed, national borders are no longer the only or necessarily the most relevant borders dividing and restricting labour mobility. Migrant labour encounters complex sets of social relations that seek to reinforce labour power as a commodity by turning migrant flows into mobile governable subjects. Indeed, the production and reproduction of differences is key to capital accumulation strategies. Impacts include the differential allocation of rights across populations. Very often, this entails being 'in one's place out of place' and 'out of place in one's place' (Casas-Cortes et al., 2015). All labour is subject to bordering, yet migrant workers' experiences are most pronounced. In particular, capital has proven particularly adept at exploiting the continuities and the gaps – the borders – between different migrant populations.

Studies on precarity, migration and borders highlights important labour trends in the global economy. They do, however, tend to suffer from the tendency toward a 'catch-all' bias in which borders and precarity are 'everywhere', leaving insufficient analytical space to unpack the nuances of particular formations of labour. The following sections address this shortcoming with attention to labour geographies and labour regimes.

LABOUR GEOGRAPHIES

The globalization of production and outsourcing of world production from North to South has necessitated theories on global space production. Capital flows, labour mobility, technology diffusion, international subcontracting chains, information networks, just-in-time production, are transnational processes that not only require intensive reconfiguration of time, but also rapid reorganization of space (Harvey, 2003). On the one hand, globalization of production is a response to competitive pressures from capital, and on the other, the need for greater flexibility in labour management and reach of capital is a reaction to strength of labour in the industrialized countries in the 1960s and 1970s (Hardt and Negri, 2000). The saturation of markets, along with the high levels of competition that introduced the process of global outsourcing, obligated firms to develop techniques and technologies to enhance or create mobility and flexibility, which also created new barriers to labour organizing. These processes have diminished workers' bargaining power at the point of production and national scale, while at the same time generated new forms of struggle on the part of feminized and racialized labour forces.

One of the most noticeable implications for labour is the rapidly diminishing trade union density since the end of the 1970s. The representational gap or its absence is inherently linked to the increasing precarization of labour and social insecurity and inequality (Standing, 2011). Thus, the lack of workplace and social representation is a central element producing, reordering, and perpetuating social, economic, and political disparity, as well as marginalization and vulnerability. Yet understanding global labour trends and particular labour regimes is not only about the paralyzing blows from inscrutable macro-scale political economic forces (Barchiesi, 2011). It is also about how workers position themselves in production systems and co-constitute the changing geographies of globalization.

Contemporary geographic labour research has coalesced around a core set of conceptual categories and theoretical concerns anchored in place, space and scale. A prominent shift has been from the 'geography of labour' which privileged the determinate role of capital in producing its social and geographical landscapes, toward 'labour geography' which foregrounds workers as active geographical agents who co-determine the mutational parameters of the global economy. Peck (2013) has argued that the ground-breaking work of Doreen Massey could be categorized as labour geography, and that intellectually and politically, Massey was a labour geographer long before the label existed. The path-breaking analysis in *Spatial Divisions of Labour* marked a paradigm shift in economic geography, from positivism and location theory to critical realism and political economy, while centring labour as a social agent (Peck, 2013).

Later iterations of labour geography began as a neo-Marxist critique of structuralist work. Much structuralist work had been concerned with the breakdown of Fordism and the Keynesian welfare state, an era in which labour in the advanced industrialized countries had an unprecedented power to shape the geographies of capitalism at the national scale. In other words the determinative role of capital in the neoliberal era had been, rightly, emphasized. Alongside emphasizing the geographies of capital and firms' management strategies, the core critique of labour geographers is that workers are too often treated as passive victims. Andrew Herod, a prominent figure in the field, offers four central tenets of an alternative approach, labour geography (Herod, 1997: 16–18, cf. Das, 2012: 21):

- The 'production of the geography of capitalism is not always the prerogative of capital' – labour plays an active role in creating this geography. In some places and times, capital is not even the most significant actor.
- The agency of workers is related to 'their desire to implement in the physical landscape their own spatial visions of a geography of capitalism' in the interest of 'their own self-reproduction and social survival'. Like capital, labour resorts to spatial fixes in its own interests. Labour's fixes may or may not coincide with those favoured by capital.
- This means that the production of space is contested, and that processes of class formation and inter- and intra-class relations are geographical processes. Workers produce landscapes in ways that increase their power and reduce that of capital.
- Workers produce scales, which impact their agency and capitals. For example, their ability to win wage increases and shorter hours help determine the size of travel to work areas and thus the urban scale. When unionized workers create regional and national contracts to equalize conditions across areas, they make it difficult for employers to play one locality against another.

A core concern with Herod and other labour geographers is that labour is not a spatially homogenous category. Scholars have developed this contention, with Coe and Jordhus-Lier (2010) identifying four distinct thematic strands within the literature, each of which is underpinned by its own bodies of theory. The first has been concerned with the collective organization of workers through labour unions and the reassertion of the potential agency of these groupings. A second has theorized the formation of geographically specific local labour markets/regimes and their ongoing regulation and segmentation, including charting the shift from welfare to workfare regimes in many nation-states and regions. The third has explored the intersections of employment relations with other facets of personal and workplace identity, including a large body of feminist work focusing on unpaid reproductive work, gendered divisions of labour and worker identities and subjectivities, among other themes. The fourth is concerned with exploring the role of the material landscape in shaping labour struggles and their outcomes, examining how workers shape landscapes, but also how landscapes influence worker action. A critique Coe and Jordhus-Lier draw from this review is the lacunae of theorization around workers' agency. Thus, they call for notions of agency that are conceptualized and fleshed out in terms of its multiple geographies and temporalities, and that the potential for worker action should always be seen in relation to the formations of capital, the state, the community and the labour market in which workers are incontrovertibly yet variably embedded.

Despite this nuanced perspective on structure-agency debates as they pertain to labour geographies, Das (2012) argues that the labour geography literature falls flat in terms of conceptualizing the materiality of labour, pointing to two tendencies. One is the emphasis on the labour *market* and on its re-regulation in the neoliberal era, usually *in abstraction* from 'the hidden abode'. Thus labour appears mainly as sellers of a commodity. Das (2012: 22) points out, drawing on Marx's *Critique of Political Economy*, that 'equating labour *market*, the realm of exchange relations, to the sphere of *production* is completely wrong. This sort of serious conceptual slippage is not atypical in labour geography and (indeed in much "critical" human geography).' He goes on to note that to the extent that labour process issues are raised, the discussion is often one-sidedly focused on identity, language and culture, leaving the materiality of the labour process under-developed. This line of critique, then, leads to a productive interface among labour geography and labour regime literatures. It is in this respect that labour geography studies are being strengthened with attention to labour regimes, the focus of the following section.

SCALING LABOUR REGIMES

The concept of 'labour regime' draws on two major theoretical sources of inspiration: Braverman's studies on control of workers at the point of production, and the regulation approach which, primarily through analysis of Fordism, focuses on periods of national-scale stabilization despite inherent crisis tendencies in capitalism (Arnold and Campbell, 2017). In its most restrictive use a labour regime is seen simply as firm-level forms of labour recruitment and use, hence primarily linked to the labour process (Mezzadri and Srivastava, 2015), defined here as how people work, how and who manages their work, the skills that are utilized, and how workers are paid. More generally, 'labour regime' conceptualizes the interconnected ways in which workers, labour processes and employment are organized and regulated in particular spaces. Bernstein (2007: 7) defines the term as the 'interrelations of (segmented) labour markets and recruitment, conditions of employment and labour processes, and forms of enterprise authority and control, when they coalesce in sociologically well-defined clusters with their own discernible "logic" and effects'.

Critical labour scholars deploying the labour regime concept often draw inspiration from Michael Burawoy's (1985) 'factory regime', which addresses specific industries' position in the wider spectrum of social relations relevant to particular forms of capital-labour relations. Burawoy's (1985) related work on 'politics of production' brought back the political and ideological effects of production regimes, emphasizing the role of the state in shaping national modes of production as well as labour politics in the workplace. Thus, Burawoy set out to understand the international as well as the national sets of relations of which the industry was part (Mezzadri and Srivastava, 2015). A core intervention of Burawoy's work is conceiving of labour regimes as hegemonic projects aimed at securing worker acquiescence to unequal and exploitative employment relations. This is a conceptualization rooted in Antonio Gramsci's analysis of the Fordist production model, in which prominent (largely male) segments of the working class were provided stable, secure and often unionized employment with relatively high wages as a means of garnering their consent to a regimented and intensified industrial labour process (Arnold and Campbell, 2017). Fordism, then, as a consent-seeking labour regime, served to bolster conditions for industrial peace as a precondition for stable capital accumulation. What was missing from Burawoy's early work was the *spatial* aspect of production, addressed above, which has become an area of concern in labour regime theorizations.

Pun and Smith (2007) explore a particular locality effect, the construction of export-orientated factories in China, and the role of the Chinese state, transnational capital and internal migrant labour in producing an exceptionally productive labour regime. In the context of China's rise as 'global factory' and its many-fold spatial fixes, they examine the linkages between the production and reproduction of labour power in this transnational labour process space. Their primary contribution to the literature is identifying a new form of labour regime, the *dormitory labour regime* in China. This labour regime reproduces low labour costs, high worker productivity, and access to extensive low-cost labour reserves that is institutionalized within a political economy between state and market. Importantly, this form of labour capture maintains access to cheap labour without a labour movement that could organize and contribute to better working conditions and labour solidarities, as the Chinese state has pre-empted grassroots mobilizations through Party domination of the trade unions. Any potential for workers' agency is tightly circumscribed by an array of structural forces. Indeed, Pun and Smith find that the systemic provision of dormitories for internal migrant labour within or around factories facilitates continuous access to labour reserves from the countryside, thus depressing wage demands and affects collective organization by workers in a particular industrial space.

Pun and Smith's dormitory labour regime conceptualization is, like many labour regime studies, about extensive, multi-scalar forms of labour regulation and control, under-analyzing workers' role in co-constituting and potentially transforming the regimes. Chan (2014) addresses this shortcoming by utilizing Coe and Jordhus-Lier's 'constrained [labour] agency' framework identified above. Chan finds that in China's manufacturing regions Chinese migrant workers have been actively participating in the shaping and reshaping of labour standards in the global factories thereby challenging global capital and the party-state regarding labour regulations. However, the political and economic constraints on labour agency should not be underestimated. He argues that the authoritarian nature of the Chinese party-state and the legacy of socialist trade unionism have impeded the rise of effective trade unionism. Furthermore, China's continued dependency on FDI (Foreign Direct Investment) and export-oriented industry in driving its economic development strategy reproduces global capital's huge leverage in influencing local labour policies.

Other labour regime studies have more explicitly centred geographical concepts including the politics of scale that have been influential in labour geography literature. Through a case study of labour agency in Yangon, Myanmar's export garment sector, Arnold and Campbell (2017) draw on Burawoy's Gramscian understandings of hegemony in labour regime analysis. They argue that the emerging labour regime in Myanmar constitutes a nascent hegemonic regulatory assemblage shaped by forces and actors at multiple scales. This assemblage involves a mix of consent-seeking discourses and labour-relations practices, including social dialogue, labour law reform, and the legalization of workers' institutional representatives, alongside coercive measures like the violent crackdowns on striking workers and implementation of repressive colonial-era laws. Discursive norms centred on building a 'clean' image for garment manufacturing have been introduced, contributing to a changing capital accumulation strategy in the sector after years of economic sanctions of the European Union and United States – the two largest apparel consuming markets globally. These relations are varyingly scaled through 'opening' to Western markets and investors and with them the discursive introduction of CSR (Corporate Social Responsibility) and other norms and development practices. In the process, the state seeks to gain worker consent at the scale of the nation – through national trade union federations and discourses of national development. Yet they assert that the hegemonic project is fraught, as workers are contesting the terms of their insertion into industrial capitalism, as well as union and other actors' efforts to scale-up the workplace into national social dialogue that contributes to notions of 'globalization with a human face'. This rejection challenges developmental assertions that workers should make 'short term' livelihood sacrifices for long-term national economic development planning – thus 'flattening' the scalar register of social and political action.

CONCLUSION

Over the past 30 years, global outsourcing and precarization of work have become central in reconfiguring labour geographies. Both trends have contributed to delimiting the collective power of labour. Indeed, economic growth has occurred alongside a reduction in workers' share of overall wealth. In the process, migrant workers have become the paradigmatic figure in the contemporary labour process. On the one hand, global capital drives migration and reshapes its patterns, directions, and forms. On the other, however, labour migration is an important factor in bringing about fundamental social transformation of both home and host countries, and is, in turn, itself a major force reshaping communities and societies.

The study of borders and states' efforts to manage mobile capital and labour focuses on classed hybridity (the multiplication of labour), and the implications of racialization, gender, and citizenship in shaping labour markets. It is an approach that recognizes global capital trends without viewing them as paralyzing blows that determine subjective experiences. Instead, the spaces that migrants traverse are mutable and multiple, offering opportunities and constraints. The spatial reorganization of labour has multiplied and fragmented its forms. In these and other debates, geographically inspired literature has been attuned to the continuous reworking and reshaping of migrant experiences. The theoretical and empirical diversity of research has made significant contributions to understanding important political, economic, and social changes.

Studies on labour geography and labour regimes offer at times more nuanced accounts of both mechanisms of labour control at the workplace and how workers co-determine the labour process and wider regimes of accumulation. At stake is developing better understanding of how labour regimes reproduce excessive share of profits to capital at labour expense, how and why workers acquiesce to their own exploitation, and the cracks and fissures that emerge in labour regimes over time – highlighting both workers' success stories as well as failures, as part of wider efforts to mitigate inequalities inherent in contemporary capitalism. Labour regime studies have, taken as a whole, usefully adopted core concerns of labour geography theorizations.

These studies include but clearly extend beyond debates within the disciplines including Geography and Sociology, rather comprising a renewed cross-disciplinary orientation around political economy.

REFERENCES

Arnold, D. and S. Campbell (2017), 'Labour regime transformation in Myanmar: Constitutive processes of contestation', *Development and Change*, **48**(4), 801–824.

Barchiesi, F. (2011), *Precarious Liberation: Workers, the State, and Contested Social Citizenship in Postapartheid South Africa*, Albany: State University of New York Press.

Bernstein, H. (2007), 'Capital and labour from centre to margins'. Keynote address at the Living on the Margins Conference, Stellenbosch, South Africa. http://urbandevelopment.yolasite.com/resources/Capital and Labour in the Margin Bernstein.pdf (accessed 19 June 2016).

Bourdieu, P. (1999), *Acts of Resistance: Against the Tyranny of the Market*, New York: New Press.

Breman, J. and M. Van der Linden (2014), 'Informalizing the economy: The return of the social question at a global level', *Development and Change*, **45**(5), 920–940.

Burawoy, M. (1985), *The Politics of Production: Factory Regimes Under Capitalism and Socialism*, London: Verso.

Casas-Cortes, M., S. Cobarrubias, N. De Genova et al. (2015), 'New keywords: Migration and borders', *Cultural Studies*, **29**(1), 55–87.

Chan, C.K-C. (2014), 'Constrained labour agency and the changing regulatory regime in China', *Development and Change*, **45**(4), 685–709.

Coe, N. and D.C. Jordhus-Lier (2010), 'Constrained agency? Re-evaluating the geographies of labour', *Progress in Human Geography*, **35**(2), 211–233.

Cohen, R. (2006), *Migration and its Enemies: Global Capital, Migrant Labour and the Nation-State*, Aldershot: Ashgate Publishing.

Das, R. (2012), 'From labour geography to class geography: Reasserting the Marxist theory of class', *Human Geography*, **5**(1), 19–35.

Hardt, M. and A. Negri (2000), *Empire*, Cambridge, MA: Harvard University Press.

Harvey, D. (2003), *The New Imperialism*, Oxford, NY: Oxford University Press.

Herod, A. (1997), 'From a geography of labour to a labour geography: Labour's spatial fix and the geography of capitalism', *Antipode*, **29**(1), 1–31.

Mezzadra, S., and B. Neilson (2013), *Border as Method, or, the Multiplication of Labour*, Durham, NC: Duke University Press.

Mezzadri, A. and R. Srivastava (2015), *Labour Regimes in the Indian Garment Sector: Capital–Labour Relations, Social Reproduction and Labour Standards in the National Capital Region: Report of the ESRC-DFID Research Project 'Labour Standards and the Working Poor in China and India'*, Research report. London: Centre for Development Policy and Research, School of Oriental and African Studies.

Peck, J. (2013), 'Making space for labour', in: D. Featherstone and J. Painter (eds) *Spatial Politics: Essays for Doreen Massey*, Oxford: John Wiley & Sons, pp. 99–114.

Pun, N. and C. Smith (2007), 'Putting transnational labour process in its place: The dormitory labour regime in post-socialist China', *Work, Employment and Society*, **21**(1), 27–45.

Samers, M. (2010), *Migration*, New York: Routledge.

Schierup, C. and M. Bak Jørgensen (2016), 'Politics of precarity: Migrant conditions, struggles and experiences', *Critical Sociology*, **42**(7–8), 947–958.

Silver, B. (2003), *Forces of Labour: Workers' Movements and Globalization Since 1870*, Cambridge, MA: Cambridge University Press.

Standing, G. (2011), *The Precariat: The New Dangerous Class*, London: Bloomsbury.

Wills, J., K. Datta, Y. Evans et al. (2010), *Global Cities at Work: New Migrant Divisions of Labour*, New York: Pluto Press.

14. Geographies of tourism in a globalizing world

Hong-gang Xu and Yue-fang Wu

INTRODUCTION

While the ongoing process of economic globalization has been made possible by the drive of capitalism to expand and grow, by the pervasiveness of new technologies in communications and transport, and by the liberalization of aviation markets (Appadurai, 1996: 1–26; Hannerz, 1996: 19), international tourism is an important beneficiary of and vehicle for its expression (Meethan, 2001). Claims from the World Travel and Tourism Council (WTTC) were made in 2003 that tourism was becoming the world's largest industry (WTTC, 2003). It was indicated by the United Nations World Tourism Organization (UNWTO) in its 'Tourism Towards 2030 – Global Overview' that international tourist arrivals were set to increase by an average of 43 million people a year between 2010 and 2030 (UNWTO, 2011). By 2030, the number was anticipated to reach 1.8 billion, meaning that in two decades' time, on a daily average, 5 million people would cross international borders for leisure, business or other purposes such as visiting friends and family, as well as four times as many tourists traveling domestically (UNWTO, 2011). Tourism has evolved into a global phenomenon – one of the most important economic sectors and social activities of our time. Besides the huge size of the tourism economy, its significant social, cultural and political effects are also significant. Apart from being a driver of development, tourism has also alternated the local ways of life and influenced the patterns of folk cultural inheritance. Tourism is often seen as a 'hyperglobalizer' (Held et al., 2000) homogenizing the world, leading to issues of authenticity, equality and sustainability. As well, the process of global tourism expansion has facilitated uneven development at the national and sub-national levels of regions and communities.

However, compared to its rapid development and pervasive influences, the current research on international tourism in a globalizing world is still lagging in addressing key aspects, for example in synthesizing approaches and paradigm shifts of geography to tourism researches, and in reflecting to the emerging trends in the Global South. To respond to the trends, patterns and complexities of the geographies of tourism in a globalizing world, the following issues are central to our review. On the one hand, it is important to understand how globalization is conceptualized in the sub-field of tourism geography, and how global tourism generates effects at national and sub-national scales. On the other hand, with the changing spatial patterns and intensified velocity of global flows of international tourism, it is critical to investigate current developments in the approaches, methodologies and paradigms in tourism geography. Since the main theoretical approaches in modernist studies of tourism have been explicitly formulated for the study of 'Western' tourism based on Eurocentric paradigmatic assumptions, the rise of tourism in emerging regions such as Asia, Latin America, Africa and the Middle

East has kept challenging the rationality and justification of traditional theories in explaining the phenomenon (Cohen and Cohen, 2015). Meanwhile, the paradigm of mobility has opened up new systematic studies of tourism in these emerging regions, where perspectives beyond Eurocentrism in contemporary tourism studies are in need.

CONCEPTUALIZING TOURISM GEOGRAPHY IN THE ERA OF GLOBALIZATION

Tourism geography is the study of tourism within the concepts, frameworks, orientations, and venues of the discipline of Geography and the accompanying fields of geographical knowledge. Although the academic focus of tourism geography is often summarized as space, place and environment, as in the subtitles of the *Tourism Geographies* journal or the text by Hall and Page (2006), its scope was broader and reflects the varieties of geographical knowledge. A spate of reflective reviews and collections on tourism geographies has occurred since 2005. This is illuminated in Hall's (2013: 603) review, that 'the growth in interest by human geographers in tourism has also been influenced by the "cultural turn" and non-representational conditions and concerns. This is also reflected in the extent to which issues of post-colonialism, performity, embodiment, and the mundane have influenced tourism geographical research, as well as the renewed focus on mobility'. Nevertheless, the 'cultural turn' in geography and elsewhere in social sciences represents only one dimension of contemporary tourism geography (Coles and Hall, 2006).

Over time, tourism geography has made significant contributions to concerns about the roles and impacts of tourism on, as is much emphasized, the physical environment (climate change and sustainability), and the restructuring of places or economies (Hannam and Knox, 2010). The diverse theoretical and methodological approaches to tourism geography, for example in the fields of spatial models, the political economy of tourism, and cultural representation of tourism, mirror the wider shifts in other social sciences, and in particular in Human Geography. It is clear that geographers have made a substantial contribution to the study of tourism in recent years (Hall and Page, 2006).

Tourism is a process which manifests itself locally and regionally and explicitly involves the construction of 'place'. As pointed out by Milne and Ateljevic (2001), the study of tourism has shown a significant potential to reveal the dialectics of production and consumption and tensions between the global and the local, which are core issues associated with social and spatial polarization. For the former, proponents of the regulation approach have advocated a perspective which gives the 'regime' of mass production and consumption, known as 'Fordism', a central role. While in the last quarter of the twentieth century it had yielded towards a more 'flexible' and dynamic pattern, categorized as 'post-Fordism', 'flexible accumulation' or 'flexible specialization' (Scott, 1989). And there is a burgeoning geographical literature addressing issues of leisure, consumption and social identity as the 'culture of consumption' further evolves and impacts the tourism industry. For the latter, the global and the local are intimately intertwined through the process of globalization, and in the past few decades, geography scholars are increasingly paying attention to global and local issues which has generated fruitful research results. Phrases such as 'global–local nexus'

(Milne and Ateljevic, 2001; Teo and Li, 2003), globalization, global meets the local (Thorns, 1997), glocalization (Salazar, 2005) and localization (Gotham, 2005) are vivid representations of the academic interest in the ways in which development at the global level has influenced tourism and the features and responses of tourism destinations under globalization.

There are controversies surrounding the relationship between the global and the local in contemporary discussions. And from the perspectives of some sociologists and geographers, globalization has brought the merging and blurring of boundaries, as well as local societies being overwhelmed by the power of capitalism (Giddens, 1990: 53). As stated by Waters (1995: 3), globalization is 'a social process in which the constraints of geography on social and cultural arrangements recede and in which people become increasingly aware that they are receding'. The global processes have disembedded local practices and re-embedded them in patterns and networks that are increasingly globalized. However, literatures on 'globalism–localism' are broad and emphasizes that localities are not merely recipients of global forces but are actively involved in their own transformation (Cooke, 1989). Thus, local cultures and identities are juxtaposed with global influences, producing unique outcomes (Massey and Allen, 1984). In tourism, global/local dynamics have been researched in the fields of cultural and heritage tourism (see for example Kearns and Philo, 1993), tourism evolution (Chang et al., 1996), and the tourism industry overall (Mowforth and Munt, 1998), which are further discussed in the following parts.

PERSPECTIVES IN TOURISM GEOGRAPHY

Spatiotemporal Dimensions

Geography is the ideal discipline to study the global tourism industry given tourism's distinct place, time, distance and activity patterns (Meyer-Arendt and Lew, 2003; Che, 2017). Butler (2015) employed a historical perspective and reviewed the origins and evolution of tourism and tourism research, and provided critical insights on the 'early roots' of tourism in order to question tourism definitions. As well, by reviewing the multiple patterns of tourism and tourist behaviour, he greatly expanded the scope of tourism research and pushed backwards the dates of ancient tourists (Butler, 2015). For example, the elites in Roman society relocating each summer to their villas in the surrounding hills to escape the heat, stench and uncomfortable conditions of the city were seen as mirroring contemporary escapism to second homes on holidays and the seasonal flight to warmer places for those living in colder climates. Butler (2015) pointed out that tourism research are best thought of as an ongoing process with varying emphases and foci at different times, beginning with factual case studies followed by a period of extensive if somewhat shallow theoretical development, and then the current situation with its paradoxes and fallacies.

Since the 1970s and 1980s, tourism geographers have sourced a variety of disciplines for relevant theoretical insights into tourism and its increasingly broad ramifications, and into globalization with its global/local complexities (Wood, 2000). For example, the mainstream examination of the impacts of global mass tourism was first based on dependency theory, which viewed tourism development in the Third World as more for the benefit of capitalist and tourism-generating countries and not for the host countries (Britton, 1982). Tourism's economic benefits in Third World countries have been acquired at considerable intangible costs. Globally leading tourism firms, with their dominant metropolitan interests, imposed a development mode that exacerbated the adverse conditions typical of dependent capitalist social formations.

Another dominant approach in this period was the Tourist Area Cycle of Evolution model (Butler, 1980), with implications for the planning and management of tourist resources in light of a continuing decline in the environmental quality and the attractiveness of many tourist areas. However, both the above approaches have been criticized for failing to incorporate local reactions and the negotiated processes between local individuals and structural forces. In the meantime, since the tourism industry operates as a link and medium between local and larger spatial scales and socio-economic and environmental systems, the core–periphery/global–local relationships are also relevant (Britton, 1991). These relationships often feature inequalities, uneven geo-political metrics and power structures, which call for more critical tourism and planning strategies (Scheyvens and Momsen, 2008).

Tourist spaces in historical urban areas are transformed and recreated under the influence of the global–local nexus. Transnational tourist spaces are demonstrated in the concentration of transnationalized tourism consumption. International hotels and accommodations, leisure firms, and transnational chains of exotic food and beverage as well as department stores are overwhelmingly dictated by globalized consumption practices through transnational travel institutions, mega hotel chains and financial service enterprises. To take Wicker Park and Logan Square in Chicago as examples, these nodes are transforming local places into internationalized tourism spaces that grow into new landscapes visualizing the territorialization of global consumption and post-industrial urbanized tourist consumption. Global tourism industries are acting as important intermediaries in understanding the interactions between global tourism and local realities. The construction of transnational spaces is carried out through a complex web of actors and social relationships at multiple scales. As well, the spaces are considered to (re)produce inequalities between local people and their communities, regions and nations.

By employing an extension of transnational theoretical perspectives and linking tourism development and transnationalism, Torres and Momsen (2005) conducted a study on issues related to tourism development in Cancun, Mexico. The concept of 'space' was used centrally to the examination of 'Gringolandia' as a phenomenon constituted by social relationships. The region was referred to by the locals as Gringolandia due to the extravagance and overbuilt nature which transformed the local space into a circus-like spectacle. The term Gringolandia not only reflected the Disneyesque quality of the spectacle mass tourism in Cancun (Torres, 2002), but also the invasion and expropriation of Mexican space by American entities.

Thus, tourism is a prominent component of the process of globalization and is a primary channel for economic and cultural exchange, but it is also shaped by globalization through the evolving system of scapes and flows. The distinction between earlier forms of internationalization and contemporary globalization is reflected in the increasing interconnections in the geographic reach of global networks of exchange. Hence, it is important to ground research on tourism in the changing relationships between individual actors and places, production and consumption, and development and sustainability in order to connect tourism geography to the wider field of geographic enquiry on globalization.

Social Dimensions

With the increase in the number of travelers seeking unique experiences beyond their regular spaces, the perspectives they employ to 'gaze at' different societies are investigated. Social theories of human geography have shed light on the perspectives of Western tourists in cross-cultural interactions, and notably theories on authenticity have been proposed in the field of tourism research to analyze the understanding of contemporary society. Researchers have argued that one of the conspicuous trends of globalized tourism is the replacement of actual authenticity with 'staged' authenticity, where local cultures and traditions are reproduced or modified for tourists' consumption. MacCannell (1973) was particularly concerned with the character of the social relationships that emerged from fascination with the 'real lives' of others. He noted that such 'real lives' could only be found in people's backgrounds and were not immediately evident. However, he also pointed out that the tourist gaze would involve an obvious and unacceptable intrusion into people's lives, where the people being observed and local tourist entrepreneurs would turn to reconstruction of people's backgrounds in a contrived and artificial manner. 'Tourist spaces' were thus organized around what MacCannell (1973) called 'staged authenticity'.

Other researchers, however, have argued that it may be wrong to suggest that the search for authenticity is the basis of tourism. Rather a sense of the unfamiliar and the extraordinary may be central to tourists' gaze. For example, 'post-tourists' found pleasure in the multiplicity of 'tourist games'. They knew that there was no authentic tourist experience, rather there was merely a series of games or narratives that could be played or followed (Feifer, 1985). Critics in more recent literature have argued that there is a neglect of other bodily experiences in tourism in favour of visual experiences. Thus, a 'performance turn' emerges. Urry and Larsen (2011) pointed out that with Goffman's (1959) dramaturgical sociology and Thrift's (2008) non-representational theory, this performative turn conceptualizes the actions of and interactions between tourism workers, tourists and locals.

Tourism geographers have further developed the authenticity theory and employed the tourist gaze paradigm in different research scenarios to investigate the increasingly complex relationships between humans and places. These theories have been developed for the analysis of the evolution of tourist products and destinations under the manipulation of international companies. The global competition and internationalization of tourism have led to standardization, commodification, historical distortion and gentrification of tourism spaces. In the meantime, by considering the typical objects of the tourist gaze, we can make sense of the wider society. By using counter-intuitive methodologies, tourism geographers are able to use differences to interrogate the 'normal' when investigating typical forms of tourism.

Political-Economic Dimensions

The roles of social, cultural, political, and economic structures that facilitate the reproduction of the organization of production, commodification, distribution, exchange, and consumption and how these structures sustain specific power constellations are of particular interest to political economy (Mosedale, 2014). However, it was also concluded by Williams (2004: 62) that 'fruitful theoretical developments in political economy have largely bypassed tourism'.

In the academic focuses on the changing structure of the tourism industry and the evolving roles in local, regional and national development processes, the model of 'Tourism and the Global–Local Nexus' proposed by Milne and Ateljevic (2001) described the composition of global tourism as including four scales: global, national, regional and local. At the global scale, tourism's development outcomes were influenced by broad-based economic changes, evolving structures of corporate governance, and the unrelenting evolutionary pressures of demographics and technological changes. It was illustrated by Mowforth and Munt (1998) that global institutions such as the IMF and World Bank played vital roles in shaping the economic environment for tourism investment and development in much of the world. Transnational corporations stood astride the global economy as the dominant form of enterprise in both the tourism industry and other sectors (Milne and Ateljevic, 2001). At the national scale, cultural traits and resource-use and economic regulations were at work, while at the regional scale, regional governments, networks and ecosystems dominated in the tourism economy. At the local scale, local governments, tourists, workers, households, and firms were at work to negotiate the complexity of the global–local nexus and dictated local development outcomes.

Tourism is viewed as a transaction process which is simultaneously driven by the global priorities of multi-national corporations, geo-political forces and broader forces of economic change, as well as the complexities of the local scale, where residents, visitors, workers, governments and entrepreneurs interact at the industry 'coal-face' (Milne and Ateljevic, 2001). Appadurai (1996), in his treatise in modernity and globalization, acknowledged that tourists and the tourism industry were part of the transnational landscape, and recognized that there was still little known about the special 'translocalities' they produced. Clearly the actors, institutions, agencies, structures, and processes associated with tourism were transnational in nature (Torres and Momsen, 2005). Torres and Momsen (2005) investigated the transnationalism of the tourism industry in Mexico. Multi-national tourism franchise agreements, international tourist flows, national government development strategies, domestic elite tourism industry investment, and migration-related livelihood strategies at the household level have played out in transnationalism in Cancun. It was pointed out that given the inherently transnational nature of tourism, transnational perspectives were important to understand the dynamics and complexities of global tourism. To theorize the relationship between the global and the local, Wood (2000) focused on cruise tourism in the Caribbean. After documenting the rapid expansion of this business, he explored three central manifestations of globalization at work in the Caribbean cruise industry: (i) the restructuring of the industry in the face of global competition, capital mobility, and labour migration; (ii) new patterns of global ethnic recruitment and stratification, including their incorporation into the product marketed to tourists; and (iii) deterritorialization, cultural theming, and simulation. Wood (2000) asserted that this 'globalization at sea' illustrated the contradictions, ambiguities, and uncharted course of contemporary globalization processes.

With the rapid changes in temporal and spatial relationships globally, and with the intensification of social and spatial mobility, identities and localities are becoming blurred, and the epistemology of tourism research is being challenged. The rational and application of traditional theories and principles is continually questioned, and under these circumstances, scholars are increasingly advocating for *new tourism* (Tribe, 2005), *hopeful tourism* (Pritchard et al., 2011) and *tourism sustainability and resilience* (Butler, 1999). There are evolving needs in understanding the epistemological and ontological foundations of tourism research, and increasing focus of the moral and ethical turn of the industry. There are also growing concerns to emphasize the development of tourism in the Global South, which is growing in significance in global tourism industries and is bringing in new perspectives and experience for the theoretical scrutiny of international tourism research, which will be illustrated in the following sections.

TOURISM IN EMERGING WORLD REGIONS

Shifting Trends in the Global Industry

International tourism is spreading more and more widely across the globe. As projected by the UNWTO (2011), besides the continual growth of international tourist arrivals worldwide in the next decades, emerging economy destinations were expected to surpass advanced economy destinations in international arrivals after 2015 (see Table 14.1). Statistics showed that international arrivals in emerging economy destinations were expected to sustain growth at double the pace (+4.4 per cent a year) of advanced economies (+2.2 per cent a year). Tourist arrivals in emerging destinations including Asia, Latin America, Central and Eastern Europe, the Eastern Mediterranean, and the Middle East and Africa would surpass 1 billion by 2030. Further indications were provided by the UNWTO that South Asia would be the sub-region with the fastest growth (+6.0 per cent a year) in international arrivals, and Asia and the Pacific would also become the outbound regions that grew the most. A large portion of the new arrivals would originate from countries in Asia and the Pacific. As a result of this, the market share of Asia and the Pacific would reach 29.6 per cent in 2030, surpassing the Americas (13.7 per cent) and ranking second to Europe (41.1 per cent).

With the rapid growth of its national economy and internationalization strategies, China has grown into the world's largest outbound tourism market since 2012, with annual tourism revenues accounting for over 13 per cent of the global tourism industry. In 2017, over 107 million outbound tourists participated in global travel, with tourism revenue reaching 115.29 billion US dollars (http://www.xinhuanet.com/2018-03/01/c_129820562.htm). The increasing significance of tourism development in the Global South is challenging the traditional perspectives in geography and tourism research, and further broadening research agendas.

Table 14.1   International tourism by UNWTO (sub)region of destination

| | International Tourist Arrivals received (million) | | | | | | | |
| | Actual data | | | | | | Projections | |
	1980	1990	1995	2000	2005	2010	2020	2030
Africa	7.2	14.8	18.9	26.5	35.4	50.3	85	134
North Africa	4.0	8.4	7.3	10.2	13.9	18.7	31	46
West and Central Africa	1.0	1.7	2.3	3.1	4.4	6.8	13	22
East Africa	1.2	2.8	5.0	6.6	8.5	12.1	22	37
Southern Africa	1.0	1.9	4.3	6.5	8.6	12.6	20	29
Americas	62.3	92.8	109.0	128.2	133.3	149.7	199	248
North America	48.3	71.7	80.7	91.5	89.9	98.2	120	138
Caribbean	6.7	11.4	14.0	17.1	18.8	20.1	25	30
Central America	1.5	1.9	2.6	4.3	6.3	7.9	14	22
South America	5.8	7.7	11.7	15.3	18.3	23.6	40	58
Asia and the Pacific	22.8	55.8	82.0	110.1	153.6	204.0	355	535
Northeast Asia	10.1	26.4	41.3	58.3	85.9	111.5	195	293
Southeast Asia	8.2	21.2	28.4	36.1	48.5	69.9	123	187
Oceania	2.3	5.2	8.1	9.6	11.0	11.6	15	19
South Asia	2.2	3.2	4.2	6.1	8.1	11.1	21	36
Europe	177.3	261.5	304.1	385.1	438.7	475.3	620	744
Northern Europe	20.4	28.6	35.8	43.7	57.3	57.7	72	82
Western Europe	68.3	108.6	112.2	139.7	141.7	153.7	192	222
Central/Eastern Europe	26.6	33.9	58.1	69.3	87.5	95.0	137	176
Southern/Mediter. EU.	61.9	90.3	98.0	132.5	152.3	168.9	219	264
Middle East	7.1	9.6	13.7	24.1	36.3	60.9	101	149

Source: UNWTO (2011).

Setting New Agendas for Future Research

As mentioned previously, localities and regions benefit from global tourism and simultaneously 'produce' tourism globalization. Tourism studies have also been recognizing the significance of the above-mentioned trends and the epistemological, theoretical and comparative issues in tourism studies on emerging regions. Critics of traditional tourism theory have argued that modernist theories based on Eurocentric paradigmatic assumptions, which root in Western modernist thoughts and reflect some neo-liberal and individualistic tendencies, are not adequate to deal with tourism from emerging regions. As noted by Cohen and Cohen (2015), the narrow concept of 'tourism' implicit in modernist tourism theories has exposed researchers to accusations of entertaining a Eurocentric (Alneng, 2002; Winter, 2009), 'post-colonialist' view of tourism (Hall and Tucker, 2004), in which the Westerner is the tourist and the non-Westerner is the 'touree', with apparently universal theoretical claims derived from this basis (Nyíri, 2006). The increasing engagement of emerging regions including Asia, Latin America, Africa and the Middle East at the domestic, intra-national and long-haul international scales is more differentiated and entangled with other forms of discretionary mobilities (a concept resembling Hall's [2005] 'voluntary temporary mobilities') (Cohen and Cohen, 2015), although the forms of movement and associated practices are becoming increasingly similar to their Western counterparts under the influences of globalization. However, the differences in the patterns of motivation and experience are still salient. For example, the new form of the 'tourist gaze' from an oriental perspective is aspired by expectations for a sight to be 'developed' rather than that of Western tourists, who value 'authenticity' (Nyíri, 2009: 165). This distinguishes carefully between the Chinese and postmodern attitudes to 'authenticity', and that unlike the 'post-tourists' of Western literature who have distracted their focus from pursuing the 'authentic', Chinese tourists seem to 'play along' because they consider participating in performances of the nation visited serious business.

With the emergence of the innovative paradigmatic and theoretical perspectives, a mobility paradigm has been proposed by some scholars as an appropriate theoretical framework for studies of tourism in emerging regions (see for example, Cresswell, 2006; Hannam et al., 2006; Urry, 2000, 2007). As Urry (2002: 256) posited, '"being on the move" has become a "way of life" for many'. The emerging mobility paradigm is challenging the 'sedentarist' approach of traditional social science research and putting forward a new 'deterritorialized' approach highlighting the pervasive conditions of mobility, fluidity and liquidity in globalization and postmodernity (see Sheller and Urry, 2006; Cresswell, 2006). As well, an interdisciplinary approach in offering the term 'lifestyle mobility' is also promoted as a theoretical lens to challenge current thinking on the intersections between travelling, leisure and migration (see also McIntyre, 2009). Lifestyle mobility has contributed to a breakdown of the binary relationship between work and leisure and has destabilized notions of 'home' and 'away'. Mobility studies imply pathways to manage the complexity associated with modern living, while future investigations can inquire deeper into the social class and power asymmetries and the significance of time and distance being reconfigured.

CONCLUSION

Globalization is often considered as a process by which events, decisions, and activities in one part of the world can come to have significant consequences for individuals and communities in other, quite distant parts (Giddens, 1990). This growing global interconnectedness and interdependency is opening up the cross-cultural production of local meanings, self-images, representations, and modes of life typical of various groups and individuals (Appadurai, 1996). Under the interactions between global and local forces, tourist destinations are endlessly (re-)invented, (re-)produced, (re-)captured and (re-)created. Thus, tourism activity not only transforms places and landscapes, but also produces meanings and representations. By employing the analytical structure of a global–local nexus, this chapter has examined the trends and developments in tourism geography in a globalizing world. The theories, principles and methodologies in geographical tourism research are reviewed through spatiotemporal, social and organizational lenses.

Tourism has offered excellent opportunities to study globalization, and pushed forward understanding of the changing world and human-place relations under the influence of globalization and modernization. However, it is also argued by the researchers that the dominating theoretical approaches have been based on Eurocentric and 'post-colonialist' perspectives, with the premise that modern tourism originally developed in Western society. Contemporary tourism development in the emerging regions and countries is challenging the existing approaches. The motivations, patterns of movement and sociocultural interactions of emerging-nation tourists are distinguished from those of Western counterparts. And the new phenomenon and experience can potentially contribute to the existing theories. Future investigations need to probe deeper into the connections between the wider social and cultural trends and patterns of discretionary travel in both Western society and emerging regions. The implementation of a wider theoretical framework of mobility is also prospective for shaping the future lens of investigation for geographers in global tourism.

REFERENCES

Alneng, V. (2002), 'The modern does not care for natives: travel ethnography and the conventions of form', *Tourist Studies*, **2**(2), 119–142.

Appadurai, A. (1996), *Modernity at Large: Cultural Dimensions of Globalization*, Minneapolis: University of Minnesota Press.

Britton, S.G. (1982), 'The political economy of tourism in the Third World', *Annals of Tourism Research*, **9**(3): 331–358.

Britton, S.G. (1991), 'Tourism, capital, and place: towards a critical geography of tourism', *Environment and Planning D: Society and Space*, **9**, 451–478.

Butler, R.W. (1980), 'The concept of a tourist area cycle of evolution and implications for management of resources', *The Canadian Geographer*, **XXIV**(1), Spring 1980, 5–12.

Butler, R.W. (1999), 'Sustainable tourism: a state-of-the-art review', *Tourism Geographies*, **1**, 7–25.

Butler, R.W. (2015), 'The evolution of tourism and tourism research', *Tourism Recreation Research*, **40**(1), 16–27.

Chang, T.C., S. Milne, D. Fallon and C. Pohlman (1996), 'Urban heritage tourism, the global–local nexus', *Annals of Tourism Research*, **23**, 284–305.

Che, D. (2017), 'Tourism geography and its central role in a globalized world', *Tourism Geographies*, **20**(1), 164-165.

Cohen, E. and S.A. Cohen (2015), 'A mobilities approach to tourism from emerging world regions', *Current Issues in Tourism*, **18**(1), 11–43.

Coles, T.E. and C.M. Hall (2006), 'Editorial: The geography of tourism is dead. Long live geographies of tourism and mobility', *Current Issues in Tourism*, **9**, 289–292.

Cooke, P. (ed.) (1989), *Localities: The Changing Face of Urban Britain London*, London: Unwin Hyman.

Cresswell, T. (2006), *On the Move: Mobility in the Modern Western World*, London: Routledge.

Feifer, M. (1985), *Going Places*, London: Macmillan.

Giddens, A. (1990), *The Consequences of Modernity*, Cambridge: Polity Press.

Goffman, E. (1959), *The Presentation of Self in Everyday Life*, Garden City, NY: Doubleday Anchor.

Gotham, K. (2005), 'Tourism gentrification: The case of New Orleans' Vieux Carre (French Quarter)', *Urban Studies*, **42**(7), 1099–1121.

Hall, C.M. (2005), 'Reconsidering the geography of tourism and contemporary mobility', *Geographical Research*, **43**(2), 125–139.

Hall, C.M. (2013), 'Framing tourism geography: note from the underground', *Annals of Tourism Research*, **43**, 601–623.

Hall, C. M. and S.J. Page (2006), *The Geography of Tourism and Recreation* (3rd edn), London: Routledge.

Hall, C. M. and T. Tucker (2004), *Tourism and Postcolonialism*, London: Routledge.

Hannam, K. and D. Knox (2010), *Understanding Tourism: A Critical Introduction*, London: Sage Publications.

Hannam, K., M. Sheller and J. Urry (2006), 'Editorial: Mobilities, immobilities and moorings', *Mobilities*, **1**(1), 1–22.

Hannerz, U. (1996), *Transnational Connections: Culture, People, Places*, London: Routledge.

Held, D., A.G. McGrew, D. Goldblatt and J. Perraton (2000), 'Global transformations: politics, economics, culture', *International Journal*, **54**(4), 218–219.

Kearns, G. and C. Philo (1993), *Selling Places: The City as Cultural Capital. Past and Present*, Oxford: Pergamon Press.

MacCannell, D. (1973) 'Staged authenticity: arrangements of social place in tourist setting', *American Journal of Sociology*, **79**, 586–603.

Massey, D. and J. Allen (eds) (1984), *Geography Matters! A Reader. Cambridge*, New York: Cambridge University Press.

McIntyre, N. (2009), 'Re-thinking amenity migration: integrating mobility, lifestyle and social-ecological systems', *Die Erde*, **140**(3), 229–250.

Meethan, K. (2001), *Tourism in Global Society: Place, Culture, Consumption*, New York: Palgrave.

Meyer-Arendt, K.J. and A.A. Lew (2003), 'Recreation, tourism and sport', in: G.L. Gaile and C.J. Willmott (eds), *Geography in America at the Dawn of the 21st Century*, pp. 524–540, Oxford: Oxford University Press.

Milne, S. and I. Ateljevic (2001), 'Tourism, economic development and the global–local nexus: theory embracing complexity', *Tourism Geographies*, **3**(4), 369–393.

Mosedale, J. (2014), 'Political economy of tourism', in: A.A. Lew, C.M. Hall and A.M. Williams (eds), *The Wiley Blackwell Companion to Tourism*, pp. 55–65, London: John Wiley & Sons.

Mowforth, M. and I. Munt (1998), *Tourism and Sustainability: New Tourism in the Third World*, London: Routledge.

Nyíri, P. (2006), *Scenic Spots: Chinese Tourism, The State and Cultural Authority*, Seattle: University of Washington Press.

Nyíri, P. (2009), 'Between encouragement and control: tourism, modernity and discipline in China', in: T. Winter, P. Teo, and T. C. Chang (eds), *Asia on Tour*, pp. 153–169, London: Routledge.

Pritchard, A., Morgan, N. and I. Ateljevic. (2011), 'Hopeful tourism: a new transformative perspective', *Annals of Tourism Research*, **38**(3), 941–963.

Salazar, N.B. (2005), 'Tourism and glocalization: "Local" tour guiding', *Annals of Tourism Research*, **32**(3), 628–646.

Scheyvens, R. and J.H. Momsen (2008), 'Tourism and poverty reduction: issues for small island states', *Tourism Geographies*, **10**, 22–41.

Scott, A.J. (1989), *New Industrial Spaces: Flexible Production Organisation and Regional Development in North America and Western Europe*, London: Pion.

Sheller, M., and J. Urry. (2006), 'The new mobilities paradigm', *Environment and Planning A*, **38**(2), 207–226.

Teo, P. and L.H. Li (2003), 'Global and local interactions in tourism', *Annals of Tourism Research*, **30**(2), 287–306.

Thorns, D.C. (1997), 'The global meets the local: tourism and the representation of the New Zealand city', *Urban Affairs Review*, **33**(2), 189–208.

Thrift, N. (2008), *Non-Representational Theory*, London: Routledge.

Torres, R. (2002), 'Cancun's tourism development from a Fordist spectrum of analysis', *Tourist Studies*, **2**(1), 87–116.

Torres, R. and J.D. Momsen (2005), 'Gringolandia: the construction of a new tourist space in Mexico', *Annals of the Association of American Geographers*, **95**(2), 314–335.

Tribe, J. (2005), 'New tourism research', *Tourism Recreation Research*, **30**(2), 5–8.

UNWTO (2011), *Tourism Towards 2030 – Global Overview*, Madrid, Spain: World Tourism Organization.

Urry, J. (2000), *Sociology beyond Societies: Mobilities for the Twenty-First Century*, London: Routledge.

Urry, J. (2002), 'Mobility and proximity', *Sociology*, **36**(2), 255–274.

Urry, J. (2007), *Mobilities*, Cambridge: Polity Press.

Urry, J. and J. Larsen (2011), *The Tourist Gaze 3.0*, London: Sage.

Waters, M. (1995), *Globalization*, London: Routledge.

Williams, A.M. (2004), 'Toward a political economy of tourism', in: A.A. Lew, C.M. Hall, and A.M. Williams (eds), *A Companion to Tourism*, pp. 61–73, Malden: Blackwell.

Winter, T. (2009), 'Asian tourism and the retreat of Anglo-Western centrism in tourism theory', *Current Issues in Tourism*, **12**(1), 21–31.

Wood, R.E. (2000), 'Caribbean cruise tourism: globalization at sea', *Annals of Tourism Research*, **27**(2), 345–370.

WTTC (2003), *Progress and Priorities 2003/04*, London: World Travel and Tourism Council.

15. Do you speak Globish? Geographies of the globalization of English and linguistic diversity[1]

Virginie Mamadouh

> I was recently waiting for a flight in Delhi, when I overheard a conversation between a Spanish UN peacekeeper and an Indian soldier. The Indian spoke no Spanish; the Spaniard spoke no Punjabi. Yet they understood one another easily. The language they spoke was a highly simplified form of English, without grammar or structure, but perfectly comprehensible, to them and to me. Only now do I realise that they were speaking 'Globish,' the newest and most widely spoken language in the world.
>
> (MacIntyre, 2005, reprinted in 2009: 239)
>
> Microsoft + English = Globish. (McCrum, 2012)

The omnipresence of English is one of the key cultural markers of globalization. The scene MacIntyre reports at the beginning of one of his 2005 columns about language in *The Times* and many similar observations (see for example Kurdi, 2006) might have been eye-openers only ten years ago. Nowadays they would probably go unnoticed. Still they remind us that while the spread of English has been associated with globalization, English has been profoundly transformed through globalization. In addition McCrum's formula underlines the role the new information and communication technologies play in the shaping of global English.

Therefore this chapter approaches the relation between the geographies of languages and globalization from the perspective of English as *the* language of globalization and its impact on linguistic diversity and will pay more specific attention to the Internet. Needless to say that there are many more aspects to the relation between globalization and languages (see Coupland, 2010 for a whole handbook devoted to the topic), and more specifically the language aspects of transnational mobility and migration (see Canagarajah, 2017 for a handbook on migration and language).

The chapter first briefly introduces the growing hegemony of English, especially in academia. It then turns to the debates regarding the evolution of English and relations between its users, including the discussion about Globish and World Englishes. Moreover it discusses the use of English and other languages on the Internet and its possible impact on offline language geographies. The conclusion offers directions for research agendas regarding the dynamics between linguistic homogenization and diversification associated with the current phase of globalization, and in the coping strategies of states and local communities.

[1]   The research leading to this chapter has received funding from the European Community's Seventh Framework Programme under grant agreement No. 613344 (MIME Project Mobility and Inclusion in Multilingual Europe).

GLOBALIZATION AND THE GLOBAL SPREAD OF ENGLISH

Alongside religious expansion, military conquest and occupation, colonization and political domination (see Ostler, 2005 for a language history of the world), an important vector of language spread has always been trade and the contacts and travels associated with trade relations. In this context, a lingua franca might emerge as language of communication. The term, for example, refers to the Mediterranean lingua franca (i.e. a free language, also known as sabir), a common linguistic code borrowing from Romance languages, Arabic, Berber languages from North Africa, that traders and sailors across the Mediterranean were using from the eleventh to the nineteenth century. Other historical and contemporary examples include Koine Greek in the Antique world, Latin in the Roman Empire and Europe in the Middle Ages and the Renaissance, Middle Low German in the Hanseatic League and the Balticum, Swahili in Eastern Africa, Cantonese (in the sense of Yue, not Cantonese proper which is the variety of Yue spoken in Guangzhou) in Southern China and South East Asia, Malay and Tagalog in South East Asia, Arabic in the Middle East, North Africa and the Sahel, Wolof in Senegal, Bambara in Mali and so on, and of course English in global trade.

Already widely spread around the world through British formal imperialism ('the sun never sets on the British Empire') and its informal imperialism (for example in South America after states obtained their independence from Spain and Portugal in the early nineteenth century), English became an even more important language of communication with the rise of the USA as a global power in the twentieth century. It began with the American intervention in the Great War and the request of American president Woodrow Wilson to negotiate the peace in Versailles in English alongside French, the diplomatic language of the time. The rise of English to its global status is further directly linked to the rise of the United States of America as a hegemonic power with the end of the Second World War and in the postwar period when the USA dominated the world economy, international relations (through the United Nations), military operations with its bases overseas, science, and popular culture with jazz music (and later pop music), Hollywood and Walt Disney. Although never really an explicit goal of British imperialism or American hegemony, the spread of English is the result of imperialism (Phillipson, 1992; Kachru, 1982).

Over 50 states and 25 non-sovereign entities now have English as the official language. English as diplomatic language has taken over French with the rise of international organizations after the Second World War, from the Universal Declaration of Human Rights (1948) to the United Nations. It is the (main) working language of most institutions of global governance, such as most UN organizations, especially those linked to economic governance and transnational infrastructure (air traffic, Internet, maritime transport, banking). This applies to many macroregional organizations too, even when their Member States have other official languages (as in the case of ASEAN).

In its cultural expression, globalization is often equaled to Americanization, sometimes symbolized by Walt Disney or McDonald's, but even more pervasively, culturally speaking, with the English language. It is shown in the borrowing of English words and expressions (especially for new technologies) in many languages across the world (including invented ones from the Dutch *beamer* for LCD projector, the German *handy* for cell phone, the French *Best of* for an anthology, the Japanese アイス *aisu* for ice cream, or the Chinese 'super app' *WeChat* which cumulates many more functions than messaging) and in the popularity of the acquisition of English as a foreign or second language.[2] First it was British English that constituted the most attractive and gratifying foreign language to learn to travel the world and to increase one's educational capital on the labour market and for prestige, later American English replaced it as the norm to emulate, although opinions still remain divided.

The global promotion of the British and/or American standard has had numerous consequences on social and economic relations (in which native speakers of English have an advantage over other speakers and get access to a wide range of jobs in teaching and communication linked to the use of correct English). The British Council was created by the British Foreign Office in the 1930s to promote the English language and culture abroad against the rise of fascism in Europe. It now has offices in over 200 cities in over 100 countries (organized in seven world regions) and is a major player in the language industry – worth almost GBP 1 billion – teaching English abroad and managing jointly with the University of Cambridge and others the International English Language Testing System (IELTS) – one of the major schemes of examination and certification of English skills across the world. The test is the only one accepted for visa requirements in the UK and Australia, and it is accepted by most academic institutions in English language countries or with English language programmes, and by various professional organizations across the world.

The main alternative certification TOEFL (Test of English as a Foreign Language) is run by an American non-profit organization called Educational Testing Services (ETS), and is also accepted in most academic institutions with language requirements for access to their English language programmes. After the Second World War, the US has also actively promoted language teaching as a soft tool (see Phillipson, 1992 for an extensive analysis of the historical evolution of English Language Teaching (ELT) and the British and American promotion of the language; see also Parmar, 2012 for the role of the Ford, Carnegie and Rockefeller foundations in the deployment of American soft power).

Next to the rising importance of English in trade and in the expanding banking and financial sector, and in the cultural sector, the role of English was greatly boosted by its hegemony in academic publishing and with the internationalization of academia. While Hamel (2007) shows that English was slightly ahead of French and German as the main language of science at the end of the nineteenth century, it was supplanted by German in the early twentieth century, peaking in the 1920s. Since the Second World War, English has become the main language of international academic exchanges, especially in the sciences. The geography of this internationalization and the impact of the growing hegemony of Anglo-American geography has been widely discussed at

---

[2]   There is no room here to elaborate on the different labels used to characterize language users. Suffice to signal that linguists and sociolinguists have critically analyzed the uses and the connotations of labels as 'mother tongue speaker' and 'native speaker', 'first language or L1 speaker', 'Second language or L2 speaker', 'foreign speaker', 'language learner' and so on, and the unduly conflation of language acquisition, identity, proficiency, and affect in these accounts. Indeed, the first language you learned from your mother might not be the one you are the most proficient in, or the one closest to your heart and to your emotions, or the one that you use to assert your identity in a given situation.

international conferences and in the pages of (English language) international geography journals (see Minca, 2000; Short et al., 2001; Garcia-Ramon, 2003; Aalbers, 2004; Belina, 2005; Kitchin, 2005 for early commentaries, Paasi, 2015; Fregonese, 2017; Van Meeteren, 2018 for recent ones) with similar debates in other disciplines (Canagarajah, 2002 from the view of Jaffna, Sri Lanka in the midst of civil war).

The hegemony of English as the language of communication at conferences and in graduate programmes, and as the language of academic publishing, is also paralleled by the dominant position of Anglo-American institutions (highly reputed departments, high-ranking academic journals and publishers, major professional associations, major annual disciplinary conferences). Academic publishers in continental Europe have switched to publishing mainly in English. International academic journals of national associations have been publishing more and more in English or even English only (without necessarily changing their name: *Tijdschrift voor Economische en Sociale Geografie* in the Netherlands, *Fennia* in Finland, *Erdkunde: Archiv für wissenschaftliche Geographie* in Germany). The resulting power relations have been criticized extensively for promoting epistemologies and theories developed in the English-speaking countries, as well as the empirical cases they study, at the expense of the circulation of the theories and the findings of scholars working at the margins (especially Belina, 2005; Fregonese, 2017 about the myth of a lingua franca).

The explosion of English language programmes in higher education is another indicator of this globalization of academia. American universities and colleges (labelled, financed, staffed and often accredited from the US) have mushroomed in the Middle East and North Africa, Central and Eastern Europe and Central Asia as well as dependences of prestigious (American) universities in the Gulf States (as a response to the combined effect of the attractiveness of their degrees and the visa restrictions after 9/11 2001). Universities in English-speaking countries (especially the UK and the US, but also Australia, Canada, Ireland and New Zealand) attract large numbers of foreign students, and other universities (in Singapore, in continental Europe) have started to offer international (read English language) programmes to compete with them and recruit international students. In Europe, the ERASMUS programme of the European Commission meant to promote student mobility within the European Communities (and later the European Union) has turned out to be a major incentive to offer English-language courses to cater to these exchange students. It was not expected from these students to learn, for example Dutch, Danish or Finnish to be able to take classes in the national language of their country of destination. And from there, it triggered the Anglicization of the whole curriculum (mostly for efficiency reasons, offering a course only once in the language understood by both local and exchange students, sometimes to boost encounters between the two groups) (see more in general Hébert and Abdi, 2013 on the internationalization of education and student mobility). Likewise, English is the working language for the applications for the research funds of the European Union and their evaluation – notwithstanding the 23 other official languages of the Union.

The economic consequences of such privileged access to knowledge and funding for native speakers is difficult to evaluate precisely but it goes without doubt that it has contributed to unequal geographies of knowledge. In the early 1990s Kaplan concluded boldly:

The English speaking nations now hold an information cartel which makes OPEC look like child's play. There are several reasons for this condition. First, petrochemical substances decreases as they are used, but information increases as it is used; second, petrochemical substances involve vast costs for exploration, exploitation, refining, and transporting, and petrochemicals are not uniformly distributed around the world while information is, and its exploitation involves much smaller incremental costs. (Kaplan, 1993: 155)

FIFTY SHADES OF ENGLISH

One of the major cultural consequences of the globalization of English is that it has prompted different perspectives on languages. After the decolonization of the mid-twentieth century, postcolonial linguists had already started to question the English ownership of the English language. The term of World Englishes (Kachru et al., 2007, see also the academic journal founded in 1985 and Jenkins, 2003, later some proposed the term Global Englishes, Pennycook, 2007; Jenkins, 2015) became popular to foreground the diversity of English varieties across the world but even more the legitimacy of these varieties. Neither the Received Pronunciation, the Queen's English, or the BBC English, nor American Standard English can still be seen as the golden standards of English they once were.

Kachru's model of the three circles of speakers has been particularly influential (Kachru, 1982, 1996) in the framing of the world of English. He distinguishes an inner circle of traditional first language speakers (the UK, Ireland, the USA, Canada, Australia and New Zealand), an outer circle where the English is institutionalized (including India, Nigeria and about fifty states, often former British colonies, plus South Africa and the Philippines), and the Expanding Circle where English is making advances as a second language for its importance for global communication (this includes countries where English is widely used in international communication, often with quite some governmental support, like in China and Chile). These categories are not fixed. Singapore has arguably moved from the Outer Circle to the Inner Circle as English becomes a home language of more and more Singaporeans, and English has become so widespread among the Dutch population in many social domains that the Netherlands could arguably be seen as part of the Outer Circle instead of the Expanding Circle and that some variety of Dutch English is emerging (Edwards, 2016) and cities like Amsterdam, or Copenhagen for that matter, are routinely described as 'effectively bilingual' (Kuper, 2018).

This expansion of the circles of English speakers has caused the speakers of English as second or third language to outnumber mother tongue speakers. This has important consequences for discussions about 'ownership' of the language and about standard norms. The demise of native speakers as the owners of English and the erosion of the hegemony of the native speakers' norm (that is the hegemonic standard norm in countries of the Inner Circle especially the UK) translates slowly into less demand for (near) native speakers' skills in job openings. Second-language speakers thus experience processes of emancipation. This happened first with the speakers of English in former colonies (hence the rich English language literature from outside Britain is not condescendingly labelled 'Commonwealth literature' anymore) and later with the speakers in the Expanding Circle. Arguably the dividend of speaking English is not reserved to native speakers (Rose, 2008; Kuper, 2018).

This prompted some sociolinguists to announce a shift from English as foreign language (EFL) to English as lingua franca (ELF), a language that belongs to everyone, and to signal the emergence of alternative norms from the interaction between users of English freed from the urge to imitate (British or American) native speakers (for example Seidhofer, 2011). Alternative labels for ELF include International English, the English language as a global means of communication, Global English, World English, Common English, Continental English or European English, Erasmus English (named after the students exchange programme of the European Commission), General English, Engas (for English as associate language).

A step further has been the promotion of Globish.[3] Although sometimes used in a pejorative way for broken English, the term Globish has been introduced as a positive label for a form or a variety of English better suited for global communication. Its promotor is the French businessman Jean-Paul Nerrière – a former vice president of IBM USA – who presented in 2004 this international auxiliary language based on English and published textbooks to learn Globish: with tips for a better pronunciation (the disconnect between sounds and spelling being a nightmare of foreign learners and the importance of intonation being notoriously difficult for French speakers), some basic vocabulary and pragmatic recommendations (avoiding British or American idioms) (Nerrière, 2004; Nerrière et al., 2005; Nerrière and Horn, 2009).

Such initiative is not particularly original: simplified languages were designed earlier in the colonial empires (there was the *français tirailleur* designed for the colonial recruits in the French army and Simplified English was created as international auxiliary language in the 1920s to compete with Esperanto). Other controlled natural languages based on English to reduce complexity and ambiguity and facilitate machine translation, are more recent: Simplified Technical English in the aerospace industry and Seaspeak designed to facilitate communication between ships and since 2001 formalized as Standard Marine Communication Phrases (SMCP). Globish is novel, however, in the sense that it is a rather bottom-up initiative from a non-native speaker of global English.

Nerrière's handbooks have been translated in many languages and the initiative got wide coverage internationally in the mid-2000s. Nerrière's initiative originates in his global experience as a businessman and observations of how non-native speakers of English communicate with each other, but is also rooted in a French debate about the role of English and its threat for (the international status of) French. Globish is meant as a tool for communication, not the language of Shakespeare and the English culture. *The Guardian* summarized it as follows:

> [a]ll this goes hand-in-hand with the Globish package deal: mass consumerism and tourism are opposite sides of the same coin. In hundreds of daily transactions, the language of Thomas Cook employees, of MasterCard call centres and the Sheraton check-out desk will be

---

[3] Globish was used earlier by Madhukar Gogate for a simplified spelling of (Indian) English. Nerrière's initiative is unrelated to that one.

Globish, the global dialect of No Logo capitalism,[4] and a kind of universal No Lingo. (McCrum, 2006)

Or as the French linguist Barbara Cassin (2017) puts it: 'it is a language of service, not a language of culture' (or a language of communication, not of identification, according to House, 2003: 559). McCrum summarized it even more succinctly:

Microsoft + English = Globish. (McCrum, 2012)

Nerrière explicitly sees Globish as a way to limit the influence of English and the power privilege of native speakers of English:

> I am helping the rescue of French and of all the languages that are threatened by English today but which will not be at all endangered by Globish. It is in the best interests of non-Anglophone countries to support Globish, especially if you like your culture and its language. (Nerrière quoted in McCrum, 2006)

McCrum later published a book on the phenomenon *Globish: How the English Language became the World's Language* (2011). The fascination with the exceptional position of English is widely shared. Many books have addressed its global status (Graddol et al., 1996; Crystal, 1997; Ostler, 2010). De Swaan (2001) for example describes a global language system in which languages are hyper-collective goods whose value increases with the number of speakers (native speakers and bilinguals building bridges with other languages), and English as the hyper-central language in this system (see De Swaan, 1993).

Even more illuminating are the two reports Graddol wrote for the British Council on *The future of English* (Graddol, 2000[1997]) and *English Next* (Graddol, 2006). Both offer a thorough assessment of the position of English and future developments from the point of view of English language teaching and more specifically Britain as a brand in that exploding market. The adaptation to changing demands for English is deemed crucial and the subtitle of the second report echoes the scope and the speed of the changes: *Why global English may mean the end of 'English as a Foreign Language'*. Indeed Graddol discusses the shift from Foreign Language (EFL) to Second Language (ESL) to Lingua Franca (ELF), to English for Young Learners (EYL) and also considers the need to anticipate the nearing peak in demand.

Other publications are much more focused on English as a threat and on the spectre of linguistic homogenization as outcome globalization. Linguists specialized in linguistic rights and human rights speak of English as a killer language and of linguistic imperialism (Phillipson, 1992), sometimes linking linguistic diversity to biodiversity (Skutnabb-Kangas et al., 2003; Maffi, 2005). Others have reviewed the linguistic geostrategies of states facing English hegemony (Maurais and Morris, 2003), and consider how linguistic justice could be achieved (Van Parijs, 2011).

Last but not least, the shift from English to Englishes has brought about an even more radical rethinking of languages, through the foregrounding of multilingualism, of other ways to deal with linguistic diversity than sequential monolingualism, i.e. the

---

[4] This refers to Naomi Klein's bestseller (Klein, 2000).

separation of different languages. Next to switching to English, other possibilities for communication across languages are made visible as viable options (Jørgensen, 2011; Grin, 2018). These include the use of other languages of wider communication, inter-comprehension (or even more broadly speaking *lingua receptiva*, when speakers develop a passive knowledge of each other's language, either because they are mutually intelligible or through exposure and study), codeswitching and codemeshing (the switching between codes/languages within a conversation) or more radically translanguaging (the use of linguistic resources from different codes to communicate), the recruitment of mediators like translators and interpreters, and the use of machine translation. The presence of other languages next to English is a timely reminder that globalization is not only a tale of English(es), it has also dramatically expanded the networks and the geographical range of many other languages, as they traveled with migrants to new localities.

LINGUISTIC DIVERSITY AND THE GLOBALIZATION OF THE INTERNET

The Internet was originally a powerful vector of Anglicization. Content was originally 100 per cent in English, both due to the origin of the network, its use in US military and international academia and technical limitations of the ASCII characters (unsuited for diacritic signs and for languages written in other alphabets or ideograms). It took some time before codes were developed to accommodate other characters and other writing systems than the Latin alphabet, and even longer before they were used in URL. In parallel, the access to Internet has exponentially increased, even if serious digital divides remain between regions, countries, social groups and generations, depending on access to infrastructure, political regulations, and digital literacy.

According to available surveys, English is still the largest language in terms of its share of users and even more in terms of content (share of websites). Nevertheless the Internet is now also a vector of linguistic diversity, even for oral languages or signs languages, since multimedia websites have eased transmission and interaction over great distances beyond writing. These specific opportunities the new media offer have turned the Internet into a powerful tool for communication between few scattered speakers of smaller languages.

Recent Internet statistics show that Chinese is catching up English in terms of Internet users, with an estimated 771 million users against 985 million, and that the process is likely to continue since the Internet penetration (the ratio between the sum of Internet users speaking a language and the total population speaking that language) is still much higher for English (70 per cent compared to 55 per cent[5]). These statistics are poor estimations: they are based on residence and foremost they allocate each person to one and only one language. They notoriously fail to take account of multilingual users and multilingual websites. Nevertheless, they provide some insights in the linguistic geographies of the virtual world. Despite the great linguistic diversity among

---

[5]   http://www.internetworldstats.com/stats7.htm (last accessed December 2017).

humans (there are about 6000 languages), the top 10 languages suffice to reach 78 per cent of the world population.[6] In terms of content, the domination of English is eroding more slowly and seems to stabilize around 52 per cent[7] followed by Russian with 7 per cent, German with 6 per cent, Spanish and Japanese (5 per cent each) and French (4 per cent). Chinese (2 per cent) still ranks much lower behind Portuguese and Italian.

Symbolically, the Internet is still an American home: the use of top domain names such as .com, .net or .edu (revealing the function of the website rather than its nationality, in these cases distinguishing between a business, a network and a university) remains largely the realm of US-based websites, by contrast to the localized domain names others use like .uk or .ca. This is a direct expression of the American hegemony on the Internet. Nevertheless, the Internet can also be a symbol and a vector of linguistic diversity.

More recently new technological development in the field of automatic translation in combination with mobile telephony and mobile apps have revolutionized language mediation and language teaching. *Google Translate* was launched in 2006 and brought machine translation to everyone (that is: providing one accesses the Internet in a territory where the state does not block Google), allegedly supporting over 100 languages (at various levels) and serving about half a billion people daily. Likewise language learning has been revolutionized by computer-assisted language learning and more specifically by apps like *Duolingo* (launched in 2011 in Pittsburgh with over 200 million registered users and providing a free website and app version) and *Babbel* (launched in 2007 in Berlin). They promote multilingualism, but at the same time reproduce the hyper-centrality of English, the language offered to learners from all other languages and the language interface offering the largest number of languages to learners.

The most popular p2p platforms have likewise participated to both the further promotion of English and the global visibility of other languages: *YouTube* and *Facebook* allowing for content in whatever language, but also *Netflix* and dubbed series. *Wikipedia*, the online grassroots encyclopedia, is available in 298 languages but the size (and the quality) of the available information varies greatly.[8] Unsurprisingly, the English version is the largest with almost 6 million articles, accounting for 12 per cent of the articles, but it is closely followed by Cebuano (the second language of the Philippines after Tagalog), Swedish, German and French. Only 13 language versions have more than a million articles and another 45 over 100 000. A few versions have less than ten articles and two African languages (Herero and Kanuri) have none apart from the main page. *Wikipedia* does not have a Globish version, but it does carry one in Simple English, based on controlled versions of English meant for foreign language speakers, especially Basic English (created as international auxiliary language in the 1920s to compete with Esperanto) and Special English (created for the broadcasting of the Voice of America at the end of the 1950s). Simple English ranks 51st between Azerbaijani and Urdu, higher than Latin, Hindi or Thai but lower than Chechen or

---

6   https://www.internetworldstats.com/languages.htm (last accessed December 2017).

7   https://w3techs.com/technologies/overview/content_language/all and https://w3techs.com/technologies/history_overview/content_language (last accessed December 2017).

8   https://meta.wikipedia.org/wiki/list_of_wikipedias (last accessed December 2017). https://en.wikipedia.org/wiki/List_of_Wikipedias#Detailed_list (last accessed December 2017).

Esperanto. It is noteworthy that the overlap between the concepts covered by different versions is limited (Hecht and Gergle, 2010): about 51 per cent of the articles of the German version are covered in the English one. Some versions were created through automatic translation by bots (computer programs that do automated tasks – in this case automatic translation) and obviously are very similar to the English source. This explains the particularly large size of the version in Cebuano and in Swedish.

The interaction of online and offline is a new field of studies and a topic of social and political debates. Geographers have explored the impact of language on our view of the world through different content analyses. Graham et al. (2014) have compared the geotagged content in different *Wikipedia* versions revealing different concentration patterns for Arabic, Persian, Hebrew or Swahili. Graham and Zook (2013) speak of geographies of uneven digital augmentations: they analyze the results of online searches in *Google Maps* for the same keywords in different languages, discussing for example the completely different options of restaurants proposed in Tel Aviv when using Arabic than when using English and Hebrew (Graham and Zook, 2013: Figure 12 a, b and c on pp. 94–95).

Arguably the Internet and its augmented reality has created new sites of language contact (i.e. sites of contact between language groups), and those interactions generally lead to linguistic change (such as borrowings, bilingualism and possibly diglossia,[9] or even death of one of the languages), and social change in the power relations between the language groups). Icelandic has recently been said to be on the verge of 'digital extinction' (Henley, 2018) because it is not isolated anymore. The younger generation grows up partly online and is immersed in an online sea of English as a consequence of the paucity of content in Icelandic (due both to the very small size of the community of speakers and the specificities of the language that makes automatic translation from English very complex). The Icelandic linguistic Eiríkur Rögnvaldsson calls it a process of 'digital minoritisation', and a recent set of studies shows that 21 European languages – now firmly institutionalized – are threatened by 'digital extinction' because they are weakly or not supported for digital services such as machine translation, speech processing, text analysis, speech and text resources.[10] The issue is that for the big players on the Internet (*Apple*, *Amazon*, *Google* etc.), providing digital support in Icelandic for example, is as expensive as in French, while business opportunities do not match up. This forces speakers of these smaller languages to use a more powerful language like English for daily online activities such as booking a flight or a hotel, searching for an address or an itinerary, watching streaming videos or looking for online educational content.

Language activists defending lesser used languages are also well aware of this challenge. In 2016 a grassroots initiative brought Frisian to *Google Translate* after a team of volunteers had provided enough corpus (i.e. translation samples) for the online translation machine to function, to boost the visibility of the language and ensure its

---

[9]   The use of two languages in a community for different function: a low variety for interpersonal relations and a high variety for formal situations: for example Swiss German and Standard German in Switzerland.

[10]   http://www.meta-net.eu/whitepapers/key-results-and-cross-language-comparison (last accessed December 2017).

online presence.[11] And they have also made sure that the home page of Google in Dutch mentions that searching is possible in Frisian too.

CONCLUSION

> Globalization's consequences for languages extend across the full range of what count as languages, their status, their corpus, and discourse forms, their spatial and sociological distribution and cultural and ideological aspects. (Lo Bianco, 2014: 315)

While English will probably consolidate its global position in the short run, it remains to be seen if it will evolve towards a global English standard (as Globish or under another label) or towards fragmentation in many World Englishes (some of them not mutual intelligible as envisioned by Mufwene, 2010: 46) in which case English might in the long run be displaced by one of her daughters (Chinglish? Hinglish? Spranglish?), like Latin was replaced by French in Europe during the Enlightenment and later by English. Alternatively it could lose ground to the language of a more powerful group, possibly Mandarin, hence the mushrooming of Chinese classes in New York and other North American cities. The impact on other languages is even more difficult to predict, and varies greatly with the demographic, economic, political and cultural weight of a specific language and the contingency of globalization processes.

From a geographical perspective, the language dimensions of globalization greatly but unevenly impact the cultural, economic, political, and social characteristics of places and states. Globalization brought English and a wide array of languages online and of migrant languages to states whose past national identities, geopolitical representations and language policies were monolingual and that might have difficulties to develop new discourses, strategies and policies to deal with this new linguistic diversity (Lo Bianco, 2005, 2014). The linguistic diversity in individual states and localities has never been so large.

Research remains particularly needed to understand these impacts and the discursive representations, the strategies and the policies developed by states (and local communities) to cope with English, with migrant languages and with the transformation both may have on the use of national (and local) languages. Initiatives to get a grip on these changes range from compulsory language tests to force newcomers to learn the national language, laws making compulsory its use in marketing and broadcasting (the *Loi Toubon* in France), attempts to protect the national language from borrowings and 'contaminations' from English, campaigns to promote one's language internationally, but also policies geared towards better and earlier access to English teaching, the Anglicization of higher education or the protection of Standard English against creolization (such as the state campaign for the use of Standard English instead of Singlish in Singapore, see Wee, 2018) and policies to support one's language in the

---

[11] Duly reported in Duchnews.nl the English language news website for expats in the Netherlands on 17 February 2016 http://www.dutchnews.nl/news/archives/2016/02/frisian-added-to-google-translate/ (last accessed 14 June 2018).

digital world. While the trajectories of specific languages, including that of English itself, remain difficult to predict, the amplification of uneven linguistic geographies is very likely.

REFERENCES

Aalbers, M.B. (2004), 'Observation. Creative destruction through the Anglo-American hegemony: a non-Anglo-American view on publications, referees and language', *Area*, **36**(3), 319–322.

Belina, B. (2005), 'Anglophones: if you want us to understand you, you will have to speak understandably!', *Antipode*, **37**(5), 853–55.

Canagarajah, A.R. (2002), *A Geopolitics of Academic Writing*, Pittsburgh: University of Pittsburgh Press.

Canagarajah, S. (ed.) (2017), *The Routledge Handbook of Migration and Language*, London: Routledge.

Cassin, B. (2017), 'The power of bilingualism'. https://conversations.e-flux.com/t/the-power-of-bilingualism-interview-with-barbara-cassin-french-philosopher-and-philologist/6252 (accessed 14 June 2018).

Coupland, N. (ed.) (2010), *The Handbook of Language and Globalization*, Malden: Wiley Blackwell.

Crystal, D. (1997), *English as a Global Language*, Cambridge: Cambridge University Press.

Edwards, A. (2016), *English in the Netherlands: Functions, Forms and Attitude*, Amsterdam: John Benjamins.

Fregonese, S. (2017), 'English: lingua franca or disenfranchising?', *Fennia*, **195**(2), 194–196.

Garcia-Ramon, M.-D. (2003), 'Globalization and international geography: the questions of languages and scholarly traditions', *Progress in Human Geography*, **27**(1), 1–5.

Graddol, D. (2000[1997]), *The Future of English? A Guide to Forecasting the Popularity of the English Language in the 21st Century*, London: The British Council.

Graddol, D. (2006), *English Next. Why Global English May Mean the End of 'English as a Foreign Language'*, London: The British Council.

Graddol, D., D. Leith and J. Swann (eds) (1996), *English: History, Diversity and Change*, London: Routledge / The Open University.

Graham, M., N. Hogan, R.K. Straumann and A. Medhat (2014), 'Uneven geographies of user-generated information: patterns of increasing informational poverty', *Annals of the Association of American Geographers*, **104**(4), 746–764.

Graham, M. and M. Zook (2013), 'Augmented realities and uneven geographies: exploring the geolinguistic contours of the web', *Environment and Planning A*, **45**(1), 77–99.

Grin, F. (ed.) (2018) *The MIME Vademecum. Mobility and Inclusion in Multilingual Europe*. Genève: MIME Project 978-2-8399-2402. http://www.mime-project.org/vademecum/ (accessed 1 July 2018).

Hamel, R.E. (2007), 'The dominance of English in the international scientific periodical literature and the future of language use in science', *AILA Review*, **20**, 53–71.

Hébert, Y. and A.A. Abdi (eds) (2013), *Critical Perspectives on International Education*, Rotterdam: Sense Publishing.

Hecht, B. and D. Gergle (2010), 'The Tower of Babel meets Web 2.0', *CHI2010*, Atlanta, Georgia.

Henley, J. (2018), 'Icelandic language battles threat of "digital extinction". *The Guardian*, 26 February 2018.

House, J. (2003), 'English as a lingua franca: a threat to multilingualism?', *Journal of Sociolinguistics*, **7**(4), 556–578.

Jenkins, J. (2003), *World Englishes: A Resource Book for Students*, London: Routledge.

Jenkins, J. (2015), *Global Englishes: A Resource Book for Students*, London: Routledge.

Jørgensen, J.N. (ed.) (2011), *A Toolkit for Transnational Communication in Europe*, Copenhagen: University of Copenhagen.

Kachru, B. (ed.) (1982), *The Other Tongue : English Across Cultures*, Urbana-Champaign: University of Illinois Press.

Kachru, B.B. (1996) 'World Englishes: agony and ecstasy', *The Journal of Aesthetic Education*, **30** (2), 135–155.

Kachru, B., Y. Kachru and C. Nelson (eds) (2007), *Handbook of World Englishes*, Malden: Blackwell.

Kaplan, R.B. (1993), 'The hegemony of English in science and technology', *Journal of Multilingual and Multicultural Development*, **14**(1–2).

Kitchin, R. (2005), 'Commentary: disrupting and destabilizing Anglo-American and English-language hegemony in geography', *Social & Cultural Geography*, **6**(1), 1–15.

Klein, N. (2000), *No Logo: No Space, No Choice, No Jobs: Taking Aim at the Brand Bullies*, London: Flamingo.

Kuper, S. (2018), 'Why Globish no longer cuts it', *Financial Times*, 13 January 2018.

Kurdi, I. (2006), 'Do you speak Globish?: How a global language tweaks English', *Arab News*, 26 December 2006.

Lo Bianco, J. (2005), 'Globalisation and national communities of communication', *Language Problems and Language Planning*, **29**(2), 109–135.

Lo Bianco, J. (2014), 'Domesticating the foreign: globalization's effects on the place/s of languages', *The Modern Language Journal*, **98**(1), 312–325.

MacIntyre, B. (2005), 'Useful language learn quick now!', *The Times*, 30 April 2005.

MacIntyre, B. (2009), *The Last Word: Tales from the Tip of the Mother Tongue*, London: Bloomsbury.

Maffi, L. (2005), 'Linguistic, cultural, and biological diversity', *Annual Review of Anthropology*, **34**, 599–617.

Maurais, J. and M.A. Morris (eds) (2003), *Languages in a Globalising World*, Cambridge: Cambridge University Press.

McCrum, R. (2006), 'So, what's this Globish revolution?', *The Guardian*, 3 December 2006.

McCrum, R. (2011), *Globish: How the English Language Became the World's Language,* New York: Norton.

McCrum, R. (2012), 'The rise of global English', *Oxford English Dictionary*. Blog entry, 16 August 2012: https://public.oed.com/blog/the-rise-of-global-english/ (accessed 25 June 2018).

Minca, C. (2000), 'Venetian geographical praxis', *Society and Space*, **18**, 285–289.

Mufwene, S. S. (2010), 'Globalization, global English, and world English(es): myths and facts', in: N. Coupland (ed.), *The Handbook of Language and Globalization*, Chichester: Wiley-Blackwell, pp. 31–55.

Nerrière, J.-P. (2004), *Parlez Globish. L'anglais planétaire du troisième millénaire*, Paris: Eyrolles.

Nerrière, J.-P., P. Dufresne and J. Bourgon (2005), *Découvrez le globish. L'anglais allégé en 26 étapes*, Paris: Eyrolles.

Nerrière, J.-P. and D. Horn (2009), *Globish the World Over: A Book Written in Globish*, Lexington, KY: International Globish Institute.

Ostler, N. (2005), *Empires of the Word: A Language History of the World*, London: HarperCollins.

Ostler, N. (2010), *The Last Lingua Franca: English until the Return of Babel*, London: Allen Lane.

Paasi, A. (2015), 'Academic capitalism and the geopolitics of knowledge', in: J. Agnew, V. Mamadouh, A. Secor and J. Sharp (eds), *The Wiley-Blackwell Companion to Political Geography*, Chichester: Wiley Blackwell, pp. 507–523.

Parmar, I. (2012), *Foundations of the American Century: The Ford, Carnegie, and Rockefeller Foundations in the Rise of American Power*, New York: Columbia University Press.

Pennycook, A. (2007), *Global Englishes and Transcultural Flows*, London: Routledge.

Phillipson, R. (1992), *Linguistic Imperialism*, Oxford: Oxford University Press.

Pimienta, D., D. Prado and Á. Blanco (2009), *Twelve Years of Measuring Linguistic Diversity in the Internet: Balance and Perspective*, Paris: UNESCO.

Rose, R. (2008), 'Political communication in a European public space: language, the Internet and understanding as soft power', *Journal of Common Market Studies*, **46**(2), 451–475.

Seidhofer, B. (2011), *Understanding English as a Lingua Franca*, Oxford: Oxford University Press.

Short, J.R., A. Boniche, Y. Kim and P. Li Li (2001), 'Cultural globalization, global English, and geography journals', *The Professional Geographer*, **53**(1), 1–11.

Skutnabb-Kangas, T., L. Maffi, and D. Harmon (2003), *Sharing a World of Difference: The Earth's Linguistic, Cultural and Biological Diversity*, Paris: UNESCO.

Swann, A. de (1993), 'The evolving European language system', *International Political Science Review*, **14**, 241–255.

Swann, A. de (2001), *Words of the World, The Global Language System*, Cambridge: Polity Press.

Van Meeteren, M. (2018), 'On geography's skewed transnationalization, anglophone hegemony, and qualified optimism toward an engaged pluralist future: A reply to Hassink, Gong and Marques', *International Journal of Urban Sciences*. DOI: 10.1080/12265934.2018.1467273.

Van Parijs, P. (2011), *Linguistic Justice for Europe and the World*, Oxford: Oxford University Press.

Wee, L. (2018), *The Singlish Controversy: Language, Culture and Identity in a Globalizing World*, Cambridge: Cambridge University Press.

16. Geographies of global production networks
Jana M. Kleibert and Rory Horner

INTRODUCTION: ECONOMIC GLOBALIZATION AND GLOBAL PRODUCTION

Under economic globalization, the organization and control over the production of goods and services often transcends the bounds of a regional or national economy. Indeed, now more than two decades ago, Gary Gereffi argued that 'globalization has eroded the traditional boundaries between nations, firms and industries' (Gereffi, 1996: 434). State-centric approaches to understanding trade, where one country is thought to trade finished products with another country, have thus become less relevant. Global lead firms have increasingly been recognized to be playing key roles in shaping and coordinating the global production and sourcing of goods for finished markets across countries, continents, and even globally. For example, major retailers or apparel companies are key actors in coordinating supply across multiple settings, despite sometimes not even possessing their own factories. In addition to these 'lead' firms, various tiers of suppliers, traders and services firms, as well as other stakeholders – including labour unions and civil society organizations – may be involved in shaping this production across national borders.

To understand the changed forms of organization, governance and development prospects resulting from increasingly complex globalized production, two related actor-centric approaches – global value chains (GVCs) and global production networks (GPNs) – have been developed. These closely related and sometimes intertwined approaches have been developed both consecutively and in parallel. Instead of providing a full review of their genealogies and idiosyncrasies (cf. Bair, 2005; Hess and Yeung, 2006; Parrilli et al., 2013; Coe and Yeung, 2015), we give a brief historical overview of their development and tease out some of their similarities and differences in relation to the focus of our intervention: their spatial dimension and engagement with territory. While drawing on both approaches, we direct our attention primarily to the latter approach, given its more explicit concern with and potential to address geographical dimensions.

The chapter proceeds as follows. In the next section, we introduce research on global value chains and global production networks which has made valuable progress in understanding new organizational arrangements as part of economic globalization. After introducing these organizational aspects, the following section focuses on the geographies of, including the articulation of territories into, transnationally networked production systems. Given that both approaches employ a *relational* perspective that foregrounds the linkages connecting different actors and regions, we argue that some aspects of the *territorial* dimension of these processes have taken the backseat in research on GVCs and GPNs. Building on vignettes from India's pharmaceuticals and the Philippines' services industry, we present a two-fold argument as to why the emerging geographies of global value chains and global production networks require more attention. First, we have a closer look at the changing multi-scalar dimensions of territoriality at which production networks operate; and second, we interrogate the territorial outcomes of network integration, further investigating the horizontal dimension of production networks, to develop a better understanding of territorial development.

KEY APPROACHES: GLOBAL COMMODITY CHAINS, GLOBAL VALUE CHAINS AND GLOBAL PRODUCTION NETWORKS

The GPN approach to understanding economic globalization is the result of a longer discussion and refinement of chain and network-based approaches to global production from the commodity chain concept in world-systems theory through to global commodity chain (GCC) and global value chain (GVC) approaches developed in the 1990s and 2000s (for a discussion and genealogy see Bair, 2005; Hess and Yeung, 2006). These relational approaches to understand economic development have a close affinity, although to some extent have remained parallel approaches with some distinct emphases (see Table 16.1). GVC research has been led most prominently by scholars at Duke University as well as at University of Sussex. Global production network research is most associated with scholars at the University of Manchester and the National University of Singapore (see Liu and Mei, 2016 for a recent bibliometric analysis of the field). The focus of all approaches has remained on the distribution of value creation activities among different types of actors and their associated implications for economic development.

The origin of the contemporary GPN approach can be found in the commodity chain concept, first developed by Terence Hopkins and Immanuel Wallerstein, as 'a network of labour and production processes whose end result is a finished commodity' (Hopkins and Wallerstein, 1986: 159). This approach started from a world-systems perspective, based on a framing of the world distinguishable by three distinct spheres: core, semi-periphery, and periphery. As opposed to concepts that took trade of products between states as the basis, this chain approach focused on one commodity and traced its production across different geographic regions. Value in this world-system was understood to be created and traded on very uneven terms, particularly between core and periphery. In illuminating the role of transnational factors and relationships, the emphasis was mostly on how integration into transnational production networks presented constraints for national development.

The commodity chains concept was given impetus, and took on a new direction, from a collection of papers (Gereffi and Korzeniewicz, 1994) arising out of a 1992 meeting of the Political Economy of the World-System section of the American Sociological Association. A key moment in the early development of the global commodity chains concept by Gary Gereffi was the re-articulation away from an interest in the power of the state in shaping global production (for example through tariffs and local content rules) towards a focus on the strategies and actions of firms (e.g. Gereffi, 1994). As opposed to the emphasis on constraints for development in the

Table 16.1 Key relational approaches to understand globalized production

Body of research	Organization & governance	Geographies of chains/ networks	Engagement with territory	Conceptual- ization of development	Disciplinary engagements	Key literature
GCC	Chain: Buyer-/ producer-driven relationships	Core–periphery (world-systems theory)	Limited, production sites as nodes. Implicit nation state reference	Value creation	Development studies, sociology, business studies	Gereffi, 1994; Bair and Gereffi, 2001
GVC	Chain: Inter-firm market, modular, relational, captive, hierarchy relationships	Global (implicitly North–South)	Limited, production sites as nodes. Implicit nation state reference	Upgrading	Organization studies, development studies	Gereffi et al., 2005; Ponte and Sturgeon, 2014
GPN	Network: Multi-scalar firm and non-firm actors and relationships	Multi-scalar (implicitly North–South)	Multi-scalar actors, Territorial embeddedness, sub-national regions	Value creation, enhancement and capture, strategic coupling	Economic geography, economic sociology	Henderson et al., 2002; Coe et al., 2004, 2008
GPN 2.0	Network: Dynamic, interplay of sector logics, context, and firm-strategies and relationships	Multi-scalar	Multi-scalar actors, Territorial embeddedness, dynamic coupling processes	Value creation, enhancement and capture, modes of strategic coupling	Economic geography, international business	Coe and Yeung, 2015

earlier world-systems-shaped approach, and shaped by economic sociology and development studies, a greater emphasis emerged on the possibilities for moving to more rewarding activities within global commodity chains – or upgrading. The global value chain framework has subsequently been developed as an elaboration of the global commodity chain framework through recognizing more organizational forms of inter-firm governance (Gereffi et al., 2005). With a greater focus on inter-firm linkages, the GVC approach thus draws more on the international business literature, drawing on Porter's (1985) conception of value chains, and less on the sociological literature.

As the 'chain' literature evolved, a group of economic geographers who have become known as the Manchester school developed the concept of global production networks (GPNs), building on, and responding to, what they saw as shortcomings of the GCC/GVC framework. This approach focuses on networks, which include both vertical and horizontal elements, in contrast to the more linear chains. Most importantly, the approach introduces an increased attention to geography, highlighting the multi-scalar nature of production networks, and including also a range of non-firm actors (for example the state, labour, consumers, civil society) in addition to firms in its analysis (Dicken et al., 2001; Henderson et al., 2002; Coe et al., 2004; Hess and Yeung, 2006; Coe et al., 2008). Economic development in the GPN approach is conceptualized as occurring through processes of value creation, enhancement and capture that may result when regions 'strategically couple' with GPNs.

More recently, the framework has been revised and advanced as GPN 2.0 by Neil Coe and Henry Yeung. In its new iteration, the framework seeks to theorize the dynamics of GPNs and identifies, on the one hand, capitalist dynamics that influence GPNs and, on the other, strategies of (firm) actors involved. Thereby, the approach seeks to reveal the causal mechanisms behind (uneven) geographic development resulting from different modes and types of regions' strategic coupling with GPNs (Coe and Yeung, 2015).

Although long recognized to have many differences, research on both GVCs and GPNs has been noted to display similarities as well (see for example Neilson et al., 2014) – such as sharing interest in the global organization of industries, notably the role of global lead firms, and also the development prospects for suppliers and regions (especially in the Global South) integrated into their chains and networks. While the organizational and spatiotemporal dimensions of GPNs are interrelated, we suggest that more analytical attention has been paid to the former as opposed to the latter. Thus, after introducing the organizational dimension of GVCs and GPNs, we will then explore the spatiotemporal dimension of GPNs.

ORGANIZATION AND GOVERNANCE OF GPNs

Initially four dimensions were identified as part of a global commodity chain – input–output structure, territoriality (geographical scope), governance structure, and institutional context (Gereffi, 1994). Gary Gereffi noted that 'although GCCs are very definitely influenced by their international, national, and local institutional settings, the main emphasis is on the organizational features of the chains' (Gereffi, 1996: 428). Indeed, the primary focus in research on GCCs and GVCs has been on the governance aspect, notably with an interest in control and distribution of value (see for example Bair, 2005: 159). Studies of GCCs and GVCs have identified various forms of governance or control that lead firms have over various suppliers. These are relation-ships which extend beyond ownership. The key initial distinction was between buyer- and producer-driven governance. Gereffi (1994) pointed to the rise of buyer-driven-ness, where leading retailers, often without owning factories, possess considerable control over a wide range of suppliers located across different geographical locations. Subsequent work on GVC governance has elaborated beyond the initial buyer-producer driven distinction to identify five key types of governance – market, modular, relational, captive and hierarchical (Gereffi et al., 2005). Standards stipulated by lead firms as well as public requirements were also identified as a crucial factor in the governance of entry barriers and setting requirements for suppliers within global production networks (Nadvi, 2008). More recently still, a conceptualization of multi-polar governance has been advanced – distinguishing between micro-governance, or linkages between individual nodes, meso-governance, or how governance travels across a chain, as well as macro-governance – in terms of the overall governance of a chain (Ponte and Sturgeon, 2014).

In addition to such GVC research on inter-firm governance, the global production network framework is conscious of intra-firm governance relations. Moreover, it recognizes a wider variety of actors who shape global production, including non-firm actors such as labour, and the state (Henderson et al., 2002; Coe et al., 2004; Coe et al., 2008; Smith, 2015; Horner, 2017). While fundamental in the functioning of GPNs, clearly these actors are not all able to wield the same power and are differentiated by their ability to influence decisions.

Understanding forms of governance in a chain has been identified as a crucial step for subsequently identifying prospects for upgrading and/or improving a firm's position in a chain (see for example Gereffi, 1999; Humphrey and Schmitz, 2002). Various types of upgrading have been identified including process (transporting inputs into outputs more efficiently), product (involving more sophisticated product lines), functional (new functions with greater skill content) and intersectoral (moving to new productive activities) (Humphrey and Schmitz, 2002). Development possibilities, emphasized as taking place through firms' upgrading, were largely understood to be related to the activities of lead firms in GPNs. A major hypothesis was that 'development requires linking up with the most significant lead firms in the industry' (Gereffi, 2001: 1622). The key, almost paradigmatic, example is the successful upgrading of East-Asian countries, whereby Japan (in the 1950s and 1960s), East Asian Newly Industrialising Economies (1970s, 1980s) and China (1990s) became leading exporters in buyer-driven chains, ultimately shifting from input assembly to Original Equipment Manufacturing (OEM) and eventually Own Brand Manufacture (OBM) (Gereffi, 1999: 37).

SPATIALITY AND TERRITORIALITY OF GPNs

Relational approaches, such as GCC/GVC and GPN analysis, are susceptible to an underappreciation of the territorial dimension. A dialectic tension has been noted between relational and territorial optics, making it 'extraordinarily difficult to conceptualize economic activity systems and territorially defined economies simultaneously' (Dicken and Malmberg, 2001: 345). While the GCC approach has usefully focused on cross-national economic organization, 'the geography ... is dealt with at a very high level of spatial aggregation, a clear reflection of the world-system ancestry of the GCC concept' (Dicken et al., 2001: 99–100) and its scalar distinction of core and periphery. In an essay on 'spatializing commodity chains', Leslie and Reimer (1999) similarly pointed to the need to understand the spatial-temporal variation of chains, space-mediating relationships between actors, and, finally, commodity chains resulting in the production of space.

GVC research, for the most part, has been interested in uncovering production relations linking suppliers in the Global South to markets in the Global North (somewhat consistent with the core-periphery aspect of world-systems thinking). The central empirical focus of such research has been on how agricultural and industrial producers in developing countries are integrated into global trade and their respective opportunities to gain from this arrangement through upgrading. The geographies of upgrading, however, have been less clearly defined and can take place at a variety of scales: the level of the national economy, the region, or the individual firm.

The GPN approach has been developed with the clear ambition to 'globalize' the conceptualization of regional development. Taking the sub-national region as its focus, such research has sought to better understand economic development opportunities in regional economies in relation to the possibilities of 'strategic coupling' with trans-national actors (Coe et al., 2004; Yeung, 2016). Stressing the (territorial) embeddedness of production, GPN scholars with a disciplinary background in (economic) geography have made an important contribution to 'bring in geography' without losing sight of power structures (Neilson and Pritchard, 2009: 36). The GPN approach has sought to employ a network perspective, which sees places as nodes, while simultaneously paying attention to the territorial spatial transformations in these places. For example, Henderson et al. argue that 'in order to understand the dynamics of development in a given place, then, we must ... focus on the flows *and* the places *and* their dialectical connections' (Henderson et al., 2002: 438, emphasis in original).

In the GPN framework, production processes are understood to be embedded in wider social and spatial relations and firms can become 'anchored' in specific places. While actors are explicitly located at multiple scales, and 'regions' are not clearly defined spatial arrangements, GPN research has largely been based on understanding regional development in places articulated into transnational networks of production organized and controlled by global lead firms, often located in the Global North. In the most recent iteration of GPN 2.0, the territoriality, defined as 'the spatial organization of global production network relations and developmental outcomes', is given more centrality. Territoriality is then differentiated between two dimensions: a vertical dimension of GPN relations, along which 'territoriality can be understood as multiple geographical scales along the global-local continuum', and a horizontal dimension (or the horizontal-territorial interface), which 'refers to the spatial form of grounding value activity know as territorial embeddedness' (Coe and Yeung, 2015: 67–74).

Although considerable progress has been made in understanding how regions are 'plugged into' these production networks, we think that a closer engagement with the geographies and territorialities of global production networks can deliver meaningful insights into economic development trajectories. Shifting geographies of global trade prompt greater consideration of two specific aspects of GPNs. First a closer look at the 'vertical' dimension of territoriality, or the geographies of GPNs in terms of the scales at which they operate (truly global or regional) and their directions (implicit assumption of North–South flows), is required. Second, a closer engagement with the 'horizontal' territorial dimension of networks, building on the concept of territorial embeddedness beyond a vaguely defined region is required to understand territorial development. We argue that we need to unpack categories of the 'global' and the 'region' and to carefully analyse the scales and geographies of global production networks and their developmental outcomes. In the next section, we will briefly discuss two empirical examples from our research to highlight these changing geographies of production.

BRIEF CASE EXAMPLES OF SHIFTING GEOGRAPHIES OF GPNs

The cases of both India's pharmaceutical industry and the Philippines' articulation into service offshoring networks are introduced below to highlight distinct aspects of the changing geography of GPNs. The first vignette makes the case for moving beyond an implicit focus and emphasis on development occurring through suppliers in the Global South serving Northern end markets. The second vignette demonstrates the need to focus on the socio-spatial outcomes of expanding global production networks.

Beyond Strategic Coupling with Northern Lead Firms: Indian Pharmaceutical Firms in Evolving Production Networks

The story of India's pharmaceutical industry (see Horner, 2014 for an overview of this case) provides a challenge to the dominant emphasis in research on GPNs on positive development outcomes occurring largely where the assets of territories in the Global South are 'strategically coupled' with global lead-firm-driven production networks. Today India's pharmaceutical industry is one of the largest in the world, especially in volume terms and has attracted worldwide attention for its role in lowering the prices of anti-retroviral drugs for people living with HIV-AIDS. Indeed, non-governmental organizations such as Médecins Sans Frontières have called the industry the 'pharmacy to the developing world' for producing large volumes of relatively low-cost generic medicines which it supplies to other developing countries. Moreover, India is increasingly a major supplier of pharmaceuticals to the Global North as well.

Much of the crucial capabilities were created among domestically owned firms during a period of 'strategic decoupling' from the multinational pharmaceutical industry in the 1970s and 1980s. After a period of decoupling from global markets, since the onset of external liberalization in India in 1991, global pharmaceutical firms have increasingly engaged with India again and have formed partnerships with Indian firms not only to get access to the Indian domestic market, but also to facilitate the supply of products manufactured in India to Northern markets. At the same time, Indian firms have formed prominent trade relationships with markets in other developing countries, outside of global lead firms' production networks. Such South–South trade comprises approximately half of India's pharmaceutical exports. Significant differences exist between these South–South trade linkages and Indian firms' participation in global lead firms' production networks oriented towards Northern end markets in terms of micro-social production/quality control, market access, and innovation practices (Horner and Murphy, 2018). Yet such trade is crucial to the supply of many generic medicines to developing countries.

This brief example presented here highlights the possibility for positive development outcomes to occur outside lead firm's GPNs and also suggests the wider significance of production networks beyond Northern end markets. Trade outside Northern lead firms' production networks has gained importance as part of a shifting geography of global trade as will be outlined below.

Beyond Manufacturing Networks: Exploring Emerging Geographies of Global Services Production in the Philippines

GVC and GPN analysis so far has been conducted on a large variety of global sectors. It remains, however, characterized by a 'productionist bias' (Coe and Yeung, 2015). Yet the GVCs and GPNs of services (stretching beyond an intermediary function) play a central role in most economies today that deserve greater attention (Kleibert, 2016a). The globalization of various services sectors has elevated the role of services as important contributors to economic development in both the Global North and South (see Beerepoot et al., 2017). Changes in information and communications technology have led to a 'tradability revolution' allowing international trade in digitally transmitted services and enabling the spatial reorganization and expansion of service value chains and production networks (UNCTAD, 2004). This new international division of service labour has a specific geography, being dependent on skilled-labour at lower-costs and often following post-colonial linkages as language-skills and cultural affinities with export-markets are decisive given the significance of emotional labour and direct client interactions (for example in call centres).

India and the Philippines have been able to strategically couple their economies with multinational firms in the information technology and business process outsourcing sector (see for example Kleibert, 2014). Whereas both countries have been rather lagging in attracting Foreign Direct Investments (FDIs) and export-oriented development in industrial production, they have been able to create high volumes of employment in the export of services. The geographies of service GPNs differ however from those of manufacturing networks not only at the scale of the country, but show also a distinct intra-national and intra-regional pattern (Kleibert, 2016b). Similar to other advanced business services, they are a very urban phenomenon, clustered in the largest urban agglomerations, and, within these, in central business districts and high-end office parks. Shedding light on these emerging labour and production geographies in services is crucial for understanding the socio-spatial outcomes of contemporary globalization dynamics.

ADVANCING GEOGRAPHIES OF GPNs

Inspired by the two examples in the previous section, we now seek to elaborate on some recent research on the shifting geographies of GPNs and outline aspects for further research.

Exploring New Geographies of Trade

Since the turn of the millennium, increasing reference has been made to a 'new geography of trade' (for example UNCTAD, 2004). Indeed, going back to world-systems theory from which the original commodity chains approach was developed, the international trade of countries in the Global South (or the 'periphery') has been assumed to involve exporting raw materials, natural resources and agricultural commodities to the Global North (or the 'core'), with manufactured goods imported in return. With shifting patterns of economic growth and end markets in the Global South (Staritz et al., 2011), including growing consumption in China, India, and Brazil, increasingly large shares of exports from developing countries go to other developing countries.

Such trade poses a challenge to the North–South orientation which implicitly underlies much research on global production networks, involving suppliers in the Global South supplying end markets in the Global North (Horner, 2016). Indeed, much of the pioneering research on global value chains has focused on, for example, retailers in North America and Europe coordinating the supply of agro-food products or textiles originally from developing countries to ultimately serve Northern end markets. Notably, given the prominence of intra-regional trade in East Asia within South–South trade, some such trade may still be oriented towards, and even controlled by firms head-quartered in, Northern end markets.

Increasingly, however, shifting geographies of trade require greater attention to new lead firms from the Global South, notably the 'rising power' economies (Sinkovics et al., 2014). Moreover, some contract manufacturers from emerging economies are also playing a more influential role. Research on how such firms seek to govern their production networks is only at a nascent stage, however (Lee and Gereffi, 2015). The influence of standards in governing such South–South trade may be distinct. Some of the earlier research on South–South value chains pointed to a lower importance of standards and thus reduced entry barriers in these markets (for example Kaplinsky and Farooki, 2011). More recently the growing significance of labour and environmental standards in rising power economies has been pointed to (Nadvi, 2014).

Another key aspect of the geography of GPNs, which warrants increasing attention as part of the shifting geography of trade, is the increasing emphasis on domestic and (supra-national) regional end markets. While Gereffi (1996: 437) had noted that the geography of trade flows can vary, and GPNs have been avowedly recognized to be multi-scalar, arguably more domestic or regionally-oriented trade has been relatively overlooked in the analysis of GPNs.

A recent special issue of *Global Networks* (Horner and Nadvi, 2018) includes a number of articles which collectively explore the changing role of the Global South in global production networks. This collection raises important issues, including the role of Southern actors as 'standard makers' (forming their own standards) as opposed to 'standard takers' (complying with Northern standards), the distinctive networks employed for Northern versus Southern end markets, governance and upgrading in South–South chains, underestimated yet significant informal South–South trading networks, the origins of regional value chains and the role of domestic market demand as a key growth opportunity. Collectively, they demonstrate considerable international-ized production and trading activity in the Global South beyond the lens of passive Southern suppliers in production networks driven by Northern-driven lead firms.

Yet arguably this is only the beginning of a new phase of analysis of global production networks in a context of more polycentric trade. Further research is required to understand, for example, the governance of these various production strands, the various standards in place, as well as how production strands oriented towards different end markets and at different scales relate to each other.

INVESTIGATING SOCIO-SPATIAL TRANSFORMATIONS RESULTING FROM GPNs

Understanding the territorial 'developmental outcomes' from integration into global production networks has been the underlying motivation for much analysis of globalization in this field. An implicit assumption has often been that positive outcomes for economic development are likely to result from integration into GPNs and subsequent upgrading within these networks. The focus on the organization of networks has, however, led to relatively limited empirical attention to investigating territorial economic impacts on places. More recent publications have given greater attention to the potential downsides (or 'dark sides') of becoming, and remaining, articulated in such networks and the simultaneous exclusions that emerge (Coe and Hess, 2010; Coe and Yeung, 2015; Werner, 2016).

A more fundamental critique argues that uneven development (and by extension socio-spatial polarization), instead of being an outcome of GPNs, can be understood to be both 'cause and consequence of these networks', linking processes of inclusion and exclusion (Werner, 2016: 457). The GPN approach has tended to be more suited to investigating the inclusions, and less attuned to understanding the territorial transformations in places. Philip Kelly has suggested that despite 'emphasizing place as central to a GPN analysis, what tends to be addressed is the place as it affects the network, not the territory per se' (Kelly, 2013: 84). Using the example of rapid industrial development in Cavite, the Philippines, his research has demonstrated how crucial territorial transformations, such as changes in landscapes and environments, households and livelihoods, can be overlooked when using a global production networks-perspective. These contributions show that a closer engagement with the horizontal-territorial interface occurring within the plugged-in 'regions' and an investigation of not only value creation, enhancement and capture, but also its distribution, are compelling avenues for future research.

The globalization of services and the increasingly complex and dispersed geographies of service delivery, alluded to above, are an interesting test case. Advanced producer services, and in particular financial services, and their production networks are now receiving increasing scholarly attention (Coe et al., 2015; Dörry, 2016; Kleibert, 2016a). The production of these services is concentrated in a select number of 'global' cities, thereby often leading to increased spatially uneven development, exacerbating urban–rural divides. Fragmentations can also occur within cities as production is organized in central business districts and special economic zones, which form globalized enclave spaces (see Kleibert, 2015). An analysis of economic development therefore requires an opening-up of the 'black-box' of the region and an inquiry into processes occurring at various interrelated scales, including the sub-national and urban scale. More sensitivity towards questions of urban and regional development and polarization, long revealed by 'global cities' researchers (see Sassen, 2001), might therefore offer a potential remedy to the slightly under-theorized spatialities and territorialities of GVCs and GPNs.

CONCLUSIONS: ADAPTING TO THE SHIFTING GEOGRAPHIES OF GPNs

Global production network analysis, and associated global commodity chain and global value chain analysis, has proven especially adept at understanding the organization and governance of trade which spans various geographic contexts under economic globalization. Such research has also deepened understanding of the possibilities for firm-level upgrading as well as regional territorial development under globalization. These relational approaches to economic production have revealed important linkages between multi-scalar actors and regions.

Research on GPNs has given more attention to geography, particularly in recognizing the multi-scalar aspects of global production as well as through seeking to understand prospects for territorial development. Two distinct vignettes presented here, of India's pharmaceutical industry and the Philippines' role in service offshoring, generate key questions for conceptualizing the geography of GPNs. This includes trade outside a South–North orientation. Moreover, the socio-spatial transformations resulting from GPN integration in countries, regions, cities and neighbourhoods, require a deeper engagement in the future.

We argue thus for a closer engagement with the geographies of production networks and value chains and an improved engagement with territory, unpacking implicit or taken-for-granted notions and categories of 'global' and 'region', with the aim of arriving at a conceptualization which fits twenty-first-century globalization and development. Such research is vital if global production network analysis is to adequately understand development possibilities within a context of both a shifting geography of global trade and increasing sub-national territorial differentiation.

REFERENCES

Bair, J. (2005), 'Global capitalism and commodity chains: looking back, going forward', *Competition and Change*, **9**(2), 153–180.
Bair, J. and G. Gereffi (2001), 'Local clusters in global chains: the causes and consequences of export dynamism in Torreon's blue jeans industry', *World Development*, **29**(11), 1885–1903.
Beerepoot, N., B. Lambregts and J.M. Kleibert (eds) (2017), *Globalisation and Services-Driven Growth: Perspectives from the Global North and South*, London and New York: Routledge.
Coe, N. and M. Hess (2010), 'Local and regional development: a global production network approach', in: A. Rodriguez-Pose, J. Tomaney and A. Pike (eds) *Handbook of Local and Regional Development*, London: Routledge, 128–138.
Coe, N. and H. Yeung (2015), *Global Production Networks: Theorizing Economic Development in an Interconnected World*, Oxford: Oxford University Press.
Coe, N., P. Dicken and M. Hess (2008), 'Global production networks: realizing the potential', *Journal of Economic Geography*, **8**(3), 271–295.
Coe, N., K. Lai and D. Wójcik (2014), 'Integrating finance into global production networks', *Regional Studies*, **48**(5), 761–777.
Coe, N., M. Hess, H. Yeung, P. Dicken, and J. Henderson (2004), '"Globalizing" regional development: a global production networks perspective', *Transactions of the Institute of British Geographers*, **29**(4), 468–484.
Dicken, P., and A. Malmberg (2001), 'Firms in territories: a relational perspective', *Economic Geography*, **77**(4), 345–363.

Dicken, P., P. Kelly, K. Olds and H. Yeung (2001), 'Chains and networks, territories and scales: towards a relational framework for analysing the global economy', *Global Networks*, **1**(2), 89–112.

Dörry, S. (2016), 'The geographies of industrialised finance: probing the global production networks of asset management', *Geography Compass*, **10**(1), 3–14.

Gereffi, G. (1994), 'The organization of buyer-driven global commodity chains: how US retailers shape overseas production networks', in: G. Gereffi and M. Korzeniewicz (eds), *Commodity Chains and Global Capitalism*, Westport, CT: Praeger, 95–122.

Gereffi, G. (1996), 'Global commodity chains: new forms of coordination and control among nations and firms in international industries', *Competition & Change*, **1**(4), 427–439.

Gereffi, G. (1999), 'International trade and industrial upgrading in the apparel commodity chain', *Journal of International Economics*, **48**(1), 37–70.

Gereffi, G. (2001), 'Shifting governance structures in global commodity chains, with special reference to the internet', *American Behavioral Scientist*, **44**(10), 1616–1637.

Gereffi, G. and M. Korzeniewicz (1994), *Commodity Chains and Global Capitalism*, Westport, CT: Praeger.

Gereffi, G., J. Humphrey and T. Sturgeon (2005), 'The governance of global value chains', *Review of International Political Economy*, **12**(1), 78–104.

Henderson, J., P. Dicken, M. Hess, N. Coe and H. Yeung (2002), 'Global production networks and the analysis of economic development', *Review of International Political Economy*, **9**(3), 436–464.

Hess, M. and H. Yeung (2006), 'Whither global production networks in economic geography? Past, present, and future', *Environment and Planning A*, **38**(7), 1193–1204.

Hopkins, T. and I. Wallerstein (1986), 'Commodity chains in the world economy prior to 1800', *Review*, **10**(1), 157–170.

Horner, R. (2014), 'Strategic decoupling, recoupling and global production networks: India's pharmaceutical industry', *Journal of Economic Geography*, **14**(6), 1117–1140.

Horner, R. (2016), 'A new economic geography of trade and development? Governing south–south trade, value chains and production networks', *Territory, Politics, Governance*, **4**(4), 400–420.

Horner, R. (2017), 'Beyond facilitator? State roles in global value chains and global production networks', *Geography Compass*, **11**(2), DOI: 10.1111/gec3.12307.

Horner, R. and J. Murphy (2018), 'South–north and south–south production networks: diverging socio-spatial networks of Indian pharmaceutical firms', *Global Networks*, **18**(2), 326–351.

Horner, R. and K. Nadvi (2018), 'Global value chains and the rise of the global south: unpacking twenty-first century polycentric trade', *Global Networks*, **18**(2), 207–237.

Humphrey, J. and H. Schmitz (2002), 'How does insertion in global value chains affect upgrading in industrial clusters?', *Regional Studies*, **36**(9), 1017–1027.

Kaplinsky, R. and M. Farooki (2011), 'What are the implications for global value chains when the market shifts from the north to the south?', *International Journal of Technological Learning, Innovation and Development*, **4**(1), 13–38.

Kelly, P.F. (2013), 'Production networks, place and development: thinking through global production networks in Cavite, Philippines', *Geoforum*, **44**, 82–92.

Kleibert, J.M. (2014), 'Strategic Coupling in "next wave cities": the role of local institutional actors and the offshore service sector in the Philippines', *Singapore Journal of Tropical Geography*, **25**(2), 245–260.

Kleibert, J.M. (2015), 'Islands of globalisation. Offshore services and the changing spatial division of labour', *Environment and Planning A*, **47**(4), 884–902.

Kleibert, J.M. (2016a), 'Pervasive but neglected: conceptualising services and global production networks', *Geography Compass*, **10**(8), 333–345.

Kleibert, J.M. (2016b), 'Global production networks, offshore services and the branch-plant syndrome', *Regional Studies*, **50**(12), 1995–2009.

Lee, J. and G. Gereffi (2015), 'Global value chains, rising power firms and economic and social upgrading', *Critical Perspectives on International Business*, **11**(3/4), 319–339.

Leslie, D. and S. Reimer (1999), 'Spatializing commodity chains', *Progress in Human Geography*, **23**(3), 401–420.

Liu, L. and S. Mei (2016), 'Visualizing the GVC research: a co-occurrence network based bibliometric analysis', *Scientometrics*, **109**(2), 953–977.

Nadvi, K. (2008), 'Global standards, global governance and the organization of global value chains', *Journal of Economic Geography*, **8**(3), 323–343.

Nadvi, K. (2014), '"Rising Powers" and labour and environmental standards', *Oxford Development Studies*, **42**(2), 137–150.

Neilson, J. and B. Pritchard (2009), *Value Chain Struggles: Institutions and Governance in the Plantation Districts of South India*, Oxford: Wiley-Blackwell.

Neilson, J., B. Pritchard and H. Yeung (2014), 'Global value chains and global production networks in the changing international political economy: an introduction', *Review of International Political Economy*, **21**(1), 1–8.

Parrilli, M.D., H. Yeung and K. Nadvi (2013), 'Local and regional development in global value chains, production networks and innovation networks: a comparative review and the challenges for future research', *European Planning Studies*, **21**(7), 967–988.

Ponte, S. and T. Sturgeon (2014), 'Explaining governance in global value chains: a modular theory-building effort', *Review of International Political Economy*, **21**(1), 195–223.

Porter, M.E. (1985), *Competitive Advantage: Creating and Sustaining Superior Performance*, New York: The Free Press.

Sassen, S. (2001), *The Global City: New York, London, Tokyo*, Princeton: Princeton University Press.

Sinkovics, R., M. Yamin, K. Nadvi and Y. Zhang (2014), 'Rising powers from emerging markets: the changing face of international business', *International Business Review*, **23**(4), 675–679.

Smith, A. (2015), 'The state, institutional frameworks and the dynamics of capital in global production networks', *Progress in Human Geography*, **39**(3), 290–315.

Staritz, C., G. Gereffi and O. Cattaneo (2011), 'Editorial', *International Journal of Technological Learning, Innovation and Development*, **4**(1/2/3), 1–12.

UNCTAD (2004), *World Investment Report 2004: The Shift Towards Services*, New York and Geneva: United Nations.

Werner, M. (2016), 'Global production networks and uneven development: exploring geographies of devaluation, disinvestment, and exclusion', *Geography Compass*, **10**(11), 457–469.

Yeung, H. (2016), *Strategic Coupling: East Asian Industrial Transformation in the New Global Economy*, Ithaca: Cornell University Press.

17. Food and globalization: 'from roots to routes' and back again

Elena dell'Agnese and Giacomo Pettenati

INTRODUCTION

Food-related practices are among the more pervasive of human life, from individual, social, cultural, political and economic points of view. They are also generally identified through regional (or national) names, implying that recipes and food habits are still mostly place-based (Farrer, 2015). Nowadays, food and globalization seem to be 'inseparable' (Nützenadel and Trentmann, 2008: 1), but long-term flows of food products and culinary ideas have been crossing the oceans for centuries (Ray, 2015). This chapter presents the more significant dimensions of the globalization of food-scapes, starting from a description of its main features and of its historical roots. The following paragraphs specifically focus on the emergence of a global food governance, on the practices of opposition to the negative externalities of the globalization of the food system, and on the relationships between food and global and local cultural identities.

Nowadays, hardly anyone is surprised that, in a supermarket in Milan (or in Moscow), one pound of bananas costs, more or less the same as one pound of apples. Bananas are tropical fruits, while apples, when in season, are usually locally produced in temperate climates. The facts that they are both available at about the same price all over the year on the same markets is just one of the many aspects of food globalization. New technologies in refrigeration, conservation and transport (Morgan et al., 2006), have apparently liberated the food supply from 'nature' and its seasons. The cost-driven upscaling of the food supply chain has now reached the global scale, and places of food production are increasingly interchangeable, competing to be on the map of global flows bringing food products from farm to fork (Wiskerke, 2009). If we followed backwards the products composing our breakfast table, welcoming the suggestion of David Harvey (1989) and using Ian Cook's 'follow the things' approach (Cook, 2006), we would discover how everything we eat represents the end of an extensive network of actors, places, flows (of goods, services and capital), and powers that constitute the framework of a global supply and value chain.

Other directly recognizable signs of food globalization are, for instance, the presence of a fast-food giant like McDonald's in about 120 countries, or the almost worldwide diffusion of certain 'global foods', like pasta and noodles, pizza, or hamburgers (Serventi and Sabban, 2003; Helstosky, 2008; Smith, 2008). Moreover, 'products of the industrial cuisine and of industrialised agriculture' are now critical elements not only in Europe or in the US, but also in the food supply of many African and Asian countries (Goody, 2013: 73).

Beyond the increasing homogenization of the world's foodscapes, there are many other, less immediately detectable, but not less relevant, aspects of food globalization. The first is the process of concentration of the sector. Corporate clusters and alliances bring together and coordinate biotechnological facilities, grain and feed trading, and meat processing, thus testifying the increasing power of food-related transnational corporations and of food retailers. According to Morgan et al. 'five major seed companies dominate world-wide' (Morgan et al., 2006: 55). So, even on the super-markets' shelves, the diversity is only apparent, since 'fresh fruit and vegetables are of standard size and colour, and the varieties on sale are very limited in number' (Morgan et al., 2006: 12; Fowler and Mooney, 1990). The cultural homogenization of some food habits at the global scale goes hand-in-hand with a dramatic increase of overeating-related diseases, even in areas where they were almost unknown, as in the case of the emerging epidemic of obesity in a growing number of low-income countries (Popkin, 2001; Prentice, 2005).

Also, the various counter-movements focused at an ethical and political problemat-ization of food can be considered as a further aspect of food globalization. They were born as a reaction against the power of international corporations with the aim of 'saving' traditional agricultural products and of promoting more socially sustainable ways of farming and trading agricultural products, but they too are acting at the transnational scale (Inglis and Gimlin, 2009; Holt Giménez and Shattuck, 2011; Larsson, 2015).

Food globalization is commonly understood as a fairly recent trend and mostly Western-led. Nowadays, the ecological support of human societies is getting 'horizon-tal' in the sense that increasingly food originates from non-local sources (Claval, 1974). In the past, on the contrary, the ecological support of human societies was supposed to be 'vertical', since people used to eat 'locally'; the ingredients of their food were produced nearby, or at the best exchanged in neighbouring regional markets. As a result, their eating practices were highly differentiated in regional terms.

A HISTORICAL PROCESS

These observations, albeit generally correct, have to be nuanced. As food historians often underline, there is a 'tendency to over-accentuate the current phase of global-ization' (Ray, 2015: 25), since, in the case of food, globalization is a very long process (Pilcher, 2005; Higman, 2011), which 'began with the invention of agriculture some ten thousand years ago in at least seven independent centres of plant and animal domestication' (Kiple, 2007: xiv). Even if, as remarked by the same Kiple (p. 24), 'we cannot be certain where the first Neolithic Revolution occurred ... as a rule pride of place is given to that large and fertile arc running from the Persian Gulf to the eastern Mediterranean and south to the Nile Valley that we call the Fertile Crescent ... a region ideally located to radiate agriculture in all directions'.

Moreover, if it is true that once staples were mostly 'locally' produced, it is also true that, even in pre-modern times, certain items could travel long distances. Sometimes, vegetables and animals followed the main routes of human interactions across the oceans as 'portable food', were cultivated in new lands, or settled spontaneously there

(Higman, 2011). In other circumstances, they were purposively transported in order to have more quantities of food available. Before the introduction of railways, ground transportation on long distances was difficult; so bulky goods were mostly carried by sea. When the Romans controlled the whole Mediterranean basin, staple foods such as grain or olive oil were imported from Sicily, Sardinia and Egypt (the so-called *provinciae frumentariae*, or 'wheat provinces') to the urban centres (Erdkamp, 2005). The shipping of food across the Mediterranean did not stop with the end of the Empire. As reported by the Arab geographer Idrisi, in Sicily, in the twelfth century, 'great quantities of pasta are made and exported everywhere, especially to Calabria and other Muslim and Christian lands. Many shiploads are sent' (Serventi and Sabban, 2003: 42). Also the Indian ocean was crossed by exchanges of staple food in pre-industrial times: rice, consumed daily in the Maldives, was imported there from the Indian subcontinent already in the fourteenth century, when Ibn Battuta visited the islands. Delicacies, like the Oriental spices, so sought after in the European Renaissance, made even longer trips to get on the tables of those who could afford the expense. When their traditional route was blocked by the Ottoman Empire, the need of finding a different seaway became one of the triggers of the Age of Discovery.

Also, the Western role in the process is often overemphasized. Even if 'the usual narrative of culinary globalization focuses on the establishment of an "Atlantic system", beginning with the Columbian exchange, followed by European plantation agriculture, and ending in American fast-food hegemony' (Farrer, 2015: 3), in the past, in each region characterized by intense trades, the crossroads of cultures favoured the exchange of products and crops of different origins. During the Han dynasty (206 BC–220 AD), for instance, sesame and onions, bound to become two essential elements of Chinese cuisine, were introduced into China along the Silk Road. If the Silk Road was one of the first culinary contact zones (Farrer, 2015), at the time of Arab conquest, the caliphate encouraged widespread trade and migration, and introduced plants of Asian origins to the west, and food crops of African roots to the east, making the Mediterranean and the Indian Ocean the main regional intersections of culinary exchanges.

So, 'culinary globalization must be recognized as an ancient and unending process that neither began nor ended with European colonization' (Farrer, 2015: 2). As Ray (2015: 23) also states: 'palatal taste can be fruitfully interrogated as an embodied archive of long-term, pre-Eurocentric globalization, especially where durable documentation is rare'. However, after Columbus' travels, an intense 'exchange between two essentially unique and separate biospheres' began (Higman, 2011), the so-called 'Columbian exchange' (Crosby, 1972). At the time, the main culinary contact zone became the Atlantic and the role of European colonizers turned out to be more and more important. The coconut, domesticated in South-eastern Asia and landed on the eastern coasts of Africa in the eleventh century, came to the Americas in the sixteenth century, following the Spaniards together with other tropical products like bananas, mangoes and sugar canes. In the opposite direction, foods from the Americas were introduced in Europe. Some of these foods were later to become so popular to be considered as typical expressions of the 'local', and later of the 'national' culture, like the 'Irish' potato. Tomato, added to pasta together with a pinch of the Asian basil, created one of the more easily recognizable symbols of Italian cuisine in the world: 'spaghetti al pomodoro' (Gentilcore, 2010).

From this point of view, a focus on food could certainly offer a fruitful historical perspective on globalization. Indeed, not only can the spatial exchange of food be considered as one of the earliest examples of globalization, but it may also be considered one of its main engines. Like coconuts and sugar canes, also bananas were brought by the Spaniards in the Americas and there were transformed into one of the main export products (Higman, 2011). More generally speaking, the desire for tropical food products, such as tea, coffee, cocoa, sugar, was one of the triggers of colonization and for the subsequent transformation of colonized lands in 'plantations'. Moreover, to obtain plantation labour, vast movements of population were carried on, mostly through the Atlantic slave trade (Ochoa, 2012).

In all cases, basic staples and exotic delicacies promoted exchanges and dominations, pushed explorers and emigrants towards new lands and new markets, crossed the oceans on the routes of the European expansions, became the objects of worldwide trades and of intercontinental territorial exploitations. Consequently, cultural historians and anthropologists have long been interested in the history of food, its production and exchange (Phillips, 2006). A pioneering publication in this respect was Sidney Mintz's book *Sweetness and Power: The Place of Sugar in Modern History* (Mintz, 1985), which opened the field, offering 'a unique analytical framework for exploring the nexus of food and globalization' (Phillips, 2006: 36). 'Mintz was one of the first to examine how a particular food, sugar, was tied into networks of production, consumption, and power, and his methodological approach utilized the emerging world systems analysis as a way to explore the macroscopic impact of Western imperialist' (Ochoa, 2012: 3). Indeed, Mintz clearly demonstrates the role of sugar in connecting European consumption, slave trade, Caribbean colonization and modern industry. As Mintz writes (1985: 175), sugar not only was 'a cornerstone of British West Indian slavery and the slave trade', it was also linked to the English industrial factory system, where it became, together with tea, the nutrition for the industrial work pause. So, the cultivation and trade in tropical products have had dramatic effects not only on the economic and political organization of the regions of productions, but also of those of consumption.

Seen from this perspective, the power of corporations in the food sector and their impact on it is not entirely a novelty as the case of bananas and of the United Fruit Company (nowadays Chiquita) demonstrates. In Central America, bananas were first grown commercially at the end of the nineteenth century, when the United Fruit Company was created. They were bound to the United States' market, where consumer tastes were formed through intensive advertising. Bananas are transportable, but require to be eaten fresh. For this reason, the demands linked to their production and transport required important investments to reconfigure the territorial settings of those regions. And the need to protect the investments required their political control, albeit in indirect forms. The power asymmetry between this domineering transnational corporation and the largest banana exporters (countries of Central America) have even led to the coinage of the label 'Banana Republics'.

THE GLOBAL FOOD GOVERNANCE

To grasp the economic aspects of food globalization, however, we need to look beyond the international circulation of food products as commodities and the transnational expansion of food-based corporations. It is also essential to take into account 'the global governance of food and food issues' (Phillips, 2006: 38) as well as the role of international trade agreements (notably the GATT (the General Agreement on Tariffs and Trade)) and of supranational organizations such as the World Trade Organization (WTO) or the Food and Agriculture Organization (FAO) of the United Nations.

The globalization of the agro-food system is strictly connected to the increased incorporation of agriculture into the global dynamics of capitalist accumulation (Le Heron, 1993; Atkins and Bowler, 2001). Its main features are: the increasing liberalization of agricultural and food products international trade, the emergence of an international division of labour in the agro-food sector and the progressively increased role of international institutions in regulating food-related issues (Goodman and Watts, 1997; Morgan et al., 2006).

According to some scholars (Friedmann, 1982, 1993; McMichael, 1992, 1994), the globe-spanning pattern of the industrialized agro-capitalist food system characterizes the current 'food regime'. It is based on the 'new global regulatory structure' lead by WTO and IMF (McMichael, 1992) and on the tension between national economic interests and transnational capital (Atkins and Bowler, 2001). The notion of 'food regime', useful in understanding the process of homogenization of production and distribution, fails – according to Goodman and Watts (1994) – to grasp the local/global specificities of the sector and the increasing contradictions.

The rise of an international governance of food production and trade is rooted in the post-Second World War new economic and political order, leading to the international debate about the strictly interconnected issues of free trade and the universal right to food.

The first frame of policies and discourses impacting on processes of food globalization is that of macro-regional and global free trade that started in 1948 with the signing of the GATT. It was intensified in 1995 with the establishment of the WTO and on the first agreements on the liberalization of agricultural products (Morgan et al., 2006). For a long time, international free trade policies excluded agricultural products, for 'the specificity of agriculture, and its centrality to questions of national social and economic organization and viability' (McMichael, 1993: 201). At the macro-regional scale, though, several free-trade agreements that also include agricultural products have been signed, such as the North American Free Trade Agreement (NAFTA) and the European Union internal market.

In the GATT framework, very intense negotiations to reach an agreement on a global free trade agreement encompassing agricultural products are still ongoing (McMichael, 1993; Morgan et al., 2006; Pritchard, 2009). Briefly, the debate about the free trade of agricultural products opposes neoliberals, considering it as the only way to deal with the market distortions caused by regional subsidies policies, and critics, worried about the privileged role that transnational firms may play in this new regulatory system (McMichael, 1993). From a geopolitical perspective, the positions are heterogeneous. Most of the developing countries (the net exporters of agricultural products ones) ask for the liberalization that would open global markets to their products (Morgan et al., 2006), but claim that possible gains would be lower than adverse effects regarding national sovereignty on development policies (Gallagher, 2008). On the other side, most developed countries are cautious in completely abandoning their policies of protection of their internal markets and production (Morgan et al., 2006; Potter, 2006; da Conceição-Heldt, 2011).

Beside nation states and intergovernmental organizations, transnational corporations play a key role in the governance of the global food system. They typically control the value chain, address food policies, steer consumption choices and monitor food security (Hendrickson and Heffernan, 2002). It is calculated, for instance, that a single corporation (Monsanto) controls about 90 per cent of the global genetically modified seeds market. Moreover, ten corporations control almost 60 per cent of the total sales of the world's main 30 retailers and around 40 per cent of the revenues earned by the leading 100 food and beverage companies at the global scale (Ziegler et al., 2011).

The second frame of policies and discourses linking food systems with globalizing forces is connected to the international efforts for the eradication of hunger and the achievement of the universal right to food, as stated in the UN Declaration of Human Rights and led by international organizations such as the World Bank and the FAO. Since the beginning of the establishment of the FAO (1945), the international cooperation in this field has been dominated by a productivistic approach, firstly aiming at helping developing decolonized countries in achieving the highest-level possible of food self-sufficiency. One of the most renowned examples of such policies was the so-called Green Revolution, which from the 1960s onwards contributed to a large increase in crop production in developing countries (Sachs, 2015), but had severe impacts in terms of environmental pollution, reduction of agricultural diversity, power inequalities, and changes in the social structure of rural areas (Shiva, 2016).

Since the 1970s, it was clear that the main problem was not the availability of food, but the inequalities in the individual and collective access to it (Sen, 1981). The main objective of food-related international cooperation policies became food security, defined as 'physical and economic access [of all people, at all time] to sufficient, safe and nutritious food to meet their dietary needs and food preferences for an active and healthy life' (FAO, 1996). According to critics (Desmarais, 2007; Shiva, 2016), the international organizations' policies against hunger and for food security played an important role in globalizing the food systems, generally considering food as a commodity and international food trade as the most viable way of achieving food security.

THE OPPOSITION TO THE GLOBALIZATION OF THE AGRI-FOOD SYSTEM

Increasingly, we can observe emerging practices of reaction and resistance to this conventional agri-food system, dominated by big companies that produce, process, and retail food on a global scale. Instead, they advocate an 'alternative food geography' (Wiskerke, 2009), mostly based on 'short food supply chains' (Renting et al., 2003), or Alternative Food Networks (Goodman et al., 2012), associated with smaller companies and a more sensitive attitude towards the environment and local societies, cultures and production systems (Watts et al., 2005; Morgan et al., 2006). The international political movement of peasants Via Campesina, which started a 'Global Campaign for Agrarian Reform' (Torrez, 2011), is an interesting example of global opposition to the dominant political and economic food system. It promotes the notion of 'food sovereignty'[1] as an alternative to the prevalent narratives on 'food security' (Desmarais, 2007). More generally speaking, this attention may lead to five different, albeit strictly connected, attitudes, supporting the practices of opposition to the globalization of the foodscape:

(1)   The preoccupation about 'food miles' and the request for locally produced ingredients. Generally speaking, there is the idea that local food is more typical, more environmentally sustainable, more traditional and fresher. Farmers' markets, veggie boxes, local foods, organic products, and Fair Trade goods are thus increasingly present in the everyday consumption (Goodman et al., 2012). At the same time, expressions like 'alternative food networks' and 'short-food supply chains' are perfused by a sort of rural idyll imaginary and advocating the necessity of 'coming home to eat' (Nabhan, 2002). So, they are used to make 'strong connections between the localization of food systems and the promotion of environmental sustainability and social justice' (Goodman et al., 2012: 11). Some critical observers highlight how these narratives and practices often confuse the scale of a food system with a desired outcome (e.g. more environmental sustainability and social equity). In this way, they fall into a 'local trap' that obscures other scalar options (even the global one), possibly more effective in pursuing some objectives (Born and Purcell, 2006).

(2)   The attention towards the environmental impacts of food production, and by the intensity of human land use, expressed by the growth of organic agriculture and by the rise of the vegan culture. The complexities of the definitions and descriptions of the different kinds of production do not represent a real guarantee of both environmentally friendly practices and social respect. As remarked by Freidberg (2003: 5), for instance, 'The "environmentally friendly" processes associated with organic farming say nothing about the highly unfriendly labour practices prevailing on many organic farms'.

(3)   The demand for 'typical' productions, endangered by the homogenization of the food products marketed by the conventional food system and by the fast-food chains. All the same, this increasing sensibility towards 'typical food' is also conveyed by the rising popularity of movements like Slow Food, and by food chains similarly promoting 'traditional' and 'high quality' food products. Examples may be represented by Eataly, a big retailer operating in Italy, Japan, and the United States (Sebastiani et al., 2013), Rosso Pomodoro, a chain of pizzerias proposing 'real' Neapolitan pizza worldwide, Obicà, an international chain of mozzarella bars, and Grom, a chain of 'hand-made', high-quality ice-cream

---

[1]   The definition of food sovereignty is that 'all peoples, nations and states are able to determine their own food producing systems and policies that provide every one of us with good quality, adequate, affordable, healthy and culturally appropriate food' (Declaration of Nyéléni, 2007 in Forum for Food Sovereignty, 2009).

sellers (Altuna et al., 2017). The Slow Food movement was created in Italy in 1987 in opposition to the ubiquity of fast-food chains, and, more generally, to 'the universal folly of the Fast Life' (Leitch, 2013), by valorizing regional products, sustainable farming and local producers. A sort of 'militant particularism' (Goodman et al., 2012) infused its initial positions; but its international success, the creation of the worldwide farmers' network Terra Madre and its growing connections with large-scale distribution retailers, like Eataly, also turned the Slow Food into a global trend (Leitch, 2013).

(4) The concern for the ethical treatment of animals, which promotes the diffusion of vegetarianism and of veganism, and challenges the modern practices of raising and slaughtering animals and the meat industry. More generally speaking, the animal advocacy movement problematizes the ethics of animal consumption, and the representation of meat consumption as 'normal, straightforward, healthy and unproblematic' (Stephens Griffin, 2017: 23), while it tries to promote a non-hierarchical conception of human and non-human animal relations.

(5) The concern for the unequal distribution of powers in the decision chain of the food system. Fair Trade was initially conceived as 'a genuine alternative to the established order of international trade and development aid' (Roozen and Van der Hoff, 2002: 238). Specifically, Fair Trade is a certification scheme aimed at creating an alternative food production system, with alternate success in economic terms, but a big appeal towards the market of ethical consumers. The focus of these practices is on the need to acknowledge adequate power and economic benefits to every part of the food chain, notably to small producers that are the weakest actors in the current globalized agro-food system. But, like the Slow Food certificate products, Fair Trade products, to make an impact, must also connect with big retailers. On the other side, food products coming from alternative food networks contribute to clean the brand of many big department stores, at a low cost (even if, as Sylla (2014: 79) clarifies, this practice is just 'greenwashing'). So, many big department stores are now also the most important sellers of organic food and Fair Trade products (see Morgan et al., 2006; Goodman et al., 2012; Colombino and Giaccaria, 2013).

In these different ways, the political meaning of food is made relevant. 'Food advocacy is a growing arena for political activism' (Counihan and van Esterik, 2013: 10). Alternative food consumption may be perceived as a source of resistance against the homogenizing effects of 'placeless', globalized, industrial modes of food provisioning and the 'McDonaldization' of regional food cultures (Goodman et al., 2012).

What we eat is not only a private matter: the kind of food we buy, its provenance, the kind of shop where we buy it, may have social, but also environmental and even political consequences. Indeed, the progressive globalization of the diet and the power of the agribusiness may have territorial impacts on consumption places, since they may cause the loss of presumed local cultural traditions, but also on production sites. Production sites may be quite far from consumption places, but all the same they may change their territorial alignment in connection with the requests of the international market. The excessive desire for certain ingredients and food products for instance may cause devastating effects on the land (eutrophication, palm oil and devastation of tropical forests, maize consumption and dust bowl), but also have territorial, and social consequences, connected with land ownership and working conditions.

So, 'our food choice has multiple implications – for our health and well-being, for economic development at home and abroad, for the ecological integrity of the global environment, for transport systems, for the relationship between urban and rural areas and ... for the very survival of democracy in poor, commodity-producing countries' (Morgan et al., 2006: 1). As remarked by Michael Pollan, eating 'is ... an ecological act, and a political act, too', and 'how and what we eat determines to a great extent the use we make of the world – and what is to become of it' (Pollan, 2006: 11). So, the general assumption is that 'eating to reinforce specific values – organic, local, vegan, fair-wage and equal exchange, child-labour free food, for example – creates market demand that could change the system of food production' (Herring, 2015: 6). Apart from ethical veganism, which aims at counteracting and combating the industrial exploitation of animals eschewing animal products, other practices of resistance are implemented through the boycott of certain multinational companies and their related products, or by the choice to consume products not only 'Made in ...' but also locally owned, and self-produced. Practices of re-territorialization through the cultivation of traditional products, in opposition to plantation farming and the consumption of imported products, may also accompany the policies for the struggle for the re-appropriation of sovereignty. In the case of Hawaii, the exponents of the various movements of resistance against global consumerism and US colonization cultivate the ancient sacred plant of the islands, the taro, and consume the resulting product, the poi, as their main staple.

FOOD AND IDENTITY

Few aspects of everyday life still appear labelled in geographical terms as what we eat. Food may acquire a role as a form of banal nationalism, which can help to 'flag the homeland' (Billig, 1995) into some of the most private spaces of our daily life, the kitchen (Ichijo and Ranta, 2016). Many different features of what we eat may perform this role, starting from names, to brands and ownership, and even recipes and ingredients. Sometimes, the symbolic usage of food in terms of national icon belongs to a not very old tradition (as Stuart Hall demonstrated with the case of 'English tea', 1997). Also 'national cuisines' are usually invented after the making of nations, creating a sort of 'imagined commensality' – as suggested by the geographers Bell and Valentine (1997), who paraphrase Anderson's popular definition of the nation as 'imagined community'. All the same, the political role of food as a symbol of the nation is patent. To illustrate, we can refer to different examples: 'naming', in the case of conflict (during the First World War, in the United States hamburgers were renamed 'Salisbury steak', while in 2003 French fries became Freedom fries, because of France's opposition to the invasion of Iraq); branding and ownership, as a form of national pride (Vegemite, a yeast-based spread, is considered a sort of Australian icon; so, when the brand was bought by a transnational corporation, it was challenged on the market by a new Australian-owned substitute, Aussie Mite); and eventually the boycott of the products of a given country (such as the Israeli owned hummus, branded Sabra) (Avieli, 2016), in order to make 'politics in the supermarket' (Stolle et al., 2005).

In these contexts, food assumes a profound symbolic and identity meaning, even more so in the construction and reproduction of a diaspora. The choice of eating a certain kind of food, or another, is a sign of the desire to cling to a particular identity, either religious, ethnic or national. So, in the contact zones of the diasporas, food is sometimes revered like the last bastion of tradition against an invasive cultural otherness, and it can be meticulously reproduced in its original recipes, at least in special occasions. Or it is creolized, in the attempt of putting together the qualities of the tradition of the origins with those of the new location (as it successfully happened with the Italo-American promotion of the 'pepperoni pizza'). Moreover, restaurants offering (partly) foreign food stuff may be accepted as a different way of globalizing the world of food since they promote culinary exchanges and they dot the global urban landscapes with signs referring to countries of origin. Those signs are both very 'typical' and very similar in different countries as Indian, Chinese or Italian restaurants across the globe amply testify.

CONCLUSION

'From roots to routes' is a metaphor, often used to underline the necessity to rethink the connection between culture and place. Instead of looking for roots, that is locality and local specificities, we should look for routes and try to understand how culture and places are made and remade by travels and intersections. If such a metaphor works for culture in general, it seems to be even more apt to speak about food and its connections with place. What we eat is the fruit of the constant and uninterrupted interaction of flows of a thousand origins, and at the same time, it appears and becomes something strongly rooted in regional terms, although today it seems to be subjected to the threats of globalization. Indeed, as remarked by Kiple (2007), food globalization is a 'movable feast', lasting since the establishment of agriculture and the subsequent diffusion of farming cultures.

Food has followed humans in all their travels, and it has offered them many reasons for travelling more. Not only it offers early examples of globalization, but also represents a trigger for colonial expansion, the slave trade, globalization itself. Nowadays, it is possible to speak about food and globalization in many different ways: in cultural terms, since global foods like pizza, pasta and hamburgers are known and consumed all over the world; in economic terms, since a small number of corporations control the global markets; in biological terms, because even seeds are worldwide controlled by a small number of corporations.

At the same time, food is also deeply rooted in culture, ethnicity, banal nationalism and identity discourses. It is one of the leading flags of regional narration and identification. So, in a world that seems to be more and more 'global', one of the key imperatives of many organic food movements and other groups and organizations' movements, mobilized to resist the global power of food corporations, is to go back 'from the routes to the roots'. Also those political movements, however, are acting at the transnational scale.

REFERENCES

Altuna, N., C. dell'Era, P. Landoni and R. Verganti (2017), 'Developing radically new meanings through the collaboration with radical circles: slow Food as a platform for envisioning innovative meanings', *European Journal of Innovation Management*, **20**(2), 269–290.

Atkins, P. and I. Bowler (2001), *Food in Society*, London: Arnold.

Avieli, N. (2016), 'The hummus wars revisited: Israeli–Arab food politics and gastromediation', *Gastronomica: The Journal of Critical Food Studies*, **16**(3), 19–30.

Bell, D. and G. Valentine (1997), *Consuming Geographies: We Are Where We Eat*, Hove: Psychology Press.

Billig, M. (1995), *Banal Nationalism*, Los Angeles and London: Sage.

Born, B. and M. Purcell (2006), 'Avoiding the local trap: scale and food systems in planning research', *Journal of Planning Education and Research*, **26**(2), 195–207.

Claval, P. (1974), *Eléments de géographie humaine*, Paris: Genin.

Colombino, A. and P. Giaccaria (2013), 'Alternative Food Networks tra locale e globale. Il caso del Presidio della razza bovina piemontese', *Rivista Geografica Italiana*, 225–240.

Cook, I. (2006), 'Geographies of food: following', *Progress in Human Geography*, **30**(5), 655–666.

Counihan, C. and Penny van Esterik (eds) (2013), *Food and Culture: A Reader*, London and New York: Routledge.

Crosby, A.W. (1972), *The Columbian Exchange: Biological and Cultural Consequences of 1492*, Westport CT: Greenwood Publishing Company.

da Conceição-Heldt, E. (2011), 'Variation in EU member states' preferences and the Commission's discretion in the Doha Round', *Journal of European Public Policy*, **18**(3), 403–419.

Desmarais, A. (2007), *La vía campesina: globalization and the power of peasants*, Hoboken: John Wiley & Sons.

Erdkamp, P. (2005), *The Grain Market In The Roman Empire. A Social, Political And Economic Study*, Cambridge: Cambridge University Press.

FAO (1996), Rome Declaration on World Food Security and World Food Summit Plan of Action. World Food Summit 13–17 November 1996, Rome: FAO.

Farrer, J. (2015), 'Introduction: traveling cuisines in and out of Asia: toward a framework for studying culinary globalization', in: J. Farrer (ed.), *The Globalization of Asian Cuisines*, New York: Palgrave Macmillan US, pp. 1–19.

Forum for Food Sovereignty (2009), *Declaration of Nyéléni* (Sélingué, Mali – February 2007).

Fowler, C. and P.R Mooney (1990), *Shattering: Food, Politics, and the Loss of Genetic Diversity*, Tuscon, AZ: University of Arizona Press.

Freidberg, S. (2003), 'Editorial not all sweetness and light: new cultural geographies of food', *Social & Cultural Geography*, **4**(1), 3–6.

Friedmann, H. (1982), 'The political economy of food: the rise and fall of the postwar international food order', *American Journal of Sociology*, **88**, 248–286.

Friedmann, H. (1993), 'The political economy of food: a global crisis', *New Left Review*, **196**, 29–57.

Gallagher, K.P. (2008), 'Understanding developing country resistance to the Doha Round', *Review of International Political Economy*, **15**(1), 62–85.

Gentilcore, D. (2010), *Pomodoro!: A History of the Tomato in Italy*, New York: Columbia University Press.

Goodman, D. and M. Watts (1994), 'Reconfiguring the rural or fording the divide? Capitalist restructuring and the global agro-food system', *Journal of Peasant Studies*, **22**, 1–49.

Goodman, D. and M. Watts (eds) (1997), *Globalizing Food: Agrarian Questions and Global Restructuring*, London: Routledge.

Goodman, D., E. DuPuis, M. Goodman and K. Michael (2012), *Alternative Food Networks. Knowledge, Practice, and Politics*, London and New York: Routledge.

Goody, J. (2013), 'Industrial food: towards the development of a world cuisine', in: Carole Counihan and Penny van Esterik (eds), *Food and Culture: A Reader*, London and New York: Routledge, pp. 72–90.

Hall, S. (1997), 'Old and new identities, old and new ethnicities', in: Anthony D. King (ed.), *Culture, Globalization and the World-System*, Minneapolis: University of Minnesota Press, pp. 41–68.

Harvey, D. (1989), 'Editorial: a breakfast vision', *Geography Review*, **3**(1).

Helstosky, C. (2008), *Pizza: A Global History*, London: Reaktion.

Hendrickson, M.K. and W.D. Heffernan (2002), 'Opening spaces through relocalization: locating potential resistance in the weaknesses of the global food system', *Sociologia Ruralis*, **42**(4), 347–369.

Herring, Ronald J. (ed.) (2015), *The Oxford Handbook of Food, Politics, and Society*, Oxford: Oxford University Press.

Higman, B.W. (2011), *How Food Made History*, Hoboken, NJ: John Wiley & Sons.

Holt Giménez, E. and A. Shattuck (2011), 'Food crises, food regimes and food movements: rumblings of reform or tides of transformation?', *The Journal of Peasant Studies*, **38**(1), 109–144.

Ichijo, A. and R. Ranta (2016), *Food, National Identity and Nationalism: From Everyday to Global Politics*, Dordrecht: Springer.

Inglis, D. and D. Gimlin (2009), 'Food globalizations: ironies and ambivalences of food, cuisine and globality', in: Inglis, D. and D. Gimlin (eds), *The Globalization of Food*, Oxford and New York: Berg, pp. 3–42.

Kiple, K.F. (2007), *A Movable Feast: Ten Millennia of Food Globalization*, Cambridge: Cambridge University Press.

Larsson, T. (2015), 'The rise of the organic foods movement as a transnational phenomenon', in: R.J. Herring (ed.), *The Oxford Handbook of Food, Politics, and Society*, Oxford Handbooks, New York: Oxford University Press, pp. 739–754.

Le Heron, R. (1993), *Globalised Agriculture*, Oxford: Pergamon.

Leitch, A. (2013). 'Slow food and the politics of "virtuous globalization"', in: Carole Counihan and Penny van Esterik (eds), *Food and Culture: A Reader*, London and New York: Routledge, pp. 409–425.

McMichael, P. (1992), 'Tensions between national and international control of the world food order: contours of a new food regime', *Sociological Perspectives*, **35**, 343–365.

McMichael, P. (1993), 'World food system restructuring under a GATT regime', *Political Geography*, **12**, 3, 198–214.

McMichael, P. (ed.) (1994), *The Global Restructuring of Agro-Food Systems*, Ithaca: Cornell University Press.

Mintz, S.W. (1985), *Sweetness and Power: The Place of Sugar in Modern History*, New York: Penguin.

Morgan, K., T. Marsden and J. Murdoch (2006), *Worlds of Food. Place, Power, and Provenance in the Food Chain*, Oxford and New York: Oxford University Press.

Nabhan, G. (2002), *Coming Home to Eat: The Pleasures and Politics of Local Foods*, New York: Norton.

Nützenadel, A. and F. Trentmann (2008), 'Introduction: mapping food and globalisation', in: Alexander Nützenadel and Frank Trentmann (eds), *Food and Globalisation. Consumption, Markets and Politics in the Modern World*, Oxford: Berg Publishers, pp. 1–19.

Ochoa, E.C. (2012), 'Political histories of food', in: Jeffrey M. Pilcher (eds), *The Oxford Handbook of Food History*, Oxford: Oxford University Press, pp. 23–40.

Phillips, L. (2006), 'Food and globalization'. *The Annual Review of Anthropology*, **35**, 37–57.

Pilcher, J.M. (2005), *Food in World History*, London and New York: Routledge.

Pollan, M. (2006), *The Omnivore's Dilemma: A Natural History of Four Meals*, New York: Penguin.

Popkin, B.M. (2001), 'The nutrition transition and obesity in the developing world', *The Journal of Nutrition*, **131**(3), 871–873.

Potter, C. (2006), 'Competing narratives for the future of European agriculture: the agri-environmental consequences of neoliberalization in the context of the Doha Round', *The Geographical Journal*, **172**(3), 190–196.

Prentice, A.M. (2005), 'The emerging epidemic of obesity in developing countries', *International Journal of Epidemiology*, **35**(1), 93–99.

Pritchard, B. (2009), 'The long hangover from the second food regime: a world-historical interpretation of the collapse of the WTO Doha Round', *Agriculture and Human Values*, **26**(4), 297.

Ray, K. (2015), 'Culinary spaces and national cuisines: the pleasures of an Indian Ocean cuisine', in: J. Farrer (ed.), *The Globalization of Asian Cuisines*, New York: Palgrave Macmillan US, pp. 23–35.

Renting, H., T. Marsden and J. Banks (2003), 'Understanding alternative food networks: exploring the role of short food supply chains in rural development', *Environment and Planning A*, **35**, 393–411.

Roozen, N. and F. Van der Hoff (2002), *L'Aventure du commerce équitable: une alternative à la mondialisation*, Paris: Éditions Jean-Claude Lattès.

Sachs, J.D. (2015), *The Age of Sustainable Development*, New York: Columbia University Press.

Sebastiani, R., F. Montagnini and D. Dalli (2013), 'Ethical consumption and new business models in the food industry. evidence from the Eataly case', *Journal of Business Ethics*, **114**: 473–488.

Sen, A. (1981), *Poverty and Famines: An Essay on Entitlement and Deprivation*, Oxford: Oxford University Press.

Serventi, S. and F. Sabban (2003), *Pasta: The Story of a Universal Food*, New York: Columbia University Press.

Shiva, V. (2016), *The Violence of the Green Revolution: Third World Agriculture, Ecology, and Politics*, Lexington, KT: University Press of Kentucky.

Smith, A.F. (2008), *Hamburger: A Global History*, London: Reaktion.

Stephens Griffin, N. (2017), *Understanding Veganism: Biography and Identity*, London: Palgrave Pivot.

Stolle, D., M. Hooghe and M. Micheletti (2005), 'Politics in the supermarket: political consumerism as a form of political participation', *International Political Science Review*, **26**(3), 245–269.

Sylla, N. (2014), *The Fair Trade Scandal: Marketing Poverty to Benefit the Rich*, Athens,. OH: Ohio University Press.

Torrez, F. (2011), 'La Via Campesina: peasant-led agrarian reform and food sovereignty', *Development*, **54**(1), 49–54.

Watts, D.C., B. Ilbery and D. Maye (2005), 'Making reconnections in agro-food geography: alternative systems of food provision', *Progress in Human Geography*, **29**(1), 22–40.

Wiskerke, J.S. (2009), 'On places lost and places regained: reflections on the alternative food geography and sustainable regional development', *International Planning Studies*, **14**(4), 369–387.

Ziegler, J., C. Golay, C. Mahon and S. Way (2011), *The Fight for the Right to Food: Lessons Learned*, New York: Springer.

18. Geographies of finance in a globalizing world

David Bassens and Michiel Van Meeteren

INTRODUCTION: FINANCIAL GLOBALIZATION

On 15 September 2008, Lehman Brothers, a big Wall Street investment bank, filed for bankruptcy. The bank had become heavily involved in home mortgage markets as packagers, sellers, and investors of mortgage debt in the United States. Unable to attract money in financial markets, Lehman Brothers was forced to cease business. This closure, however, proved to be only the beginning of a banking crisis that would almost immediately spread across the Atlantic through deep connections between European and US banks. Yet, this still did not end the story. In the European Union (EU), as in the US, uncertainty about exposure to US mortgage markets led to a credit crunch, a near to complete freezing up of the credit flows between banks. As many banks were financed via these short-term schemes, the credit crunch generated acute liquidity problems, and the potential threat of collapse for European banks. Subsequently, EU governments stepped in with guarantees or fresh capital injections in an effort to save the financial system from the collapse of 'systemic' banks, most of which had a Northern European signature. This move stabilized the northern 'core' of the EU. However, by 2009 it had become clear that the initial uncertainty about US mortgage market exposure had morphed into financial market uncertainty about the ability to pay of the net-debtor states in what was soon dubbed 'the periphery' of Europe (or GIPSI: Greece, Ireland, Portugal, Spain, and Italy). What started as a mortgage market crisis in the US had travelled to become a European banking crisis and, in the end, took hold of Europe as a sovereign debt crisis from which the EU still had to recover a decade after the crisis emerged.

The above sketch of the inception and spread of the North Atlantic Financial Crisis (Jessop, 2015) is emblematic of a condition of financial globalization, a process characterized by the growing range and depth of financial relations enshrined in a global financial system that supersedes the international state system. The magnitude of the global financial system is hard to fathom due to its offshore character and the incomplete geographic coverage of indicators developed by key regulatory institutions. Still, as per 2017, the most recently available information indicates that cross-border bank claims amounted to 79 trillion US dollars (BIS, 2017), while derivatives markets and the murky shadow banking system amount to as much as 544 trillion US dollars (BIS, 2016) and 137 trillion US dollars (FSB, 2015) respectively. Even if we consider these as modest estimates (excluding foreign direct investments and portfolio investments), the size of the global financial system is by now about ten times the size of the global economy, which was estimated at 74 trillion US dollars in 2015 (World Bank, 2017). While the cross-border character of finance is nothing new, the concept of financial globalization captures the tremendous growth of these financial exchanges in recent decades. Moreover, financial *globalization* signals the difference from earlier efforts to contain and regulate what was formerly known as the *international* financial system as financial globalization largely escapes the financial and monetary governance capacities of the state system (Strange, 1994).

The main purpose of this chapter is to provide an overview of the main drivers of financial globalization. What is it about finance that allows its globalization to take place so rapidly and deeply? In the next section we contend that a main answer lies in the inherent capacity of finance to overcome space and time constraints. Finance appears to produce space, yet, co-constitutively, geographic-institutional variegation feeds back into how finance operates. Hidden beneath this abstract mechanism of time–space compression (Harvey, 1989) lie concrete georgraphies of finance that are to an important extent co-evolutionary with pre-existing geographies of knowledge, law, and power. As the resulting variegation raises important analytical questions, the third section provides an overview of how financial geographers study this process from the perspectives of international financial centres, firms, flows, and products. The fourth section examines the history of post-1945 European financial integration to illustrate how European states have sought ways to contain and benefit from financial global-ization, a process that, paradoxically, has been fundamental in deepening uneven development on the continent. The final section concludes with a research agenda and a plea to give space a central focus in future studies of finance.

THE CO-CONSTITUTION OF FINANCE AND SPACE

In capitalism, the decision about 'where to put your money' is usually determined by the ability of capital to generate a surplus. As the amount of capital in the world increases, the surplus-generating opportunities needed for stable accumulation increases at a compounding rate. In absolute terms, the 2014 world economy would have to grow 333 times more than in 1970 to sustain a normal rate of accumulation (six billion in 1970, two trillion in 2014, see Harvey, 2014: 228). Somewhat counter-intuitively, a capitalist crisis emerges when more money is around than can be profitably invested, frustrating the system's need for ongoing expansion. The result is a 'wall of money', restlessly and sometimes desperately seeking normal returns (Fernandez and Aalbers, 2016), logically resulting in (trading of) financial instruments to reap future profits and to defer risks to others – a condition called 'financialization' (Arrighi, 1994[2010]). In order to gain a surplus, capital needs to become mobile to switch easily from one circuit of value to another, yet not all forms of capital can do this with the same ease. Capital, a process of value in motion, appears in three forms, which display varying geographical mobilities: the productive, commodity, and financial capital forms. Harvey (2013: 37–40) proposes the metaphor of a butterfly to illustrate the capital accumulation cycle that links the three capital moments. Productive capital (e.g. factories, infrastructure, and machinery) is hardly mobile (like an egg or a chrysalis), while commodity capital (i.e. sellable goods and services) is only mobile to a limited extent (like a caterpillar). Finance capital (i.e. credit), however, is free like a butterfly to relocate and become part of a different circuit. Hence the pivotal role of finance capital for globalization: only in its financial form can capital rapidly move around to become part of the most profitable circuit. This capacity to become mobile has enabled time–space compression with financial products being traded globally on the nanosecond.

The structural tendencies described above can result in many different concrete outcomes, depending how actors act upon them (Van Meeteren and Bassens, 2018). Globally operating finance has crystallized into a particular structure: money does not flow everywhere with the same ease, yet sticks to certain places 'like mercury' (Clark, 2005). At the same time, investment opportunities are not simply waiting to be discovered, but are socially constructed by financial intermediaries who produce geographical narratives (i.e. economic geographies) on where to invest next (Lee, 2002). Concrete geographies thus hinge upon how actors in particular places are able to (re)combine knowledge, legal, and power resources (Van Meeteren and Bassens, 2016). First, gaining from financial globalization requires knowledge about profitable investment opportunities. For finance to 'work' across the borders of different political economies, different kinds of expertise have to be combined. The ability to interpret financial information is not ubiquitous, but instead tends to be concentrated in a limited number of places, where actors can marshal agglomerated resources and an advanced division of labour of specialized service firms and their networks (Bassens and Van Meeteren, 2015). Combined, these resources enable interpretation of information flows, innovation in new markets and product areas, and the representation of financial power (Amin and Thrift, 1992). Consequently, the more complicated and advanced financial products – those that are most attractive in the context of the 'wall of money problem' – tend to be only accessible from the larger, better-connected financial centres (Clark and O'Connor, 1997). Moreover, access to knowledge is not only a question of 'being there', one also needs to be socialized in the profession to have the cognitive repertoire and the social and cultural resources to access this market information.

Second, global finance is crucially mediated by legal geographies. Financial assets are contracts concerning claims on future revenues, the value of which depends on their legal vindication, that is, the plausibility that the contract will be honoured at a later date (Pistor, 2013). The globalization of finance has coincided with the emergence of an expanding and deepening set of free trade rules (e.g. embodied in the World Trade Organization or regional blocks such as the EU) that seek to competitively (re)regulate, and not much deregulate, cross-border financial exchanges. However, the world is anything but a singular legal space and different jurisdictions remain marked by different legal regimes. It is the law that helps to create, segment, and demarcate a mosaic of markets across the world (Christophers, 2015). Moreover, legal regimes overlap, creating innumerable loopholes for legal arbitrage, once territories with slightly different legal regimes are combined (Sassen, 2013). Intermediaries, such as big global law firms, do not only help facilitate actors to traverse these boundaries and exploit the loopholes, but they are also active in changing legal spaces to their and their clients' advantage (Faulconbridge and Muzio, 2015).

Third, power relations not only mediate the way global finance operates, but also who gains from its practices. Concerning the former, the relations between financial actors are influenced by the relative position of the nation-states from which they operate in the inter-state system. Even though financial sector interests have become to a degree disembedded from 'territorial' interests of their hosting nation-states, the way the global financial system is structured is still reminiscent of power relations between states. For instance, the period of American political-economic hegemony after the Second World War was matched by the growing globalization and influence of its private financial institutions. Similarly, the growing weight of China in the global economy currently translates in the growing weight and range of its state-related banks and sovereign wealth funds (Haberly, 2011). Concerning the spoils of financial globalization, it appears these are distributed unevenly between the too-big-to-fail banks, and the wider public (be it states or taxpayers), the latter having borne most of the cost of systemic failure in 2008. Also within the financial sector, some animals are more equal than others: when push comes to shove, some legal claims are more likely to be honoured. Generally speaking, laws regarding financial resolution tend to be flexible and negotiable when the core of the financial system benefits and remorseless when the powerless or the periphery have to pay up (Pistor, 2013: 317). Wójcik (2013) shows how banks in the dyad London–New York, largely responsible for the financial crisis of 2008, went back to reaping exuberant profits in no time. Part of the power of these financial intermediaries is structural in nature, as collectively they act as obligatory passage points for all those – including citizens of capitalist states – in need of accumulation for their private or collective household pension and schemes.

ANALYZING FINANCIAL GEOGRAPHIES

The discussed mechanisms of financial globalization come together in four analytical perspectives utilized by financial geographers who study the global financial system.

First, studies of International Financial Centres (IFCs) analyze the agglomeration of financial activities. A key debate here regards to what extent contemporary IFCs are a path-dependent result of long-term historical development. The study of the rise and fall of European financial centres (see e.g. Arrighi, 1994[2010]; Cassis, 2006) reveals that cycles of economic expansion and decline coincide with geographical shifts in the leading IFCs. Financial geographers are interested in how these histories provide a heritage of place-based institutions and regimes (King, 1990). While path-dependencies matter, the ever-expanding role of Information and Communication Technologies (ICT) has helped producing a global, electronically wired financial system, in which many second- and third-tier financial centres have seen their position weakened (Engelen, 2007). Yet, while many popular and policy debates interpret IFC formation as a zero-sum game with harsh implications for smaller centres, geographers have offered compelling evidence that financial intermediation depends on a spatial division of labour. Different IFCs and offshore financial centres perform complementary and niche roles in the global financial system (Lee and Schmidt-Merwede, 1993; Van Meeteren and Bassens, 2016) in both mature (Faulconbridge, 2004), and emerging (Lai, 2012) markets.

Second, financial geographers have studied the spatial organization of financial intermediaries. As these firms are the enablers of financial globalization, it is relevant to inquire about the spatial strategies of financial intermediaries and an array of supporting producer services. How do these firms organize global work and govern their global activities as part of their business model? World city network research has systematically mapped out the connectivities of financial centres, based on the insertion of financial, accounting, management consultancy, advertising, and law intermediaries in global urban networks (Taylor and Derudder, 2016). Changing ownership structures – due to mergers and acquisitions in these sectors (Zademach and Musil, 2014) – in turn affect the positionality of IFCs that host these intermediaries in the world city network. Other studies focus on the governance geographies of institutional investors such as pension funds (Hebb and Wójcik, 2005), sovereign wealth funds (Dixon and Monk, 2014), niches such as Islamic finance (e.g. Bassens et al., 2011), private equity and venture capital (Martin et al., 2002), and financial technologies that may disrupt the financial sector (Hendrikse et al., 2018). Financial geography also offers in-depth qualitative studies of the narratives on the importance of global expansion of financial institutions, which started to flourish in a context of relaxing anti-trust rules and shareholder capitalism in the financial sector (Van Meeteren and Bassens, forthcoming). Ultimately, the firm-based perspective raises questions about the extent to which financial intermediaries are not only the executers, but also targets of corporate financialization.

A third perspective deals with the spatialities and power relations of credit flows and the financial products and techniques that allow capital to move across borders. Unfortunately, financial geographies, in the strict sense of mapping financial flows, are uncommon, despite the availability of official data on Foreign Direct Investment (FDI) and cross-border bank holdings. Still, geographers have used these data sources to understand cross-border interdependencies between banking systems (Bassens et al., 2013) or to disclose the footprint of the offshore system (Haberly and Wójcik, 2015). The fundamentally geographical nature of the global financial crisis of 2008 gave an impetus to this line of inquiry as a global web of financial obligations ground to a standstill with disastrous effects on economies and societies. Financial geographers have subsequently become interested in the techniques that have allowed the crisis to spread between different areas of the financial system. The technique of the bundling and repackaging of assets known as 'securitization' emerged as key in this context. Its geographies of variegated implementation and diffusion have come to constitute an important point of attention (Aalbers et al., 2011).

Fourth, financial geographers have studied the territorially variegated outcomes of financial globalization. The global financial system is embedded in formal (e.g. legal systems, regulatory frameworks) and informal institutions (e.g. relating to business cultures) that are context-bound (Peck and Theodore, 2007). Authors conceptualize these differences as a process of unfolding variegated financialization, which plays out differently depending on the context (see e.g. Fernandez and Aalbers, 2016). Typically, this is examined at the national scale where particular finance–society relations are hypothesized. Often, geographers make an ideal–typical distinction between bank-based regimes and capital-market based regimes, even though that distinction has increasingly blurred in contemporary economies (Hardie and Howarth, 2013). Geographers working from this institutional perspective have been interested in grasping the changes in national financial systems under globalization, for instance mapping cross-border stock listings (Wójcik, 2011). Alternatively, geographers have wondered how (national) markets are socially constructed as an investment category, under the banner of emerging markets research (see Bassens, 2012; Heinemann, 2016).

This construction of geographically articulated markets raises fundamental questions about how these markets relate to political-economic scales such as the EU, a question to which we will now turn.

ILLUSTRATION: EUROPEAN FINANCIAL INTEGRATION

The incessant drive for finance capital's spatial expansion presents state actors, such as politicians and regulators, with a conundrum. While state actors seek to stabilize their economies in spite of the crisis-prone volatilities of the global financial system, they also have a desire to encapsulate financial globalization to their benefit. This necessitates a scalar fix (Cox, 2002): the construction of a scale that can withstand the geographical upscaling and acceleration of the global financial system. A key example of such 'scalar fixing' is the deepening of European financial and monetary integration over the past four decades. In order to ensure the political-economic stability of the present-day EU, it became necessary to shield the European economy from currency volatility after the collapse of the 1944–1973 transatlantic Bretton Woods regime (Strange, 1994). Bretton Woods was largely designed by the United States to direct the power of financial globalization into re-building the defeated economies of the Second World War into a pro-US geopolitical constellation. Central mechanisms to achieve such orchestration were the restriction of capital movements and pegging currencies to the dollar (Varoufakis, 2015[2011]). In Northern Europe, this created spaces in which national financial and banking systems could flourish relatively shielded from foreign competition (Lee and Schmidt-Merwede, 1993). Meanwhile, tensions in the Bretton Woods system helped to lay the foundations of the contemporary global financial system with a crucial role cut out for the City of London and its Commonwealth offshore satellites (Roberts, 1995). When the Bretton Woods regime finally collapsed in 1973, the European refuge from the volatility of financial globalization came to an end and immediately raised the stakes of a European scalar fix of financial integration. From 1973 onwards, an increasing number of regulations and institutions have been issued to promote free and frictionless movement of capital within what in 1993 became the EU (Maes, 2007). The North Atlantic Financial Crisis of 2008 accelerated this process considerably as the power of the European Central Bank was strengthened, movements towards further financial integration were initiated, and the pressure on member states that 'treaded outside the commonly agreed' budgetary rules were increased (Matthijs and Blyth, 2015).

However, the struggle to construct a European scalar fix in fact has exacerbated uneven development within the continent (Bassens et al., 2013). A growing series of EU initiatives towards monetary, banking and capital market union have diminished legal fragmentation within the EU. This homogenization of legal geographies is predicted to affect the division of labour between IFCs across Europe (Faulconbridge et al., 2007). Although some regional specializations – for instance Vienna as a centre of East-European finance (Zademach and Musil, 2014) – have emerged, generally European financial integration has strengthened the position of London, Paris, and Frankfurt at the expense of smaller centres, such as Amsterdam and Brussels (Derudder et al., 2011). Although the more intricate knowledge geographies found in the larger

IFCs have become more important, local knowledges have not lost their relevance altogether. Nevertheless, core–periphery relations within Europe have become strained as financial integration has mostly benefited expansion by core banks into the periphery during successive waves of mergers and acquisitions. In 2015, only three out of the 20 largest European commercial banks came from the GIPSI countries: Banco Santander and BBVA in Spain and Unicredit in Italy (Goddard et al., 2016: 113), none were headquartered in Eastern Europe. Moreover, IFCs in the periphery are predominantly conduits channelling in capital coming from the EU's Northern members (Karreman, 2009). After the inception of the Euro and until 2008 banks from core countries were key in financing credit-led economic development in the GIPSI countries (Bassens et al., 2013). Pre-crisis geographical expansion also came with the use of new techniques such as securitization to keep assets of the books, increase bank leverage, and churn out higher profits (Wainwright, 2015). Lingering national legal differences pertaining to securitization (Thiemann, 2012), however, left banks unevenly exposed to the crisis, producing a banking crisis geography skewed to Northern Europe. Yet, as the crisis morphed into its sovereign debt phase, GIPSI countries were ultimately the ones facing liquidity issues, as refinancing their debt with core banks had become increasingly challenging. Since then, debt asymmetries have augmented the political dominance of the EU's core countries over the periphery (Varoufakis, 2015[2011]). This is reflected in the way in which the Eurozone crisis is understood in the various EU jurisdictions with representatives of the Northern countries likely to 'blame the periphery' for their own troubles (Engelen et al., 2011). Moreover, some EU member states have more influence on how things are organized at the EU scale than others, giving rise to ever-louder accusations that the scalar fix of European financial integration boils down to a form of financial imperialism (Hadjimichalis, 2011). Greece, for instance, has been repeatedly denied the budgetary flexibility by the EU institutions that was granted to Germany in 2003 to pay for its reunification in the 1990s (Heipertz and Verdun, 2004; Varoufakis, 2015[2011]: 201–203). In sum, despite the EU Institutions' enduring rhetoric of economic convergence, the political-economic exercise in containing and benefitting from financial globalization through fixing it at the EU scale has in reality produced a fundamentally uneven financial geography within its spatial confines.

CONCLUSION

This chapter has highlighted the fundamentally co-constitutive nature of finance and space. While finance capital breeds globalization in its search for profitable outlets, a global financial system has emerged with a traceable spatial footprint. It is the geography of this socio-spatial structure that financial geographers aim to pin down to understand its variegated entanglements with polities, economies, and societies across the globe. To do so in the current conjuncture is intriguing because of at least three processes of ostensible disruption. First, we are entering a period where the two states harbouring the leading IFCs – US and Wall Street, UK and The City – are renegotiating their interaction with the global economy while new geo-economic players are entering the scene. At the time of writing, the US president Donald Trump is proclaiming a future of stringent economic protectionism, whereas UK's Prime Minister Theresa May is executing the Brexit from the European Union. Meanwhile, besides being recipients of finance capital, key emerging economies such as China are by now also instigators of financial globalization themselves, with an important role for sovereign investments into Europe and the US. Second, current geo-financial shifts run parallel with rapid technological developments in the financial industry, a process referred to as the rise of Financial Technology (FinTech). The digitization of finance carries the promise of a more decentralized financial power outlay as new players beyond the big bank shortlist might rejuvenate smaller IFCs. Simultaneously, counter currents in the banking world seem to suggest that FinTech may very well be a strategic asset in maintaining existing banking oligopolies in a limited number of IFCs (Hendrikse et al., 2018). Third, despite the confluence of a growing wall of money and citizen dependence on accessing the debt-driven global financial system for their social reproduction (Graeber, 2011), the current finance-dominated form of capitalism has failed to produce an inclusive and redistributive system. Together, these three developments raise serious questions about the spatial reproduction of the global financial system, the analysis of which, we may hope, can draw on the apparatus developed in this contribution.

REFERENCES

Aalbers, M.B., E. Engelen and A. Glasmacher (2011), '"Cognitive closure" in the Netherlands: mortgage securitization in a hybrid European political economy', *Environment and Planning A*, **43**(8), 1779–1795.

Amin, A. and N. Thrift (1992), 'Neo-Marshallian nodes in global networks', *International Journal of Urban and Regional Research*, **16**(4), 571–587.

Arrighi, G. (1994[2010]), *The Long Twentieth Century. Money, Power and the Origins of our Times*, London and New York: Verso.

Bassens, D. (2012), 'Emerging markets in a shifting global financial architecture: the case of Islamic securitization in the Gulf region', *Geography Compass*, **6**(6), 340–350.

Bassens, D. and M. Van Meeteren (2015), 'World cities under conditions of financialized globalization: towards an augmented world city hypothesis', *Progress in Human Geography*, **39**(6), 752–775.

Bassens, D., B. Derudder and F. Witlox (2011), 'Setting Shari'a standards: on the role, power and spatialities of interlocking Shari'a boards in Islamic financial services', *Geoforum*, **42**(1), 94–103.

Bassens, D., M. Van Meeteren, B. Derudder and F. Witlox (2013), 'No more credit to Europe? Cross-border bank lending, financial integration and the re-birth of the national scale as a credit scorecard', *Environment and Planning A*, **45**(10), 2399–2419.

Bank for International Settlements (BIS) (2016), 'Statistical release OTC derivatives statistics at end-June 2016', accessed on: 7 March 2017, https://www.bis.org/.

Bank for International Settlements (BIS) (2017), 'Summary of consolidated statistics, by nationality of reporting bank', accessed on: 7 March 2017, https://www.bis.org/.

Cassis, Y. (2006), *Capitals of Capital: The Rise and Fall of International Financial Centres 1780–2009*, Cambridge, MA: Cambridge University Press.

Christophers, B. (2015), 'The law's markets: envisioning and effecting the boundaries of competition', *Journal of Cultural Economy*, **8**(2), 125–143.

Clark, G.L. (2005), 'Money flows like mercury: the geography of global finance', *Geografiska Annaler: Series B, Human Geography*, **87**(2), 99–112.

Clark, G.L. and K. O'Connor (1997), 'The informational content of financial products and the spatial structure of the global finance industry', in: K.R. Cox (ed.), *Spaces of Globalization. Reasserting the Power of the Local*, New York/London: The Guilford Press, pp. 89–114.

Cox, K.R. (2002), '"Globalization", the "regulation approach", and the politics of scale', in: A. Herod and M.W. Wright (eds), *Geographies of Power, Placing Scale*, Malden: Blackwell Publishing, pp. 85–114.

Derudder, B., M. Hoyler and P.J. Taylor (2011), 'Goodbye Reykjavik: international banking centres and the global financial crisis', *Area*, **43**(2), 173–182.

Dixon, A. and A. Monk (2014), 'Frontier finance', *Annals of the Association of American Geographers*, **104**(4), 852–868.

Engelen, E. (2007), '"Amsterdamned"? The uncertain future of a financial centre', *Environment and Planning A*, **39**(6), 1306–1324.

Engelen, E., R.P Hendrikse, V. Mamadouh and J.D. Sidaway (2011), 'Turmoil in Euroland: the geopolitics of a suboptimal currency area?', *Environment and Planning D: Society and Space*, **29**, 571–583.

Faulconbridge, J.R. (2004), 'London and Frankfurt in Europe's evolving financial centre network', *Area*, **36**(3), 235–244.

Faulconbridge, J.R. and D. Muzio (2015), 'Transnational corporations shaping institutional change: the case of English law firms in Germany', *Journal of Economic Geography*, **15**(6), 1195–1226.

Faulconbridge, J.R., E. Engelen, M. Hoyler and J.V. Beaverstock (2007), 'Analysing the changing landscape of European financial centres: the role of financial products and the case of Amsterdam', *Growth and Change*, **38**(2), 279–303.

Fernandez, R. and M.B. Aalbers (2016), 'Financialization and housing: between globalization and varieties of capitalism', *Competition and Change*, **20**(2), 71–88.

Financial Stability Board (FSB) (2015), 'Global shadow banking monitoring report 2015', accessed on: 7 March 2017, http://www.fsb.org/.

Goddard, J., D.G. McKillop and J.O.S. Wilson (2016), 'Ownership in European banking', in: T. Beck and B. Casu (eds), *The Palgrave Handbook of European Banking*, London: Palgrave Macmillan, pp. 103–134.

Graeber, D. (2011), *Debt: The First 5000 Years*, New York: Melhouse House Publishing.

Haberly, D. (2011), 'Strategic sovereign wealth fund investment and the new alliance capitalism', *Environment and Planning A*, **43**(8), 1833–1852.

Haberly, D. and D. Wójcik (2015), 'Regional blocks and imperial legacies: mapping the global offshore FDI network', *Economic Geography*, **91**(3), 251–280.

Hadjimichalis, C. (2011), 'Uneven geographical development and socio-spatial justice and solidarity: European regions after the 2009 financial crisis', *European Urban and Regional Studies*, **18**(3), 254–274.

Hardie, I. and D. Howarth (2013), *Market-Based Banking and the International Financial Crisis*, Oxford: Oxford University Press.

Harvey, D. (1989), *The Condition of Postmodernity*, Cambridge MA/Oxford: Blackwell.

Harvey, D. (2013), *A Companion to Marx's Capital vol 2*, London and New York: Verso.

Harvey, D. (2014), *Seventeen Contradictions and the End of Capitalism*, Oxford: Oxford University Press.

Hebb, T. and D. Wójcik (2005), 'Global standards and emerging markets: the institutional-investment value chain and the CalPERS investment strategy', *Environment and Planning A*, **37**(11), 1955–1974.

Heinemann, T. (2016), 'Relational geographies of emerging market finance: the rise of Turkey and the global financial crisis 2007', *European Urban and Regional Studies*, **23**(4), 645–661.

Heipertz, M. and A. Verdun (2004), 'The dog that would never bite? What we can learn from the origins of the Stability and Growth Pact', *Journal of European Public Policy*, **11**(5), 765–780.

Hendrikse, R., D. Bassens and M. Van Meeteren (2018), 'The Appleization of finance: Charting incumbent finance's embrace of FinTech', *Finance and Society*, Early View.

Jessop, B. (2015), 'Hard cash, easy hard credit, fictitious capital: critical reflections on money as a fetishized social relation', *Finance and Society*, **1**(1), 20–37.

Karreman, B. (2009), 'Financial geographies and emerging markets in Europe', *Tijdschrift voor Economische en Sociale Geografie*, **100**(2), 260–266.

King, A.D. (1990), *Global Cities, Post-Imperialism and the Internationalization of London*, London and New York: Routledge.

Lai, K. (2012), 'Differentiated markets: Shanghai, Beijing, and Hong Kong in China's financial centre network', *Urban Studies*, **49**, 1275–1296.

Lee, R. (2002), '"Nice maps, shame about the theory?" Thinking geographically about the economic', *Progress in Human Geography*, **26**(3), 333–355.

Lee, R., and U. Schmidt-Merwede (1993), 'Interurban competition? Financial centres and the geography of financial production', *International Journal of Urban and Regional Research*, **17**(4), 492–515.

Maes, I. (2007), *Half a Century of European Financial Integration: From the Rome Treaty to the 21st Century*, Brussels: Mercatorfonds.

Martin, R., P. Sunley and D. Turner (2002), 'Taking risks in regions: the geographical anatomy of Europe's emerging venture capital market', *Journal of Economic Geography*, **2**(2), 121–150.

Matthijs, M. and M. Blyth (eds) (2015), *The Future of the Euro*, Oxford: Oxford University Press.

Peck, J. and N. Theodore (2007), 'Variegated capitalism', *Progress in Human Geography*, **31**(6), 731–772.

Pistor, K. (2013), 'A legal theory of finance', *Journal of Comparative Economics*, **41**(2), 315–330.

Roberts, S. M. (1995), 'Small place, big money: the Cayman Islands and the international financial system', *Economic Geography,* **71**(3), 237–256.

Sassen, S. (2013), 'When territory deborders territoriality', *Territory, Politics, Governance,* **1**(1), 21–45.

Strange, S. (1994), 'From Bretton Woods to the casino economy', in: S. Corbridge, R. Martin, N. Thrift (eds), *Money, Power, and Space,* Blackwell: Oxford, pp. 49–62.

Taylor, P.J. and B. Derudder (2016), *World City Network* (2nd edn), London and New York: Routledge.

Thiemann, M. (2012), '"Out of the shadows?" Accounting for special purpose entities in European banking systems', *Competition & Change,* **16**(1), 37–55.

Van Meeteren, M. and D. Bassens (2016), 'World cities and the uneven geographies of financialization: unveiling stratification and hierarchy in the world city archipelago', *International Journal of Urban and Regional Research,* **40**(1), 62–81.

Van Meeteren, M. and D. Bassens (2018), 'Chasing the phantom of a "global end game": the role of management consultancy in the narratives of pre-failure ABN AMRO', in: Hoyler, M., Parnreiter, C. and Watson, A. (eds), *Global City Makers: Economic Actors and Practices in the World City Network,* Cheltenham, UK and Northampton, MA, USA: Edward Elgar Publishing, pp. 170–191.

Varoufakis, Y. (2015[2011]), *The Global Minotaur: America, Europe and the Future of the Global Economy,* London: Zed Books.

Wainwright, T. (2015), 'Circulating financial innovation: new knowledge and securitization in Europe', *Environment and Planning A,* **47**(8), 1643–1660.

Wójcik, D. (2011), *The Global Stock Market: Issuers, Investors, and Intermediaries in an Uneven World,* Oxford: Oxford University Press.

Wójcik, D. (2013), 'The dark side of NY-LON: financial centres and the global financial crisis', *Urban Studies,* **50**(13), 2736–2752.

World Bank (2017), 'GDP (current US$)', accessed on: 7 March 2017, http://data.worldbank.org/.

Zademach, H. and R. Musil (2014), 'Global integration along historical pathways', *European Urban and Regional Studies,* **21**(4), 463–483.

19. Geographies of health in a globalizing world
Mark Rosenberg

INTRODUCTION

Global outbreaks of contagious diseases speak to the imagination. The outbreak of SARS (Severe Acute Respiratory Syndrome) in November 2002 made the public aware, arguably for the first time, of how globalization affects human health regardless of whether one lives in an affluent neighbourhood in a rich country in the developed world, or a squatter settlement in a poor country in the developing world (Smith, 2006; Wallis and Nerlich, 2005). Since the SARS outbreak, there have been other global health scares, most notably the H1N1 (Influenza A virus subtype H1N1/09) outbreak of 2009, the Ebola outbreak of 2013 and the Zika outbreak of 2015, which have reinforced the views of the public that some aspects of globalization are a threat to human health. At the extreme, an apocalyptic view of how globalization affects human health has been created by a global film industry and the popularity of movies such as *Outbreak* (1995), *Doomsday* (2008), *The Happening* (2008), *Carriers* (2009), and *The Thaw* (2009). There is, however, much more to the geographies of health in a globalizing world than focusing only on emerging (e.g. SARS) and re-emerging (e.g. H1N1) diseases. To present some of these, this chapter is divided into six parts: the meaning of globalization; health and the environment; the movement of people; the movement of economic activities; health care as a global economic activity; and some concluding comments.

THE MEANING OF GLOBALIZATION

There are many definitions of globalization. The one used to guide the discussion that follows comes from Voisey and O'Riordan (2001: 26) who define globalization as 'a process of primarily economic, but also social and political change that encompasses the planet, resulting in greater homogeneity, hybridization and interdependence ... of money, people, images, values and ideas that has entailed smoother and swifter flows across national boundaries'. Homogeneity places an emphasis on the sameness of phenomena regardless of where one lives (e.g. how local foods are replaced by the ubiquity of fast food whether one lives in the United States, China or Ghana). Hybridization is the merging of phenomena (e.g. the incorporation of traditional medical practices such as acupuncture with Western biomedical practices in some health care systems). Interdependence is the growing acceptance that actions in one part of the world have consequences and effects on other parts of the world (e.g. pollution generated in the United States not only crosses borders into Canada and

Mexico but contributes to overall global warming). The above definition of globalization is also about 'smoother' and 'swifter' flows at the international and global scale (e.g. how an outbreak of a highly infectious disease can rapidly and easily spread around the world in hours and days now).

What is also useful in the Voisey and O'Riordan (2001) definition of globalization is their emphasis on various aspects of globalization depending on the scale of analysis. At the international scale, Voisey and O'Riordan (2001) focus on technological advance and the growth of the information technology sector. Politically and economically, they draw attention to international agreements (e.g. the North American Free Trade Agreement [NAFTA]), multinational corporations and international government institutions (e.g. World Trade Organization) and the role they play in facilitating globalization. At the national scale, states either promote or fight against the growing ubiquity of Western culture and capitalism to the detriment of local cultures and economic systems. With the diminution of local cultures and economic systems also comes a loss of diversity and the disappearance of local distinctiveness and community. At the individual scale, people feel the loss of control over their own lives and question the ability of national, state/provincial/regional and local governments to act in the best interests of the citizenry. Ultimately, globalization emphasizes individual consumption over notions of the collective good.

Since Voisey and O'Riordan (2001) proposed their definition of globalization much has changed about the world. The '9/11 and the war on terror', global economic crisis in 2008, the conclusion of the Millennium Development Goals (MDGs) and their replacement with the Sustainable Development Goals (SDGs), the Brexit vote and the election of Donald Trump as president of the United States stand as contradictions to the general trends of globalization. Each of these examples and many others stand as significant counter evidence to the inexorable trajectory of globalization that some have predicted. Even so, there can be no doubt that processes that we call globalization in terms of creating swifter flows of people, technology, flora and fauna have profound direct and indirect effects on health and the environment, the movement of people, the movement of economic activities and health care as a global economic activity.

HEALTH AND THE ENVIRONMENT

At the global scale, climate change affects human health directly and indirectly. According to the Intergovernmental Panel on Climate Change (IPCC, 2014) indirect impacts on human health will result from water shortages and the impacts that water shortages and global warming have on food security. The direct impacts will result from how climate change exacerbates health issues that already exist. For example, through stratospheric ozone depletion, the rates of skin cancer are growing globally at an alarming rate (Lens and Dawes, 2004). The number of new cases of melanoma of the skin in developed countries was 191 100, placing it eighth among the top 10 sites of cancer for men and ninth for women in 2012 (Torre et al., 2015). Global warming is contributing to desertification and land degradation which in turn leads to the destruction of agroecosystem productivity and ultimately, to malnutrition, starvation and mass movements of people who can no longer support themselves (see below).

Global warming is also expanding the zones in which disease vectors (e.g. mosquitoes) can survive. For example, between 300 and 500 million people were estimated to have malaria at the end of the 1990s. Martens et al. (1999) demonstrated how mosquitoes of the genus *Anopheles* will spread as global warming continues putting an additional 260 to 320 million people at risk for *P. falciparum* and from 100 to 200 million for *P. vivax* (two types of malaria) by 2080.

At national and regional scales of analysis, air pollution (which also contributes to global warming) contributes directly to growing rates of asthma, other respiratory diseases (e.g. chronic obstructive pulmonary disease) and other non-respiratory diseases (e.g. heart disease) (see for example Ailshire et al., 2017; Jerrett et al., 2005; Romero-Lankao et al., 2013). While air pollution comes from many sources, the underlying driver is the rapid growth of urban populations fuelled by population growth, rural to urban migration and transportation and energy needs within cities. Particularly in the rapidly growing cities of Africa, Asia and South America, urban infrastructure cannot keep up with population growth resulting in the growing size and number of slums and shanty towns where the lack of clean water and proper sanitation result in disease outbreaks (e.g. diarrhoeal diseases and cholera) and vector-borne diseases (e.g. malaria and dengue). While improved housing and safe water and sanitation facilities are the preferred long-term solutions but seem highly unlikely in many parts of the developing world, local small-scale interventions have a role to play in the short-term (e.g. Tumwebaze and Mosler, 2015). The Tumwebaze and Mosler example also highlights the tension between efforts to resolve health issues at higher levels of government up to the global scale through efforts like the Millennium Development Goals and the more recent Sustainable Development Goals and resolving issues at the local level.

THE MOVEMENT OF PEOPLE

The movements of people alluded to above, whether as refugees from violence (e.g. refugees from the wars in Iraq and Syria), environmental refugees (e.g. being driven off the land as the result of drought), searching for a better way of life (e.g. rural to urban movement of working age Chinese people giving up an agricultural way of life for a factory job) to the continental migration of economic refugees (e.g. from central America to the United States or from north Africa to Western Europe) are underpinned by the pervasive forces of globalization that favour high-income countries over lower- and middle-income countries. 'From 2000 to 2015, high-income countries received an average of 4.1 million net migrants annually from lower- and middle-income countries' (United Nations, 2015a: 11). Notwithstanding the nativist and nationalist push-back against immigration and refugees taking place in many developed countries (USA TODAY, 2016), developed countries have mainly benefited from international migration (OECD, 2014) at the expense of the developing world (see below).

While documented immigration is mainly about people seeking a better way of life in high-income countries, it often comes at a very high cost to low-income countries in the form of the loss of the most educated and talented people. In terms of global health, this has meant the loss of physicians, nurses and other health care workers in low income countries who find employment in high-income countries (Blacklock et al., 2014; Connell, 2014; Connell et al., 2007; Walton-Roberts, 2015).

Beyond documented migration, one of the most powerful reflections of the forces of globalization and its destructive powers on local economies is refugee migrations. These types of mass migrations place many of those who make them at risk every step of the way. For example, of the approximately 1.8 million refugees who entered Europe by sea between 2014 and 2017, UNHCR estimates that another 15 500 died or are missing (UNHCR, 2018). Once refugees arrive in developed countries, their health challenges continue because of the physical and mental trauma that result from their past and present experiences (Grove and Zwi, 2006).

Increasingly, in many countries, people are also moving from rural areas to urban areas at historically unprecedented rates. 'In 2007, for the first time in history, the global urban population exceeded the global rural population' and 'by 2050, the world will be one-third rural (34 per cent) and two-thirds urban (66 per cent), roughly the reverse of the global rural–urban population distribution of the mid-twentieth century' (United Nations, 2015b: 7). The movement of working age people from rural to urban agglomerations has created slums and shanty towns in cities throughout the developing world creating public health challenges because of the lack of resources and the limitations of governments at all levels to provide basic services (e.g. clean water and sanitation).

Lastly, globalization involves improved connectedness that allows people to easily travel from city to city, country to country and continent to continent through air travel. In 2017, over 4 billion people travelled on airlines affiliated to the International Air Transport Association (IATA, 2017). What this means for global health can be contrasted between the time it took the 1918 Spanish Influenza Epidemic to spread around the world in contrast to the time it took SARS in 2002 to spread around the world. It is estimated that the former took about two years (Billings, 1997) while the latter took seven days to spread from Hong Kong to the United States, eight days to spread to Canada, nine days to spread to Australia, and 14 days to the United Kingdom (WHO, 2018). The very connectedness of travel is also a key mechanism for the potential of a global pandemic that might have untold health consequences sometime in the future.

The above examples also highlight other intersections between globalization and health, particularly the heightened importance of international institutions and agreements as responses to the crises that the movement of people generates. For example, the International Health Regulations (IHR) have been given new and expanded meaning as legally binding rules for reporting and controlling the spread of highly infectious diseases among all member countries of the World Health Organization (WHO). Now that the developed world feels threatened by the possibilities of global pandemics from existing diseases (e.g. influenza) and emerging diseases (e.g. SARS), more efforts are being made to find vaccines (Sandberg et al., 2010), but these global efforts remain problematic in terms of which health threats are the focus of vaccine development and anti-vaccine movements (Craddock, 2007; Freimuth and Jamison, 2017; Ward, 2016).

THE MOVEMENT OF ECONOMIC ACTIVITY

Closely linked to the movement of people is the movement of economic activities. One of the key aspects of globalization is the removal of international trade barriers to enable the free flow of goods and services. One of the outcomes of increasing the free flow of goods and services is that global corporations have moved their factories to countries in the developing world where they pay lower wages and there are fewer environmental restrictions placed on the factories. The result has been the movement of 'dirty' industries (e.g. the production of iron and steel) from developed countries to developing countries. The health benefits to developed countries have been the decline in air and water pollution in places that once housed mills and factories that produced smoke and effluence. Paradoxically, many of those same places now face health costs related to mental health, alcohol and substance abuse linked to unemployment and underemployment because well-paid jobs in factories and mills have been replaced by permanent unemployment or work in low-paid service sector employment. For example, Caldbick et al. (2014) link globalization, precarious employment and health. They note how precarious workers often lack health benefits and are thus at greater risk. In another study of Canada's three largest cities, Labonté et al. (2015) show how globalization in the form of changing labour markets, housing and social protection measures led to the paradox of both improved living standards but also deteriorating health. In the developing countries, jobs in 'dirty industries' have meant higher relative wages but also occupational health problems linked to increased air, water pollution and worker safety. Overall the health costs are high, resulting from lower standards of worker safety and higher rates of health problems in developing countries.

The forces of globalization are also creating new economic activities linked to health and dependent on the ability of people to easily travel between countries. One example is the development of medical tourism. People from developed countries who are dissatisfied with their health care systems because they cannot get a service that they want (e.g. a hip replacement) when they want it, or they want a service that is not offered by their health care system or when they cannot afford to pay for a service that they want in their country, are travelling to countries like India (Crooks et al., 2011) or countries in the Caribbean to receive health care. Since the medical tourism facilities are priced for this clientele from developed countries, the services often are unaffordable for the local population, while they are offered by local health care workers that give up positions in facilities that serve the local population, therefore reducing further the access to health care.

HEALTH AND HEALTH CARE AS GLOBAL INDUSTRIES

Health and health care are now global industries. In their annual report on the largest corporations in the world, the consultancy firm PricewaterhouseCoopers (PWC, 2018) identified 16 global corporations that provide health services, health insurance, develop pharmaceuticals or deliver pharmaceuticals as their core businesses in the global top 100 list of companies by market capitalization in 2017. Twelve global corporations that deliver food, beverages (soft drinks and/or alcohol) and tobacco products as their core businesses could also be found in the global top 100 list of companies by market capitalization in 2017 (PWC, 2018). In other words, 38 out of the 100 largest global corporations affect health and health care through their decision-making, the goods and services that they produce and the strategies that they use to convince governments and people to consume their goods and services.

One obvious example of the importance of health and health care as global industries is the role that global pharmaceutical companies play, not only in the development of new drugs but also for whom the drugs are developed, who their target markets are and how the drugs are priced and marketed differently in various parts of the world. How the global pharmaceutical industry chooses to invest in research and development of drugs that have minor health benefits but enormous markets in the developed world (e.g. drugs to enhance males' sexual prowess) while ignoring the development of drugs that might have enormous health benefits (e.g. an effective malaria vaccine) at a low price and would benefit people in the developing world, is well-documented (Robinson, 2001). Blume and Zanders (2006) describe the interplay among the global pharmaceutical industry, international bodies (e.g. WHO) and nation states as 'vaccine politics', to show the consequences when a country that questions global trends in favour of a response that they deem better meets national needs. They document the debate that took place in the Netherlands over what pertussis vaccination to use and the pressure to conform to international standards in the light of the concerns of some parents about the potential side-effects. Another example of how the global pharmaceutical industry favours high-income countries over low-income countries is the research by Fisher et al. (2015). They created a database for the period 2006 to 2011 based on the clinical trials initiated to address diseases mainly found in high-income countries compared to low-income countries. They found that the pharmaceutical industry was 3.46 times more likely to be developing a drug for high-income countries in contrast to a drug for low-income countries. What should not, however, be ignored in the behaviour of the global pharmaceutical industry is that wealth dictates who has access to prescription drugs whether they live in a wealthy developed country or a poor developing country.

A particularly controversial aspect of the global pharmaceutical industry is their ability to influence international agreements in their favour and in favour of developed countries. The focus of some research has been on the World Trade Organization and Trade Related Intellectual Property Rights (TRIPS) agreement that has resulted in the growth of market share by the global pharmaceutical industry in developing countries and the ability of the global pharmaceutical industry to influence changes in local regulations in their own interests (Gabe et al., 2015). For example, the Indian government hoped to grow basic research in pharmaceuticals by encouraging the global pharmaceutical industry to carry out clinical trials of drugs. While clinical trials grew, basic research in pharmaceuticals did not (Sariola et al., 2015). In another study carried out in India, Aivalli et al. (2018) found that even though the quality of generic and branded drugs was the same, patients' perceptions of quality favoured patent drugs. They also found that negative perceptions of generic drugs and promotion of branded drugs by the pharmaceutical companies influenced prescriber behaviour and affected people's trust in the public health care system.

The global food, beverage, tobacco and alcohol industries are much like the global pharmaceutical industry in searching for profits even if the health outcomes of their products are questionable. A critical examination of global food and beverage companies shows how they are contributing to the global obesity epidemic. The health outcomes can be charted in terms of the increase in obesity-related diseases taking place globally (WHO, 2000). It is widely recognized that the growth in obesity in developed countries (e.g. Mokdad et al., 1999; Prentice and Jebb, 1995) is now repeating itself in developing countries with similar negative health effects (e.g. Monteiro et al., 2004; Bauman et al., 2012). At least part of the global obesity epidemic can be linked to the success of global food and beverage companies to encourage the consumption of high calorie, high sugar content and low nutritional value foods and beverages ('fast foods'). In a case study from Australia, Baker et al. (2017) analyse how the power of actors, the power of ideas, the political context and issue characteristics come together in the debates among governments, civil society and the global food and beverage industry. In the case of the global food and beverage industry, they finance national associations (e.g. the Australian Food and Grocery Council) to argue for voluntary self-regulation in contrast to the 'nanny state', and how they draw on the power of ideas and political context to draw support from the rural workforce that is heavily involved in food production. Similarly, global tobacco and beverage companies have also been successful in encouraging the growing consumption of tobacco and alcohol products in developing counties in spite of global efforts (e.g. the 2003 WHO Framework Convention on Tobacco Control). Although the WHO Framework Convention on Tobacco Control (FCTC) commits member countries to reduce tobacco use, the global tobacco companies make every effort to convince the public to continue to smoke and weaken government efforts to reduce tobacco consumption (Mamudu et al., 2008). In a study of countries in Southeast Asia many of which have among the highest rates of tobacco consumption in the world, Barraclough and Morrow (2010) show how governments legitimize tobacco production and cooperate with the multinational tobacco corporations while also promoting reductions in tobacco consumption. What Barraclough and Morrow ultimately advocate is that governments need to disinvest from the tobacco industry and end the promotion of tobacco.

CONCLUSIONS

The goal of this chapter was to argue that we need to see health in a globalizing world through various lenses beyond the effects of climate change on health. Even the discussion of climate change and health needs to be expanded to other environmental issues including the impacts on water, land, air and extreme events and the various spatial scales at which environmental change is affecting health of people, flora and fauna both directly and indirectly. Globalization is also about the movement of people, economic activities and health and health care as global industries. The net health benefits of the movement of people, economic activities and the health and health care industry generally favour the countries of the developed world over the countries of the developing world. What is also noted is that the health of vulnerable groups regardless of where people live is affected by globalization. To improve health in a globalizing world, the challenges that people and nations face is how to take advantage of the forces of globalization while reducing the disadvantages to the people and nations who are most at risk and have the fewest resources to combat the emerging, re-emerging and chronic health diseases of the twenty-first century.

REFERENCES

Ailshire, J., A. Karraker and P. Clarke (2017), 'Neighbourhood social stressors, fine particulate matter air pollution, and cognitive function among older US adults', *Social Science & Medicine*, **172**, 56–63.

Aivalli, P.K, M.A. Elias, M.K. Pati, S. Bhanuprakash, C. Munegowda, Z.C. Shroff and P.N. Srinivas (2018), 'Perceptions of the quality of generic medicines: implications for trust in public services with the local health system in Tumkur, India', *BMJ Global Health*, **2**, e000644. doi:10.1136/bmjgh-2017-000644.

Baker, P., T. Gill, S. Friel, G. Carey and A. Kay (2017), 'Generating political priority for regulatory interventions targeting obesity prevention: an Australia case study', *Social Science & Medicine*, **177**, 141–149.

Barraclough, S. and M. Morrow (2010), 'The political economy of tobacco and poverty alleviation in Southeast Asia: contradictions in the role of the state', *Global Health Promotion*, **17**, 40–50.

Bauman, A.E., R.S. Reis, J.F. Sallis, J.C. Wells, R.J.F. Loos and B.W. Martin (2012), 'Correlates of physical activity: why are some people physically active and others not?' *The Lancet*, **380**, 258–271.

Billings, M. (1997), *The Influenza Pandemic of 1918*, https://virus.stanford.edu/uda/ (accessed on 3 January 2018).

Blacklock, C., A.M. Ward, C. Heeneghan and M. Thompson (2014), 'Exploring the migration decisions of health workers and trainees from Africa: a meta-ethnographic synthesis', *Social Science & Medicine*, **100**, 90–106.

Blume, S. and M. Zanders (2006), 'Vaccine independence, local competences and globalisation: lessons from the history of pertussis vaccines', *Social Science & Medicine*, **63**, 1825–1835.

Caldbick, S., R. Labonté, K.S. Mohindra and A. Ruckert (2014), 'Globalization and the rise of precarious employment: the new frontier for workplace health promotion', *Global Health Promotion*, **21**, 23–31.

Connell, J. (2014), 'The two cultures of health worker migration: a Pacific perspective', *Social Science & Medicine*, **116**, 73–81.

Connell, J., P. Zurn, B. Stilwell, M. Awases and J-M. Braichet (2007), 'Sub-Saharan Africa: beyond the health worker migration crisis?' *Social Science & Medicine*, **64**, 1876–1891.

Craddock, S. (2007), 'Market incentives, human lives, and AIDS vaccines', *Social Science & Medicine*, **64**, 1042–1056.

Crooks, V.A., L. Turner, J. Snyder, R. Johnston, and P. Kingsbury (2011), 'Promoting medical tourism to India: messages, images and the marketing of international patient travel', *Social Science & Medicine*, **72**, 726–732.

Fisher, J.A., M.D. Cottingham and C.A. Kalbaugh (2015), 'Peering into the pharmaceutical "pipeline": investigational drugs, clinical trials and industry priorities', *Social Science & Medicine*, **131**, 322–330.

Freimuth, V.S. and A.M. Jamison (2017), 'Determinants of trust in the flu vaccine for African Americans and Whites', *Social Science & Medicine*, **193**, 70–79.

Gabe, J., S. Williams, P. Martin and C. Coveney (2015), 'Introduction – pharmaceuticals and society: power, promises and prospects', *Social Science & Medicine*, **131**, 193–198.

Grove, N.J. and A.B. Zwi (2006), 'Our health and theirs: forced migration, othering, and public health', *Social Science & Medicine*, **62**, 1931–1942.

International Air Transport Association (IATA) (2017), *Fact Sheet. Industry Statistics. December 2017*, Montreal: IATA.

International Panel on Climate Change (IPCC) (2014), *Climate Change 2014: Synthesis Report*. Contribution of Working Groups I, II and III to the Fifth Assessment Report of the Intergovernmental Panel on Climate Change [Core Writing Team, R.K. Pachauri and L.A. Meyer (eds)]. Geneva: IPCC.

Jerrett, M., M. Buzzelli, R.T. Burnett and P. F. DeLuca (2005), 'Particulate air pollution, social confounders, and mortality in small areas of an industrial city', *Social Science & Medicine*, **60**, 2845–2863.

Labonté, R., E. Cobbett, M. Orsini, D. Spitzer, T. Schrecker and A. Ruckert (2015), 'Globalization and the health of Canadians: "Having a job is the most important thing"', *Globalization and Health*, **11**(1), 19.

Lens, M.B. and M. Dawes (2004), 'Global perspectives of epidemiological trends of cutaneous malignant melanoma', *British Journal of Dermatology*, **150**(2), 179–185.

Mamudu, H.M., R. Hammond and S. Glantz (2008), 'Tobacco industry attempts to counter the World Bank report *curbing the epidemic* and obstruct the WHO framework convention on tobacco control', *Social Science & Medicine*, **67**, 1690–1699.

Martens, P., R.S. Kovats, S. Nijhof, P. de Vries, M.T.J. Livermore, D.J. Bradley, J. Cox and A.J. McMichael (1999), 'Climate change and future populations at risk of malaria', *Global Environmental Change*, **9**(S1), S89–S107.

Mokdad, A.H., M.K. Serdula, W.H. Dietz, B.A. Bowman, J.S. Marks and J.P. Koplan (1999), 'The spread of the obesity epidemic in the United States, 1991–1998', *Journal of the American Medical Association*, **282**, 1519–1522.

Monteiro, C.A., E.C. Moura, W.L. Conde and B.M. Popkin (2004), 'Socioeconomic status and obesity in adult populations of developing countries: a review', *Bulletin of the World Health Organization*, **82**, 940–946.

Organisation for Economic Co-operation and Development (OECD) (2014), 'Is migration good for the economy?', *Migration Policy Debates*. Paris: OECD.

Prentice, A.M. and S.A. Jebb (1995), 'Obesity in Britain: gluttony or sloth?', *British Medical Journal*, **311**, 437–439.

PWC (2018), *Global Top 100 Companies by Market Capitalisation*. https://www.pwc.com/gx/en/audit-services/assets/pdf/global-top-100-companies-2017-final.pdf (accessed on 3 January 2018).

Robinson, J. (2001), *Prescription Games: Money, Ego, and Power inside the Global Pharmaceutical Industry*, Toronto: M&S.

Romero-Lankao, P., H. Qin and M. Borbor-Cordova (2013), 'Exploration of health risks related to air pollution and temperature in three Latin American cities', *Social Science & Medicine*, **83**, 110–118.

Sandberg, K.I., S. Andresen and G. Bjune (2010), 'A new approach to global health institutions? A case study of new vaccine introduction and the formation of the GAVI Alliance', *Social Science & Medicine*, **71**, 1349–1356.

Sariola, S., D. Ravindran, A. Kumar and R. Jeffery (2015), 'Big-pharmaceuticalisation: clinical trials and contract research organisation in India', *Social Science and Medicine*, **131**, 239–246.

Smith, R.D. (2006), 'Responding to global infectious disease outbreaks: lessons from SARS on the role of risk perception, communication and management', *Social Science & Medicine*, **63**, 3113–3123.

Torre, L.A., F. Bray, R.L. Siegel, J. Ferlay, J. Lortet-Tieulent and A. Jemal (2015), 'Global cancer statistics, 2012', *CA: A Cancer Journal for Clinicians*, **65**(2), 87–108.

Tumwebaze, I.K. and H-J. Mosler (2015), 'Effectiveness of group discussions and commitment in improving cleaning behaviour of shared sanitation users in Kampala, Uganda slums', *Social Science & Medicine*, **147**, 72–79.

UNHCR (2018), Operational Portal refugees situations. http://data2.unhcr.org/en/situations/mediterranean (accessed on 2 January 2018).

United Nations, Department of Economic and Social Affairs, Population Division (2015a), *World Population Prospects: The 2015 Revision, Key Findings and Advance Tables*, New York: United Nations, Working Paper No. ESA/P/WP.241.

United Nations, Department of Economic and Social Affairs, Population Division (2015b), World Urbanization Prospects: The 2014 Revision (ST/ESA/SER.A/366).

USA TODAY (2016), 'Anti-immigration groups rally across Europe amid ongoing tensions'. https://www.usatoday.com/story/news/world/2016/02/06/anti-immigration-groups-rally-across-europe-amid-migrant-tensions/79929396/ (accessed on 16 January 2018).

Voisey, H. and T. O'Riordan (2001), 'Globalization and localization', in: T. O'Riordan (ed.) *Globalism, Localism & Identity*, London: Earthscan, pp. 25–42.

Wallis, P. and B. Nerlich (2005), 'Disease metaphors in new epidemics: the UK media framing of the 2003 SARS epidemic', *Social Science & Medicine*, **60**, 2629–2639.

Walton-Roberts, M. (2015), 'International migration of health professionals and the marketization and privatization of health education in India: From push–pull to global political economy', *Social Science & Medicine*, **124**, 374–382.

Ward, J.K. (2016), 'Rethinking the antivaccine movement concept: a case study of public criticism of the swine flu vaccine's safety in France', *Social Science & Medicine*, **159**, 48–57.

World Health Organization (WHO) (2000), *Obesity: Preventing and Managing the Global Health Epidemic*, Geneva: WHO.

World Health Organization (WHO) (2018), *Summary of probable SARS cases with onset of illness from 1 November 2002 to 31 July 2003.* http://www.who.int/csr/sars/country/table2004_04_21/en/ (accessed on 3 January 2018).

20. Digital media
Paul C. Adams

INTRODUCTION

Digital communication is fundamental to processes of globalization. It encourages cross-border payments via *PayPal* and *Bitcoin*. It disseminates information in some 250 different languages through Wikipedia. It facilitates the sharing of amateur videos through *YouTube* and *Vimeo*. It fosters international sociality via *Skype* and *WhatsApp*. It supports global travel through *Travelocity*, *Lonely Planet*, and *TripAdvisor*. It rebalances geopolitics through interstate hacking and the manipulation of social media. In these ways, as well as through ordinary web search, email, use of mobile media, credit card transactions, and the carrying of locationally aware devices, digital communications weave globalization processes into daily practices. The taken-for-granted awareness of the world which is achieved through the use of digital media may partly explain why the term 'globalization' has fallen by 70 per cent in online searches since 2004.[1] Globalization has become less salient precisely as it has become more pervasive and multifaceted. Digital media draw people into international, interstate, and cross-border networks but we now focus more on the new media than the globalization processes they facilitate.

If we ask how digital media contribute to globalization we must be careful from the outset to distinguish between several processes that are interlinked and too often confused. First are global infrastructures that permit communications to move around the world at an ever increasing volume and rate. More communication technologies are available every year, even in remote locations, indicating a global process of material diffusion. Second, we can consider the worldviews accessed through digital communications. Media are key to globalization insofar as they permit people to look and listen to remote locations. Third is the phenomenon of mediatization, which folds global forces and processes into everyday life, transforming our lives in ways that go beyond overt revelations of globalization. This dimension of globalization involves technological diffusion and long-distance perspectives but goes beyond their separate globalizing influences by reworking daily practices in ways that respond to global influences. We consider each of these dimensions of globalization in turn, then see how they come together in the case of Rwanda.

---

[1] As measured by Google Trends, the popularity of 'globalization' as a search term ranged between 50 and 100 in 2004 but now ranges between 15 and 27, indicating that it has fallen in popularity by about 70 per cent over that period.

GLOBALIZING MEDIA TECHNOLOGIES

The most common way in which people access the global infrastructure of information and communication technology (ICT) is through the use of a cell/mobile phone. Worldwide, the number of functioning mobile cellular phones is over seven billion and hence is approaching parity with the human population. These devices have found their way into rural, otherwise undeveloped areas where they appear out of place to observers from wealthy countries accustomed to thinking of the rest of the world as 'behind'. Driving this classic example of *leapfrog* technological diffusion, cell phone service is now available in some form to 95 per cent of the global population, and mobile broadband, to more than four-fifths of the global population, although far fewer can actually afford broadband service (ITU, 2016: 1; Graham and De Sabbata, 2014).

Cell phone access is generally measured as a percentage of population, a statistic known as penetration, but what is actually calculated is the ratio of registered subscriber identity module (SIM) cards to the total population. By this measure, many places appear rather puzzlingly to have more cell phones than people: cell phone penetration is around 110 per cent in the Americas and the Arab states, 120 per cent in Europe, and 140 per cent in the Commonwealth of Independent States (CIS) (Comer and Wikle, 2017: 167). The numbers are so high because 'each device (i.e. mobile phone, cellular computer modem, cellular-capable tablet computer) authenticates to a network through a separate SIM, and it is quite possible for one person to have several such devices' (Comer and Wikle, 2017: 165). There are contingencies in each situation that are missed by the statistic of 'penetration'. These contingencies affect the mix of technologies that are actually being used and how easy it is for a person to access phone services. For example in sub-Saharan Africa, the percentage of the population with access to a cell phone is higher than what is indicated by relatively low penetration statistics because of both informal and commercial arrangements that support the sharing of phones (Comer and Wikle, 2017: 166). In short, although global comparisons are rendered tricky because of local contingencies, the fact that crude measurements of access are at or above 100 per cent in many places suggests how important the cell phone has become to global communications.

Turning to the Internet we see a picture that is both similar and different. The general distinction between information haves and have nots has often been discussed, with some 20 per cent of the population going online in the least developed countries and more than 80 per cent using the Internet in the developed countries (ITU, 2017). The digital access gap is particularly noticeable in Africa where Straumann and Graham have identified an 'archipelago of disconnection' (Straumann and Graham, 2016) stretching from Mauritania to Madagascar and including 28 countries in sub-Saharan Africa where less than 10 per cent of the population has used the Internet at all in the past year. Aside from sub-Saharan Africa, the 'archipelago' includes Iraq, Turkmenistan, Afghanistan, Cambodia, Burma and Papua New Guinea.

Age-based variation complicates this picture. Worldwide, 71 per cent of youths and young adults (persons aged 15–24) go online while only 48 per cent of older adults do so (ITU, 2017). The gap between old and young is most striking in developing countries where some 26 percentage points separate the age groups, with 67 per cent of younger persons and 41 per cent of older persons regularly going online (ITU, 2017).

A gender divide is evident as well, particularly in the least developed countries, where 12.5 per cent of women use the Internet as opposed to 18 per cent of men, in effect making the men half again as likely to go online, versus developed countries where the percentage of both men and women who go online is about 80 per cent (ITU, 2016, 2017).

Access to the Internet can mean different things depending on whether one has a state-of-the-art, high bandwidth connection, a connection that works tolerably well except for streaming video, or a connection that takes many minutes to download any image larger than a thumbnail. In Western Europe, North America, Australia, Japan, Singapore and South Korea, between a quarter and two-fifths of the population have broadband Internet access, benefiting from data transmission rates of 2 Mbit/s or greater, while in developing countries high speed service is scarce to non-existent (ITU, 2015: 5). Likewise, nine out of ten people living in developed countries have mobile broadband subscriptions whereas in developing countries less than half of the population has such service (ITU, 2016: 4).

Whether using cell phones or computers, international communications depend on two complementary infrastructures: fibre optic cables and communications satellites. The services carried on this infrastructure include telephone calls and the various applications of the Internet, but other types of data are also conveyed in these infrastructures including television, radio, teleconferencing, and internal company communications on privately leased links. While satellites are more practical for certain kinds of long-distance communications, particularly to and from remote rural areas, fibre optic cable is superior in terms of security and bandwidth while costing less in many cases. Consequently satellites relay about one-tenth of the global data flow while optical fibres carry most of the rest (Warf, 2017a: 123). We will consider each of these systems in turn.

Strands of quartz crisscross the world's oceans and continents, carrying terabytes of data each second at the speed of light – ideas and innovations, financial transactions, internal company directives, entertainment products, and interpersonal exchanges. The net result of greater access to such communications is a multiplication of the social characteristics that led to greater digital access in the first place, a cyclical relationship between connectivity and global power which spurs further connectivity (Warf, 2017a). The geographical arrangement of fibre optic cables provides a clear example of what Massey (1993) refers to as 'power-geometries', intensifying the connectivity of places that already enjoy greater wealth and global influence while creating a strategic disadvantage for places that are lagging. Working against this logic, however, is a force toward global communication integration in the form of an oversupply of digital communication capacity. Immediately after a frenzy of cable installation in the 1990s the percentage of fibre optic cables that were 'dark' or unutilized rose to over 50 per cent; oversupply lowered the cost of digital transmission which in turn stimulated markets for digital services, particularly those accessed by cell (mobile) phone (Warf, 2017a: 123). The low cost has helped ensure that digital communication services are increasingly international even if they are still profoundly uneven.

Satellites find greatest commercial use in the transmission of television programming, but also relay broadband Internet and radio communications. While commercial communications account for 38 per cent of all communications satellite activity, other uses that are also quite important include earth observation (remote sensing), government communications and research and development (R&D), which together make up another 45 per cent, with navigation, military surveillance, scientific research, meteorology, the non-profit sector, and space observation accounting for the remaining 20 per cent (SIA, 2017). Like fibre optic cables, satellites are now deployed and maintained by private companies. The satellite communication system began under the supervision of a consortium of state governments called Intelsat and despite its privatization and commercialization the name Intelsat remains. In addition to operating 59 communications satellites and more than 80 per cent of all public earth stations that communicate with these satellites, Intelsat serves international coordination and regulation functions through non-profit facets of its operations (Warf, 2017b: 153–154). The company's legacy as an intergovernmental organization is also preserved in the price structure which evens out costs between areas with high and low utilization of satellite services. This lowers barriers for entry into the global telecommunications market in poorer, less connected, parts of the world; certain remote rural areas and isolated islands are globally connected as a result of this policy, somewhat mitigating the tendency towards geographical concentration of communication power (Warf, 2017b: 154).

The global digital infrastructure now includes some 2 000 communications satellites. About a quarter of these are in a geostationary orbit that holds the satellite in a fixed position exactly 35 786 kilometres above the earth's equator. Oddly, the growth of global communications flows has turned these orbital parking spaces into a scarce resource. Satellites are now close enough to each other to cause problems when additional satellites are placed in this orbit unless a minimum distance is carefully maintained, and the dominance of premium parking spots by powerful countries on a first-come-first-served basis mirrors the uneven geopolitical and economic power on the earth's surface (Warf, 2017b: 152).

Because a satellite in geostationary orbit is visible from some 40 per cent of the earth's surface, the technology is inherently border-transcending. 'States that seek to restrict imports of foreign media—including much of the Muslim world and former Soviet bloc—have typically found it impossible to assert national controls over global flows of information beamed from above' (Warf, 2017b: 152). According to one source Chinese authorities attempt to control this border-crossing flow by confiscating privately installed satellite dishes and replacing them with government-approved equipment (*Tibetan Review*, 2009). Digital communication activities of other sorts leave traces, and users of interpersonal and social media expose themselves to oppression: 'the same technology that assists social movements in their pursuit of democracy also assists states and supra-state organizations in their subversion and surveillance of these movements' (Croucher, 2004: 30). Thus digital communications infrastructure presents contradictory forces, working against geographical isolating forces yet accentuating globally uneven concentrations of knowledge. It challenges existing power centres while at the same time creating new forms of vulnerability for those who take part in contesting apparatuses of autocratic power.

As data flow faster, more frequently, and farther among the world's digitally connected populations, the implications of disconnection are increasing. Rather than merely remaining stagnant, the disconnected populations are falling further behind with regard to economic competitiveness, cultural integration and geopolitical power. But this geography of disconnection is complexly scaled. 'Certain parts of South Africa and Thailand (especially privileged spaces such as cities and tourist destinations) are well wired while their hinterlands remain cut off, and parts of the developed world such as Appalachia are struggling to maintain meaningful digital connections' (Zook et al., 2004: 156). The challenge of uneven communication access must therefore be considered at all scales from regional (sub-Saharan Africa) to national (North Korea) to local (poor parts of cities around the world).

GLOBALIZING WORLDVIEWS THROUGH DIGITAL MEDIA

In light of the ideas presented thus far it would be easy to assume that digital media have radically transformed how people think about their place in the world. The global *worldview* is not actually a particularly new phenomenon, however. There have been several phases in the evolution of the globalizing imagination, as captured by Denis Cosgrove in *Apollo's Eye* (2001). As communication media, maps that supposedly represented the world have existed since antiquity. These were not global maps in a modern sense, but earnest efforts to show the known part of the world and infer the rest: cartographic gestures towards a global outlook. The period in human history when it was possible to think of oneself as living in a global society extends back only a few hundred years. Fifteenth-century literary and scientific culture brought a flowering in international communications via printed material and handwritten letters and this 'Republic of Letters' evaded censorship and state suppression by listing fictitious places of publication, including Cosmopolis. More than just an ideal, Cosmopolis imagined the world as a society for the purpose of liberating communication.

Globalization via digital media offers the potential to work towards the cosmopolitan vision, manifested now as a virtual globe and envisioned in terms of 'networks of infinite individual points linked across invisible channels over a frictionless surface' (Cosgrove, 2001: 236). Thus, there is a long history of the global vision, but digital media have helped people to envision the world holistically as captured by Manuel Castells' term 'space of flows' (Castells, 1989) and, with somewhat more nuance, Arjun Appadurai's 'ideoscapes', 'mediascapes', 'technoscapes', 'ethnoscapes', and 'finance-scapes' (Appadurai, 1996). Their works stand as a persistent reminder that digital media create their own international communications spaces or landscapes. Each 'scape' not only connects places differently and has its own high spots and low spots (places that drive global communication flows and places driven by such flows), but it also looks at the world differently, providing its own particular vantage points on the world, and by extension, on globalization.

Euclidian distance and topological connection are now competing with each other as complementary logics structuring the world:

> The need to click a lot to get on to a website that is actually physically nearby is perceived as it being more remote and less accessible than a site that, just one click away, is physically more than thousands of miles away. The time cost is so small that virtual access is practically invariant with physical distance. Access in virtual space, therefore, follows logical links rather than physical paths. (Devriendt et al., 2008: 15)

Digital media do not just give everyone more global interaction. In terms of experience, digital globalization means the near can be distant and the distant can be near, as experience is subsumed by a digital media logic. If the most recent round of globalization promised a kind of flattening of the world (Friedman, 2005) which was supposed to even out economic divergences between places, the reality of globalization has often been the cruel accentuation of inequality and accelerated concentration of power in the hands of those at the 'switching points' that control flows of information and social exchange across multiple scales (Devriendt et al., 2008: 19; Crang, 1999).

Despite these cautionary comments, it is also true that developing regions are coming to dominate the Internet, with twice as many users residing in the developing world as compared to those in the developed world (ITU, 2015). This preponderance can be expected to eventually shift the worldviews offered on the Internet to better reflect realities of the developing world, even if services such as *Wikipedia* and *Google* currently promulgate the views of developed countries much more than the views of developing countries (Ballatore et al., 2017).

When considering such flows, however, 'an ostensibly global digital service is differentiated by multiple layers of industrial and regulatory contingency' (Lobato, 2017: 183). In Thailand, local regulatory practices may lead to the blurring or 'pixellation' of scenes in foreign films showing smoking, drinking and violence (Lobato, 2017). South Korean medieval dramas meet standards of decency in Iran, Iraq, the UAE and Turkey, where short skirts can run afoul of state censors, and have accordingly been shown during prime time in these countries (Mundy, 2013). South Korean television programs also generate social media responses that are of great interest to the producers, who 'typically shoot only the first few episodes of a drama before it goes to air ... [then] develop the plot according to the audience reaction, which is tracked on social media and discussion boards on shows' fan pages' (Mundy, 2013). Whether producers attend only to local social media responses or tailor shows to an international audience will mark the extent to which these entertainment products are merely exported or more thoroughly globalized.

In short, digital media do not offer a single worldview but many contrasting globalized visions. One concern is that insofar as worldviews need not be shared by distant populations or even neighbours, the global perspectives offered by digital media may do more to pull people apart than to bring them together. This potential for people to become isolated, cognitively encapsulated in 'echo chambers' (Quattrociocchi et al., 2016) or 'filter bubbles' (Pariser, 2011) depends, as well, on daily media-dependent practices.

MEDIATIZATION AS A GLOBAL PROCESS

Mediatization is the process, or rather set of processes, through which mediated communications become integrated into daily practices, altering the functions of organizations and the nature of social institutions, and ultimately affecting personal identity. Mediatization is separate from the mere proliferation of communication technologies and the increase in access to such technologies. It entails a transformation in ways of doing things that incorporates media into social and material processes, including processes of group coordination and individuation. The globalization of digital communication involves mediatization rather than merely the diffusion of communication technologies and the proliferation of mediated worldviews.

Using an on-board global positioning system (GPS) to find one's way through a strange city or country is an illustrative example. The device seems to merely 'know' where one is; the stream of communications between the device and several satellites orbiting the earth disappears into the background of daily practice. A driver is more likely to notice the oddness of the electronic voice emanating from the device than the fact that a 5-kilometre drive is now guided from tens of thousands of kilometres above. But the synthesized voice in the automobile interior and the radio transmissions streaming to and from the distant satellites are both digital communications, and their joint coordination of the daily practice of navigation provides a perfect illustration of mediatization. The two types of communication (digital flows over tens of thousands of kilometres and acoustic flows over less than a metre) converge to rework geographical knowledge, awareness, practices and routines. Long distance data flows blend into a person's environment becoming integral parts of the holistic experience we call 'place' (Adams, 2017).

Mediatization must be distinguished from mediation – the conveyance of particular communications on or through particular media. It is trickier to comprehend the deep and wide-reaching implications of mediatization. A mediatized activity is one that is 'to a greater or lesser degree performed through interaction with a medium, and the symbolic content and the structure of the social and cultural activities are influenced by media environments which they gradually become more dependent upon' (Hjarvard, 2007: 3). In the example of driving with the aid of a GPS device, a driver may forget strategies for getting from here to there, heuristics incorporated through the use of a paper map. Someone dependent on GPS is likely to have a hard time locating him or herself 'on the map' if the device quits working or contains bad data. In this and many other realms of activity, mediatization reworks life according to what Altheide and Snow (1979) described as a 'media logic'. More recently Zook and Graham (2007) coined the term DigiPlace to indicate 'the use of information ranked and mapped in cyberspace to navigate and understand physical places'. Mobility is achieved with increasing reference to DigiPlace in a transformation that becomes internalized as the loss of particular skills and the gain of others. This internalization of a new spatial logic is emblematic of the more general process of mediatization.

Another example of mediatization is the practice of conducting an Internet 'search', which reworks the process of seeking information about the world. The same resource presents different facets: someone in Africa who searches on *Google* for information about their own capital is likely to see that capital city through remote lenses of distant websites, whereas this is not the case when Europeans look for information about their own capital cities (Ballatore et al., 2017: 1212). So while the practice of using the Internet involves the user in distant locations, this effect is not the same in every place. Here again we can see a geography of unevenness. Less intensely connected places are at the same time more distantly connected when it comes to digital information flows. It is as if the conduits tend to be either long and thin or short and thick. Insofar as the thin information on the Internet relative to the Global South is heavily skewed towards problems and solutions formulated in and for the Global North, merely improving access to this information is not in itself a solution to many of the problems in the Global South.

Part of mediatization is the process of media convergence. What used to be distinctly different media – television, cinema, and the Internet – have merged in the form of streaming video applications such as *Netflix*, a platform accessible throughout much of the world that can be used to access television programs and movies. Technically, *Netflix* is available in every country except China, North Korea, Crimea and Syria (although in practical terms 3.0 mbps is a lower limit for uninterrupted streaming, which excludes much of the world from the *Netflix* service area) (Lobato, 2017: 180). Beyond these technical limitations, there is also a kind of geographical filtering that occurs through the need to pay online service fees via credit card, which favours 'the global urban middle classes' (Lobato, 2017: 181).

Digital communications contribute to mediatization whether we are thinking of new media, social media, mobile telecommunications, locationally aware devices, the Internet of Things, voice over IP protocols such as *Skype*, crowd-sourced information, or the Internet. The distinctions between these forms of communication are breaking down in the general trend of media convergence (Jenkins, 2004). Digital media convergence is underlined on a daily basis by actual events that unsettle previously local communications through the intrusion of long-distance, border-crossing communications that simultaneously cross between the familiar types of communication. When a customer talks on the telephone with a company representative it may seem to be a person-to-person communication, but in fact the service employee's phone line is linked to a computer database that calls up intimate details about the customer many thousands of kilometres away (Aneesh, 2015), surprisingly rich records in cross-referenced databases that have been variously identified as 'data profiles' (Curry, 1997), 'data shadows' (Graham, 2014), 'capta shadows' (Kitchin and Dodge, 2011), and 'digital droppings' (Deibert, 2013). The service personnel trained to use databases as leverage in phone interactions are themselves reconfigured to suit the global digital environment they inhabit, undergoing training in a 'neutral accent' and receiving a pseudonym that will supposedly let them interface seamlessly with English speakers in Australia, the United States, and the UK (Aneesh, 2015). The same dynamic stimulates the growth of French-language call centres in Morocco, Tunisia, Senegal, and Mauritius (Moriset and Bonnet, 2005). Still more media are drawn in if one considers how a particular digital 'object' can lead to heterogeneous sharing practices in social media. For example, a viewer of a popular television series could use *Facebook* to share a link to a parody on *YouTube*, which appropriates the tune and lyrics of a popular song, and in turn prompts that person's 'Facebook friend' to post a link to another video, a news story, or a stand-up comedy routine. These links create a hybrid media network with digital products of various types and origins loosely joined together through *Facebook*'s social media platform. Most important to this discussion, globalization occurs through processes of media convergence: a German bus driver passing time online in a café with wireless data (Wi-Fi) service can discover a Japanese music group, while in the process of chatting online about a scientific discovery in Australia, following news about a political event in the Middle East, or checking in on a hectic day at the New York Stock Exchange. The links one follows in a hybrid, convergent media network are likely to be international and interregional; they cross borders but in doing so they demonstrate spatial concentration and unevenness rather than the evenness and universality often implied by the term 'global'.

CASE STUDY

Rwanda provides an interesting case study. The country qualifies as 'least developed' and is in the poorest eighth of the world as measured by gross domestic product (GDP) per capita. At the same time, it has implemented an ICT policy framework called 'Vision 2020' that 'aims at making Rwanda a middle-income economy by the year 2020' (Lwakabamba, 2005: 214). Adopted in 2000, Vision 2020 led to what was hailed as 'tremendous progress' in the country's ICT development (Lwakabamba, 2005: 215). This ICT policy was ostensibly 'among the best on the continent' (Etta, 2005: 296), judged in terms of explicit state government efforts to promote the communication sector. However, latest estimates of Internet penetration show Rwanda lagging behind neighbours such as Kenya, Uganda and Tanzania with its percentage online still stuck in the single digits (CIA, 2017).

Foster and Graham (2017) examine Rwandan tea production and perhaps surprisingly find numerous examples of digital communications incorporated into the production of this commodity. First, new ways of selling tea leaves via online auctions are leading to fragmentation of the market and, in a seeming paradox, the integration of tea production under the control of large companies in a new, mediatized production system. A second facet of mediatization that is reworking the production of tea is the development of new procedures for assessing the value of each tea shipment. Automated weighing of tea has brought new levels of accuracy and precision which facilitates pricing but at the same time undermines the role of local cooperative associations. Third, and most subtly, mediatization invades the Rwandan tea market as algorithms reinforce social power relations, producing greater efficiencies that increase profits far from the site of production but do little to reduce precarity near to the place of production. In these ways, digital communications are translated into Rwandan tea through their incorporation into producing, financing, marketing and distributing the product. More generally, digital communications are reworking Rwanda's situation within global economic flows, in terms of its embeddedness, value, and power (Foster and Graham, 2017: 82).

The expectations of some observers that digital communications would help overcome productivity imbalances between developed places, developing places, and least developed places have proven largely unfounded in sub-Saharan Africa. Research by Ojanperä et al. (2017) shows that not only are traditional forms of knowledge generation (e.g. books and research articles) unusually sparse throughout the region but so are new forms of knowledge generation such as top-level domain names and collaborative computer programming. In fact, 'the geographies of collaborative coding and domain registrations are more uneven than the spatial distribution of academic authoring' (Ojanperä et al., 2017: 48). Even though improved access to digital technology is promoted as a 'path for the region to move away from reliance on agriculture and extractive industries and towards a focus on the quaternary and quinary sectors (in other words, the knowledge-based parts of the economy)' (Graham and

Foster, 2014), ordinary digital services are out of the financial reach of many in the area. Broadband subscription costs range as high as 100 per cent of the average annual income (Graham and De Sabbata, 2014) and the proliferation of Internet cafés appears thus far to cater more to travellers than to locals (Warf, 2013: 38). Thus while globalization is being furthered by digital communications, sub-Saharan Africa, and Rwanda in particular, offer glimpses of how a familiar power geometry is perpetuated in three basic ways: through digital communications infrastructure, through worldviews subsumed to a media logic, and through mediatized lifestyles.

REFERENCES

Adams, P.C. (2017), 'Place', in: D. Richardson (ed.) *The International Encyclopedia of Geography: People, the Earth, Environment, and Technology*, Vol. X, Hoboken, NJ: Wiley-Blackwell and Washington, DC: American Association of Geographers, 5073–5085.

Altheide, D. and R. Snow (1979), *Media Logic*, Beverly Hills, CA: Sage.

Aneesh, A. (2015), *Neutral Accent: How Language, Labor and Life Became Global*, Durham, NC and London: Duke University Press.

Appadurai, A. (1996), *Modernity at Large: Cultural Dimensions of Globalization*, Minneapolis, MN: University of Minnesota Press.

Ballatore, A., M. Graham and S. Sen (2017), 'Digital hegemonies: the localness of search engine results', *Annals of the American Association of Geographers*, **107**(5), 1194–1215.

Castells, M. (1989), *The Informational City: Information Technology, Economic Restructuring, and the Urban-Regional Process*, Oxford, UK and Cambridge, MA: Basil Blackwell.

CIA (2017), *CIA World Factbook*. https://www.cia.gov/library/publications/resources/the-world-factbook (accessed on 30 October 2017).

Comer, J.C. and T.A. Wikle (2017), 'The geography of mobile telephony', in: B. Warf (ed.) *Handbook on Geographies of Technology*, Cheltenham, UK and Northampton, MA, USA: Edward Elgar Publishing, 162–177.

Cosgrove, D. (2001), *Apollo's Eye: A Cartographic Genealogy of the Earth in the Western Imagination*, Baltimore and London: Johns Hopkins University Press.

Crang, P. (1999), 'Local-global', in: P. Cloke, P. Crang and M. Goodwin (eds) *Introducing Human Geographies*, New York and Oxford: Oxford University Press, 24–34.

Croucher, S.L. (2004), *Globalization and Belonging: The Politics of Identity in a Changing World*, Lanham, MD: Rowman & Littlefield.

Curry, M.R. (1997), 'The digital individual and the private realm', *Annals of the Association of American Geographers*, **87**(4), 681–699.

Deibert, R.J. (2013), *Black Code: Inside the Battle for Cyberspace*, Toronto, ON: Signal/McClelland & Stewart.

Devriendt, L., B. Derudder and F. Witlox (2008), 'Cyberplace and cyberspace: two approaches to analyzing digital intercity linkages', *Journal of Urban Technology*, **15**(2), 5–32.

Etta, F. (2005), 'Policy matters: recommendations for responsible policy making', in: F.E. Etta and L. Elder (eds) *At the Crossroads: ICT Policy Making in East Africa*, Nairobi, Kenya: East African Educational Publishers, 295–298.

Foster, C. and M. Graham (2017), 'Reconsidering the role of the digital in global production networks', *Global Networks*, **17**(1), 68–78.

Friedman, T.L. (2005), *The World is Flat: A Brief History of the Twenty-First Century*, New York: Farrar, Straus and Giroux.

Graham, M. (2014), 'Internet geographies: data shadows and digital divisions of labour', in: M. Graham and W.H. Dutton (eds) *Society and the Internet: How Networks of Information and Communication are Changing our Lives*, Oxford: Oxford University Press, 99–116.

Graham, M. and S. De Sabbata (2014), 'Broadband affordability', Map, https://commons.wikimedia.org/wiki/File:Broadband_Affordability.png. Oxford: Oxford Internet Institute (accessed on 27 October 2017).

Graham, M. and C. Foster (2014), 'Geographies of information inequality in sub-Saharan Africa', Paper presented at the Network Inclusion Roundtable, Bangalore, India (October 2014). Oxford Internet Institute, University of Oxford.

Hjarvard, S. (2007), 'Changing media, changing language: the mediatization of society and the spread of English and Medialects', Paper presented to the 57th ICA Conference, San Francisco, CA, 23–28 May.

ITU (2015), 'ICT facts and figures 2015', ICT Data and Statistics Division, Telecommunication Development Bureau. Geneva, Switzerland: International Telecommunication Union.

ITU (2016), 'ICT facts and figures 2016', ICT Data and Statistics Division, Telecommunication Development Bureau. Geneva, Switzerland: International Telecommunication Union.

ITU (2017), 'ICT facts and figures 2016', ICT Data and Statistics Division, Telecommunication Development Bureau. Geneva, Switzerland: International Telecommunication Union.

Jenkins, H. (2004), 'The cultural logic of media convergence', *International Journal of Cultural Studies*, 7(1), 33–43.

Kitchin, R. and M. Dodge (2011), *Code/Space: Software and Everyday Life*, Cambridge, MA: The MIT Press.

Lobato, R. (2017), 'Streaming services and the changing global geography of television', in: B. Warf (ed.) *Handbook on Geographies of Technology*, Cheltenham, UK and Northampton, MA, USA: Edward Elgar Publishing, 178–192.

Lwakabamba, S. (2005), 'The development of ICTs in Rwanda: pioneering experiences', in: F.E. Etta and L. Elder (eds) *At the Crossroads: ICT Policy Making in East Africa*, Nairobi, Kenya: East African Educational Publishers, 213–224.

Massey, D. (1993), 'Power geometry and a progressive sense of place', in: J. Bird, B. Curtis, T. Putnam, G. Robertson and L. Tickner (eds) *Mapping the Futures: Local Cultures, Global Change*, London and New York: Routledge, 59–69.

Moriset, B. and N. Bonnet (2005), 'La géographie des centres d'appels en France', *Annales de Géographie*, 641(1), 49–72.

Mundy, S. (2013), 'South Korean soap operas hook foreign audiences', *Financial Times* (13 November 2013). https://www.ft.com/content/948719ae-4858-11e3-a3ef-00144feabdc0 (accessed on 16 October 2017).

Ojanperä, S., M. Graham, R.K. Straumann, S. De Sabbata and M. Zook (2017), 'Engagement in the knowledge economy: regional patterns of content creation with a focus on sub-Saharan Africa', *Information Technologies & International Development*, 13: 33–51.

Pariser, E. (2011), *The Filter Bubble: What the Internet is Hiding from You*, London: Penguin.

Quattrociocchi, W., Scala, A. and C.R. Sunstein (2016), 'Echo Chambers on Facebook', Discussion Paper No. 877. Cambridge, MA: Harvard Law School.

SIA (2017), 'State of the satellite industry report', Satellite Industry Association. Alexandria, VA: Bryce Space and Technology, http://www.sia.org/wp-content/uploads/2017/10/SIA-SSIR-2017-full-2017-10-05-update.pdf (accessed on 13 November 2017).

Straumann, R.K. and M. Graham (2016), 'Who isn't online? Mapping the "archipelago of disconnection"', *Regional Studies, Regional Science*, 3(1), 96–98.

Tibetan Review (2009), 'Satellite receivers replaced to tighten information control' *Tibetan Review* (1 June 2009).

Warf, B. (2013), *Global Geographies of the Internet*, Dordrecht: Springer.

Warf, B. (2017a), 'Fiber optics: nervous system of the global economy', in: B. Warf (ed.) *Handbook on Geographies of Technology*, Cheltenham, UK and Northampton, MA, USA: Edward Elgar Publishing, 113–125.

Warf, B. (2017b), 'Eyes in the sky: satellites and geography', in: B. Warf (ed.) *Handbook on Geographies of Technology*, Cheltenham, UK and Northampton, MA, USA: Edward Elgar Publishing, 148–161.

Zook, M.A. and M. Graham (2007), 'Mapping DigiPlace: geocoded Internet data and the representation of place', *Environment and Planning B: Planning and Design*, 34(3), 466–482.

Zook, M., M. Dodge, Y. Aoyama and A. Townsend (2004), 'New digital geographies: information, communication, and place', in: S.D. Brunn, S.L. Cutter and J.W. Harrington (eds) *Geography and Technology*, Dordrecht: Kluwer Academic Publishers, 155–176.

21. Patterns and dynamics of globalization of cultural industries

Robert C. Kloosterman and Rosa Koetsenruijter

CHALLENGING THE CORE

Fashion design from Nigeria (Abrams, 2013), films and contemporary art from China (Kharchenkova, 2017; Zhang and Yajuan, 2018) and TV series from Brazil, Mexico, South Korea and India (Kloosterman and Koetsenruijter, 2016). Emerging economies are now home to globally competitive cultural industries (UNCTAD, 2010, 2013; Lipovetsky and Serroy, 2013; EY, 2015). These 'industries' focus on products for which the selling argument is their symbolic or aesthetic value. While (both academic and public) opinions differ regarding what ought to be included, generally they tend to comprise performing and visual arts, design (from jewellery to fashion, furniture and buildings), gaming, and audio-visual content production such as music, films and television programmes (UNCTAD, 2010).[1] According to a recent estimate by Ernst & Young (EY, 2015: 5), the worldwide turnover of these activities amounts to some 2,250 billion US dollars and the total employment is nearly 30 million jobs, making them 'major drivers of the economies of developed as well as developing countries'.

Although Western countries, and the US in particular, are still very important in the production and consumption of these industries there is no doubt that a fundamental shift regarding the geography is occurring. Former peripheral countries are now challenging those in what used to be the core in the global division of labour by selling all kinds of commodified culture across borders on a large scale (Lipovetsky and Serroy, 2013).

This fundamental shift is, as we will argue below, linked to wider processes of globalization. However, the defining characteristic of products of cultural industries – their symbolic or aesthetic qualities – also makes them, in principle, much more context-specific than many other goods and services. Cultural distance, then, may hamper the appreciation of, for instance, a TV series set in India or a girl group singing in Korean among Western or Latin-American audiences. Moreover, it might also be the case that governments (or other powerful authorities) want to preserve what they see as *their* national culture and for example ban content from abroad which they see as at

---

[1] We use the term *cultural industries* and not *creative industries* as the former denotes a more narrowly defined group of activities focused on commodifying and producing culture in one way or another, while the latter comprises a much broader set of activities. Even regarding the more focused category of cultural industries there are still many definitions to be found in the literature which is partly to blame for the lack of reliable aggregate data (EY, 2015: 13). Here we stick, in principle, to the core cultural industries but the data from the literature that we present may comprise other activities as well.

odds with this. They might also actively promote their own cultural industries through subsidies. Cultural goods and services even form a formal exception in the overarching framework of international trade agreements. The importance of cultural diversity (on a nation state basis) is formally acknowledged by UNESCO (2005) in the *Convention on the Protection and Promotion of the Diversity of Cultural Expressions* which explicitly states that nation states are allowed to exempt cultural goods and services from international agreements aimed at promoting free trade. One of the 'key players' in the construction of this agreement was China aiming to protect its own cultural industries (Vlassis, 2016: 485–486).

Below, we offer a brief overview of processes of globalization of cultural industries after 1945. This overview is, inevitably sketchy and anything but exhaustive. This has to do, of course, with the limited space. It is also caused by the relative dearth of suitable and rigorous data on cultural industries on a more global scale. Although there exists quite a broad consensus (at least among geographers) about the underlying common element of the rather varied and disparate activities which make up cultural industries – namely the crucial role of symbolic and aesthetic aspects of the products – this has not been translated (yet) into anything like a uniform system of classification accepted by statistical bureaus across the globe. Data on cultural industries, then, are patchy, inconsistent and often more or less buried in larger categories. The reports produced by UNCTAD (2010, 2013) and Ernst & Young (EY, 2015) testify to these difficulties in constructing trends encompassing emerging economies.

What we do have, however, is an increasing number of case studies of particular cultural industries in specific countries and cities. This field of studies has expanded rapidly after the pioneering publications by, among others, Andy Pratt (1997) (the cultural industries production systems in London and the South East of the UK), Allen Scott (2000, 2004) (the geography of the image industries in general and the film industry in Hollywood specifically) and Dominic Power (2003) (the cultural industries in Nordic countries), which showed that cultural industries have become important drivers of employment and economic growth in advanced urban economies. Many studies have followed suit, increasingly also of cases in emerging or developing economies which allow us to get an, albeit partial, glimpse of processes of globalization concerning cultural industries.

We will first outline the characteristics of cultural industries. This is followed by an overview of the patterns and dynamics of the globalization of cultural industries, how they have become international, and where the largest exporters among the cultural industries are located. We then zoom in on two particular cultural industries to illustrate various patterns of globalization. The film industry represents a case of cultural industries which creates products with an evident explicit narrative content rooted in the place or country of production. This content is both lingual and visual which, in principle, impacts on the possibilities of films to travel across (cultural and political) borders. The case of architectural design, by contrast, elucidates the possibilities and constraints of non-verbal cultural industries. The concluding section will reflect on the wider implications of research on cultural industries.

CULTURE AS A COMMODITY

Culture is at least as old as mankind. Cultural industries, however, are much more recent. Ever more culture, which first 'went on in the private sphere has been annexed by the market, by capitalism' (Sassoon, 2006: xxi). Theodor Adorno pointed to the fundamental characteristic of the 'culture industry' namely the production of 'art' as a consumption good or in other words turning culture into a commodity (Adorno, 2001: 40). According to Power and Scott (2005: 3), cultural industries refer to those economic activities 'concerned in one way or another with the creation of products whose value rests primarily on their symbolic content and the ways in which it stimulates the experiential reactions'. Again, one could say that turning culture into sellable products goes far back in time – painters in the Dutch Golden Age of the seventeenth century, for instance, definitely produced for markets (Rasterhoff, 2012).

Contemporary cultural industries, however, differ in organization, scale of production, size of production series, market orientation, and, often use of technology from their predecessors (Scott, 2000). They form an integral part of the currently dominant production system – so-called 'cognitive-cultural capitalism' – in which knowledge is *the* strategic input, value chains tend to be vertically disintegrated and firms are inserted in dense networks partly based on proximity. Like other cognitive-cultural activities, cultural industries tend to cluster in cities and within cities in particular areas to reap the agglomeration economies of matching (labour pool), sharing (networks, dedicated formal and informal infrastructure from educational institutions to meeting places) and learning (knowledge spill-over fostering innovations). Agglomeration economies typically provide increasing returns or set in motion positive feedback processes which strengthen an existing cluster by attracting more capital, specialized labour, extending the dedicated infrastructure, and generating more localized knowledge. The clusters of film making in Hollywood (Scott, 2005), fashion design in New York (Rantisi, 2004) and Milan (Jansson and Power, 2010; D'Ovidio, 2015), and music production in New York, Nashville and Los Angeles (Florida and Jackson, 2010) are cases in point of such path-dependent trajectories of an agglomerated production of cultural industries. Although in many ways similar to other cognitive-cultural activities, cultural industries also display specific characteristics which have significant implications for the extension and intensification of cross-border flows of cultural goods and services.

First, there is a structurally rising demand for products of cultural industries as they usually have a high income elasticity of demand. They are part and parcel of the evolving post-Fordist consumer cultures in which consumer markets are not just fragmented along the vertical dimension of price but also horizontally along lines of lifestyles forming a myriad of rather distinct niches (Scott, 2000, 2004, 2012; Mansvelt, 2005). Cultural goods and services are essential in expressing lifestyles (Currid-Halkett, 2017) and related niches. A dialectical relationship between cultural industries and consumer market fragmentation evolves, driving up both the demand for cultural goods and services as well as further enhancing market fragmentation. In addition, these processes are also intimately intertwined with developments on the supply side as firms have to be able to cater to these niches and also to rapidly adjust their output to meet shifts in demand by using flexible specialization production methods (Scott, 2000). This rising demand also drives a push for aestheticization of products more in general, thereby considerably broadening the scope for cultural industries as well as blurring boundaries with other economic activities as ever more products become imbued with symbolic value (Lipovetsky and Serroy, 2013: 27). This trend has been reinforced by changes in communication technologies with mobile telephones and digital media expanding the range of cultural experiences (UNCTAD, 2010: 50).

Second, in cultural industries there tends to be a much stronger connection between place and products than in other more mundane economic activities. This connection is not just in the sense of the in-place clustering of cultural industries, but also in a more symbolic (and dialectical) way as fashion from Paris or films from Seoul illustrate (Shim, 2006). In these cases, the place of origin confers a particular quality or even aura to a product which cannot easily be copied by firms elsewhere (Molotch, 2002).

Third, and partly related to the former, their selling point is, by definition, their symbolic value and this is inevitably deeply intertwined with aspects of a local and national cultural identity – much more than is the case with, for example, electronic devices or cars. This obviously holds in a very fundamental way for those products in which (a significant part of) the symbolic value is expressed in a particular (national) language, but it also holds for say music or cinematographic styles (Lorenzen and Mudambi, 2013). The cultural specificity may, on the one hand, hamper its export potential as (in particular more subtle) meanings may get lost in translation (Appadurai, 1990; Deinema, 2012). On the other, because of their apparent intimate relationship with national identity, cultural industries are often seen to be an integral part of an accepted national culture. They touch upon 'identity issues because cultural expressions – distributed by cultural industries – are usually components of a national, regional or local identity and many actors are increasingly worried on cultural dominance' (Vlassis, 2016: 481). Frequently, cultural industries are, then, the target of conscious state policies to support them and to fend off foreign competition which may be seen as a threat to the avowed national virtues. In many countries the cultural industries – notably the audio-visual sector – are in one way or another regulated and protected. This protection has been often aimed against American cultural hegemony and what has been seen as the corroding effects of McDonaldization (Osterhammel and Peterson, 2003; Ritzer, 2008; Barraclough and Kozul-Wright, 2011; Jacques, 2012).

Processes of globalization regarding cultural industries are, therefore, intrinsically linked to structural shifts in the production system and society at large. They are also impacted by state policies aimed at preserving national culture and, hence, domestic cultural industries. Next, we will delve a bit deeper into these relationships by looking at cultural industries from a more general perspective of globalization in the past three decades.

CONVERGENCE OF CULTURAL INDUSTRIES

One could argue that the United States were the crucible of modern cultural industries which transformed and commodified culture into products aimed at large audiences (Sassoon, 2006). Before 1920 'Americans were still minor players in the global culture' (Sassoon, 2006: 605). The US first and foremost imported culture (mostly from Europe) and even showed a disregard for international agreements on intellectual property rights (Sassoon, 2006: 605). After the First World War, however, American films, music (especially jazz) and books were exported to Europe (Sassoon, 2006: 935). In the US, culture was much less directed at the elite and instead – by use of new technologies (films and recorded sound) – catered to the emerging domestic mass market. American culture was not just popular but also pluralistic: 'There was something for everyone. American films and novels could be populist and anti-rich, or could celebrate the glamour' (Sassoon, 2006: 943).

After the Second World War, the United States were clearly the hegemon, projecting its economic (including financial), military and political power on the global stage. It was also the dominant cultural force exporting films, music, television series to other (mostly Western) countries (EY, 2015: 55). The American cultural industries exploited their Fordist mass-production methods and their scale of production developed to supply their large domestic market which they now could use to meet the growing demand in the emerging European consumer societies amidst post-war reconstruction (Lipovetsky and Serroy, 2013). They could rely on a highly diverse pool of talent which could tap into, among others, Afro-American, Italian, Irish, English, German, and Jewish (including artists who fled Nazi-dominated Europe) sources of inspiration for music, films, TV series and novels. In addition, American cultural industries also benefited from the more general attractiveness of the US as, apparently, a rich, democratic and open society ahead of the rest. Elvis Presley, James Dean, Marilyn Monroe, and Lucille Ball together with Coca Cola, Chevrolet, and McDonald's were all part of the soft power of the United States (cf. Vlassis, 2016).

The hegemonic position of the US in cultural industries, however, has always been incomplete and contested. Those local and national cultures in countries (and regions) which have been at a relatively large distance in cultural terms from the hegemonic power, proved quite resistant to globalization and showed quite different responses to American culture after the Second World War than, for instance, the United Kingdom, the Netherlands or even France which were more culturally proximate (Osterhammmel and Peterson, 2003). The cultural proximity of these Western European countries might have been the reason for the fact that their cultural industries were easily influenced by American hegemony. Cultural industries in (semi-)peripheral countries in which the vernacular language, musical vocabulary, and established role models played an important role were, thus, at least to some extent, protected against foreign competition on their domestic markets. In addition, Communist governments in countries behind the Iron Curtain actively tried to keep out Western and in particular American culture (Judt, 2005; Applebaum, 2013). Notwithstanding these caveats, the US clearly dominated global markets of cultural goods and services, notably of film, music, and TV programmes before 1980. Accordingly, between 1985 and 1990 the revenues from major US film studios from foreign film theatre film rentals (i.e. films that were rented by foreign cinemas from US studios) increased from 735 to 1650 million US dollars (11 per cent point increase). The domination of US culture in foreign countries in that period is also evident from the US shares of film theatre box office in 1983: 30 per cent in Italy, 50 per cent in Germany, 45 per cent in France and even 92 per cent in the UK (Roper Center, 1992).

One could argue about the starting point, but there is a broadly shared view that sometime after 1980, a fundamentally new phase of globalization set in. This phase of 'hyper-globalization' is characterized by a radical shift in the global division of labour intertwined with far-reaching changes in the production system of notably urban economies (Castells, 2011; Scott, 2008, 2012; Mason, 2016; Baldwin, 2016). Global linkages extended and intensified after 1980 as value chains were carved up into separate parts located in different parts of the world (the second unbundling, Baldwin, 2016). This was enabled by a worldwide wave of the trade liberalization which occurred, including the opening up of China and India (cf. Frieden, 2006; Straubhaar, 2010; Kloosterman et al., 2015) and, notably after 1990, the rapid advances in digital storage, processing and communication technologies (ICT).

More specifically, we have witnessed the rise of cities in what used to be the periphery – especially in Southeast Asia, but not only there – as important global nodes in conjunction with a shift towards production systems dominated by high-tech, high-finance, high-concept and high-craft activities which compete on global markets by using highly skilled, specialized labour in combination with digital processing and communication technologies (Scott, 2012). With this shift, a rapid expansion of urban middle classes in Asia (according to the report by EY, 2015: 20, Asia already has some 525 million middle class cultural consumers) and elsewhere in the Global South took place (Milanovic, 2016). As happened earlier in the US and Europe, these middle classes significantly pushed up demand for cultural goods and services thereby creating ever more opportunities for cultural industries (EY, 2015). Also, just like their counterparts in the West, they used these cultural goods and services to express and fit with their lifestyles.

How have these interrelated changes of this phase of hyper-globalization affected the global division of labour which before 1980 was characterized by the dominance of the West in general with notably French and Italian films and ditto fashion design, British pop music and first and foremost the US in a broad array of cultural industries? Addressing this question can only be tentative, patchy and highly selective given the, on the one hand, large number of cultural industries and countries and cities, and on the other, the lack of aggregate data (cf. UNCTAD, 2010: 136). What we offer here are empirical observations of the development of key cultural industries from a global perspective using specific angles based on the main elements of the current processes of globalization. First, we describe in broad terms the relative importance of countries and regions in terms of producing and exporting cultural goods and services mainly based on UNCTAD and UNESCO publications, than we present a tentative explanation for the observed patterns.

Recently, Ernst & Young (EY, 2015) carried out a study commissioned by the International Confederation of Societies of Authors and Composers (CISAC – the body representing authors' societies worldwide). The study aims to assess the global significance of cultural and creative industries (CCI) by looking at it on a global scale. The study presents data on the revenues and employment for five global regions: Africa and the Middle East, Asia-Pacific, Europe, Latin America, and North America (see Table 21.1). As reliable data are hard to come by, the study used 'a bottom-up approach' combining several sources, and the writers are very much aware about the shortcomings of the figures they present – notably ignoring informal economic activities and piracy (EY, 2015: 13). Notwithstanding these caveats, the data can still be used to get a more general impression of the significance of cultural industries on a global scale.

Table 21.1   *Total revenues, employment and average labour productivity in cultural industries (advertising, architecture, books, gaming, music, movie, newspapers and magazines, performing arts, radio, TV, and visual arts) in five global regions, 2013*

	Total revenues in US$ (billion)	Share of total global revenues	Total employment (millions of jobs)	Share of total global employment	Average labour productivity (in 1000 US$ per worker) (revenue/ employment)
Africa and the Middle East	58	3	2.4	8	19.3
Asia-Pacific	743	33	12.7	43	22.5
Europe	709	32	7.7	26	22.1
Latin America	124	6	1.9	7	20.7
North America	620	28	4.7	16	22.1

Source: EY (2015), *The First Global Map of Cultural Industries* (various pages).

On this high level of spatial aggregation, there is not a hegemonic global region with regard to cultural industries. Asia-Pacific and Europe are more or less in the same league, North America is not far behind (this also holds for the labour productivity, which is between 22 100 and 22 500 US dollars per year). Latin America and Africa are clearly far behind in revenues, employment and also labour productivity. These recent data, then, show a more polycentric pattern with three, rather equally powerful global regions.

The data presented in Table 21.1 are first and foremost about the size of the cultural industries in the five global regions. Globalization is, by definition, about cross-border linkages and, hence, about exporting cultural goods and services. Table 21.2 shows the ranking of individual countries according to the export value of what UNCTAD (2010: 132) has called creative goods. These goods consist of art crafts, audio visuals, design, new media, performing arts, publishing, and visual arts, and, hence, refers partly to a different set of economic activities than the cultural industries of Table 21.1. In addition, the export value does not necessarily measure the value added in the country since, with many value chains being unbundled in global production networks, the cultural or creative part can nowadays be located in a very different country than the one where the final product is assembled and where the (gross) export value is being counted. An Armani suit, for instance, may be designed in Milan, Italy; the fabrics may come from Java; the actual sewing occurs in Bangladesh, Turkmenistan, Slovakia, Turkey and/or Romania. It is quite likely that the export value is registered in the country where the final assemblage takes place whereas much of the value added and captured occurs in the country of design (Brun et al., 2008; Toste, 2013). The export values, then, on the one hand, are lacking in reliability when it comes to where the actual symbolic or aesthetic value has been created (and probably captured as well). In addition, it is important to note that the production of many of the export-oriented cultural products from developing countries are commissioned by often large firms (e.g. the Walt Disney Company outsourcing the production of merchandizing) from developed countries. On the other hand, however, they do provide an important indicator of which countries are currently key players in the global production networks of creative and cultural goods. Moreover, being inserted in global audio-visual production networks, has also allowed Indian firms doing post-production tasks for American film companies to move upstream and use these distribution channels to sell their own products on foreign markets (Mukherjee, 2008: 186).

Table 21.2   *Top 20 countries according to export value (in millions of US$) of creative goods (art crafts, audio visuals, design, new media, performing arts, publishing, visual arts) and market shares in % in 2008, and growth rates 2003–2008*

	Country	Export value 2008	Market share 2008	Growth rate 2003–2008
1	China	84,807	20.8	16.9
2	United States	35,000	8.6	13.3
3	Germany	34,408	8.5	14.7
4	China, Hong Kong SAR	33,254	8.2	6.3
5	Italy	27,792	6.8	9.7
6	United Kingdom	19,898	4.9	6.5
7	France	17,271	4.2	10.2
8	Netherlands	10,527	2.6	11.6
9	Switzerland	9,916	2.4	13.5
10	India	9,450	2.3	15.7
11	Belgium	9,220	2.3	6.7
12	Canada	9,215	2.3	-0.9
13	Japan	6,988	1.7	14.7
14	Austria	6,313	1.6	8.5
15	Spain	6,287	1.5	4.9
16	Turkey	5,369	1.3	15.0
17	Poland	5,250	1.3	14.9
18	Mexico	5,167	1.3	9.1
19	Thailand	5,077	1.2	10.3
20	Singapore	5,047	1.2	6.0

Source: UNCTAD (2010: 162).

The data shown in Table 21.2 do indeed point to an emergence of non-Western countries as global players and, arguably, to a new phase in the globalization of cultural industries characterized by a more polycentric pattern on the level of individual countries (cf. Lipovetsky and Serroy, 2013: 28). No less than seven of the top twenty exporters can be found outside North America and Europe. China, workshop of the world, is clearly leading the pack – even more so if one adds the export value of Hong Kong SAR to Mainland China. The export data thus confirm the finding in the EY report (2015) of 'a very rapid rise of the Asian-Pacific Rim' with besides China and Hong Kong, also Japan, India, Thailand and Singapore. Still, we can observe the continuing importance not just of the US, but also of (Western) European countries which traditionally have been important cultural exporters: Germany, Italy, United Kingdom, France, and Spain, but also smaller countries such as the Netherlands, Switzerland, Belgium, and Austria. Entrepreneurs in the CCI in Europe can tap into the rich cultural legacy of Europe and its 'unique concentration of heritage and arts institutions' (EY, 2015: 16). On a more negative note, the data also highlight the presence of Mexico as the sole representative of Latin America (which could be related to its NAFTA membership) and the complete absence of countries in Africa and the Middle East as significant global exporters of creative goods. The production of cultural goods and services in those regions might be too small to express in US dollars or even monetary terms in general and remain most likely informal, which could possibly explain the absence from formal statistics. Still, the trend towards a polycentric pattern of cultural industries on a global scale is evident and can be corroborated by many other sources. *The First Global Map of Cultural Industries* (EY, 2015) also looks beyond the level of the global regions and points to global centres of excellence in cultural industries.

When it comes to what UNCTAD (2010: 191) calls 'creative services' (consisting of (architectural services; personal, cultural and recreational services; advertising services; research and development; audio-visual services), the dominance of the former core is still very much present as the developed economies accounted for 83 per cent of total exports of creative services. This dominance is even more pronounced if we look at main exporters of audio-visual services (see Table 21.3). The US is by far the largest exporter of these services and still holds a hegemonic position especially in film (see below).

The same caveats as mentioned above also apply to these data. In addition, some countries (e.g. India and Mexico) did not report figures for 2008. It may also be the case that these data are quite outdated when it comes to Southeast Asia. The recent emergence of South Korea as a global powerhouse of popular culture – selling K-pop, TV soaps, films as well as fashion and other forms of design to youth not just in Asia but increasingly also in other parts of the world (Shim, 2008; EY, 2015: 45) – is, for instance, probably only scarcely reflected in the figures for 2008.

Table 21.3   Top 15 countries according to export value of audio-visual and related services (in millions of US$) in 2008

1	United States	13,598
2	United Kingdom	2,520
3	Canada	2,102
4	France	1,167
5	Germany	1,065
6	Hungary	949
7	Spain	751
8	Belgium	540
9	The Netherlands	498
10	Argentina	447
11	China	418
12	Italy	414
13	Norway	275
14	Russian Federation	261
15	Republic of Korea	208

Source: UNCTAD (2010: 369).

The emerging multipolar global geography of cultural industries can also be observed at the level of individual cities. Large East Asian cities – Hong Kong, Singapore, Taipei, Seoul, Beijing, Shanghai and Tokyo – are joining the ranks of European and North American cities as global cultural centres (Skórska and Kloosterman, 2012; Caset and Derudder, 2017). These cities show that indeed 'Culture and the city, especially the global city, are close companions' (Pratt, 2012: 266).

CASES OF GLOBALIZATION: THE FILM INDUSTRY AND ARCHITECTURAL DESIGN

The globalization of the cultural industries could have been exemplified by a variegated array of particular cultural industries. However, due to limited space in this chapter we have chosen to illustrate the complex processes of cultural globalization through the case of the film industry and that of architecture. The film industry represents a case of products with an evident explicit narrative content typically rooted in the place or country of origin. This content is both lingual and visual which, in principle, impacts on the potential of films to travel across (cultural and political) borders. In addition, films are targeted at – often mass – audiences of consumers. The case of architectural design, by contrast, elucidates the possibilities and constraints of globalization of non-verbal cultural industries which depend on the commissioning of small numbers of usually institutional (e.g. municipalities) or corporate actors (e.g. global firms).

US Domination of the Global Film Market

Wiedemann (2011: 251) has deemed the film industry the 'crown jewel of the audiovisual sector' while Lorenzen (2008) has indicated that it is the largest of the cultural industries in terms of revenue. Crane (2014: 365–366) additionally points out that the fact that the film industry has been the target of various national cultural policies implies that films are perceived as having considerable symbolic and cultural value. Both this economic and cultural significance makes the film market a strategic site for the investigation of the globalization of cultural industries. And despite the ever increasing (online) availability of (free of charge) films due to technological advances, the global film industry continues to grow, both in terms of films produced and tickets being sold worldwide (Crane, 2014: 366).

Notwithstanding the rise of other centres of film production such as India, Nigeria, South Korea and China – both India and Nigeria have since 2011 surpassed the US in terms of films produced per year (UIS, 2013) – and despite technological advances that have allowed for the ever increasing availability of (free of charge) films online, the US film industry still dominates the global market (Banerjee, 2002; Crane, 2014; Fu and Govindaraju, 2010; EY, 2015; Vlassis, 2016). In the early 1900s, the US film industry, centred in Los Angeles (Hollywood), emerged as one of the world's most prominent centres of film production, and has since become the dominant locale of film production in the world due to three interrelated organizational and economic factors (Scott, 2002; van Elteren, 2003; Crane, 2014; Lipovetsky and Serroy, 2013). First, Scott (2002) alludes to the concentration of talent and economic resources dedicated to film production in Hollywood which allowed companies to benefit from agglomeration economies. Second, according to van Elteren (2003), the US market in general offered profitable economies of scale which ensured that cultural exports could be sold against relatively low rates well below the production costs of companies in other nations. Third, the widespread and effective distribution system resulted in the effective exclusion of foreign films from the US market, as well as ensuring the successful export of American films abroad (Scott, 2002; Crane, 2014). The export was also actively promoted by the US government which used its hegemonic position to put pressure on countries to open their border for American films (Judt, 2005: 232). These films were also attractive according to Tony Judt (2005: 232): 'What made American films so appealing, beyond the glamour and lustre that they brought to the gray surroundings in which they were viewed was their "quality". They were well-made, usually on a canvas far beyond the resources of any European producer.'

Starting in the late 1950s however, the US film industry has seen organizational changes related to, amongst others, technological changes (smaller cameras which allowed for outdoors shooting), the upsurge of TV and hence TV series, the 1948 Paramount decree (which severely bounded the possibilities for studios to control the whole production chain) and wider processes of globalization (Storper, 1989). While US corporations such as the Walt Disney Company, Warner Bros, Universal and Paramount still dominate film distributions, the vertically integrated production system has transformed into a more vertically disintegrated chain wherein smaller and independent companies are increasingly playing a bigger role (Scott, 2002; Lampel and Shamsie, 2003; Lorenzen, 2008). These smaller and independent companies spread all over the world (many US companies now outsource their production of animation to Asian countries due to higher quality and lower costs) cannot however benefit from the economies of scale that were created in the US since the large conglomerates still act as gatekeepers in the global film market through dominating distribution. Furthermore, the US have attempted to remain in its dominant position by producing 'transnational' films that could be watched in any cultural setting, and as such trying to offset cultural barriers to the consumption of audiovisual products (Crane, 2014).

Challenging the US

Although the US still dominate the global film market, especially in terms of distribution, since the 1980s we can observe that developing nations have entered this global market. Supported by national cultural policies in the form of tariffs, tax credits, subsidies and quotas, countries such as China, Japan, South Korea and India have developed a strong domestic film sector in terms of films produced yearly, the generation of jobs and revenue. Aside from India, whose Bollywood films have attained global status and are consumed all over the world, these countries have not yet succeeded in fully reaching the global market in terms of distribution and consumption, which could be attributed to linguistic, cultural and economic entry barriers. However, these national film sectors have succeeded in distributing films regionally to diasporic populations and more culturally proximate nations and sub-regions (Straubhaar, 2010; Mukherjee, 2008; Wiedemann, 2011).

Thus, although the organizational and economic structure of the US film industry has evidently changed since the 1950s, which has allowed for other firms and film producing countries to enter the stage, these other countries have not yet been successful in actually challenging the American dominance of the global film market (EY, 2015). The sheer size of the US film industry generating significant economies of scale (also in distribution), its concentration of talent (also from abroad) in and around Hollywood setting in motion agglomeration economies, its competence (notably in creating recognizable genres as, for example, westerns, thrillers, and sci-fi films), and the 'star system' in combination with winner-takes-it-all outcomes (Lipovetsky and Serroy, 2013: 210) are crucial components in the path-dependent reproduction of the American dominance.

Architectural Design

Whereas the film industry is organized around very large firms, architectural design shows a very different format. The largest architectural firm in the world, Gensler from the US, employs 2570 architects and has a fee income of 1–1.5 billion US dollars whereas Walt Disney Corporation has 195 000 employees and a turnover of more than 55 billion (Arch Daily, 2017; Wikipedia, n.d.). Economies of scale are hard to generate in architectural design in general. This is even more true for the high-end innovative architectural design practices which are mainly responsible for high-profile projects in (global) cities (Kloosterman, 2008; McNeill, 2009). Below, we briefly focus on these strong-idea architectural practices as they are strongly focused on creating something aesthetically unique which distinguishes them from their strong-service and strong-delivery practices counterparts (Kloosterman, 2010).

These strong-idea or boutique practices tend to be quite small – typically organized around one signature architect whose name is often a brand in itself as exemplified by Norman Foster, Zaha Hadid, Rem Koolhaas and Toyo Ito (Kloosterman, 2010). Their brand recognition is key in getting commissions, attracting talent and delivering projects on a global scale (Ren, 2011: 34). This form of branding also explains why barriers of entry for these global architectural practices are relatively high – notwithstanding their small size – as architects need to accumulate high levels of symbolic capital to carve out a name for themselves. This symbolic capital can then be transformed into economic capital when large corporations or cities commission them to design an iconic building, bridge, station or stadium (McNeill, 2009).

Another difference with the film industry is the ease with which architectural codes travel across borders. Although these boutique firms are strongly embedded in local production milieus, languages of design styles (for example modernism or postmodernism) originating in the West have become globally accepted (Kloosterman, 2008; McNeill, 2009; Ponzini and Nastasi, 2011). What has emerged then is a divergence between 'the production and the consumption of architectural design; cities with robust construction markets ... have yet to become equally competitive production sites of innovative architectural design' (Ren, 2011: 34).

Design Society, the new design museum in Shenzhen which opened its doors in December 2017, exemplifies key aspects of the globalization of high-end architectural design. It is located in one of the fastest growing urban regions in China and it is part of a conscious strategy by the Chinese government to move from 'Made in China' to 'Created in China'. The complex, late-modernist building itself has been designed by the Japanese architect Fumihiko Maki, whereas part of the display is designed by British architects Sam Jacobs Studio and another part by MVRDV from the Netherlands (Heathcote, 2017). So, on the one hand, *Design Society* is very much part of the giant construction site and the ensuing booming market for architectural services that China and other fast-growing regions in Asia have become in recent decades (EY, 2015). On the other, however, it shows that when an aspiring global city in China wants to make statement with an iconic building, foreign 'starchitects' are commissioned and Western styles are preferred.

CONCLUSIONS

The global landscape of production and consumption of creative goods and services has fundamentally changed in the past few decades. Although reliable data for a more comprehensive overview are still lacking, there is no doubt about the emergence of a polycentric or multipolar world of cultural industries. Allen Scott's prognostication (Scott, 2000: 211) 'for a global but polycentric and multi/faceted system of cultural production/straddled by large multinational corporations – to make its appearance over the course of the twenty-first century' has already materialized. Emerging economies are not just about churning out toys, cars, laptops or mobile phones, but increasingly also producing and, moreover, exporting series, music, films, design, fashion and visual and performance arts. It seems that it will only be a matter of time until these countries spawn their own multinational cultural corporations.

Especially larger cities in Asian countries are becoming home to competitive clusters of cultural industries. After 1990, a fast rise in the number of advertising agencies, architectural practices, and design studios has occurred in emerging economies. Moreover, we can also observe an enlargement and deepening of the broader field of cultural production with, for instance, the establishment of design and fashion schools, the increasing number of design and fashion magazines as well as art and design biennales and fashion weeks in countries like China, Turkey, and South Korea (Lipovetsky and Serroy, 2013: 239). South Korea has pioneered building such infrastructures as part of a conscious state policy to boost the export potential of its cultural industries (Shim, 2006, 2008) and China is now following suit with the explicit mentioning of cultural industries in its 11th Economic Plan (EY, 2015).

Because of the high explicit symbolic content of their products and the subsequent potential for cultural and political contestation, processes of globalization of cultural industries tend to display their own specific spatial patterns and dynamics. Creative goods or services with a relatively high culturally specific content typically cater first and foremost to global regions sharing a set of broader cultural notions or cultural proximity – such as East Asia or the West (Straubhaar, 2010).

To what extent China and other, mainly Asian, countries will be able to appeal to audiences which are at a greater cultural distance than those in their own global region remains to be seen. There are clear exceptions to this rule – Bollywood films from India, strong-idea architectural design from Japan, Super Mario from Japan and, increasingly K-pop from South Korea. It might even be the case that globalization of cultural industries is entering a new phase with a marked increase in the crossing of the borders of the global region enabled by a combination of technology (digitization and social media) and demographics (diaspora populations).

The role of governments in promoting vibrant cultural industries is not limited to creating infrastructures. To develop globally competitive and innovative cultural industries, a strong protection of intellectual property rights is important (Wiedemann, 2011; EY, 2015). To what extent the freedom of expression is also a necessary condition for the realization of this potential remains a moot point (Wiedemann, 2011; Jacques, 2012). Many emerging economies seem to struggle with creating and maintaining an open, tolerant creative climate which allows innovative, eccentric and even deviant forms of expression. The fact that the recent, critically acclaimed erotic psychological thriller *The Handmaiden* (based, interestingly, on Sarah Waters' novel *Fingersmith*) was made in South Korea seems to offer a glimmer of hope.

REFERENCES

Abrams, M. (2013), 'Lagos: global fashion hotspot', *Financial Times*, 1 November 2013. https://www.ft.com/content/e1668b92-31b0-11e3-817c-00144feab7de (accessed 3 December 2017).

Adorno, T.W. (2001), *The Culture Industry*, London: Routledge.

Appadurai, A. (1990), 'Disjuncture and difference in the global cultural economy', *Theory, Culture And Society*, **7**(2), 295–310.

Applebaum, A. (2013), *Iron Curtain; The Crushing of Eastern Europe*, London: Penguin Books.

Arch Daily (2017), 'The World's 20 Largest Architecture Firms'. https://www.archdaily.com/870842/the-worlds-20-largest-architecture-firms (accessed 12 December 2017).

Baldwin, R. (2016), *The Great Convergence; Information Technology and the New Globalization*, Cambridge, MA/London: The Belknap Press of Harvard University Press.

Banerjee, I. (2002), 'The locals strike back? Media globalization and localization in the new Asian television landscape', *Gazette: The International Journal for Communication Studies*, **64**(6), 517–535.

Barraclough, B. and Z. Kozul-Wright (2011), 'Voice, choice and diversity through creative industries: towards a new development agenda', in: B. Barraclough and Z. Kozul-Wright (eds), *Creative Industries and Developing Countries; Voice, Choice and Economic Growth*, London/New York: Routledge: 3–36.

Brun, A., F. Caniato, M. Caridi, C. Castelli, G. Miragliotta, S. Ronchi, A. Sianesi and G. Spina (2008), 'Logistics and supply chain management in luxury fashion retail: empirical investigation of Italian firms', *International Journal of Production Economics*, **114**(2), 554–570.

Caset, F., and B. Derudder (2017), 'Measurement and interpretation of "global cultural cities" in a world of cities', *Area*, **49**(2), 238–248.

Castells, M. (2011), *The Rise of the Network Society: The Information Age: Economy, Society, and Culture (Vol. 1)*, Hoboken: John Wiley & Sons.

Crane, D. (2014), 'Cultural globalization and the dominance of the American film industry: cultural policies, national film industries, and transnational film', *International Journal of Cultural Policy*, **20**(4), 365–382.

Currid-Halkett, E. (2017), *The Sum of Small Things: A Theory of the Aspirational Class*, Princeton: Princeton University Press.

Deinema, M. (2012), *The Culture Business Caught in Place. Spatial Trajectories of Dutch Cultural Industries, 1899–2005*, Amsterdam: University of Amsterdam (Dissertation).

D'Ovidio, M. (2015), 'The field of fashion production in Milan: a theoretical discussion and an empirical investigation', *City, Culture and Society*, **6**(2): 1–8.

EY (2015), *The First Global Map of Cultural Industries*. https://en.unesco.org/creativity/sites/creativity/files/cultural_times._the_first_global_map_of_cultural_and_creative_industries.pdf (accessed 22 November 2017).

Florida, R. and S. Jackson (2010), 'Sonic city: the evolving economic geography of the music industry', *Journal of Planning Education and Research*, **29**(3), 310–321.

Frieden, J. (2006), *Global Capitalism, Its Fall and Rise in the Twentieth Century*, New York/London: W.W. Norton and Company.

Fu, W.W. and A. Govindaraju (2010), 'Explaining global box-office tastes in Hollywood films: homogenization of national audiences' movie selections', *Communication Research*, **37**(2), 215–238.

Heathcote, E. (2017), 'Created in China'. *Financial Times, FT Weekend*, 9 December/10 December 2017: 12.

Jacques, M. (2012), *When China Rules the World*, London: Penguin Books.

Jansson, J., and D. Power (2010), 'Fashioning a global city: global city brand channels in the fashion and design industries. *Regional Studies*, **44**(7), 889–904.

Judt, T. (2005), *Postwar; A History of Europe Since 1945*, New York: The Penguin Press.

Kharchenkova, S.S. (2017), *White Cubes in China. A Sociological Study of China's Emerging Market for Contemporary Art*, Amsterdam: University of Amsterdam (Dissertation).

Kloosterman, R.C. (2008), 'Walls and bridges: knowledge spillover between "superdutch" architectural firms', *Journal of Economic Geography*, **8**(4): 545–563.

Kloosterman, R.C. (2010), 'Building a career: labour practices and cluster reproduction in Dutch architectural design', *Regional Studies*, **44**(7): 859–871.

Kloosterman, R.C. and R. Koetsenruijter (2016), 'New kids on the block? The emerging global mosaic of cultural production', in: B. Lambregts, N. Beerepoot and J. Kleibert (eds), *Globalisation and New Patterns of Services Sector-Driven Growth*, London: Routledge: 142–163.

Kloosterman, R.C., N. Beerepoot and B. Lambregts (2015), 'Service-sector driven economic development from a historical perspective', in: B. Lambregts, N. Beerepoot and Robert C. Kloosterman (eds), *The Local Impact of Globalization in South and Southeast Asia: Offshore Business Processes in Services Industries*, Oxford/New York: Routledge: 17–28.

Lampel, J. and J. Shamsie (2003), 'Capabilities in motion: new organizational forms and the reshaping of the Hollywood movie industry', *Journal of Management Studies*, **40**(8), 2189–2210.

Lipovetsky, G. and J. Serroy (2013), *L'Esthétisation du monde; Vivre à l'âge du capitalisme artiste*, Paris: Gallimard.

Lorenzen, M. (2008), 'On the globalization of the film industry. Creative encounters', Working paper #8. http://www.cbs.dk/content/download/77698/1040003/file/8-Lorenzen-Globalization_Film_Industry-08.pdf (accessed 27 June 2018).

Lorenzen, M., and R. Mudambi (2013), 'Clusters, connectivity and catch-up: Bollywood and Bangalore in the global economy', *Journal of Economic Geography*, **13**(3), 501–534.

Mansvelt, J. (2005), *Geographies of Consumption*, Los Angeles/London: Sage.

Mason, P. (2016), *Postcapitalism: A Guide to Our Future*, London: Macmillan.

McNeill, D. (2009), *The Global Architect: Firms, Fame and Urban Form*, London: Routledge.

Milanovic, B. (2016), *Global Inequality; A New Approach for the Age of Globalization*, Cambridge, MA: The Belknap Press of Harvard University Press.

Molotch, H. (2002), 'Place in product'. *International Journal of Urban and Regional Research*, **26**(4), 665–688.

Mukherjee, A. (2008), 'The audio-visual sector in India', in: B. Barraclough and Z. Kozul-Wright (eds), *Creative Industries and Developing Countries; Voice, Choice and Economic Growth*, London/New York: Routledge: 174–192.

Osterhammel, J. and N.P. Peterson (2003), *Globalization; A Short History*, Princeton/Oxford: Princeton University Press.

Ponzini, D. and M. Nastasi (2011), *Starrchitecture; Scenes, Actors and Spectacles in Contemporary Cities*, Turin: Umberto Allemandi.

Power, D. (2003), 'The Nordic "cultural industries": a cross-national assessment of the place of the cultural industries in Denmark, Finland, Norway and Sweden', *Geografiska Annaler: Series B, Human Geography*, **85**(3), 167–180.

Power, D. and A.J. Scott (2005), 'A prelude to cultural industries and the production of culture', in: D. Power and A.J. Scott (eds), *Cultural Industries and the Production of Culture*, London: Routledge: 3–15.

Pratt, A.C. (1997), 'The cultural industries production system: a case study of employment change in Britain, 1984–91', *Environment and Planning A*, **29**(11), 1953–1974.

Pratt, A.C. (2012), 'The cultural economy and the global city', in: B. Derudder, M. Hoyler, P.J. Taylor and F. Witlox (eds), *International Handbook of Globalization and World Cities*, Cheltenham, UK and Northampton, MA, USA: Edward Elgar Publishing: 265–274.

Rantisi, N.M. (2004), 'The ascendance of New York fashion', *International Journal of Urban and Regional Research*, **28**(1), 86–106.

Rasterhoff, C. (2012), *The Fabric of Creativity in the Dutch Republic. Painting and Publishing as Cultural Industries, 1580–1800*, Doctoral dissertation, Utrecht University.

Ren, X. (2011), *Building Globalization: Transnational Architecture Production in Urban China*, Chicago: University of Chicago Press.

Ritzer, G. (2008), *The McDonaldization of Society (5)*, London: Pine Forge Press.

Roper Center (1992), 'The Export of American Culture', *The Uniting of America*, **3**(4), 117–120.

Sassoon, D. (2006), *The Culture of the Europeans; From 1800 to the Present*, London: HarperCollins.

Scott, A.J. (2000), *The Cultural Economy of Cities: Essays on the Geography of Image-Producing Industries*, London: Sage.

Scott, A.J. (2002), 'A new map of Hollywood: the production and distribution of American motion pictures', *Regional Studies*, **36**(9), 957–975.

Scott, A.J. (2004), 'Hollywood and the world: the geography of motion-picture distribution and marketing', *Review of International Political Economy*, **11**(1), 33–61.

Scott, A.J. (2005), *On Hollywood: The Place, the Industry*, Princeton: Princeton University Press.

Scott, A.J. (2012), *A World in Emergence; Cities and Regions in the 21st Century*, Cheltenham, UK and Northampton, MA, USA: Edward Elgar Publishing.

Shim, D. (2006), 'Hybridity and the rise of Korean popular culture in Asia', *Media, Culture & Society*, **28**(1), 25–44.

Shim, D. (2008), 'The growth of Korean cultural industries and the Korean wave', *East Asian Pop Culture: Analysing the Korean Wave*, **1**, 15–32.

Skórska, M.J. and R.C. Kloosterman (2012), 'Performing on the global stage: exploring the relationship between finance and arts in global cities'. http://www.lboro.ac.uk/gawc/rb/rb412.html (accessed 15 June 2018).

Storper, M. (1989), 'The transition to flexible specialisation in the US film industry: external economies, the division of labour, and the crossing of industrial divides', *Cambridge Journal of Economics*, **13**(2), 273–305.

Straubhaar, J. (2010), 'Chindia in the context of emerging cultural and media powers', *Global Media and Communication*, **6**(3), 253–262.

Toste, S. (2013), 'Armani Supply & Production Chain'. Presentation, https://prezi.com/wuh67ycqravm/armani-supply-production-chain/ (accessed 10 December 2017).

UIS, Unesco Institute for Statistics (2013), Emerging markets and the digitalization of the film industry; An analysis of the 2012 UIS International Survey of Feature Film Statistics. UIS Information Paper No. 14. Montréal: Unesco Institute for Statistics.

UNCTAD (2010), 'Creative Economy Report 2010, United Nations Conference on Trade and Development'. http://unctad.org/en/docs/ditctab20103_en.pdf (accessed 15 June 2018).

UNCTAD (2013), *Creative Economy Report 2013. Special Edition: Widening Local Development Pathways*, United Nations Conference on Trade and Development. http://www.unesco.org/culture/pdf/creative-economy-report-2013.pdf (accessed 6 December 2017).

UNESCO (2005), *Convention on the Protection and Promotion of the Diversity of Cultural Expressions*, Paris: UNESCO.

van Elteren, M. (2003), 'US cultural imperialism: today only a chimera', *SAIS Review*, **23**(2), 169–188.

Vlassis, A. (2016), 'Soft power, global governance of cultural industries and rising powers: the case of China', *International Journal of Cultural Policy*, **22**(4), 481–496.

Wiedemann, V. (2011), 'Promoting creative industries: public policies in support of film, music, and broadcasting', in: B. Barraclough and Z. Kozul-Wright (eds), *Creative Industries and Developing Countries; Voice, Choice and Economic Growth*, London/New York: Routledge: 251–274.

Wikipedia (n.d.), 'The Walt Disney Company'. https://en.wikipedia.org/wiki/The_Walt_Disney_Company (accessed 13 December 2017).

Zhang, X. and L. Yajuan (2018), 'Concentration or deconcentration? Exploring the changing geographies of film production and consumption in China', *Geoforum*, **88**, 118–128.

22. Globalization and mega-events: thinking through flows

Martin Müller and Christopher Gaffney

INTRODUCTION

If we understand globalization as the coming together of people and things from different parts of the planet for a common purpose, mega-events such as the Olympic Games and the Football World Cup are its quintessential expressions. The people involved in staging a mega-event immediately come to mind: athletes and coaches from different nations, spectators from around the world, security personnel, journalists, VIPs. The activities of these people are broadcast to billions of viewers around the world, hammering home that, yes, the Christ the Redeemer statue stands indeed in Rio de Janeiro. Capital from international broadcasters and global corporations bankrolls mega-events. NBC, BBC and others compete for broadcasting rights and Coca-Cola, Panasonic, Dow Chemical and other blue chips sign sponsorship deals to get public exposure of their brands.

A systematic and agnostic way of thinking through globalization and mega-events is through the concept of flows. Metaphors make a difference in how we conceive space and spatialities (Simonsen, 2004) and flows have a number of specific characteristics. First, they are dynamic. Flows can increase and decrease, change direction and intensity, volume and speed. This is in contrast, for example, to the notion of 'circulation', which suggests both a circular movement and a measure of containment and order. Second, flows are material, encompassing a diversity of things that can move or be mobilized: people, money, images, knowledge, words, objects, technology, air, water. Third, flows have no origin and no end point; they are pure movement. This directs attention not just to the result of flows coming together (to the event), but to the before and after, and to the fixed pathways and conduits that facilitate and condition their movement (Santos, 1996: 274–279). Some scholars go so far as to elevate flows to the status of a foundational concept, where place, space, and scale only exist as a function of flows (Deleuze and Guattari, 1997; Shields, 1997; Urry, 2002).

For understanding the global nature of mega-events, flows come with several conceptual advantages. They allow examining the movement of images, knowledge, people, and capital – all of which are indispensable for mega-events (see Figure 22.1) – without changing register or concept. Flows cut across scales, obviating the need to speak of global and local as distinct or even opposed scales and focusing instead on the temporary articulation of arrangements in certain contexts – or de- and reterritorializations, as Deleuze and Guattari (1987) would have it. With mega-events, there is never something that is purely local, as media transmit every second of the event to the farthest corner of the world. But at the same time, there is never something that is purely global, as mega-event hosts adapt knowledge, enact policies, and deploy capital

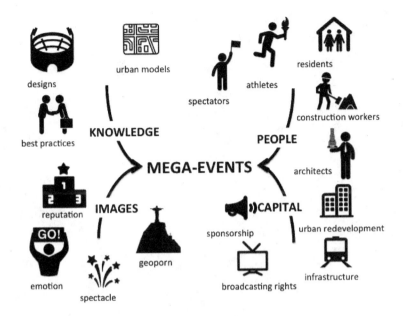

Figure 22.1   Flows in the organization of mega-events

from elsewhere in their specific contexts. Finally, flows reflect that mega-events are inherently unstable, itinerant phenomena. Not only do they move from place to place every few years, but their very organization in one place relies on many short-term fixes and workarounds, as plans go wrong and crises strike. Each unexpected change necessitates a shifting of flows to accommodate new realities.

Drawing on the available literature, the following sections examine the characteristics and patterns of the four main flows constituting mega-events today: people, images, capital, knowledge.

MEGA-EVENTS AS FLOWS

People

People first come to mind when thinking of flows that comprise mega-events. After all, the competition between athletes from different countries around the world is the main attraction of events such as the Football World Cup. But while important and visible, athletes are, in fact, the smallest group of people involved in staging a mega-event, as Figure 22.2 demonstrates. Spectators are by far the largest physically present group, indicating the importance of the global consumption of spectacle. Together with the representatives of the media, spectators and athletes are the drivers of the globalization of mega-events. All three groups choose, indeed are eager, to attend mega-events, often travelling long distances and spending large sums of money.

Sources: Rio 2016 Organizing Committee (athletes, temporary service workers); Brazilian Tourist Board (foreign spectators); Comitê Popular da Copa e Olimpíadas do Rio de Janeiro (people displaced); Brazilian Ministry of Sports (security personnel); International Olympic Committee (number of volunteers); O Globo (media; construction workers).

Figure 22.2 Size of people flows at the Rio 2016 Olympic Games

There are a number of other, less visible human flows that are shaped and directed by the globalizing forces of mega-events. These flows are indispensable for preparing and staging events, but are often overlooked. For construction workers, mega-events bring coveted jobs. Workers often migrate large distances, also crossing country borders, to build stadia, transport infrastructure, or housing. The World Cup 2022 in Qatar has been at the centre of a global campaign for workers' rights and is but the most egregious example of the exploitation of migrant workers that accompanies many mega-events, whether in Beijing, Sochi, or Rio (Amnesty International, 2015; Buchanan, 2013; Carter, 2017). Mega-events stimulate these labour flows because they lead to a temporary increase of construction activity and create jobs that can often not be filled from the local labour market.

Volunteers, security personnel and temporary service workers are also part of the temporary labour flows. They can number several hundreds of thousands, but many of them are just employed for the duration of the event. Here again, the temporary surge in demand for labour stimulates flows, mostly from within the host country, to meet this demand (Horne and Manzenreiter, 2017).

A final human flow that is perhaps the most important one for the production of the event but, paradoxically, also the least visible and least well documented is that of people displaced. The size of construction for mega-events makes displacement almost inevitable. Displacement can happen through two mechanisms: through direct displacement, by moving people from the sites of construction projects (Porter et al., 2009), or through indirect displacement, for example through gentrification, which replaces poorer residents with wealthier ones and is a common side effect of mega-events (Gaffney, 2016; Watt, 2013). The size of these flows can reach staggering dimensions: for Rio 2016 an estimated 77,200 people were directly displaced (Comitê Popular, 2015: 20) and more than one million for the Olympic Games in Beijing in 2008 (COHRE, 2007: 154; on Beijing see also Shin, 2009). While displacement usually does not make for flows over a long distance (often people settle not far from where they used to live), it is the most disruptive of the flows because it results in a permanent move. While most other flows of people for mega-events involve temporary physical movement, displacements result in profound and permanent changes to people's lives.

Capital

The flows of financial capital that accompany mega-events have grown in conjunction with the global reach and size of mega-events. The combined value of broadcasting contracts for the Winter Olympic Games in Sochi 2014 and Rio de Janeiro 2016 exceeded USD 4 billion, while global sponsorship income ran to USD 1 billion (IOC, 2016). Despite the rising income from sponsorship and broadcasting rights, the majority of the capital to bankroll mega-events still comes from the public. This is true for neoliberal economies in the West, but even more for state-led economies such as in China, Russia, and Brazil (Preuss, 2004). While income from the private sector typically covers the operational costs, capital investments for infrastructure or urban redevelopment remain with the state.

Event boosters argue that the public financing of the event will stimulate flows of tourists, images, news, and money that can only be obtained through hosting. The logics of hosting the event are to invest public wealth with the goal of inserting a city or country ever more into global circuits of accumulation. It is through this process of accelerated globalization, the argument goes, that a city or country should risk its public resources in order to attract global flows of capital – which will benefit the population as a whole (Harvey, 1989; Lauermann and Davidson, 2013). Considered from a purely financial standpoint, however, research shows that there is typically a negative return on investment, indicating a large-scale public subsidy for private profit and public spectacle (Matheson, 2009; Zimbalist, 2015).

In order to develop a more complete understanding of how capital flows function in relation to the mega-event, we should also consider flows of political and symbolic capital. Politicians hope to accrue political capital from mega-events (Burbank et al., 2001). The presidents of FIFA and the IOC are accorded the status of heads of state, even though they are the elected representatives of small, private Swiss-based NGOs. Neither organization has more than 500 full-time employees, yet commands the world's attention for their events. In local and national contexts, mayors, ministers, even queens and presidents are keen to be associated with mega-events and be present at opening ceremonies, final matches, and preparatory meetings that have all of the protocols accorded to the highest ranking diplomatic missions.

Images

The Olympic Rings are among the most recognizable of icons and global brands pay top dollar to associate with them. For individuals, taking selfies in front of the Olympic Rings and sharing them on *Facebook*, *Instagram* or *Snapchat* has become a naturalized element of the mega-event experience. It is the global flow of images that creates this symbolic capital and structures global consciousness of events and the places where they occur (Roche, 2000). While contemporary mega-events are still dependent on the production of physical spaces for their realization, they are primarily mediated spectacles, distant events consumed by an ever-growing audience through increasingly diverse media.

The production and circulation of images is at the core of the current mega-event business model, and the monopoly rights holders of the event charge dearly so that companies can gain exclusive access. The laws and regulations that regulate the flow of images protect 'stakeholders' against ambush marketing, establish 'clean sites' within cities that are free of competitors' advertisements, and prevent athletes and spectators from capitalizing on their participation in the event. Of course, the exchange of selfies, production of *Instagram* photos, and *Facebook* updates from live sites are means through which individuals accumulate symbolic capital through the circulation of images. Yet when we consider the official production of images associated with mega-events we are presented with an aseptic vision of a city that has been produced for global consumption (Broudehoux, 2007; Gruneau and Horne, 2016; Waitt, 1999).

Knowledge

Global knowledge flows are a relative latecomer to mega-events. In the 1990s, at a time when talk of the knowledge society and knowledge as the critical resource for companies began to figure prominently in corporate and national strategies, knowledge was still not considered a major resource for organizing mega-events, let alone one that should be systematically managed, stored, or transferred. 'Knowledge transfer [between host cities of the Olympic Games] operated on a relatively unsophisticated basis, with future host cities sending numerous persons to literally look over the shoulders of the officials involved in staging one or another event' (Cashman and Harris, 2012: 4). The basic setup of mega-events as one-time occasions taking place in different geographic and cultural contexts with a different team of people every couple of years, however, offers a clear case for the mobilization of the experience and knowledge from previous hosts; it avoids reinventing solutions that have already been found elsewhere and it lends legitimacy to policies if other cities have applied them successfully before.

Work on global knowledge flows in mega-events has gained in currency together with the focus on policy mobilities in geography (McCann, 2011). It is therefore comparatively recent and a budding area of research. The travelling of global expertise and policy models in the bidding and planning process of mega-events has attracted particular attention. The importance of policies and experience from elsewhere in bidding for and hosting mega-events is well-documented for cities such as Manchester (Cook and Ward, 2011), Rio de Janeiro (Silvestre, 2013), London (Allen and Cochrane, 2014), Vancouver (Temenos and McCann, 2012), and Sochi (Müller, 2015).

Global knowledge flows come on the back of the other flows discussed before. Consultants who advise cities in bidding and hosting spread models of 'best practice' and bring their own network of contacts. Variously called, 'the Olympic caravan' (Cashman and Harris, 2012), 'Gamers' or 'mega-event gypsies', these experts travel from mega-event to mega-event and contribute to the emergence of a global mega-events industry. Capital flows play an important part in this, as knowledge turns into 'policy commodities' that are traded on the global market for urban and planning knowledge (Lauermann, 2014). Finally, images such as architectural design or planning models contribute to the mobilization of knowledge. The iconic status of stadia such as Beijing's Bird's Nest or Munich's Allianz Arena has increased the demand for designs of global starchitects as cities have increasingly sought to build brands and communicate through images (Ren, 2008).

A CLOSER LOOK AT THE 2014 FIFA WORLD CUP AND 2016 OLYMPIC GAMES IN RIO DE JANEIRO

While every mega-event is an apt case study for examining the convergence of global flows in a specific site, Rio de Janeiro is particularly interesting because it hosted the 2014 FIFA World Cup and the 2016 Summer Olympics. An examination of the flows contained and conditioned by these two mega-events reveals the dynamism of the events themselves and their role in reshaping their host cities and countries.

Rio's quest for the Olympics began in the mid-1990s with the contracting of a Barcelona-based consulting firm that proposed a strategic plan for Rio that included a bid for the 2004 Olympics. Even though that bid failed, the flows of knowledge from Spain to Brazil were established and would figure heavily in the eventual fashioning of Rio's successful 2016 Olympic bid (Gusmão de Oliveira and Gaffney, 2010). Similarly, the much-publicized security paradigm that was implemented in Rio de Janeiro in the years leading up to the events, the Police Pacification Units (UPP – Unidades de Polícia Pacifcadora), was the fruit of knowledge exchanges between the government of the state of Rio de Janeiro with the municipal government of Medellín, Colombia, and with the Brazilian-led UN force in Haiti, MINUSTAH.

The flows of documents from FIFA and the IOC to the organizing committees in Brazil and three levels of government, conditioned the implementation of large-scale infrastructure projects for both events. Once the Brazilian government legalized hosting contracts, a series of knowledge exchanges and international observer missions helped to make the events a physical and social reality. For instance, the Brazilian World Cup organizing committee went to South Africa to observe their handling of security and event management during the 2010 World Cup. In the local context, these exchanges took the form of trade shows for stadiums, transportation, weapons manufactures, marketing agencies, and sports business interests. During the events themselves, these exchanges accelerated in the 'national houses' established by governments to showcase their major tourist attractions and industries. In addition to the acceleration and vectorization of capital flows of the event, an intertwined system of personal and professional relations functioned to consolidate the business opportunities opened by the spectacle. Global sponsors pay hundreds of millions to FIFA and the IOC in exchange for VIP and hospitality packages in which to conduct business, or expand networks of patronage. In this way, a localized event coalition of actors intersects with globalized agents to use the mega-event as a way to maximize exchange, flows, and accumulation possibilities.

The added stress that the flows of athletes, national delegations and hundreds of thousands of tourists bring to the urban infrastructure could easily overwhelm systems if provisions for increasing or changing flow capacity are not addressed. To this end, it is not uncommon for cities to declare public holidays on days of peak demand so that fixed infrastructures can handle the increased demand. For the 2014 World Cup, Rio de Janeiro declared a public holiday for each day that a match was played at the Maracanã stadium. This measure removed the normal flow load from the system and replaced it with more predictable, syncopated flows. While this kind of intervention is useful for the realization of the event, it is a form of dispossession as residents are prevented from freely using the systems that condition their urban movements.

One of the biggest challenges in hosting mega-events is to create transportation solutions for the event that will also serve the long-term needs of the city. In this sense, the building of mass-transport systems under the time pressures of an event will supposedly help to stimulate urban developments that would have otherwise taken many more years. Rio de Janeiro's transportation interventions for mega-events are examples of this tendency.

All of the transportation interventions undertaken in the city of Rio de Janeiro were orientated around the flows determined by the location of the four Olympic clusters. All levels of Brazilian government contributed to the financing of these projects, by far the most expensive elements of Brazil's BRL 40+ billion (ca. USD 12 billion) Olympic budget. By concentrating investment in these areas, the state contributed directly to a process of real-estate valorization within the Olympic zones. In addition to removing people from their homes and businesses through compulsory purchase orders, the city of Rio de Janeiro engaged in forced removals.

In the local context and in conjunction with unchecked real-estate speculation (Gaffney, 2016), the implementation of these specific transportation lines resulted in the dislocation of lower income residents from the nominal Olympic zones to the far west and north, where their displacement was justified through the provision of those self-same transport lines. In the words of one major real-estate developer, 'if the poor can't afford to live here, they can take the BRT [Bus Rapid Transit]' (Watts, 2015). In effect, the daily flows of the city were rearticulated through an urban planning agenda that was largely determined by the demands of the mega-event that happened to articulate with a coalition of vested interests that were using the events as a rent-seeking mechanism. While it is also true that FIFA and the IOC are also rent-seeking organizations, their capacity to extract monopoly rents is predicated upon the willingness of their hosts to collude and collaborate. In the local context, the mega-events serve to benefit the coalitions of actors that bid, build, and stage them, capturing flows of public money at the same time that they condition future urban development.

As with every mega-event, Rio de Janeiro's event coalition articulated these substantial urban interventions behind closed doors and presented them to the IOC in the form of the Rio 2016 candidature dossier. Once the two parties signed a legally binding hosting contract in 2009, the federal government stepped in with subsidized loans and the city and state of Rio de Janeiro assumed significant debt as the cost of the infrastructure projects spiralled upwards. The Rio state government declared bankruptcy one month before the Olympics in order to guarantee federal monies for covering last-minute preparations. The public risk for implementing these projects was justified with the familiar refrains of post-event prosperity (a.k.a. 'legacy'), but in the year leading up to the 2016 Olympics, Brazil found itself in the midst of its worst-ever economic crisis, and the international flows of capital that made the country the darling of investors had slowed to a trickle.

CONCLUSION

Mega-events are among the most complex social phenomena of our time and no researcher can hope to understand the totality of any one event or its multifarious impacts on cities, populations, and political economy. What we have suggested here is to approach mega-events from the perspective of flows. This perspective takes away notions of fixity and stability, allowing us to read mega-events as a constantly unfolding web of relationships that provide key insights into processes of globalization. Within this perspective, we would highlight three elements of flow dynamics that will merit our attention in the coming years: extensity, intensity, and velocity.

The extensity of mega-event flows can be measured by examining patterns, nodes, and networks over time. For instance, when an event bid is under construction, a local coalition of actors will typically employ one of a handful of specialist global public relations firms. In the local context, these actors have specific interests that are tied to the real estate, construction, sports, and media sectors – each of which has an identifiable geographic footprint. The public relations firms are tasked with increasing the extensity of the city's flows by capturing the mega-event. If they are successful, a range of new actors begins to visit the city, bringing ever more flows and expanding the network of actors. These flows do not increase all at once, nor are they sustained indefinitely, nor do they only move in one direction. Rather, these flows move in a staccato rhythm, reaching their highest point of extensibility during the event itself: approximately one billion people watched the 2014 World Cup final. Measuring these flows has become increasingly complicated with electronic and social media, yet there are increasingly sophisticated tools for researchers to evaluate quantitatively the variegated flows of the event cycle.

But even if we were able to measure effectively the extent of mega-event flows at different scales, we would also need to consider their intensity. For instance, a tourist who visited Rio de Janeiro for the Olympics might have had an intense and memorable experience, but the impact of her visit on the city itself will have faded rather quickly: this is not an intense flow. On the other hand, if the same woman started an import–export business as part of her visit, then the flow of goods, capital, and people may become more sustained over time. The kinds of flows that are most intense for contemporary mega-events are associated with the media, which is the dominant force in their promotion, production, and consumption. Media, of course, mobilizes billions of dollars and global corporations depend on advertising to sell their products. During the event itself, the intensity of flows is matched by its extensiveness, though this too can be highly variable. For instance, the Summer Olympic media flows to Nepal (population 27 million) are not as intensive or extensive as those to Jamaica (population 2.7 million).

The velocity of global flows is strongly correlated to their intensity, but measures the degree to which the mega-event alters the status quo ante for a city. London is a hub of global tourism and finance, capital of a once-sprawling empire with one of the world's busiest airports. Images of London are well-known globally and the city's reputation as a cosmopolitan centre of learning and culture is well-established. London's flows did not significantly increase in the years leading up to the Olympic Games, but its media exposure (image and text flow) jumped during the Games themselves, before returning to pre-Games levels. Mega-events therefore do not necessarily lead to an acceleration of all flows, but speed up certain flows for a certain period of time.

As with most processes of globalization, global flows are not even or equal. In hosting mega-events there are primary flows, such as those of various forms of capital, that induce secondary flows – of service workers or displaced people – as a response. It would be misleading not to consider the flows of displaced people, workers and volunteers as part of how globalization is interwoven with mega-events. Although many of these flows do not cross national borders, they are a necessary requirement for the global flows of capital, athletes, media, and spectators to be possible in the first place. Displaced people make way for stadiums, security personnel protect athletes, volunteers contribute free labour to FIFA and the IOC, who profit from the event. A focus on flow and flux should therefore not seduce us into forgetting about the different characteristics of flows: whether they are voluntary or forced, whether they serve consumption or production and whether they are desirable or undesirable.

REFERENCES

Allen, J., and A. Cochrane (2014), 'The urban unbound: London's politics and the 2012 Olympic Games', *International Journal of Urban and Regional Research*, **38**(5), 1609–1624.

Amnesty International (2015), *Promising little, delivering less: Qatar and migrant labor abuse ahead of the 2022 Football World Cup*, London: Amnesty International. Retrieved on 15 June 2018 from https://www.amnesty.org/download/Documents/MDE2215702015ENGLISH.PDF.

Broudehoux, A.-M. (2007), 'Spectacular Beijing: the conspicuous construction of an Olympic metropolis', *Journal of Urban Affairs*, **29**(4), 383–399.

Buchanan, J. (2013), *Race to the Bottom: Exploitation of Migrant Workers Ahead of Russia's 2014 Winter Olympic Games in Sochi*, New York: Human Rights Watch.

Burbank, M., G. Andranovich and C.H. Heying (2001), *Olympic Dreams: The Impact of Mega-events on Local Politics*, Boulder: Lynne Rienner.

Carter, T.F. (2017), 'Pulling back the curtain: on mobility and labour migration in the production of mega-events', in: N. B. Salazar, C. Timmerman, J. Wets, L. Gama Gato and S. Van den Broucke (eds), *Mega-event Mobilities*, Abingdon: Routledge, pp. 16–32.

Cashman, R. and R. Harris (2012), *The Australian Olympic caravan from 2000 to 2012: A Unique Olympic Events Industry*, Petersham: Walla Walla Press.

COHRE (2007), *Fair Play for Housing Rights*, Geneva: COHRE.

Comitê Popular (2015), *Olimpíada Rio 2016: os jogos da exclusão*, Rio de Janeiro: Comitê Popular da Copa e Olimpíadas do Rio de Janeiro.

Cook, I.R. and K. Ward (2011), 'Trans-urban networks of learning, mega events and policy tourism: the case of Manchester's Commonwealth and Olympic Games projects', *Urban Studies*, **48**, 2519–2535.

Deleuze, G. and F. Guattari (1987), *A Thousand Plateaus* (B. Massumi, Trans.), Minneapolis: University of Minnesota Press.

Deleuze, G. and F. Guattari (1997), 'City/state'. *Zone*, **1–2**, 195–199.

Faulhaber, L. and L. Azevedo (2015), *SMH 2016: remoções no Rio de Janeiro Olímpico* (1st edn), Rio de Janeiro: Mórula Editorial.

Gaffney, C. (2016), 'Gentrifications in pre-Olympic Rio de Janeiro', *Urban Geography*, **37**(8), 1132–1153.

Gruneau, R. and J. Horne (eds) (2016), *Mega-Events and Globalization: Capital and Spectacle in a Changing World Order*, Abingdon: Routledge.

Gusmão de Oliveira, N. and C. Gaffney (2010), 'Rio de Janeiro e Barcelona: Os Limites do Paradigmo Ollímpico', *Biblio 3W*, **XV**(895(17)). Retrieved on 15 June 2018 from http://www.ub.edu/geocrit/b3w-895/b3w-895-17.htm.

Harvey, D. (1989), 'From managerialism to entrepreneurialism: the transformation of urban governance in late capitalism', *Geografiska Annaler B*, **71**(1), 3–17.

Horne, J. and W. Manzenreiter (2017), 'The production of the spectacle: conceptualising labour and global sports mega-events', in: N.B. Salazar, C. Timmerman, J. Wets, L. Gama Gato, and S. Van den Broucke (eds), *Mega-event Mobilities*, Abingdon: Routledge, pp. 33–51.

IOC (2016), *IOC Marketing Fact File*, Lausanne: International Olympic Committee.

Lauermann, J. (2014), 'Competition through interurban policy making: bidding to host megaevents as entrepreneurial networking', *Environment and Planning A*, **46**(11), 2638–2653.

Lauermann, J. and M. Davidson (2013), 'Negotiating particularity in neoliberalism studies: tracing development strategies across neoliberal urban governance projects', *Antipode*, **45**(5), 1277–1297.

Matheson, V. (2009), 'Economic multipliers and mega-event analysis', *International Journal of Sport Finance*, **4**(1), 63–70.

McCann, E.J. (2011), 'Urban policy mobilities and global circuits of knowledge: toward a research agenda', *Annals of the Association of American Geographers*, **101**(1), 107–130.

Müller, M. (2015), '(Im-)Mobile policies: why sustainability went wrong in the 2014 Olympics in Sochi', *European Urban and Regional Studies*, **22**(2), 191–209.

Porter, L., M. Jaconelli, J. Cheyne, D. Eby and H. Wagenaar (2009), 'Planning displacement: the real legacy of major sporting events', *Planning Theory & Practice*, **10**(3), 395–418.

Preuss, H. (2004), *The Economics of Staging the Olympics: A Comparison of the Games, 1972–2008*, Cheltenham, UK and Northampton, MA, USA: Edward Elgar Publishing.

Ren, X. (2008), 'Architecture and nation-building in the age of globalization: construction of the national stadium of Beijing for the 2008 Olympics', *Journal of Urban Affairs*, **30**, 175–190.

Roche, M. (2000), *Mega-events and Modernity*, London: Routledge.

Santos, M. (1996), *A Natureza do Espaço*, São Paulo: Hucitec.

Shields, R. (1997), 'Flow as a new paradigm', *Space and Culture*, **1**(1), 1–7.

Shin, H.B. (2009), 'Life in the shadow of mega-events: Beijing Summer Olympiad and its impact on housing', *Journal of Asian Public Policy*, **2**, 122–141.

Silvestre, G. (2013), *An Olympic City in the Making: Rio de Janeiro Mega-event Strategy 1993–2016*, Lausanne: IOC Olympic Studies Centre.

Simonsen, K. (2004), 'Networks, flows, and fluids – reimagining spatial analysis?', *Environment and Planning A*, **36**(8), 1333–1337.

Temenos, C. and E. McCann (2012), 'The local politics of policy mobility: learning, persuasion, and the production of a municipal sustainability fix', *Environment and Planning A*, **44**(6), 1389–1406.

Urry, J. (2002), *Global Complexity*, Malden, MA: Polity Press.

Waitt, G. (1999), 'Playing games with Sydney: marketing Sydney for the 2000 Olympics', *Urban Studies*, **36**, 1055–1077.

Watt, P. (2013), '"It's not for us": Regeneration, the 2012 Olympics and the gentrification of East London', *City*, **17**(1), 99–118.

Watts, J. (2015), 'The Rio property developer hoping for a $1bn Olympic legacy of his own'. *The Guardian*. Retrieved on 15 June 2018 from http://www.theguardian.com/sport/2015/aug/04/rio-olympic-games-2016-property-developer-carlos-carvalho-barra.

Zimbalist, A.S. (2015), *Circus Maximus: The Economic Gamble Behind Hosting the Olympics and the World Cup*, Washington, DC: Brookings Institution Press.

PART IV

GEOGRAPHIES OF PLACES

23. Climate change, Gaia and the Anthropocene

Simon Dalby

GLOBALIZATION'S GEOLOGICAL DIMENSION

Until recently most discussions of globalization did not include geological matters beyond concerns about how mining, resources and fossil fuel supplies affected economic development in various regions of the world. Mostly, globalization has been about political changes and challenges to traditional notions of sovereignty. It has been about economic growth and contentious discussions of the consequences of trade agreements. Globalization has included environmental themes, but often as a side effect, an afterthought to the supposedly all-important matter of economic growth. It is something to be dealt with under the rubric of sustainable development. It often involves technical fixes to the more obvious pollution problems, and innovative production methods using new techniques and products that provide markets for global corporations, an *Environmentalism of the Rich* in Peter Dauvergne's (2016) terms. Ecological modernization, as these processes are often termed (Mol, 2003), are just one more part of the progress of corporate power driven by technological innovation. More recently questions of climate adaptation and development have impinged on the discussion, particularly in terms of energy supplies for states in the Global South and the possibilities of food shortages or more extreme events disrupting economic life.

The sense of living in a single place, a matter of 'globality' is tied up with both the growth of the global brands and their presence globally because of enhanced communication, satellite TV broadcasts, the Internet and the rise in global sporting events. We are all in at least some senses participants in these global performances, citizens of the world up to a point. Very little of this discussion has seriously engaged with the global scale of physical and biological transformations that have been caused by the global economy which ties ever larger urban populations into a world-wide set of commodity chains, transportation links and complicated logistical arrangements. Many of these are at the expense of poor and marginalized populations in the Global South, who suffer the slow violence of dispossession and polluted and degraded lands (Nixon, 2011).

But the relative silence about the larger context in which globalization has taken place, and the transformations it has set in motion, is changing. Climate change disruptions in particular have increasingly caused alarm among political elites, not least the national security consequences to the United States in particular of disruptions and the fear of 'climate refugees' as a looming political problem (CNA, 2014). That said, the economic response to fear of food shortages and climate disruptions has in part accelerated processes of globalization as states and corporations try to ensure supply sources across the globe in the face of potential future disruptions (Funk, 2014).

This in turn has accelerated the processes of rural transformation that are a key part of globalization, and a matter of dramatic environmental change that is part of the climate change story too on what ecologist George Woodwell (2016) simply calls a plundered planet.

Many of these themes are not new. Some of them are part of the discussion of the Gaia Hypothesis that dates from the 1970s. Earlier fears of a world transformed by both nuclear war and population outstripping food supplies on a polluted world fed into the formulation of what became the environmental movement. But as the later sections of this chapter elaborate, what is new in this discussion is the growing awareness of how humanity is changing a dynamic planet, so much so that earth system scientists are now discussing matters in terms of a new geological epoch, the Anthropocene. Globalization is much more than an economic, political and cultural matter; its processes are changing the biophysical context for life itself (Bonneuil and Fressoz, 2016). As such it is posing profound questions of who will decide what kind of a planet we produce for future generations, and which species will reside with us now that globalized humanity has become a geological scale force shaping the planetary future.

THE GAIA HYPOTHESIS

Concerns about our global fate and the consequences of the economic growth are not new. Discussions of the significance of the nuclear bomb, and the thermonuclear weapons that followed in the 1950s shaped a new sense of endangered globality and the need for global governance (van Munster and Sylvest, 2016). This was in part a matter of the direct fears of nuclear war and the direct destruction that would result if these bombs were used to destroy major cities. But related to this was a concern about nuclear fallout, the radioactive particles that are lofted into the atmosphere by nuclear explosions, and then dispersed by winds around the world. Tracking the course of nuclear fallout from bomb tests in the 1940s and 1950s led both to widespread fears of global contamination and to a growing appreciation of the interconnectedness of the earth's ecological systems (Masco, 2015). These concerns also led to a campaign in the 1950s and 1960s to stop nuclear testing precisely because it was becoming clear that strontium and other radioactive isotopes were showing up in humans around the world. In part this provided the impetus of the partial nuclear test ban treaty of the early 1960s which caused the United States and the Soviet Union to move their nuclear tests underground so that most radioactive particles remained buried.

Research into the effects of weapons and fallout coupled to the development of the ability to launch satellites into orbit, and use their remote sensing capabilities to measure earth in new ways, as well as provide much more accurate weather forecasts, enhanced geophysical research. Once coupled to an understanding of the mechanisms of plate tectonics that caused continental drift, a dynamic interconnected view of the planet emerged which has become the basis for more recent investigations into what is now called earth system science. The 1960s discussions of environmental crises and food shortages related to famine in particular were dramatically enhanced by the NASA photographs of the earth from space, images that suggested a unique but vulnerable 'blue marble' against a backdrop of inhospitable black empty space. This in turn fed into a larger discussion of the limits to growth (Meadows et al., 1972) and the first United Nations Conference on the Human Environment in Stockholm in 1972 where the unofficial report to the meeting posed environmental matters as a case of their being *Only One Earth* (Ward and Dubos, 1972).

The space programme also raised the question of whether the earth really was unique in terms of being the only planet that had life. Probes to Mars in the 1970s were in part justified by the question of whether there was life there too. One of the consultants on this question was James Lovelock, a British scientist and inventor who developed instruments to, among other things, remotely measure ultra violet radiation on earth. Posing the question as to what biological byproducts life might generate, which a probe to Mars should look for, raised the larger issue of what traces life had left elsewhere. Clearly oxygen is one such substance, and its absence in the Martian system, suggested to Lovelock that life similar to what is on earth was unlikely to be found on the surface of Mars. Needless to say this was not the answer NASA wanted, given that this undercut an important rationale for going to Mars, but the argument led to what became the Gaia Hypothesis.

Very crudely summarized the argument is that if viewed from space then the very odd atmosphere this planet has suggests life has to be present. The atmosphere is odd in the sense that it is clearly very far from chemical equilibrium. That being so some mechanism must be generating the very high levels of highly reactive oxygen in the atmosphere, and Lovelock's simple conclusion was the obvious one that life itself partly shapes the planetary system (Lovelock, 1979). Further than this his research went on to suggest that life itself has kept the planetary system broadly within conditions that facilitate life. It is loosely a self regulating system, one that as a result of chemical and biological feedback mechanisms, has kept temperature in particular, and such things as the salinity of ocean waters, within a convenient range for life to exist. The atmosphere currently keeps the planet warmer than it would be without the presence of trace greenhouse gases. Anthropogenic climate change is a problem precisely because burning fossil fuels at the rate the global economy is doing is dramatically changing the composition of the key gases, carbon dioxide and methane, that effectively regulate planetary temperature.

GAIA AND PLANETARY MANAGEMENT

In using the term Gaia, suggested to him by novelist William Golding, an ancient Greek term for the earth, and for a goddess who brings order from chaos, as well as the root of the modern English term geo as in geology and geography, Lovelock had a term that suggested that the planet was in some sense alive. In further talking about the physiology of the planet the imputation was of a single organism, analogous to an animal or even a human that requires medicine in present times (Lovelock, 1988, 1991). This was a hugely evocative suggestion, one that simultaneously inspired many environmentalists to use the term in their advocacy campaigns and annoyed many scientists because the metaphor often implied intentionality or conscious control of the system by some entity; a suggestion that was much more theological than natural science usually allows. Nonetheless biologist Norman Myers ([1984]1993) edited

Gaia: An Atlas of Planet Management – a popular, vividly illustrated, volume that used the metaphor of Gaia to put numerous environmental problems and their potential solutions into a global change framework just as the notion of 'sustainable development' was becoming popular in the 1980s (World Commission on Environment and Development, 1987). The second edition of the Gaia Atlas appeared the year after the Earth Summit in Rio de Janerio in 1992, warning of the need for a transition to much more sustainable modes of living, a new earth ethic, if disastrous collapse of key earth systems was to be avoided.

More recently James Lovelock has returned to the Gaia theme arguing in strident terms that just such a disastrous collapse is becoming more likely as the trajectory of globalization continues apace. In *The Revenge of Gaia* (Lovelock, 2006) he warns that the earth is effectively fighting back against the depredations inflicted on her. The earth has a fever induced by human activity and this warming might be disastrous for humanity: 'we are now so abusing the Earth that it may rise and move back to the hot state it was in fifty-five million years ago, and if does most of us, and our descendants, will die' (Lovelock, 2006: 1). Lovelock is never one to mince words. He notes the objections to considering the world as alive, but insists this framework allows the seriousness of the current situation for humanity to be clearly understood. Three years later at the age of ninety he issued what he called a final warning about *The Vanishing Face of Gaia* (Lovelock, 2009). Focusing on Britain in particular he suggested that rising temperatures, energy shortages and numerous refugees from parts of the world will be the future. Dealing with these problems may need a wartime mobilization, rationing, and the suspension of democracy. Internationalism, he suggests, will not solve the problem and island havens, such as Britain, will have to adapt as best they can, not least because large parts of the planet may degenerate into warring fiefdoms.

This dystopic vision of the future turned out to be not quite his 'final' word on the subject. Five years later he returned to these themes reflecting on a life in science and on the possibility that humanity might become the intelligent part of Gaia, and find ways to avoid the worst excesses of global warming. Doing so requires us to recognize both that life is a key player in the earth system and that humanity is involved in a process of accelerated evolution whereby we have changed from being mobile, middle-sized, grassland-dwelling apes into hive-dwelling beings who have dramatically altered their habitat. In terms of other species our artificial habitats, that we frequently call cities, are a new evolutionary form of living. Rejecting calls to try to geo-engineer the planet by such things as injecting sulphate aerosols into the atmosphere to cool the planet (see Hamilton, 2013), he suggests instead that we concentrate on air conditioning our hives, leaving the larger biosphere to adapt as it will. Clearly we do not understand its operation anything like well enough to intervene in ways that will be guaranteed to be effective; humanity is in for *A Rough Ride to the Future* (Lovelock, 2014), but thinking carefully about what we need to do to adapt to our new circumstances gives us a reasonable chance of surviving, along with Gaia herself.

THE ANTHROPOCENE

Such concerns about dramatically altered future circumstances for humanity have heightened the attention paid to the growing science of earth systems, focused on climate change in part, but on a wider array of issues related to the physical transformations of the earth's atmosphere, biosphere and the oceans too. Fears that humanity is so disrupting its global habitat that we may be transcending various boundaries within which we know we can fairly safely live, have triggered a discussion of a 'safe operating space' for humanity (Rockström et al., 2009; Steffen et al., 2015). Much of this has crystallized around the idea that the cumulative changes that industrialization has wrought, have led to new global circumstances that require us to recognize that we live in a new period in the earth's history. In August 2016, a couple of months prior to the Paris Agreement on Climate Change coming into force in November that year, a working group of geologists concluded that there was indeed enough evidence of dramatic recent change in the earth system to declare that we were now living in a new geological epoch, what is now called the Anthropocene (Carrington, 2016).

Geologists are concerned to find a sedimentary marker, a suitable location for a golden spike to mark the end of the Holocene, the period of earth history which stretches back about 12 000 years to the end of the last ice age. One of the obvious markers in the sediments that have accumulated round the planet in the last few generations is the presence of artificial radioactive isotopes, nuclear fallout from the atmospheric bomb tests carried out with such enthusiasm by American and Soviet nuclear weapons designers in the 1950s in particular. Another potential marker is the emergence of 'plastiglomerates', combinations of plastic waste and other detritus that, when washed up on beaches around the world is starting to form what will eventually be a new form of sedimentary rock. Fly ash from high temperature carbon fuel burning is also widely dispersed, and, in addition, a product of the combustion that is causing climate change; this too works well as a marker of the new epoch. The rapid decline in global biodiversity caused by habitat loss, environmental disruption, pollution, fishing and hunting, leading to the sixth global extinction event in the planet's history is another obvious phenomenon that justifies a specification of present times as a novel geological period (Kolbert, 2014).

Lovelock frequently criticized conventional science for its failure to understand the dynamism and interconnectedness of the world. The assumption that life was a superficial phenomenon on a separate rocky substrate, and hence biology and geology could be dealt with separately has, now, with the emergence of earth system science been rendered much more difficult to sustain. The separation has never made much sense to paleontologists in particular. Limestone, coal and petroleum are products of the decay of previous life forms; life turned into rock quite literally. Now with the use of fossil fuels, humanity is – through the processes of combustion – turning rocks back into air, and in the process generating the greenhouse gases that are heating the planet. We use plenty of limestone too as building material, in concrete and fertilizer. These interconnections make it very evident that the earth is not a given habitat for humanity, instead it is constantly being remade by biological, chemical and physical processes. The point about the Anthropocene is that humanity has become a larger part of these processes, large enough so that we have already left a clearly discernable assemblage of things in the sedimentary record of the planet, one that will be there long after humanity has either died off or evolved into some as yet unknowable future entity.

The implications of this are profound for how we think about globalization. The Anthropocene makes it clear that traditional ideas of an environment, 'out there' parts of which humanity would be well advised to 'preserve' in some relatively unchanging state, has to be transcended by an understanding of humanity as a large-scale agent in remaking many parts of the earth system. Related to that is the insight drawn from geological investigations into the recent history of the planetary system, which makes it clear that the last 12 000 years of the Holocene constitute a highly unusual period of relative climatic stability. Whereas the preceding few hundred thousand years were marked by an ice age where glacial periods were interrupted by short periods of relatively warm climate (the so called 'inter-glacials') which repeatedly quickly gave way to more ice, the Holocene has been markedly different with its relatively stable climate. Now, with greenhouse gas levels already far higher than those for at least the last million years, humanity is living in a climate system that has no analogue state in the recent history of the planet. Our actions have already postponed the next ice age and possibly the one after that, and how we collectively shape the next stage of the global economy matters greatly in terms of what kind of climate system our descendants will experience.

STAGES OF THE ANTHROPOCENE

The question of when the Anthropocene might be said to have begun sheds light on the discussion of globalization from a new perspective. It does so because the consequences of human activities are seen to matter much more than has been recognized until recently. One of the most interesting arguments is that humanity may have inadvertently shaped its own future conditions back in the early stages of agriculture. William Ruddiman's (2005) investigations of historical levels of methane in the interglacial periods of the past led him to wonder if the higher levels of that gas in the mid Holocene period might have been enough to prevent the world being glaciated once again, as had been the pattern at the end of previous inter-glacials. Might the clearing of forests and the emergence of agriculture have generated just enough methane and carbon dioxide to prevent the return of the ice? If so then humanity has been changing the climate for a much longer time than recent concerns with industrial combustion as a source of greenhouse gases have considered. That said, the invention of agriculture did not generate greenhouse gases on anything like the scale that contemporary globalization does.

A second suggestion points to the colonization of the Americas by Europeans and the huge die off of human population that followed the introduction of smallpox and other diseases to those continents (Lewis and Maslin, 2015). The elimination of most of the population lead to rapid reforestation which, by the early seventeenth century, had slightly reduced the levels of carbon dioxide in the atmosphere, and might be responsible for the worst of the 'little ice age' in that century, one marked by climate disruptions, widespread food shortages and political instabilities in Asia as well as

Europe. The colonization of the Americas is easy to understand as the beginning of globalization when the economics of the planet became fully linked into what we call the modern world system. Until recently historians have not paid much attention to the ecological consequences of colonization, but that has changed and now we understand this as a profoundly disruptive process for natural systems, one that may even have caused climate change at the beginning of what we now understand as modernity (Hornborg et al., 2007).

The initial discussion of the start of the Anthropocene often suggested that the emergence of the steam engine, and the beginning of widespread use of fossil fuels for industrial power and locomotion, was the key moment (Steffen et al., 2007). The industrial revolution changed things, most obviously by beginning the, at first slow, and then accelerating use of fossil fuels which has caused much of the growth of greenhouse gases in the atmosphere. Steam engines both increased the productive capabilities of industry and greatly accelerated the economic interconnections around the world in the latter part of the nineteenth century. This in turn set what we now know as globalization in motion and simultaneously began the process of altering trace gases in the atmosphere. It is important to note that this period also saw the dramatic extension of cultivated landscapes and rapid deforestation of many places, as modern commercial agriculture became a global phenomenon thereby accelerating the human transformation of the terrestrial biosphere. It also gave rise to working-class movements based in the coalfields and transportation sectors that through their struggles for better conditions and political rights for all, greatly expanded the remit of democratic rule (Mitchell, 2011).

However, most recent discussions of the Anthropocene have focused on the period since the Second World War, now frequently called the Great Acceleration (McNeill and Engelke, 2016), when the global economy expanded very rapidly and the automobile culture of the consumer society became something close to a global aspiration. This has been driven (pun intended) by the expansion of the use of petroleum both as a fuel for cars and for commercial vehicles, airplanes and ships that have linked places together, greatly enhanced urbanization while simultaneously changing the atmosphere. The preoccupation with economic growth and the promise that a larger economy will at least eventually benefit everyone, has impelled ever larger financial institutions to invest globally and generated complicated international arrangements that are often collectively called globalization, even when they might perhaps be better termed neoliberalism, given the priority given to economic market measures over all other considerations.

The politics of this period have increasingly been matters of expert knowledge, technological innovation and the governance of international trade by obscure procedures and agreements, all of which are much less subject to veto by national populations, at least until the Brexit vote in Britain in 2016. The rise of the anti-globalization movement in the 1990s reflected this political reality as activists, union members and environmentalists tried to find ways to challenge at least some of the worst inequalities and environmental damage done by an increasingly neoliberal, market-driven set of governance priorities. Climate change itself has generated numerous new market mechanisms, green development funds, carbon offsets, ecological additionalities and certification procedures, which at least by the time of the Paris Agreement late in 2015 had failed to curb the seemingly inexorable increase in carbon dioxide and methane in the atmosphere.

THE FUTURE AS GEOSTORY

All of which has raised the crucial questions of what comes next. Are there modes of governance, perhaps something called earth system governance (Biermann, 2014) that can effectively grapple with the gap between what science suggests is necessary to keep the biosphere in something approximating the Holocene conditions that have been so conducive to humanity, and the current trajectories of ecological damage and climate change (Galaz, 2014)? Is a sustainable earth the next possible stage of the Anthropocene, and if so, what needs to be done by which institutions to shape a future that avoids James Lovelock's nightmare of a world analogous to what it was fifty-five million years ago when there were no ice caps on earth? Is a future of constrained fossil fuel energy use compatible with any forms of democracy? Can people reimagine themselves as citizens participating in collectively shaping the future, or are we stuck with our identities as consumers merely choosing between various brands of commodities that are presented to us as ecological friendly or somehow 'green'?

Above all else, can we abandon traditional modern notions of a nature, or environment 'out there' separate from our economic production processes? Can we think about production, and futures as something we make rather than as potential disasters caused by processes that, while predictable, are beyond human control? Is it possible to think of ourselves as earth, as participating in Gaia, and as the supposedly intelligent part of it, beginning to shape it so that humanity and the planet both flourish? Can the implicit cultural promise of globalization to make us all citizens of one world – rather than of particular competing sovereign territories – be fulfilled? And can this be done quickly enough to prevent disruptive environmental change that is beyond the coping capabilities of existing institutions? Indeed can we invent flexible arrangements that transcend the political logjams of the present in time to make the future in some sense sustainable? These are the questions the Anthropocene poses.

Bruno Latour (2017) has recently pondered these issues in a philosophical mode suggesting that the Anthropocene requires us to abandon the distinction between the natural and social sciences. Geology and history have now effectively been merged in the Anthropocene discussion. History that ignores the changing material context that has shaped, and been shaped by human action is not helpful. Neither can we consider the next stages of the geology of the planet without focusing on the decision-making process of humanity. Humanity is not the only force shaping the future, but in planetary terms, apart from tectonic forces, we are now the largest ecological force moving around more material than natural processes do now. We have become geomorphological agents as well as determining which species will succeed where and what the future climate likely will be (Ellis, 2011). Hence, we need to think in terms of 'geostory', combining the stories of planet and people into one narrative rather than separate modes of thinking.

All this may be pointless if a major asteroid strike or a huge volcanic event happens in coming decades and disturbs the climate. Assuming that it does not, then the rich and powerful parts of humanity are primarily shaping the future of the planetary system. They are doing so by how they choose to organize production and which energy systems they use to power the factories and cities of the next few decades. Whether land is understood as something to continue using for the monoculture production of single crops, or as a series of complex ecologies necessary to simultaneously buffer extreme events, act as a carbon sink and provide a food supply for urban humanity, matters in terms of both practical survival for rural peoples and as a matter for long-term climate trends.

The notion of geostory requires, Latour (2017) suggests, us to rethink ourselves as part of Gaia, perhaps to understand Gaia as now in a state of civil war over what kind of a future will be produced (see Conway, 2015). This has profound implications for the key conceptual distinctions that European imperialism and modernity was based on (Hamilton et al., 2015). The discussion of globalization has made clear that states can no longer pretend to be autonomous entities unaffected by activities elsewhere. Sovereignty is constrained, not absolute. The Anthropocene makes it clear that this is not just a matter of politics or economics, but that the future shape of the planetary system is dependent on an intelligent appreciation of our circumstances, ones increasingly of our own making.

The Paris Agreement on climate change is at best a first step in coming to terms with this realization. The process of getting to the agreement was painfully slow, and the plans for periodic 'stocktaking' exercises do not make for the kind of rapid progress that James Lovelock and most climate scientists think is necessary (Anderson, 2015). The compromise between a binding agreement and voluntary nationally determined contributions to the overall enterprise of keeping global temperatures within a few degrees of the pre-industrial levels, is altogether inadequate in terms of current fossil fuel consumption trajectories, but at last, with only a few notable exceptions, political leaders from all the major states are beginning to recognize that globalization is a profoundly physical process, not just one that is a matter of economics and politics somehow separate from the larger planetary context where those processes play out.

Above all, now it is slowly dawning on policy makers that unrestricted economic growth is not the solution to all human ills, and the assumption that growth is facilitated by wholesale appropriation of natural resources regardless of the long-term consequences, is quite literally unsustainable. The sense of the global revealed by nuclear fears at the beginning of the Great Acceleration is now being complemented by a much more sophisticated understanding of how interconnected the planet as a living entity actually is. Gaia, has it seems come of age, and globalization has come to mean very much more than it did in the immediate aftermath of the Cold War, now that we are finally starting to understand the significance of living in the Anthropocene.

REFERENCES

Anderson, K. (2015), 'Duality in Climate Science', *Nature Geoscience*, **8**, 898–900.
Biermann, F. (2014), *Earth System Governance: World Politics in the Anthropocene*, Cambridge, MA: MIT Press.
Bonneuil, C. and J-B. Fressoz (2016), *The Shock of the Anthropocene*, New York: Verso.

Carrington, D. (2016), 'The Anthropocene Epoch: Scientists Declare Dawn of Human Influenced Age', *The Guardian*, https://www.theguardian.com/environment/2016/aug/29/declare-anthropocene-epoch-experts-urge-geological-congress-human-impact-earth (accessed 15 June 2018).

CNA Military Advisory Board (2014), *National Security and the Accelerating Risks of Climate Change*, Alexandria, VA: CNA Corporation.

Conway, P. (2015), 'Back down to earth: reassembling Latour's anthropocenic geopolitics', *Global Discourse*, **6**, 43–71.

Dauvergne, P. (2016), *Environmentalism of the Rich*, Cambridge, MA: MIT Press.

Ellis, E.C. (2011), 'Anthropogenic transformation of the terrestrial biosphere', *Philosophical Transactions of the Royal Society A*, **369**, 1010–1035.

Funk, M. (2014), *Windfall: The Booming Business of Global Warming*, New York: Penguin.

Galaz, V. (2014), *Global Environmental Governance, Technology and Politics: The Anthropocene Gap*, Cheltenham, UK and Northampton, MA, USA: Edward Elgar Publishing.

Hamilton, C. (2013), *Earthmasters: The Dawn of the Age of Climate Engineering*, New Haven: Yale University Press.

Hamilton, C., C. Bonneuil, and F. Gemenne (eds) (2015), *The Anthropocene and Global Environmental Crisis: Rethinking Modernity in a New Epoch*, Abingdon: Routledge.

Hornborg, A., J.R. McNeill and J. Martinez-Alier (eds) (2007), *Rethinking Environmental History: World System and Global Environmental Change*, Lanham, MD: Altamira.

Kolbert, E. (2014), *The Sixth Extinction: An Unnatural History*, New York: Henry Holt.

Latour, B. (2017), *Facing Gaia: Eight Lectures on the New Climatic Regime*, Cambridge: Polity.

Lewis, S.L. and M.A. Maslin (2015), 'Defining the Anthropocene', *Nature*, **519**, 171–180.

Lovelock, J.E. (1979), *Gaia: A New Look at Life on Earth*, Oxford: Oxford University Press.

Lovelock, J.E. (1988), *The Ages of Gaia: A Biography of our Living Earth*, Oxford: Oxford University Press.

Lovelock, J.E. (1991), *Gaia: The Practical Science of Planetary Medicine*, Oxford: Oxford University Press.

Lovelock, J.E. (2006), *The Revenge of Gaia*, London: Allen Lane.

Lovelock, J.E. (2009), *The Vanishing Face of Gaia: A Final Warning*, London: Allen Lane.

Lovelock, J.E. (2014), *A Rough Ride to the Future*, London: Allen Lane.

Masco, J. (2015), 'The age of fallout', *History of the Present: A Journal of Critical History*, **5**(2), 137–168.

McNeill, J.R. and P. Engelke (2016), *The Great Acceleration: An Environmental History of the Anthropocene Since 1945*, Cambridge, MA: Harvard University Press.

Meadows, D.H., D.L. Meadows, J. Randers, and W.W. Behrens III (1972), *The Limits to Growth*, New York: Universe Books.

Mitchell, T. (2011), *Carbon Democracy: Political Power in the Age of Oil*, London: Verso.

Mol, A. (2003), *Globalization and Environmental Reform: The Ecological Modernization of the Global Economy*, Cambridge, MA: MIT Press.

Myers, N. (ed) ([1984]1993), *Gaia: An Atlas of Planetary Management*, New York: Doubleday.

Nixon, R. (2011), *Slow Violence and the Environmentalism of the Poor*, Cambridge, MA: Harvard University Press.

Rockström, J., W. Steffen, K. Noone, Å. Persson, F.S. Chapin, III, E. Lambin, T.M. Lenton, M. Scheffer, C. Folke, H. Schellnhuber, B. Nykvist, C.A. De Wit, T. Hughes, S. van der Leeuw, H. Rodhe, S. Sörlin, P.K. Snyder, R. Costanza, U. Svedin, M. Falkenmark, L. Karlberg, R.W. Corell, V.J. Fabry, J. Hansen, B. Walker, D. Liverman, K. Richardson, P. Crutzen and J. Foley (2009), 'Planetary boundaries: exploring the safe operating space for humanity', *Ecology and Society*, **14**(2): 32.

Ruddiman, W.F. (2005), *Plows, Plagues, and Petroleum: How Humans Took Control of Climate*, Princeton, NJ: Princeton University Press.

Steffen, W., P. Crutzen and J.R. McNeill (2007), 'The Anthropocene: are humans now overwhelming the great forces of nature?', *Ambio*, **36**(8), 614–621.

Steffen, W., K. Richardsson, J. Rockström, S.E. Cornell, I. Fetzer, E.A.M Bennett, R. Biggs, S.R. Carpenter, W. de Vries, C.A. de Wit, C. Folke, D. Gerten, J. Heinke, G.M. Mace, L.M. Persson, V. Ramanathan, B. Reyers and S. Sörlin (2015), 'Planetary boundaries: guiding human development on a changing planet', *Science*, **347**(6223): 10.1126/science.1259855.

van Munster, R. and C. Sylvest (eds) (2016), *The Politics of Globality Since 1945: Assembling the Planet*, London: Routledge.

Ward, B. and R. Dubos (1972), *Only One Earth*, Harmondsworth: Penguin.

Woodwell, G. (2016), *A World to Live In: An Ecologist's Vision for a Plundered Planet*, Cambridge, MA: MIT Press.

World Commission on Environment and Development (1987), *Our Common Future*, Oxford: Oxford University Press.

24. Globalization and the incremental impact on the security and defense sector

Soul Park[1]

INTRODUCTION

Globalization is a multifaceted phenomenon that has become a prominent feature of world politics in the post-Cold War era. In particular, the impact of globalization on the realm of international security has garnered heightened interest in the aftermath of the 9/11 terror attacks (Hughes, 2009: 34). While some have argued that greater interconnectedness and interactions have become a force for peace especially amongst the more industrialized states (Gartzke, 2007), globalization's link to darker aspects of political violence has evoked a sense of insecurity in the contemporary political context – 'globalization's shadow' (Devetak, 2008). Today, transnational forms of violence – criminal networks, terrorist organizations and private military contractors – have emerged to become an integral component in how we understand the new security environment. Moreover, such global trends have gone hand-in-hand with recent attempts to broaden the conception of security studies in world politics (Baldwin, 1997).

Globalization continues to be a difficult concept to define as scholars have focused on different aspects of it to explain various social and political phenomena (Scholte, 2008). Within the subfield of international relations (IR), globalization is broadly described as an:

> [O]ngoing *process* in which the world moves toward an integrated global society and the significance of national borders decreases. It thus calls into question the distinction between domestic and foreign relations. In this view, the living conditions of people and local communities have changed through globalization; distant events of all sorts have immediate consequences not only for states but for individuals' daily lives. (Emphasis in original, Zurn, 2013: 402; Held et al., 1999: chapter 1)

It is the broadening and deepening of interdependence of not just nation-states but other societal and transnational actors through growing cross-border interactions, linkages and political reorganizations (Keohane and Nye, 1989; Cerny, 1995). The contemporary process of globalization has brought about new forms of interconnectedness and the emergence of new security players in the international realm. In comparison to past periods, globalization includes features, such as transnational modes of production and greater flow of goods and services, that are deemed relatively unprecedented and novel (Chapter 7 in this *Handbook*).

[1] I would like to thank Kimberly Peh for providing invaluable research assistance.

The nature of globalization has challenged the traditional notions of national security and broadened our conceptualization of international threats in the post-Cold War era. According to the traditional IR paradigms, international security has been broadly conceived as a realm in which rational unitary actors strive to survive in an anarchical world (Waltz, 2000; Mearsheimer, 2001). Within the national security state framework, territorial states are the prime actors when it comes to political violence and the use of force. The international system remains a self-help world mired in security competition and inter-state relations have largely been framed in terms of military issues, such as great power wars and territorial defense (Ripsman and Paul, 2005).

Proponents of globalization have contested this traditional model in two fundamental ways. First, they argue that it has foundationally altered the way in which we conceptualize national security, especially the autonomy and capability of the state in the international system. The emergence of non-state actors in the globalized world have posed increasing pressures to the state-centric model. As such, it has produced a complex global marketplace for security services in which states do not have control of the use of force like they once used to. Second, a particular feature of globalization – the emergence of global networks of production (see Chapter 15 in this *Handbook*) – has brought about changes to the global arms industry; similar to other realms, weapons procurement and development are now conducted in an internationally collaborative manner. As a result, scholars argue that a self-sustaining autarkic defense industry is becoming much less feasible for a sovereign state due to the trends brought about by globalization.

I analyze the two themes in greater detail and parse out the causal logic underpinning the arguments put forth by the globalization theorists. In the process, I show that while the contemporary changes warrant greater attention they still come with qualifications. Recent developments have certainly broadened and deepened the interconnections between globalization and national security in unique ways. Nevertheless, states still remain the prime actors when it comes to the use of force. In all, the impact of globalization on issues of defense and national security remains an incremental process. Globalization is broadening rather than radically altering the state-centric model of international security. The rest of this chapter describes the two shifts that globalization has brought about in the realm of defense and armaments – the rise of the global marketplace for security services and the global production of arms.

GLOBAL MARKETPLACE FOR SECURITY SERVICES

The first major change that we witness is in the composition of the global marketplace for security services in the international system. Globalization has impacted the traditional notion of security and has fostered the re-conceptualization of the role and limits of the nation-state, the power balances between states captured in terms of relative material capability, and the environment in which it operates (Kirshner, 2006: 6). In particular, the more recent economic, political and societal interdependence has brought about two interrelated shifts to the global marketplace for security services: (1) the broadening of the modes of conflict; and (2) the emergence of new security opportunities and actors.

The first major shift is the broadening of the modes of conflict. With the rise of globalization, increasing cross-border interactions have resulted in the growth of transnational resource bases and networks for political mobilization and contestation (Adamson, 2005). As such, globalization has fostered a process of de-territorialization (or supra-nationalization) of the state and has also provided non-state actors with the transnational base to initiate political violence. Such trends have led to the reconfiguration of 'social space away from, and beyond notions of, delineated territory, and *transcends* existing physical and human borders imposed upon social interaction' (emphasis in original, Hughes, 2009: 37). Furthermore, globalization has affected what is referred to as 'the geography of conflict' as political violence has become more prevalent in regions where sovereignty is weakest – failed states – and the international system has witnessed an increase in the number of low-intensity conflicts, such as civil wars and counter-insurgency operations (Gleditsch et al., 2002). Subsequently, different types of political violence can no longer be fully captured by the national security framework in the post-Cold War era.

Greater transnational linkages have increased the risks of conflict between different types of groups both across and within states, especially the ones characterized by issues of identity and ethnicity (Gleditsch, 2007; Olzak, 2011). Such issues have served as a source of domestic political mobilization and, at times, a tool to weaken state capacity for certain segments of society (Neuman, 2010). As ethnicity has emerged as a new source of conflict, the trend has moved from one of instrumental to that of expressive violence (Coker, 2002: 40). Subsequently, conflicts that have often been regarded as residuals of the Cold War great power dynamics are argued to be better understood within the broad framework of globalization: '[t]he process of globalization carries implicit homogenization tendencies and messages, which in combination with the "borderlessness" of the globalization phenomenon elicits a cultural pluralist response' (Cha, 2000: 394–395). The rise of religious extremism in recent times in response to the diffusion of liberal democracy represents one example of such tendencies.

Second, globalization has created a complex marketplace of global security that accommodates and accounts for the new opportunities and the emergence of non-state (or transnational) actors in the international system. The conclusion of the Cold War signaled the end of the stable bipolar world. And as the US–Soviet Union superpower rivalry seemed to have come to a close, so did a period often referred to as 'hypermilitarization' (Singer, 2001/02: 193). Downsizing of military personnel and equipment in its aftermath, the changing character of warfare (Kaldor, 1999), and the sophistication of weapons technology has led to the commercialization (or privatization) of the military industry (Shearer, 1998: 26–31; Neuman, 2010: 110–111). Consequently, globalization and security issues have become more intertwined in recent years and, as a result, this process has led to the blurring of private and public sectors.

Such a transformation has created 'increasingly complex conditions under which international actors exercise power' (Kay, 2004: 11). States traditionally held the monopoly of the legitimate use of violence within its recognized territory (Tilly, 1985). In the contemporary period, the issue of global governance and the struggle for the control of the use of force has become highly contested. This trend is especially evident in weak states where sovereignty is contested and has led to increased political violence (Avant, 2006: 129). Thus, non-state actors, often neglected in the traditional literature, have become prominent players in understanding the security dynamics in the globalized world of today; in recent years we are witnessing 'a proliferation of military, quasi-military and increasingly militarized civilian actors beyond the traditional state-based armies or military forces' (Cheeseman, 2008: 34).

One key example is the link between globalization and the rise of transnational terrorism. Greater levels of interconnectedness of goods, services and people have served as critical conditions for the rise of transnational terrorism and associated illicit networks: '[t]he current wave of international terrorism ... not only is a reaction to globalization but is facilitated by it' (Cronin, 2002/03: 30). These terrorist organizations exist and thrive within the networks of globalization or the 'nexus of networks' (Russell, 2008: 3). Another example would be the illicit trade (or trafficking) of small arms and light weapons (SA/LW) to non-state actors both on the 'black' and 'grey' markets. The network form of economic organization has served as the main vehicle for the spread of these weapons to various groups competing for political influence. An estimated 1.4 million SA/LW are under possession by non-state actors/militias while criminal gangs hold anywhere between 2 to 10 million (Bourne, 2013). However, networks of globalization are also available to states and are not the sole purview of transnational actors as the literature often portrays it to be. As a matter of fact, governments have also been active in similarly exploiting the network system to counter such threats by relying on more 'flexible' modes of response (Eilstrup-Sangiovanni, 2005).

Another example would be the proliferation of Private Military Companies (PMCs) in the post-Cold War era, such as Dyncorp International, Aegis Defence Services and Academi.[2] Viewed in a longer historical context, the use of force by non-state actors or mercenary activities is well documented and is perhaps as old as war itself (Thompson, 1994). PMCs oftentimes do not qualify the legal orthodox definition of mercenaries and usually tend to cover a 'grey' area today (Shearer, 1998: 16–21). Nevertheless, by embodying a 'corporate' nature that is well-suited to the contemporary business world, they have exploited the global security services market which is estimated to be in excess of 100 billion US dollars annually (in 2010) and projected to grow (Singer, 2008: 78). Activities by PMCs in African states in the 1990s as well as in Iraq and Afghanistan the past decade have caught the attention of the public. For instance, in the summer of 2007, there were more private contractors (190 000) than US troops in Iraq (160 000) (Cotton et al., 2010).

Their very existence and the role such companies play in the global security market have not only been viewed with moral pessimism but also challenged the state-centric theories of IR (Pingeot, 2012: 18–21; Mandel, 2002). By making a diverse array of security provisions more accessible to various domestic and international clients, these PMCs are able to provide an extension of foreign-policy tools to great powers, particularly the US. In other words, military services, ranging from combat and

---

[2] This view goes parallel with the argument that strategy is now being more globalized (Guehenno, 1998/99).

military training to support provision, are now being internationally outsourced to global corporate entities, similar to other realms of globalization.

While PMCs have come in to exploit the market opportunities created by the disjuncture between the international system and national interest of states, their goals and business interests, for the most part, remain largely aligned to those of states. While they thrive in regions of the world where sovereignty and the control of the use of force are contested, many of these companies work for their respective governments and are considered an extension of their foreign policies (Avant, 2006: 113–116). PMCs vie for government contracts and work side-by-side with the national armed forces. For instance, Military Professional Resources Incorporated's (MPRI) involvement in the Bosnian War in the 1990s is a plain example of private contractors working alongside governments rather than against it. Thus, though they may be run in the private sector, the question as to whether they are able to run independently without the support of national governments still looms. In cases where these companies acted against the national interests of the state, governments quickly stepped in and regulated their actions.[3]

In all, advances of globalization have led to changes in how we conceptualize the marketplace for global security services; greater interplay between the public and private sectors pertaining to the provision of security has rendered today's security environment much more complex. In particular, the mode of conflict has now been broadened to include different levels of political violence and not just inter-state wars. Similarly, the new opportunities and the subsequent emergence of transnational actors have changed the composition of the global security market both in terms of supply and demand. The increasing privatization of security has also seen many of the security provisions now being provided by corporate entities rather than national governments. For example, Halliburton has provided various non-combat services to US troops as part of the Logistics Civil Augmentation Program (LOGAP). Yet, these actors have, by and large, conducted their businesses in parallel to state interests rather than challenging them.

GLOBAL PRODUCTION OF ARMS

The second key shift we observe is the changes that are occurring to the defense industry – the rise of the globalization of arms production. As discussed above, globalization has brought about changes to both the interests and capabilities of states while rendering the security environment much more complex than it once used to be. Yet, global arms trade in the post-Cold War era has not shown a decline. On the contrary, much of the regions outside of Western Europe have shown an increase in military expenditure (Tan, 2010). In other words, there seems to be more continuity rather than radical change when it comes to the development and production of arms and weapons.

---

[3] Controversial companies such as Executive Outcomes and Sandline in the 1990s have closed down operations soon thereafter due to restrictions and pressures from their respective governments in South Africa and Great Britain.

What has changed in the last couple of decades is weapons technology itself, as it has become much more sophisticated and costlier to produce advanced weapons in the modern era (Skons and Wulf, 1994). Technological development, a key aspect of globalization, has led to an increase in both the unit cost of production of the weapons system itself and the subsequent fixed ratio costs of successive projects associated with it (Kirkpatrick, 2004). As a result, an indigenous, single-country model of the defense industry in face of globalization currents is no longer sustainable, given that arms production and development has internationalized and, thus, taken a much more transnational character (Bitzinger, 1994: 170).[4] The process of globalization has fostered a model of arms production that is based on international intra-firm division of labour and growing integration of international production networks (Brooks, 2005). Collaboration of arms production between firms based in different countries has served as a means through which states have attempted to consolidate their defense industries.

What is now much more evident is that economic motivations have emerged as a key determining factor for the rise of this global production of arms (Bitzinger, 1999). To tackle some of the fiscal pressures due to military downsizing in the aftermath of the Cold War, governments have loosened their grip on weapons production and have relied on a trans-border process of production and procurement. Some measures include licensed production, co-production and co-development programs between firms from different countries (Taylor, 1990; Brooks, 2005).[5] This move away from an autarkic model is a trend that began in the mid-1980s and has intensified ever since. For instance, according to a study conducted by the US Department of Commerce in the early 1990s, '15,000 companies were identified at the subcontractor level, with 11,638 companies still serving as active suppliers to the prime contractors' for the production of just three weapons systems – the High-Speed Anti-Radiation (HARM) missile, the Mark-48 Advanced Capability (ADCAP) torpedo, and the VERDIN communication system (quote in Brooks, 2005: 89–91, 90). Today, internationalization of ownership and of supply chain via subcontracting and the pursuit of technological cooperation has become the new norm (Hoyt and Brooks, 2003/04: 131–132). It has also served as a more efficient and cost-effective means through which governments have tried to maintain the military and defense capacity in the post-Cold War era.

The move towards global production has resulted in two broad changes for the defense sector. First, the dividing line between foreign and domestic defense firms has become somewhat blurred in recent times (Guay, 2007). In addition to cross-border acquisitions and mergers between international firms, the weapons production chain normally integrates companies from multiple countries into the process. Subsequently, it is argued that advances of globalization have challenged the once dominant autarkic

---

4 Internationalization refers to 'a broad concept covering a range of different company activities involving actors in more than one country. These activities can take different forms, including foreign trade, foreign investment, cross-border mergers and acquisitions, international subcontracting, international licensing, international joint ventures, and international inter-firm agreements' (Skons and Wulf, 1994: 44).
5 Licensed production refers to 'the transnational sale or transfer of the rights to manufacture a weapon system that was originally developed in the supplier's country' and co-development programs refer to joint efforts by firms in different countries to develop weapons systems (Bitzinger, 1994: 175–176; Brooks, 2005: 82–83).

model of arms production for sovereign states; no longer can the production of weapons be conceived as occurring solely in one country. Second, the lines of demarcation between firms in the civilian and defense sectors have also become further blurred with the global diffusion of dual-use technology (Bansak, 2011).

Increasingly, advances made in one sector affect the other as the military industry has undergone greater sophistication in the past few decades. As Stephen Brooks notes, 'this dynamic was the most important driving force behind the globalization of U.S. defense-related production during the 1970s and 1980s' (Brooks, 2005: 84–85). This is especially evident in what has been classified as weapons of mass destruction (WMD), such as nuclear and biochemical weapons. Materials utilized in developing these weapons often pose a dual-use dilemma, where 'the facilities, material, and knowledge used for peaceful purposes such as bio-medical research and pharmaceutical production can also be used for hostile purposes such as biological warfare and bioterrorism' (Koblentz, 2010: 115–116). The process of globalization has exacerbated the dual-use dilemma further as the spread of biochemical and other lethal materials and know-hows have become even more accessible to a wider set of transnational actors due to the changing security environment. Such rapid pace of development has also heightened fears of the spread of WMD by non-state actors through illegal networks and black markets (Hastings, 2012). Similarly, greater accessibility further signals the possibility of theft or illicit trafficking by potentially hostile non-state actors. As reported by the International Atomic Energy Agency's Incident and Trafficking Database factsheet, there have been 2 889 confirmed incidents of unauthorized criminal activities and instances of theft from 1993 until 2015 (IAEA ITDB, 2016).

The effects of the global network on the defense sector have certainly brought about structural changes in arms production and development. Yet, such changes have not had a transformative effect on the global arms industry, or, at minimum, the impact has been uneven across different regions. With defense expenditures close to 600 billion dollars in 2016, the US, as the unipolar power in the international system, continues to provide the biggest and most lucrative market for the global arms industry. The overall size of US military expenditure is estimated to be more than three times that of the next military spender, China, and roughly equal to the next 14 countries combined (Taylor and Karklis, 2016). The US also owned a 34 per cent share of the global arms exports from 2013 till 2017 (Wezeman et al., 2018). Similarly, the US defense companies dominate the global market with seven of the top ten firms being based in the US (Guay, 2007: 7–9). Companies such as Lockheed Martin, Northrop Grumman, and Raytheon serve as prime contractors in the chain of weapons production and essentially produce monopolies that highly favor the home market and help the US wield greater influence abroad (Caverley, 2007). For instance, the US defense industry has undergone an expansion in the post-9/11 era with the war in Iraq and Afghanistan proving to be an opportunity for the private contractors. Overall profits of the defense industry rose, from an estimated US$144.6 (2001) to US$295 billion (2006), as more government contracts became available with the US pursuing an aggressive foreign policy under the Bush administration (Skons, 2008: 242–243).

Moreover, collaboration and cooperation of armament production and development at the regional or transatlantic level have not materialized as expected; the globalization process on weapons procurement seems to have hit a stumbling block in the past decade or so. For instance, the European defense industry remains fragmented and collaborative programs on weapons procurement have not taken off like in the economic sphere (Bitzinger, 2010). Furthermore, despite the globalizing effects on the production of armaments, even small- and medium-powers have come up with means through which they have managed to sustain their autonomy while being able to produce weapons at competitive prices (Devore, 2013). Rather than being dictated to by the forces of globalization, states have re-aligned their priorities and re-adjusted to the global market; the so-called second-tier arms producing states, such as South Korea, France, and Israel, have been guided by 'techno-nationalism' and 'appear willing to pay the premium for achieving high levels of autarky in arms acquisition' (Bitzinger, 2010: 218; Serfati, 2001). In the process, they have not only remained globally competitive but have also retained a large portion of their military autonomy.

In short, the technological innovation and information revolution brought about by globalization, have forced states to seek external opportunities for the development of its arms industry. Faced with global market pressures and international structural constraints in the aftermath of the Cold War, the arms industry has resorted to the transnational production and development of weapons technology. International collaborations between multinational firms have blurred both the foreign–domestic and civilian–defense divides that were regarded to have been the norm in the past. However, the trends within the global network of production still highlight the importance of states and the role they continue to play in the arms industry. Regional collaboration of weapons procurement has been much more incremental than originally projected as states have adjusted to the changing dynamics of globalization. This is well captured in the Netherland's decision to collaborate in the US-led Joint Strike Fighter jet program to develop and test the F-35-Lightening II rather than join a more European-based initiative (Mamadouh and Van der Wusten, 2011).

CONCLUSION

In all, the process of globalization has resulted in incremental changes when it comes to issues of national security. Developments in the global security environment and changes to the defense sector in the post-Cold War era all exhibit elements of both change and continuity. Greater interconnectedness and linkages have led to changes in the global market of security services with the broadening of the modes of conflict and the emergence of non-state threats and security providers. However, such changes in the global security provision do not signal the end of the national security framework in international politics. States continue to play a foundational role in the globalized security environment in the contemporary era. Similarly, the global network of production has altered the development and production of weapons technology; a trans-border process of production through inter-firm alliances and international subcontracting is much more prevalent today than in the past. While the global arms industry has undergone changes, states have by and large retained their self-sustaining mode of production in face of such globalizing trends. Globalization has not led to a convergence of the defense sector. Rather, it seems to be an uneven market as the US continues to dominate the defense sector both in terms of market share and weapons development. Moreover, military modernization, undertaken by many developing nations, largely seems to indicate the ability of states to readapt and readjust to the changing context of globalization in contemporary world politics.

Consequently, calls for the end of the centrality of nation-states in security affairs seem to be premature, or at least in need of qualifications, at this stage. Globalization may have resulted in the rise of new transnational actors (and associated threats) but these actors co-exist along with nation-states on the international stage; that is, non-state actors have not replaced states when it comes to national security issues. Similarly, in the defense sector, the development and production of weapons, while done much more globally than before, exist primarily in a market created by states; the network of production supplements the national security state rather than providing an alternative to it. Furthermore, the process of globalization has impacted different states in different ways, and states continue to pursue much more traditional security policies than commonly portrayed within the globalization literature. Thus, the national security framework still provides a baseline approach in understanding the global marketplace of security and dealing with the burdens of military modernization in the post-Cold War era. Sustained globalization can eventually reshape the national security establishments and the nation-state in the long run. However, given the context of international anarchy and the current developments related to national security, states should continue to play a foundational role in the contemporary globalized world.

REFERENCES

Adamson, F. (2005), 'Globalisation, Transnational Political Mobilisation, and Networks of Violence', *Cambridge Review of International Affairs*, **18** (1), 31–49.

Avant, D. (2006), 'The Marketization of Security: Adventurous Defense, Institutional Malfunction, and Conflict', in: Jonathan Kirshner (ed.), *Globalization and National Security*, London: Routledge, pp. 105–142.

Baldwin, D. (1997), 'The Concept of Security', *Review of International Studies*, **23** (1), 5–26.

Bansak, K. (2011), 'Biodefense and Transparency: The Dual-Use Dilemma', *Nonproliferation Review*, **18** (2), 349–368.

Bitzinger, R. (1994), 'The Globalization of the Arms Industry: The Next Proliferation Challenge', *International Security*, **19** (2), 170–198.

Bitzinger, R. (1999), 'Globalization in the Post-Cold War Defense Industry: Challenges and Opportunities', in: Ann Markusen and Sean Costigan (eds), *Arming the Future: A Defense Industry for the 21st Century*, New York: Council on Foreign Relations Press, pp. 305–333.

Bitzinger, R. (2010), 'Globalization Revisited: Internationalizing Armaments Production', in: Andrew Tan (ed.), *The Global Arms Trade: A Handbook*, London: Routledge, pp. 208–220.

Bourne, M. (2013), 'International Trafficking in Weapons', in: Philip Reichel and Jay Albanese (eds), *Handbook of Transnational Crime and Justice* (2nd edn), Los Angeles, CA: Sage, pp. 85–100.

Brooks, S. (2005), *Producing Security: Multinational Corporations, Globalization, and the Changing Calculus of Conflict*, Princeton: Princeton University Press.

Caverley, J. (2007), 'United States Hegemony and the New Economics of Defense', *Security Studies*, **16** (4), 598–614.

Cerny, P. (1995), 'Globalization and the Changing Nature of Collective Action', *International Organization*, **49** (4), 595–625.

Cha, V. (2000), 'Globalization and the Study of International Security', *Journal of Peace Research*, **37** (3), 391–403.

Cheeseman, G. (2008), 'Globalization and Military Force(s)', in: Richard Devetak and Christopher Hughes (eds), *The Globalization of Political Violence: Globalization's Shadow*, London: Routledge, pp. 27–45.

Coker, C. (2002), 'Globalization and Insecurity in the Twenty-First Century: NATO and the Management of Risk', *Adelphi Paper 345*, International Institute for Strategic Studies, Oxford: Oxford University Press, 2002.

Cotton, S., U. Petersohn, M. Dunigan, Q. Burkhart, M. Zander-Cotugno, E. O'Connell and M. Webber (2010), *Hired Guns: Views About Armed Contractors in Operation Iraqi Freedom*, Santa Monica, CA: RAND.

Cronin, A.K. (2002/03), 'Behind the Curve: Globalization and International Terrorism', *International Security*, **27** (3), 30–58.

Devetak, R. (2008), 'Globalization's Shadow: An Introduction to the Globalization of Political Violence', in: Richard Devetak and Christopher W. Hughes (eds), *The Globalization of Political Violence: Globalization's Shadow*, London: Routledge, pp. 1–26.

Devore, M. (2013), 'Arms Production in the Global Village: Options for Adapting to Defense-Industrial Globalization', *Security Studies*, **22** (3), 532–574.

Eilstrup-Sangiovanni, M. (2005), 'Transnational Networks and New Security Threats', *Cambridge Review of International Affairs*, **18** (1), 7–13.

Gartzke, E. (2007), 'The Capitalist Peace', *American Journal of Political Science*, **51** (1), 166–191.

Gleditsch, K.S. (2007), 'Transnational Dimensions of Civil War', *Journal of Peace Research*, **44** (3), 293–309.

Gleditsch, N.P., P. Wallensteen, M. Erikson, M. Sollenberg and H. Strand (2002), 'Armed Conflict, 1945–99: A New Dataset', *Journal of Peace Research*, **39** (5), 615–637.

Guay, T. (2007), *Globalization and Its Implications for the Defense Industrial Base*, Carlisle, PA: US Army War College.

Guehenno, J. (1998/99), 'The Impact of Globalisation on Strategy', *Survival*, **40** (4), 5–19.

Hastings, J. (2012), 'The Geography of Nuclear Proliferation Networks: The Case of AQ Khan', *Nonproliferation Review*, **19** (3), 429–450.

Held, D., A. McGrew, D. Goldblatt, and J. Perraton (1999), *Global Transformations: Politics, Economics and Culture*, Cambridge: Polity Press.

Hoyt, K. and S. Brooks (2003/04), 'A Double-Edged Sword: Globalization and Biosecurity', *International Security*, **28** (3), 123–148.

Hughes, C. (2009), 'Beyond Interdependence: Globalization, State Transformation and Security', in: Geoffrey Till, Emrys Chew and Joshua Ho (eds), *Globalization and Defense in the Asia-Pacific: Arms across Asia*, London: Routledge, pp. 34–49.

International Atomic Energy Agency (IAEA ITDB) (2016), 'IAEA Incident and Trafficking Database (ITDB): Incidents of nuclear and other radioactive material out of regulatory control – 2016 Fact Sheet', https://www-ns.iaea.org/downloads/security/itdb-fact-sheet.pdf (accessed 19 June, 2018).

Kaldor, M. (1999), *New and Old Wars: Organized Violence in a Global Era*, London: Polity.

Kay, S. (2004), 'Globalization, Power, and Security', *Security Dialogue*, **35** (1), 9–25.

Keohane, R. and J. Nye (1989), *Power and Interdependence* (2nd edn), Glenview: Scott Foresman.

Kirkpatrick, D. (2004), 'Trends in the Costs of Weapon Systems and the Consequences', *Defence and Peace Economics*, **14** (3), 259–273.

Kirshner, J. (2006), 'Globalization and National Security', in: Jonathan Kirshner (ed.), *Globalization and National Security*, London: Routledge, pp. 1–34.

Koblentz, G. (2010), 'Biosecurity Reconsidered: Calibrating Biological Threats and Responses', *International Security*, **34** (4), 96–132.

Mamadouh, V. and H. Van der Wusten (2011), 'The Footprint of the SF/F-35 Lightning II Military Jet in the Netherlands: Geopolitical and Geo-Economic Considerations in Arms and Arms Production', *L'Espace Politique* 2011-3, http://espacepolitique.revues.org/2124 (accessed 25 June 2017).

Mandel, R. (2002), *Armies without States: The Privatization of Security*, London: Lynne Reinner.

Mearsheimer, J. (2001), *The Tragedy of Great Power Politics*, New York: W.W. Norton.

Neuman, S. (2010), 'Power, Influence, and Hierarchy: Defense Industries in a Unipolar World', *Defence and Peace Economics*, **21** (1), 105–134.

Olzak, S. (2011), 'Does Globalization Breed Ethnic Discontent?' *Journal of Conflict Resolution*, **55** (1), 3–32.

Pingeot, L. (2012), *Dangerous Partnership: Private Military and Security Companies and the UN*, New York: Global Policy Forum.

Ripsman, N. and T.V. Paul (2005), 'Globalization and the National Security State: A Framework for Analysis', *International Studies Review*, **7** (2), 199–227.

Russell, J. (2008), 'WMD Proliferation, Globalization and International Security: Whither the Nexus and National Security', in: James Wirtz and James Russell (eds), *Globalization and WMD Proliferation: Terrorism, Transnational Networks and International Security*, London: Routledge, pp. 1–13.

Scholte, J.A. (2008), 'Defining Globalisation', *The World Economy*, **31** (11), 1471–1502.

Serfati, C. (2001), 'The Adaptability of the French Armaments Industry in an Era of Globalization', *Industry and Innovation*, **8** (2), 221–239.

Shearer, D. (1998), 'Private Armies and Military Intervention', *Adelphi Paper 316*, International Institute of Strategic Studies, London: Oxford University Press.

Singer, P. (2001/02), 'Corporate Warriors: The Rise of Privatized Military Industry and Its Ramification for International Security', *International Security*, **26** (3), 186–220.

Singer, P. (2008), *Corporate Warriors: The Rise of the Privatized Military Industry*, Ithaca: Cornell University Press.

Skons, E. (2008), 'The US Defence Industry after the Cold War', in: Andrew T.H. Tan (ed.), *The Global Arms Trade: A Handbook*, London: Routledge, pp. 235–249.

Skons, E. and H. Wulf (1994), 'The Internationalization of the Arms Industry', *The Annals of the American Academy of Political and Social Science*, **535** (1), 43–57.

Tan, A. (2010), 'The Global Arms Trade', in: Andrew Tan (ed.), *The Global Arms Trade: A Handbook*, London: Routledge, pp. 3–12.

Taylor, A. and L. Karklis (2016), 'This Remarkable Chart Shows How U.S. Defense Spending Dwarfs the Rest of the World', *The Washington Post*, https://www.washingtonpost.com/news/worldviews/wp/2016/02/09/this-remarkable-chart-shows-how-u-s-defense-spending-dwarfs-the-rest-of-the-world/?utm_term=.37dbc4a74d7d (accessed 26 April 2017).

Taylor, T. (1990), 'Defence Industries in International Relations', *Review of International Studies*, **16** (1), 59–73.

Thompson, J. (1994), *Mercenaries, Pirates, and Sovereigns: State-Building and Extraterritorial Violence in Early Modern Europe*, Princeton: Princeton University Press.

Tilly, C. (1985), 'War Making and State Making as Organized Crime', in: Peter Evans, Dietrich Rueschemeyer, and Theda Skocpol (eds), *Bringing the State Back In*, Cambridge: Cambridge University Press, pp. 169–190.

Waltz, K. (2000), 'Structural Realism after the Cold War', *International Security*, **25** (1), 5–41.

Wezeman, P., A. Fleurant, A. Kuimova, N. Tian and S. Wezeman (2018), 'Trends in International Arms Transfers, 2017', *SIPRI Fact Sheet*, https://www.sipri.org/sites/default/files/2018-03/fssipri_at2017_0.pdf (accessed 19 June 2018).

Zurn, M. (2013), 'Globalization and Global Governance', in: Walter Carlsnaes, Thomas Risse and Beth Simmons (eds), *Handbook of International Relations* (2nd edn), London: Sage, pp. 401–425.

25. Regions and clusters and the global economy
Franz Tödtling, Arne Isaksen and Michaela Trippl

INTRODUCTION

The post-war period has been characterized by a strong growth of economic inter-dependencies at a global level (Dicken, 2015). National and regional economies and their industrial clusters are challenged by this development and forced to reposition themselves in the new global economy. Some regions – particularly core areas – have undergone successful innovation-based reconfiguration processes, while many old industrialized and peripheral regions have often lost employment and parts of their economic base. In this chapter, we discuss some conceptual approaches to regional aspects of globalization, types of regions and modes of innovation. We also provide examples of clusters located in different geographical contexts and investigate how they cope with innovation challenges and place-specific innovation barriers.

CONCEPTUAL CONSIDERATIONS

Definition of Key Terms

By 'globalization' we understand a shift of socio-economic interdependencies and institutional arrangements towards the global scale (Swyngedouw, 2004). Beside an increase of trade relations, growing financial flows, direct investment, transnational corporations (TNCs) and their organizational links, we find globally configured technology links and knowledge flows (Dicken, 2015).[1] There are also sociocultural interdependencies such as global migration and mobility, and rapidly increasing Internet-based social networks. Finally, there is a growing importance of global institutional arrangements and organizations such as WTO, IMF, and the World Bank (Dicken, 2015).

Regions represent one of several geographical scales involved in the process of globalization. We regard regions as sub-national territories for production, consumption and regulation. Due to factors such as specialized infrastructure (e.g. for transport, communication and education), trust, social capital and local culture, regions may constitute a synergetic space for the development of firms, entrepreneurship and innovation (Scott and Storper, 2003; Healy and Morgan, 2012).

---

[1] These global economic interdependencies have grown strongly over the past decades. For the past few years, however, one could observe a slowing down of this process (see, for example, the *Economist* from 28 January 2017).

Clusters are spatial concentrations of specific industries and related activities. Due to their specialization and synergetic complementarities (such as specialized suppliers, educational bodies and knowledge organizations) they may provide dynamic advantages for firms in global competition by improving factor and demand conditions, related industries, firm strategy and rivalry (Porter, 2008). Clusters are both locally rooted and globally linked (Maskell and Malmberg, 2007).

KEY CONTRIBUTIONS AND APPROACHES

Various scholars have analyzed the role of regions and clusters in the global economy, highlighting accompanying agglomerative tendencies (Scott and Storper, 2003), scale-shifts of power (Swyngedouw, 2004), and the relation between the positioning of regions in global production networks (GPNs) and their potential gains or losses from globalization (Coe and Hess 2011; see also Chapter 15 in this *Handbook*). Contrary to views that regions become less relevant as spaces for production and regulation due to globalization (Cairncross, 1997), regional economies can be regarded as synergetic spaces of physical and relational assets that have gained in importance (Morgan 2004), although the level of synergy may vary between regions as we will show. Regions are thus an essential dimension of the development process, not only in the more advanced countries but also in less-developed parts of the world. But it would be fundamentally mistaken to equate globalization with a simple spreading out of economic activity (Scott and Storper, 2003). On the contrary, globalization has been accompanied by increasing agglomerative tendencies in many different areas of the world. Cities appear as privileged sites for economic growth because they economize on capital-intensive infrastructure and permit significant economies of scale. Furthermore, there are dynamic backward and forward linkages of firms, the formation of large and diverse local labour markets and localized relational assets promoting learning and innovation effects (OECD, 2006).

Erik Swyngedouw (2004) suggests that the alleged process of globalization should be understood as a process of 'glocalization'. He argues that the pre-eminence of the 'global' in much of the literature marginalizes an ongoing socio-spatial struggle in which the re-configuration of spatial scale is a key arena. 'Glocalization' describes a scale shift of power both upwards to the supranational level and down to the sub-national level. Governments at various scales are challenged to align their social and economic policy. In response to the threat of (hyper)mobile capital, Swyngedouw (2004) argues that regional and national states try to initiate or restore a fertile entrepreneurial culture, create more flexible labour markets and minimize environmental and social regulation.

Coe et al. (2004) highlight the nexus between regional development and global production networks (GPN). They focus on the dynamic 'strategic coupling' of such networks and regional assets, an interface mediated by a range of institutional activities across different geographical scales. The argument is that regional development depends on the ability of regions and their clusters to tie into GPNs to stimulate processes of value creation, enhancement and capture. The creation of value involves efforts of regional institutions in attracting value-added activities by improving the skills of the local workforce, promoting start-up firms and supplier networks, facilitating venture capital formation and encouraging entrepreneurial activities. Value enhancement essentially includes knowledge transfer and industrial upgrading. Value capture concerns the issues of power and control that are critical for the distributional aspects of regional development within GPNs. Coe and Hess (2011), furthermore, stress the multi-scalarity of the forces and processes underlying regional development and argue against privileging one particular geographical scale over the other.

More recent studies look at innovation processes from a local and global perspective. A large body of work has been dealing with the 'globalization of innovation' (Bunnel and Coe, 2001). This literature stresses the role of non-local and international knowledge linkages, emphasizing the global scale in the innovation process. Scholarly work on GPNs (Yeung and Coe, 2015) and more recently, studies of global innovation networks (Cooke, 2013) have provided insights into the global organization and coordination of production and innovation activities and the co-creation, exchange and diffusion of knowledge across geographical space. This literature has challenged older versions of cluster and regional innovation systems approaches that have seen innovation as a result of regional knowledge interactions, disregarding distant and global knowledge interdependencies.

Nowadays, innovation systems are seen as open, internationally and globally connected phenomena. The regional innovation system (RIS) approach (Asheim and Gertler, 2005) emphasizes that in the globalizing economy, RIS should be understood as institutional contexts for innovation within globally configured knowledge and production networks. The innovation capacity of clusters and RIS is shaped by complex multi-scalar institutional settings and organizations (local, regional, national and international), and respective knowledge flows (Bathelt, 2011). Since different types of industries rely on different kinds of knowledge that vary in their transferability across space, one might expect that the patterns of innovation and the geography of knowledge sourcing activities differ significantly between types of industries (Asheim et al., 2011). In some industries innovation activities are strongly based on analytical and codified knowledge, which can be easily transmitted across space. Knowledge sourcing and innovation networks in such sectors tend to be dispersed or global in nature. Other industries rely more on synthetic and symbolic knowledge, which is more tacit in nature. Such knowledge is more often drawn from the local and regional levels (Asheim et al., 2011). Furthermore, regions differ in the local availability of knowledge (see the following section). Regions which host only a few firms and knowledge organizations will rely more on global knowledge flows than regions that are well populated with such sources (Tödtling et al., 2011).

To summarize, in the globalizing economy regions can serve as synergy spaces for achieving competitive advantages, improving innovation and policy coordination. Globalization affects regions differently, where in particular agglomerations and metropolitan regions represent key centres for economic development and innovation (Scott and Storper, 2003; Simmie, 2005). This implies that globalization shows distinct core and periphery patterns at the global, macro-regional (e.g. EU, North America) and national levels (Dicken, 2015). Some metropolitan regions serve as key nodes in global corporate networks, GPNs and innovation networks (Scott and Storper, 2003). Clusters and whole regional economies may benefit from 'strategic coupling'-processes within such networks (Coe and Hess, 2011), whereas 'de-coupled' or less innovative and competitive regions may get left behind and decline. Clusters can serve as tools for achieving competitive advantages through specialization, knowledge flows and complementarities (Porter, 2008). Our core argument, however, is that globalization enhances the need for innovation in regional economies in order to bring forward new products, processes, or new industry paths. But not all regions have the same preconditions for innovation (Simmie, 2005; Tödtling and Trippl, 2005), and also industries differ in their innovation pattern and in local–global innovation interactions as will be demonstrated below. Table 25.1 summarizes the local and non-local dimensions and factors of regional development in the globalizing economy.

Table 25.1   Local and non-local factors of regional development in a globalizing economy

Dimensions	Local factors	Non-local factors
Firms	Indigenous SMEs Industrial clusters Intra-regional markets Venture capitalists	TNCs and global corporations Business and financial networks Distant global markets Global production networks
Labour	Skilled and unskilled workers Locally mobile labour	Skilled experts and technologists Transnational migrants Transnational business elites
Innovation	Spill-over effects Tacit knowledge Infrastructure and assets	Intra- and inter-firm innovation networks Technological licensing Strategic alliances
Institutions	Conventions and norms Local growth coalitions Local authorities Development agencies	Global standards and practices National agencies and authorities Inter-institutional alliances Supranational organizations

Source: Own modification based on Coe et al. (2004: 471).

GLOBAL CHALLENGES AND INNOVATION CONDITIONS FOR DIFFERENT TYPES OF REGIONS AND INDUSTRIES

Which kind of challenges do regions face in the process of globalization? And which potential for innovation and adjustment can be observed in different types of regions and clusters? Isaksen and Trippl (2017a) suggest that both place-specific factors (such as relevant infrastructure, institutional settings, social capital and knowledge spill-overs) and factors at higher spatial scales (such as markets, global firms, GPN, strategic alliances and innovation networks) have an impact on firms' innovative activities and performances (Table 25.1). As argued below, the endowment with and the interplay between those factors vary strongly across regions and clusters.

TYPES OF REGIONS

Regions face different kinds of challenges and threats as well as potential benefits in the globalizing economy. Metropolitan regions and large cities seem to be more capable than other region types of coping with sectoral shocks, to innovate and to develop new growth paths. Metropolitan areas and advanced technology regions often have organizationally thick and diversified regional innovation systems (RIS) (Isaksen and Trippl, 2017a). These include a large number of firms in different industries and a critical mass of knowledge and support organizations that stimulate experimentation and innovation across a wide range of economic and technological fields. Due to well-developed transport and ICT infrastructures, firms in such regions may find it easier to couple with global production and innovation networks. Metropolitan regions, however, may also face problems such as substantial immigration of poor and unskilled people, high unemployment, social segmentation, and high demand on social infrastructure and related expenses due to globalization (OECD, 2006).

Old industrial regions in comparison face more severe problems since they are often specialized in vulnerable (labour- or capital-intensive) industries, which have seen relocation to and competition from emerging countries. Old industrial and specialized technology regions are characterized by organizationally thick and specialized RIS which host strong firms and clusters in a limited number of industries (Trippl and Tödtling, 2008). Knowledge and support organizations are often well-tuned to the region's narrow industrial base and further contribute to the dominant specialization. They are often 'locked into' their existing industrial and technology paths.

Peripheral regions tend to face similar problems, but in addition they are disadvantaged by a weakly developed transport, ICT and knowledge infrastructure and their firms face difficulties in tying into GPNs or innovation networks. They often have rather thin RIS characterized by a low number of innovative companies, few knowledge and support organizations and only weakly developed clusters (Isaksen, 2015).

INDUSTRY SPECIFICITIES AND INNOVATION MODES

Besides regional differences there are innovation specific modes and local–global patterns of innovation that differ by industries and clusters. Jensen et al. (2007) distinguish between the STI (Science, Technology and Innovation) and the DUI mode (Doing, Using and Interacting) of innovation. The STI mode relies on the exploitation of codified scientific and technical knowledge and the application of scientific methods and models. Investment in R&D and employment of scientifically trained workers are of key importance for firms and linkages to universities and other research organizations help to access relevant scientific knowledge. Both the sourcing of global scientific knowledge and the reliance on local 'sticky' knowledge seem to matter. Codified, scientific knowledge is highly mobile in geographical space and is transferred within international scientific communities, through scientific journals and other channels. Local knowledge is often transferred in the form of spin-offs from universities and research organizations or through branching from existing industries and knowledge bases and is much less mobile (Boschma and Frenken, 2011).

The DUI mode, in contrast, relies on experience-based know-how and more informal learning. Innovation activities are based on knowledge acquired on the job in the context of problem-solving activities. Often, novel combinations of existing knowledge are applied for which skilled workers, learning by doing and using, and learning by interacting are of central importance. The DUI mode is based on synthetic and symbolic knowledge that is more often tacit in nature when compared with the analytical knowledge base of STI industries. Synthetic and symbolic knowledge is more difficult to transfer across space, however. Knowledge, then, is exchanged through interactive forms such as user-producer relationships, labour mobility and collaboration with technology centres. Whereas the STI mode tends to prevail in research-intensive industries and clusters such as biotechnology, microelectronics or nanotechnology, the DUI mode dominates in more traditional industries as well as in cultural and creative industries (Isaksen and Karlsen, 2011). However, no 'pure' forms exist. Firms and industries hardly rely exclusively on just one of the two innovation modes but often integrate both modes. Empirical studies suggest that the combination of STI and DUI modes and of different knowledge bases is particularly favourable for innovation (Jensen et al., 2007).

Firms draw knowledge from various sources and geographical scales by using different kinds of knowledge linkages (Isaksen and Trippl, 2017a). Important knowledge carriers are the recruitment of qualified employees as well as of graduates from universities; cooperation with customers, suppliers, and research organizations; knowledge spill-overs in informal settings; the use of publications, the Internet and electronic discussion forums; and the purchase of knowledge that is embodied in equipment and services. These knowledge linkages are applied on different geographical scales. Knowledge flows through labour mobility and informal spill-overs benefit from geographical proximity, whereas knowledge embodied in purchased goods and services, and knowledge acquired from publications and the Internet have a more global character (Bathelt, 2011). Trade fairs allow establishing temporary geographical proximity between actors.

LOCAL–GLOBAL INNOVATION INTERACTIONS: EXAMPLES FOR TYPES OF REGIONS AND CLUSTERS

In the following we will investigate the challenges, innovation patterns and local–global innovation interdependencies for selected industries and clusters in the three different types of regions. The industry examples will cover both the STI and the DUI mode of innovation. The evidence shows that both innovation modes have relevance for all three types of regions. This implies that also in specialized industrial and in peripheral regions one can find clusters that innovate along the STI mode. However, the patterns of innovation and knowledge interdependencies differ between the regional contexts as the following analysis demonstrates.

Diversified Metropolitan Regions

Diversified metropolitan regions are characterized by a heterogeneous industrial structure and a large variety of knowledge and supporting organizations that facilitate experimentation, R&D and innovation in a wide range of economic and technological fields (Scott and Storper, 2003). All kinds of knowledge bases and innovation modes, and a variety of industrial clusters, are well developed in such areas. The preconditions for facing competition in the global economy seem to be good in general. Metropolitan regions are often centres of radical innovation and of cluster growth in science- and technology-based industries (Simmie, 2005). This is due to a broad variety of universities and research organizations, knowledge spill-overs, labour mobility, diverse suppliers of inputs, and access to a large local market. These factors facilitate the combination of different types of knowledge and the emergence of new industries.

Metropolitan regions also host many DUI clusters. They are in particular key centres for cultural and creative industries such as advertising, music, fashion, design, film and television, new media and computer games. Such industries tend to cluster in metropolitan areas, while their products and services are often sold on the global market (Power and Scott, 2011). Metropolitan regions offer a fertile ground for these industries to emerge, grow, and innovate. These regions provide advantages of knowledge diversity, enabling firms to get access to ideas and complementary competences from a broad variety of sources. Geographical proximity supports the organization of temporary projects, which are common for these industries. Factors such as social diversity, openness and tolerance attract creative people and talent (Florida, 2003). Knowledge linkages in such industries and regions are often (but not exclusively) of a local nature (see Sinozic and Tödtling, 2015, for the new media cluster in Vienna). This reflects the large diversity of knowledge assets in such regions and the limited transferability of 'symbolic knowledge' that plays an important role for innovation in these industries (Asheim et al., 2011). Symbolic knowledge is highly place-specific and depends on sociocultural contexts. However, also non-local and even global knowledge connections and networks of co-productions and creative partnerships matter (Power and Scott, 2011).

Specialized Industrial Regions

Such regions have organizationally thick and specialized RIS hosting well-established clusters in a limited number of industries. Strong industrial specialization is often reinforced by knowledge and support organizations that are well tuned to the region's narrow industrial base. Such configurations are typical for old industrial areas and younger technology regions. Although these regions are well endowed with firms and specialized knowledge organization, they lack complementary knowledge and industrial variety that could drive regional renewal. Knowledge links to sources outside the region (such as R&D and innovation collaborations) are thus of vital importance for accessing complementary knowledge assets.

Old industrial regions are usually not associated with a high innovation performance but rather characterized by industrial continuity (Isaksen and Trippl, 2016), moving along well-established technological paths. In many cases, this has led to negative lock-in and decline (Hassink, 2010). These regions are home to traditional industries such as coal and steel, chemicals, textiles and shipbuilding that have experienced strong global competition and employment losses in the past decades. Specialization in one or few of those industries is often backed up by narrowly specialized knowledge and support organizations that are too much directed to the region's industrial base (Tödtling and Trippl, 2005). While many of these regions have suffered from the decline of their mature industries, some have been successful in stimulating economic rejuvenation by employing innovation-based restructuring activities. Several studies point to a modernization of mature industries based on the injection of new scientific and technological knowledge, reflecting a combination of DUI and STI innovation modes. There are various examples of such renewal processes like the introduction of new materials and technologies in the metal industry in Styria (Austria), and the integration of electronics, control and communication technologies in the industrial machinery sector in the Finnish region of Tampere (Trippl and Tödtling, 2008). Such an innovation-oriented restructuring of traditional sectors may take two directions. First, when firms are shifting to high-value added segments of the market, such as the move from mass-produced, standardized steel products towards steel specialities for niche markets in the case of the Styrian metal cluster (Trippl and Tödtling, 2008). Second, when firms diversify into new but related industries where they can make use of their accumulated knowledge and competences. Examples are branching processes from old steel and engineering clusters into environmental technology industries in the Ruhr area in Germany and in Upper Austria (Tödtling et al., 2014). Usually the knowledge assets required for such innovation and transformation processes can hardly be found within the regional context only but are often accessed through global knowledge linkages.

Younger knowledge-based industrial regions are often home to high-tech clusters. Such regions have excellent research universities within a restricted number of scientific fields. Technology-based clusters in such locations are therefore more dependent on global knowledge links than those in larger and more diverse metropolitan regions. The emergence of high-tech clusters may be driven by spin-offs from the universities and R&D organizations that may lead to cluster growth in specialized fields (Patton and Kenney, 2010). Such spin-offs are accompanied by significant knowledge flows through labour mobility, spill-overs, and research collaboration with the incubator organizations. There is a danger, however, that knowledge sourcing becomes too narrowly focused on those local knowledge organizations, hampering a more dynamic cluster development (Karlsen et al., 2011). It thus seems important to combine those narrow knowledge assets with complementary knowledge as well as with other resources (such as investment and risk capital) that often may come from global sources.

Peripheral Regions

Peripheral regions usually have organizationally thin RIS that have only a few innovative companies, a low number of knowledge and support organizations and only a few or weakly developed clusters (Tödtling and Trippl, 2005). Industries are often resource-based (e.g. related to agriculture and food, forestry, wood and paper, furniture and mining). Other sectors that innovate mainly along the DUI mode are found in these areas, such as machinery, mechanical engineering, construction and services. Due to a lack of knowledge organizations within the region, firms often have to source knowledge from non-local sources (Isaksen and Trippl, 2017a). This requires, however, that regional firms have the absorptive capacity to identify, acquire, and use such external knowledge. An illustrative example for innovation challenges in peripheral regions is provided by Isaksen (2015), who shows how smelter and aluminium firms in Lister (Norway) compensate the disadvantages of a thin regional innovation system by building up strong expertise internally and by maintaining extra-regional knowledge linkages with multinational corporations.

However, there are also cases of more technology-intensive sectors in such regions such as electronics, ICT and software that innovate along the STI mode. In general, peripheral regions lack good conditions for initiating new growth paths in science-based industries due to the absence of higher education institutes, weak clusters and low levels of local knowledge circulation. A recent study of the growth of electronics and software industries in two peripheral regions in Norway and Austria shows that the inflow of new knowledge through the arrival of research institutes, firms, and skilled labour from outside can initiate new industrial paths (Tödtling et al., 2011; Isaksen and Trippl, 2017b).

CONCLUSION AND RESEARCH AGENDA

The core argument made in this chapter is that in the globalizing economy regions and clusters can serve as synergetic space for achieving competitive advantage. However, regions and clusters have different roles in globalization processes. Some regions, in particular metropolitan regions and advanced technology regions, are dynamic economic cores and often include command centres in global production networks. Specialized industrial and peripheral regions are more often on the producing end of global networks and are then engaged in strategic coupling of regional assets in such networks. Globalization tendencies also enhance competition in many industries and thus the need of firms, clusters and regions to embark on innovation-based restructuring strategies. This chapter demonstrates that the preconditions for and challenges of such endeavours and the resulting pattern of local–global innovation interactions may vary widely, depending on the type of region under consideration as well as on industry-specific learning modes. Diversified metropolitan regions with a wide variety of knowledge assets can rely much on local knowledge circulation for innovation purposes. Firms in specialized industrial regions and peripheral regions have fewer opportunities to combine different local knowledge which forces them to source more knowledge from national and global sources.

However, more research is required to grasp in a more nuanced way how metropolitan, specialized and peripheral areas can successfully reposition themselves by modernizing existing STI and DUI clusters and by growing new industries. Particular attention should be given to enhancing our understanding of the reconfiguration of knowledge linkages at the local, national and global levels that underpin such economic renewal and innovation processes. The possibilities of regions to create new growth paths will also depend on the role of firms and clusters in global production and innovation networks. It may be relevant to distinguish between regions that contribute in global networks mainly with inherited production factors, such as natural resources, and regions that have more created production factors, such as skilled labour, knowledge organizations and social capital. An important question, then, concerns if such a division would reveal important differences in the opportunities of regions to take advantage of global investments and knowledge links for economic restructuring. Deepening our knowledge in these regards would also be important in order to inform innovation and industrial policies at various spatial scales.

REFERENCES

Asheim, B.T. and M.S. Gertler (2005), 'The geography of innovation: regional innovation systems', in: J. Fagerberg, D.C. Mowery and R.R. Nelson (eds), *The Oxford Handbook of Innovation*, Oxford: Oxford University Press, 291–317.

Asheim, B.T., J. Moodysson and F. Tödtling (2011), 'Constructing regional advantage: towards state-of-the-art regional innovation system policies in Europe?', *European Planning Studies*, **19**(7), 1133–1139.

Bathelt, H. (2011), 'Innovation, learning and knowledge creation in co-located and distant contexts', in: A. Pike, A. Rodriguez-Pose and J. Tomaney (eds), *Handbook of Local and Regional Development*, Abingdon, Oxon: Routledge, 149–161.

Boschma, R. and K. Frenken (2011), 'Technological relatedness, related variety and economic geography', in: P. Cooke, B. Asheim, R. Boschma, R. Martin, D. Schwartz and F. Tödtling (eds), *Handbook of Regional Innovation and Growth*, Cheltenham, UK and Northampton, MA, USA: Edward Elgar Publishing, 187–197.

Bunnel, T.G. and N.M. Coe (2001), 'Spaces and scales of innovation', *Progress in Human Geography*, **25**(4), 569–589.

Cairncross, F. (1997), *The Death of Distance*, Cambridge, MA: Harvard University Press.

Coe, N.M. and M. Hess (2011), 'Local and regional development – a global production network approach', in: A. Pike, A. Rodriguez-Pose and J. Tomaney (eds), *Handbook of Local and Regional Development*, Abingdon, Oxon: Routledge, 28–138.

Coe, N.M., M. Hess, H.W. Yeung, P. Dicken and J. Henderson (2004), '"Globalizing" regional development: a global production networks perspective', *Transactions of the Institute of British Geographers*, **29**, 468–484.

Cooke, P. (2013), 'Global production networks and global innovation networks: stability versus growth', *European Planning Studies*, **21**(7), 1081–1094.

Dicken, P. (2015), *Global Shift, Mapping the Changing Contours of the World Economy*, London: Guilford Publications, 7th edn.

Florida, R. (2003), 'Cities and the creative class', *City & Community*, **2**(1), 3–19.

Hassink, R. (2010), 'Locked in decline? On the role of regional lock-ins in old industrial areas', in: R. Boschma and R. Martin (eds), *The Handbook of Evolutionary Economic Geography*, Cheltenham, UK and Northampton, MA, USA: Edward Elgar Publishing, 450–468.

Healy, A. and K. Morgan (2012), 'Spaces of Innovation: learning, proximity and the ecological turn', *Regional Studies*, **46**(8), 1041–1053.

Isaksen, A. (2015), 'Industrial development in thin regions: trapped in path extension?', *Journal of Economic Geography*, **15**(3), 585–600.

Isaksen, A. and J. Karlsen (2011), 'Organisational learning, supportive innovation systems and implications for policy formulation', *Journal of the Knowledge Economy*, **2**(4), 453–462.

Isaksen, A. and M. Trippl (2016), 'Path development in different regional innovation systems: a conceptual analysis', in: M.D. Parrilli, R.D. Fitjar and A. Rodriguez-Pose (eds) *Innovation Drivers and Regional Innovation Strategies*, London: Routledge, 66–84.

Isaksen, A. and M. Trippl (2017a), 'Innovation in space: the mosaic of regional innovation patterns', *Oxford Review of Economic Policy*, **33**(1), 122–140.

Isaksen, A. and M. Trippl (2017b), 'Exogenously led and policy-supported new path development in peripheral regions: analytical and synthetic routes', *Economic Geography*, **93**(5), 436–457.

Jensen, M.B., B. Johnson, E. Lorenz and B.Å. Lundvall (2007), 'Forms of knowledge and modes of innovation', *Research Policy*, **36**(5), 680–693.

Karlsen, J., A. Isaksen and O.R. Spilling (2011), 'The challenge of constructing regional advantages in peripheral areas: the case of marine biotechnology in Tromsø, Norway', *Entrepreneurship & Regional Development*, **23**(3), 235–257.

Maskell, P. and A. Malmberg (2007), 'Myopia, knowledge development and cluster evolution', *Journal of Economic Geography*, **7**(5), 603–618.

Morgan, K. (2004), 'The exaggerated death of geography: learning, proximity and territorial innovation systems', *Journal of Economic Geography*, **4**(1), 3–21.

OECD (2006), *OECD Territorial Reviews: Competitive Cities in the Global Economy*, Paris: OECD.

Patton, D. and M. Kenney (2010), 'The role of the university in the genesis and evolution of research-based clusters', in: D. Fornahl, S. Henn and M.-P. Menzel (eds), *Emerging Clusters. Theoretical, Empirical and Political Perspectives on the Initial Stage of Cluster Evolution*, Cheltenham. UK and Northampton, MA, USA: Edward Elgar Publishing, 214–238.

Porter, M.E. (2008), *On Competition. Updated and Expanded Edition*, Cambridge, MA: Harvard Business School Publishing.

Power, D. and A.J. Scott (2011), 'Culture, creativity, and urban development', in: A. Pike, A. Rodriguez-Pose and J. Tomaney (eds), *Handbook of Local and Regional Development*, London and New York: Routledge, 162–171.

Scott, A. and M. Storper (2003), 'Regions, globalisation, development', *Regional Studies*, **37**(6/7), 579–593.

Simmie, J. (2005), 'Innovation and space: a critical review of the literature', *Regional Studies*, **39**(6), 789–804.

Sinozic, T. and F. Tödtling (2015), 'Adaptation and change in creative clusters: findings from Vienna's new media sector', *European Planning Studies*, **23**(10), 1975–1992.

Swyngedouw, E. (2004), 'Globalisation or "glocalisation"? Networks, territories and rescaling', *Cambridge Review of International Affairs*, **17**(1), 28–48.

Tödtling, F. and M. Trippl (2005), 'One size fits all? Towards a differentiated regional innovation policy approach', *Research Policy*, **34**(8), 1203–1219.

Tödtling, F., L. Lengauer and C. Höglinger (2011), 'Knowledge sourcing in "thick" and "thin" regional innovation systems: comparing ICT firms in two Austrian regions', *European Planning Studies*, **19**(7), 1245–1276.

Tödtling, F., C. Höglinger, T. Sinozic and A. Auer (2014), 'Factors for the emergence and growth of environmental technology industries in Upper Austria', *Mitteilungen der Österreichischen Geographischen Gesellschaft*, **156**, 115–140.

Trippl, M. and F. Tödtling (2008), 'Cluster renewal in old industrial regions – continuity or radical change?', in: C. Karlsson (ed.), *Handbook of Research on Clusters*, Cheltenham, UK and Northampton, MA, USA: Edward Elgar Publishing, 203–218.

Yeung, H. and N. Coe (2015), 'Toward a dynamic theory of global production networks', *Economic Geography*, **91**(1), 29–58.

26. World cities and globalization

Ben Derudder

CONCEPTUALIZATION

The idea of a global economy articulated through cities is longstanding (cf. Braudel, 1992), but the specific suggestion of a set of 'world cities' having a strategic role in the coordination and control of the global economy is relatively new. In spite of some earlier contributions and a variegated genesis of the concept (cf. Hymer, 1972), contemporary research on world cities is most commonly traced back to a handful of writings in the 1980s. The theoretical raw materials of this then-emerging literature are most clearly articulated in John Friedmann's (1986) 'World City Hypothesis', in which he identifies world cities as those centres from which the 'new international division of labour' created by multinational enterprises was being organized and controlled. In the 1990s, a more encompassing language of globalization emerged as the conceptual backdrop for much social sciences research, and the world cities research agenda blossomed as one of the possible takes on how globalization and cities intersected. This resulted in a more varied conceptual and empirical literature (alongside a more complex terminological toolkit, see Taylor and Lang, 2005), in which – alongside Friedmann's (1986) seminal paper – Saskia Sassen's (1991) *The Global City* arguably came to serve as a key reference point for the literature at large.[1]

The unambiguous connection between writings on world cities and on globalization implies that the former literature has been a major laboratory for new theories, methodologies, and intellectual frameworks developed in the latter literature. Major examples include vigorous and wide-ranging debates on the potential of post-structuralist thinking on (e.g. Smith, 2003; Smith and Doel, 2011), and post-colonial readings of globalization (e.g. Robinson, 2002; Roy, 2009), as well as on the modified role and relevance of national states under conditions of contemporary globalization (e.g. Therbörn, 2011; Brenner, 2004). Importantly, debates on the cities/globalization-nexus have moved well beyond the world city terminology, with new frameworks such as 'planetary urbanization' (Brenner, 2013) and a more encompassing language of globalizing cities (Keil et al., 2017) being developed. Nevertheless, in spite of the term 'world city' (and what it conceptually engenders) frequently being contested and purposely bypassed, it has clearly become a major rallying point in social sciences research on globalization in general and its urban-geographical dimensions in particular.

---

[1] Although 'global cities' and 'world cities' (alongside other concepts such as 'global city-regions') tend to – and are sometimes even specifically constructed to – cover different concepts, here I will not attempt to discriminate between both terms: I will stick to 'world cities' as a common signifier for the literature at large.

Drawing out key approaches/methodologies in the world city literature is not an easy task. This is because casual references to a 'world city paradigm' (cf. Friedmann, 1995) and especially 'world city theory' (cf. Short et al., 2000) tend to overstate the coherence of the academic literature invoking that term (see also Brenner and Keil, 2006; Van Meeteren et al., 2016). Instead, following Saey's (2007) assessment of the disparate roots of world cities research, it seems more apt to think of different perspectives unevenly coming together in this literature, and which thus most certainly do not add up to a paradigm in a Kuhnian (1970) sense. Nonetheless, it seems fair to state that the most important (in terms of output) of the different threads running through the world city literature is a political-economic approach that aims to (1) understand urban-economic geographies of (capabilities for) capitalist command and control on a global scale; after which (2) these changing geographies of command and control are critically related to socio-spatial changes within the cities involved. Sassen's *The Global City* (1991) is a case in point, as the book fundamentally deals with (1) how the rise of producer services economies catering to a global economy in a de facto limited number of cities is related to global power structures, after which (2) attention is paid to the growth and changing nature of inequality in these cities. Given the relative dominance of the political-economic literature that uses the term world city to examine urban-economic geographies of exercising control on a global scale, in the remainder of this chapter I largely restrict myself to this particular take on the literature.

However, even within this take on world cities diversity reigns. Indeed, neither Sassen's (1991) book nor its subsequent use as the input to other research can be taken to be a wholesale characterization of 'the' world city literature, which is characterized by a pronounced pluralism. This is evident from the wide variety of research topics, epistemologies, and methodologies marshalled in the literature. As Van Meeteren et al. (2016) highlight, world city research encompasses both model-based approaches (e.g. Mahutga et al., 2010) and qualitative research methods (e.g. Beaverstock, 2005) for understanding cities within contemporary globalization. Some conduct detailed empirical studies on specific (sets of) cities (e.g. Lai, 2012), while others advance a more theoretical approach (e.g. Brenner, 1998). Furthermore, some authors emphasize structure (e.g. Alderson and Beckfield, 2004), while others underscore agency (e.g. Watson and Beaverstock, 2014); some focus on issues of global capitalist expansion (e.g. Wójcik, 2013), while others have a strong interest in methods for analysis and visualization (e.g. Neal, 2011).

Additionally, and in line with the two tiers of attention in Sassen's (1991) book, certain scholars primarily try to understand the remit of the notion of (capabilities for) global capitalist command and control (Allen, 2010), while others mainly deal with socio-spatial changes within cities (Van der Waal and Burgers, 2009) and/or the politics of world city-formation 'on the ground' (Golubchikov, 2010). Importantly, within this diversity there are also plenty of authors who critically examine the applicability of world city frameworks to less-obvious cases (e.g. Grant and Nijman, 2002). And finally, world city research is still a fast-developing field of literature (Acuto and Steele, 2013; Keil et al., 2017). For example, the second edition of *World City Network* (Taylor and Derudder, 2016) is not simply an update of a variety of empirical patterns in the analysis of the global connectivity of world cities, but represents a comprehensive revision that involves engaging with new developments, critiques, refined and improved methods, as well as a broader appreciation of how the research relates to other urban and globalization studies. In light of this diversity and continuing developments, Acuto (2011) and Van Meeteren et al. (2016) propose to think of the world city research agenda as being produced by an 'invisible college'. An invisible college refers to the sociological formation of a group of authors in a particular research field who constitute a social circle, but have varying degrees of involvement on the basis of diverging research interests. This circle has an (informal) stratification, and is characterized by internal disagreements, debate, and openness to internal mavericks and criticism (cf. Crane, 1969).

SPATIOTEMPORAL DIMENSIONS

In spite of the marked diversity in the literature, political-economic research on world cities seems premised on two key observations. First, the increasingly worldwide (re)distribution of economic activities necessitates strategic control functions that are found in a limited number of locations: globalization in its various guises has led to increased levels of geographical complexity, and this calls for control points to ensure the smooth functioning of the global system. In other words: world cities contain a disproportionate number of strategic agents in the global system (e.g. headquarters of multinational corporations and international institutions, specialized and international-ized business firms). Second, this practice of strategic control is accomplished through the capacity of these world city agents to network across space. In addition to developments in logistics and airline connectivity facilitating the flows of goods and people (Ducruet et al., 2010), from the 1970s onwards advances in two distinct technologies – computers and communications – combined to create a new enabling infrastructure for global organization that has been instrumental for world city-formation. As the infrastructures and technologies became more pervasive and sophis-ticated, this global infrastructure implied that spatial organization became increasingly conceptualized through networks (in which interaction is defined by quasi-simultaneity) to the detriment of territoriality (in which interaction is defined by proximity). Envisaging world cities in the context of a 'network society' particularly gained momentum in the 1990s, when influential authors such as Manuel Castells (1996) claimed that globalization processes are basically all about transnational processes operating through numerous networks, with networks between world cities as a prime example of this logic.

In any case, key contributions to the world cities literature are loosely united in their observation that cities such as New York and London derive their importance from a privileged position in networks of capital, knowledge, and people. Bassens and Van Meeteren (2015: 752) recently proposed re-theorization of the 'world city archipelago (as) an obligatory passage point for the relatively assured realization of capital'. Furthermore, and as already suggested, they commonly refer to the effects of the increasing internationalization of trade and production, give some weight to the increased service intensity in all sectors of economic life, and identify the relevance of recent evolutions in transport and technology. However, at the same time, it is quite obvious that such a sweeping approach rings rather hollow. Teasing out the spatio-temporal dimensions of world cities is thus best achieved by more concretely exploring how key contributors position themselves with respect to (i) the main function of cities; (ii) the key agents in the formation of world cities; (iii) the global structure that thus emerges; and (iv) the impacts on the ground. In the remainder of this section, I will focus on the first two points; the latter two will be dealt with in the next sections. That is, here I will briefly single out the spatiotemporal dimensions of world city-formation by exploring the major vantage points of what are arguably the two most formative contributions in this literature, i.e. those by Friedmann (1986, 1995; with Wolff, 1982) and Sassen (1991, 2000).

Friedmann (1986) frames the emergence of world cities in the context of a major geographical transformation of global production and the ensuing complexity in the organizational structure of multinational enterprises. The increased economic-geographical complexity associated with contemporary globalization, he argues, requires a limited number of control points, and world cities are deemed to be such points. A world city is a spatially integrated economic and social system at a given location or metropolitan region and can therefore even refer to polycentric urban regions such as the Kansai Region or Randstad Holland (cf. Kloosterman and Musterd, 2001). Friedmann (1986, 1995) gives theoretical body to his framework for research by (implicitly) subsuming it under Wallerstein's (1979) world-systems analysis, hence the title of Knox and Taylor's (1995) *World Cities in a World-System*. This implies a vision of globalization that emphasizes the presence of a capitalist system where capital accumulation involves hierarchical and spatial inequalities based on the concentration of relatively monopolized and therefore high-profit production in a limited number of core zones. World cities are deemed to be the key organizational nodes in the networks of power and dominance that (re)produce global inequality. Although no systematic analysis of the global geography of world cities was attempted, Friedmann (1986) provided a schematic (and now almost iconic) outline of what this geography looked like, differentiating between primary and secondary cities in core and semi-peripheral countries (e.g. New York primary/core; Sydney secondary/core; Sao Paulo primary/semi-periphery; Bangkok secondary/semi-periphery). Although Friedmann (1986) points out that the economic variable of command-and-control is likely to be the decisive, world cities can also be seen as the locales from which other forms of global power are projected, for example, geopolitical and/or ideological-symbolical control. Miami's control position over Central America is a case in point (e.g. Grosfoguel, 1995; Nijman, 2011), as is the position of Los Angeles in the production of global consumer identities (e.g. Hoyler and Watson, 2013) or Berlin in defining 'global culture' (Skórska and Kloosterman, 2012).

In *The Global City*, Saskia Sassen (1991) presents an alternative take on the role of world cities in the global economy, and does so by focusing upon the attraction of producer service firms to major cities that offer knowledge-rich and technology-enabled environments. In the 1980s and 1990s, many such service firms followed their global clients to become important multinational enterprises in their own right, albeit that service firms tend to be more susceptible to the agglomeration economies offered by major city locations. Sassen thus advocates a shift of attention to the advanced servicing of worldwide production, which implies a shift in focus from formal command power in the global economy to 'the practice of global control: the work of producing and reproducing the organization and management of a global production system and a global market-place for finance' (Sassen, 1995: 63–64). Through their transnational, city-centred spatial strategies, producer service firms have created worldwide office networks covering major cities in most or all world regions, and it is exactly the myriad of connections between these service complexes that gives way to a global urban system centred on world cities. Although, not unlike Friedmann, Sassen herself has not attempted a systematic mapping of the global geography of world cities, she suggests a break away from rigid core/periphery thinking by stating that the geography of world cities will likely cut across established North/South divides.

Some critics have questioned the Friedmann/Sassen (and their followers) take on world cities. They argue that it is both naïve and problematic to (even if implicitly) trace the causal dynamics of world city-formation back to a combination of economic globalization and changes in transportation and communication technologies. Massey (2007) emphasizes that the primary task of global/urban researchers should be to focus on the politics of the *how* and *why* of, say, the emergence of the New York–London axis as the organizational backbone of the global economy. To frame this critique, it is useful to think of research on world cities as part of broader social-scientific narratives on (the resurgence of) cities as key territorial units within contemporary globalization. Key examples of such city-centred narratives of an increasingly globalized economy and society are Edward Glaeser's (2011) *The Triumph of the City* and Benjamin Barber's (2013) *If Mayors Ruled the World*, in which straightforward and normative arguments are presented for an unavoidable and imminent urban future. In spite of their disparate roots and despite covering very different topics, such books tend to share an undercurrent of the objective need for a new strategic economic and political role for cities against a backdrop of economic deregulation, liberalization, and privatization, thus de facto considering this neoliberal backdrop as an undisputable given.

From this vantage point, world city research in its most uncritical form, with its focus on recounting the global centrality of cities, can also be cast as naively describing, and therefore subscribing to, neoliberal globalization. It is therefore no surprise that critics such as Michael Peter Smith (1998; see also Massey, 2007) primarily and dismissively label world cities (alongside globalization) as a social construct, arguing that research should focus on critically dissecting the genesis and the adoption of the concept more than anything else. The uneven and partial adoption of world city research in urban policy frameworks would point to the relevance of such a critical perspective. For example, world city league tables have become popular amongst policy-makers and global firms alike, with the image being consumed and internalized, largely stripped of its critical antecedents (for a brilliant parody, see http://syruptrap.ca/2014/10/vancouver-ranked-the-most-city-in-the-world/). Being aware of the potential performativity of knowledge has certainly become a major point of attention for world city researchers. However, in tune with Van Meeteren et al. (2016), I believe it is useful to not be too much stifled by this: neoliberal urban development in general and processes analyzed under the world city rubric in particular would very likely have occurred anyway, and stopping researching it may simply mean we understand it less well.

ORGANIZATIONAL DIMENSIONS

World cities are (re)produced by means of a number of underlying organizational structures. The most visible and obvious of these are the material bases on which the connectivity of world cities is built, that is, high-quality (tele)communication, transport and logistics networks. Cities such as London, New York, and Singapore not only harbour some of the world's most well connected airports, but are anchor points for extensive fibre-optic-based telecommunications networks. Bel and Fageda (2008), for example, show how the availability of nonstop intercontinental flights has a large influence on the location of large firms' headquarters across urban areas, confirming the importance of transport infrastructures and tacit information exchanges between world cities. Meanwhile, Malecki (2012) reveals how the geography of high-end Internet infrastructures is clearly tied to the geography of world cities. Examining key dimensions of the infrastructures needed for high-speed communication, the pattern that emerges is consistent with rankings of world cities, with London, Amsterdam, Frankfurt, and San Francisco standing out. Meanwhile, Hong Kong, Tokyo, Singapore and Sydney are key nodes in Asia-Pacific, while Sao Paulo emerges as the best-linked location in Latin America alongside Miami, which functions as the hub for Central and South America.

Another key organizational dimension of world cities is their centrality in migration flows. These migration flows are tied to a broader social restructuring in world cities, as they are often claimed to be defined by polarized class structures comprised of 'transnational elites' and a 'permanent underclass' (Friedmann and Wolff, 1982: 322). Immigrants comprise a large and significant component of both the transnational elites (cf. Sklair, 2001) and a permanent underclass (cf. Mingione, 1996) in world city class structures. This wide-ranging observation has been conceptually and empirically elaborated in subsequent research on world cities. Although the complex nature of social restructuring has cast doubt over the emergence of a straightforward polarized and dual class structure (see Hamnett, 1994; Fainstein, 2001; Van der Waal, 2015), rising patterns of socio-economic inequality have been recorded in world cities, and the nexus with migration has been confirmed (cf. Timberlake et al., 2012). At the transnational elites end, Beaverstock (2004) explores how the expatriation of managerial elites embodies a major globalization strategy of service firms in world cities. He finds that expatriation represents a deliberate organizational strategy to develop, manage, and diffuse idiosyncratic knowledge between world cities, the major objectives of which are to service the client and increase profitability and market share. Thus the large number of high-skilled migrants from all over the world in London's financial district is instrumental to keep London running. Yet, what is at least equally instrumental is the vast army of low-skilled workers engaged in industries that 'cater to the privileged classes for whose sake the world city primarily exists' (Friedmann, 1986: 73). Wills et al. (2010) have documented this for London, noting that London's expanded low wage economy is dominated by migrant workers leading to the emergence of what may be termed as a 'migrant division of labour'. As an illustration, in 2001, it was estimated that 46 per cent of London's low wage jobs were occupied by foreign-born workers (Spence, 2005: 61). This reliance on migrant workers is even more significant in certain sectors such as cleaning where the numbers of foreign-born workers rose from 40 per cent in 1993–1994 to almost 70 per cent in 2004–2005.

An increasingly important organizational dimension of world cities is their de facto emergence in terms of policy frameworks and networks. Against the backdrop of an increasingly urbanizing world, Benjamin Barber's (2013) *If Mayors Ruled the World* claims that nation states are poorly equipped to tackle global challenges such as terrorism and climate change or poverty. According to Barber, the world's cities should/will thus lead the way to a new urban millennium of renewed democracy and sustainability. Although few social scientists share Barber's somewhat naïve excitement about the liberating potential of a political/policy world centred on cities, it certainly does point to a consensus about the rising influence of cities in politico-organizational dimensions of globalization. Research on urban policy mobilities (McCann and Ward, 2011), for example, has highlighted the meaning of cities in the making of global space(s). Using the example of the C40 Cities Climate Leadership Group (C40), which brings together mayors of megacities to address the issue of climate change, Barthold (2018) illustrates how cities increasingly engage in networks aimed at 'global policy-learning' through the exchange of experiences, best practices and innovations. She shows how C40 is functioning both (1) as a network of cities that is pooling and multiplying the particular authority, fiscal competency and technical expertise that distinguishes cities from other scales of climate governance; and (2) as collective actor in global environmental governance as it provides city governments with access and influence on global decision-making procedures and international policy circles that non-state actors usually lack.

CASE STUDY: STRATIFICATION WITHIN A GLOBAL GEOGRAPHY OF WORLD CITIES

The variegated, and sometimes contentious nature of world city writings implies that the very selection of a case study implies choosing a particular conceptual and empirical vantage point that cannot be taken to stand for 'the' literature. Here I draw on some of my own work with Peter Taylor and other colleagues in the context of the Globalization and World Cities (GaWC) research network (for a recent outline, see Taylor and Derudder, 2016). In this research, we explore what one of the possible global geographies of world cities looks like. To this end, we map how strongly cities are inter-connected in the global economy by drawing on Sassen's (1991) recognition of producer service firms as the major agents of world city-formation.

Producer service firms provide the financial, professional and creative means of servicing transnational corporations in their global pursuits. These activities range from resourcing capital, to navigating multiple jurisdictions, and to developing generic marketing strategies. These are largely accomplished through face-to-face meetings requiring work to be carried out in offices across all major cities where their clients have important interests. The skyscraper cityscapes that epitomize the world city are the 'knowledge factories' that make globalization possible through myriad inter-city communications. Thus, to map world city-formation we investigate the offices of producer service firms across leading cities, information that is (mostly) readily available on firms' websites where they describe their office network to both impress potential clients and attract potential employees. From this information on offices in different cities, a data matrix is created arraying firms (the agents) against cities (where the agents do their work). From this matrix, estimates of knowledge flows (advice, direction, plans, strategies and so on) can be generated by applying a tailored network model. The outcome is a description of global inter-city flows of financial, professional and creative knowledges that define the organizational framework on which globalization is built.

In our latest application of the model, based on a data gathering carried out in the summer of 2016, we garnered information on the size and extra-locational information of the offices of 175 service firms across 707 cities (for operational details, see Taylor and Derudder, 2016: Chapter 4). Applying our model produces a global geography of world cities as proxied through their connectivity in the office networks of major producer services firms. This global geography is shown in Figure 26.1. The cartogram illustrates the 100 most connected cities to solve the problem of illustrating a very uneven distribution of cities across the world. In the cartogram, each city is given its own equal space in approximately its correct relative position, with darker shades reflecting stronger connectivities. Cities are indicated by intuitive two-letter codes, e.g. 'NY' for New York and 'JB' for Johannesburg. The first geographical result is that there is indeed a worldwide pattern of interconnected world cities, albeit clearly an uneven one. At its simplest, the cartogram reproduces the old North–South divide: higher connected cities tend to be in the North and lower connected cities in the South, with the Western Pacific Rim firmly bucking this trend. But the global geography is, of course, much more complicated: this simple interpretation is only a trend with many lower connectivity cities in the North and some higher connectivity cities in the South. Figure 26.1 illustrates clearly the three leading zones of the core of the global economy through the lens of cities: northern America, Western Europe and parts of Pacific Asia. However, this is not a homogeneous core, the three zones have very different histories associated with their trajectories to core status.

The oldest, indeed original, core zone is Western Europe and this is reflected in two features. First, this region has more world cities than the other regions. And second, there is a wide variety of levels of connectedness amongst the region's cities, ranging from the likes of London and Paris to the likes of Manchester and Hamburg. In other words, in this region, there is a mixture of cities of varying importance all linking into the world city network. This is the opposite of Pacific Asia in which the connectivity levels of the cities is generally top heavy. In spite of booming and now widespread urbanization, few of these cities in the most recent of the core zones are strongly connected. Thus this region has far less world cities than Western Europe, although the number increases if we add Australasian cities and Auckland to create a Western Pacific Rim region. The third core zone, northern America (i.e. USA and Canada), is in between the other two in numbers of world cities identified. However, in this case the range of levels of connectedness is very similar to Western Europe with numerous less important cities such as Philadelphia and Minneapolis joining the world city network. But there is a regional difference in that in northern America the more connected cities tend to be in the east and west of the region, leaving the centre bereft of well-connected cities apart from the major exception of Chicago.

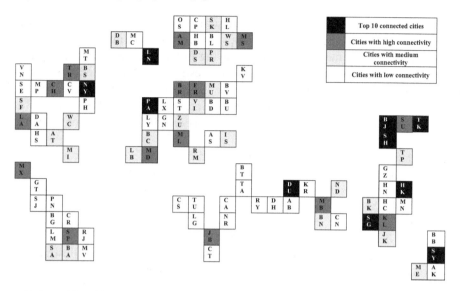

Source: Taylor and Derudder (2016).

Figure 26.1 A global geography of world cities

Beyond the core there are no regions with concentrations of highly connected cities. In Eastern Europe (the former Communist states), the most common pattern is for capital cities to take on the world city role. Having lost its political and economic distinctiveness, this region seems to have become an appendage to the Western European core, albeit that unlike Western Europe there is only one well connected city (Moscow) and that there is much more concentration with generally one (capital) city per state. The same can be said for Latin America with respect to northern America where again capital cities dominate although in this case Sao Paulo, despite being neither a former nor current Brazilian capital city, has become a highly connected world city in its own right. Together with Mexico City, this city stands out in Latin America. This pattern is also found in South Asia and the large North African/West Asian region, where Mumbai and Dubai as non-capital cities have become highly connected world cities. Sub-Saharan Africa has few cities on this particular map, but it does sport a clear regional leading city in terms of connectivity: Johannesburg.

As indicated, the above case study does not represent what world city research is about. Yet, this mapping of cities in globalization can be used as a starting point to make analytical connections to a number of other pertinent world city research agendas. For example, these connectivity measures can be used as the input to analyses of how, and to what degree, world city-formation is linked to rising inequality in general or migrant division of labour in particular (e.g. Timberlake et al., 2012). It also allows exploring how and to what degree this global geography is related to other geographies of global inequality (e.g. Mahutga et al., 2010; Alderson et al., 2010). But above all, it can be used as a backcloth for in-depth analyses of the making of world cities. Lai's (2012) research provides a clear-cut example of this. Her qualitative research into the Chinese space-economy helps us understand the very distinctive world city pattern in China discernable in Figure 26.1, with Shanghai and Beijing – combined with the special administrative region Hong Kong – dwarfing the rest of the 100-or-so Chinese cities with more than 1 million inhabitants in a 'tri-primate' world city pattern of global connectivity. Lai (2012) shows how this pattern is co-produced by the sheer size of the national market, making it difficult to operate from a single city; its historical and ongoing political divisions, with Hong Kong still operating as a quasi-autonomous area in financial and economic terms; boosterist visions of severe inter-city competition among some of China's urban elites, which have led to the active pursuit of world city status in Shanghai; and functional divisions of labour as the Chinese political system imposes a context where producer services need to be near the centre of political decision-making in Beijing irrespective of commercial opportunities in Shanghai or Hong Kong.

CONCLUSION AND RESEARCH AGENDAS

Research into world cities has been one of the major ways in which (urban) geography has become rooted in globalization debates in the social sciences from the 1980s onwards. A straightforward interpretation of the success of this research field is that it directly confronts the commonplace observation that contemporary urban life cannot be properly understood without making some sort of reference to globalization. This has led to a wide-ranging literature that produced a number of new and exciting insights. However, two points to be emphasized when appraising the success – in terms of cross-referencing – of the literature invoking the world city concept.

First, to a large extent, the literature lacks a central paradigm and often offers not much more than a very broad background to frame more concrete research into very different topics. Indeed, the most apt definition of world city research is simply that what those researchers using the term world city do (Van Meeteren et al., 2016): the praxis of doing research defines the evolving research subject and prevails over a rigid definition or conceptualization of what a world city is, how it should be researched, and how one should interpret the results. Even the most widely cited world city contributions – Friedmann (1986), Sassen (1991), and Taylor (2004) – are best understood as specific building blocks within an increasingly diverse literature on cities in globalization.

Second, even the most basic conceptual and empirical building blocks continue to be contested, and the continued popularity of the term therefore also somewhat paradoxically stems from it having become a rallying point for radically different approaches towards understanding globalized urbanization (e.g. Roy, 2009). Indeed, a large number of references to the world city literature stem from publications lamenting the provincialism of world city research, which putatively narrates and possibly normalizes the urban experience of the Londons and New Yorks of this world.

The need for a more diverse literature on cities in globalization that stretches well beyond the most widely cited world city contributions in ontological, epistemological and geographical terms cannot be disputed (for a more encompassing overview of 'doing global urban studies', see Hoyler and Harrison, 2018; Keil et al., 2017).

However, the central tenets of the key world city contributions still offer a powerful analytical framework to help understanding globalization through an (albeit specific) urban lens. From this perspective, on-going world city research agendas that aim to refine, deepen and extend our understanding of how (1) specific cities are key platforms in the organization of a globalized economy/society; and (2) how this impacts socio-spatial changes within those cities remain instrumental to sharpen our understanding of contemporary urban geographies. Some notable examples are contained in a recent special section of *Environment and Planning A* (Hoyler and Harrison, 2018) on the topic. Neal (2017), for example, proposes refined mappings of the global geography of world cities by developing formal statistical methods for identifying well-connected cities and thus help identifying cases that warrant more detailed investigation. Meanwhile, Krijnen et al. (2017), observing a tendency in the world literature to superimpose distinctions between high- and low-skilled labour and between North and South (see earlier), use the case of Beirut to marshal the need for a more plural conceptualization of professionals and elites to include expatriate or transnational service workers. This critical probing and extension of some key ideas is also apparent in Kleibert (2017; see also Kleibert and Horner, Chapter 16 in this *Handbook*), who, based on empirical research on producer service firms in Manila, argues that the existence of linkages as per Figure 26.1 does not automatically lead to an increased command and control position. Instead, and in line with Parnreiter (2014), the attraction of lower-end services may lead to cities such as Manila and Mexico City being articulated in a dependent way in the geography of world cities. And finally, one key emerging research agenda has been to diversify our understanding of world cities by revealing the global geographies of other corporate (Wall and Van der Knaap, 2011; Martinus et al., 2015; Krätke, 2014; Sigler and Martinus, 2017) and institutional (Chow and Loo, 2015; Derudder and Taylor, 2018) world city agents, articulated through institutional and industry-specific factors, grounded in diverse histories and path-dependent trajectories.

REFERENCES

Acuto, M. (2011), 'Finding the global city: An analytical journey through the "Invisible College"', *Urban Studies*, **48**(14): 2953–2973.

Acuto, M. and W. Steele (2013), *Global City Challenges: Debating a Concept, Improving the Practice*, New York: Palgrave Macmillan.

Alderson, A.S. and J. Beckfield (2004), 'Power and position in the world city system', *American Journal of Sociology*, **109**(4): 811–851.

Alderson, A.S., J. Beckfield and J. Sprague-Jones (2010), 'Intercity relations and globalisation: the evolution of the global urban hierarchy, 1981–2007', *Urban Studies*, **47**(9): 1899–1923.

Allen, J. (2010), 'Powerful city networks: more than connections, less than domination and control', *Urban Studies*, **47**(13): 2895–2911.

Barber, B.R. (2013), *If Mayors Ruled the World: Dysfunctional Nations, Rising Cities*, New Haven: Yale University Press.

Barthold, S. (2018), 'Greening the global city. The role of C40 cities as actor in global environmental governance', in: Oosterlynck. S. et al. (eds), *The City As Global Political Actor*, London: Routledge, in press.

Bassens, D. and M. Van Meeteren (2015), 'World cities under conditions of financialized globalization', *Progress in Human Geography*, **39**(6): 752–775.

Beaverstock, J.V. (2004), '"Managing across borders": knowledge management and expatriation in professional service legal firms', *Journal of Economic Geography*, **4**(2): 157–179.

Beaverstock, J.V. (2005), 'Transnational elites in the city: British highly-skilled inter-company transferees in New York city's financial district', *Journal of Ethnic and Migration Studies*, **31**(2): 245–268.

Bel, G. and X. Fageda (2008), 'Getting there fast: globalization, intercontinental flights and location of headquarters', *Journal of Economic Geography*, **8**(4): 471–495.

Braudel, F. (1992), *Civilization and Capitalism, 15th–18th Century: The Structure of Everyday Life*, Oakland: University of California Press.

Brenner, N. (1998), 'Global cities, glocal states: global city formation and state territorial restructuring in contemporary Europe', *Review of International Political Economy*, **5**(1): 1–37.

Brenner, N. (2004), *New State Spaces*, New York: Oxford University Press.

Brenner, N. (2013), *Implosions/Explosions: Towards a Study of Planetary Urbanization*, Berlin: Jovis.

Brenner, N. and R. Keil (2006), *The Global Cities Reader*, London: Routledge.

Castells, M. (1996), *The Rise of the Network Society*, Oxford: Blackwell.

Chow, A.S. and B.P. Loo (2015), 'Applying a world-city network approach to globalizing higher education: conceptualization, data collection and the lists of world cities', *Higher Education Policy*, **28**(1): 107–126.

Crane, D. (1969), 'Social structure in a group of scientists: a test of the "invisible college" hypothesis', *American Sociological Review*, **34**(3): 335–352.

Derudder, B. and P.J. Taylor (2018), 'Central flow theory: comparative connectivities in the world city network', *Regional Studies*, **52**(8): 1029–1040.

Ducruet, C., C. Rozenblat and F. Zaidi (2010), 'Ports in multi-level maritime networks: evidence from the Atlantic (1996–2006)', *Journal of Transport Geography*, **18**(4): 508–518.

Fainstein, S.S. (2001), *The City Builders: Property Development in New York and London, 1980–2000*, Lawrence: University Press of Kansas.

Friedmann, J. (1986), 'The world city hypothesis', *Development and Change*, **17**(1): 69–83.

Friedmann, J. (1995), 'Where we stand: a decade of world city research', in: Knox, P.L. and Taylor, P.J. (eds), *World Cities in a World System*, Cambridge: Cambridge University Press, pp. 21–47.

Friedmann, J. and G. Wolff (1982), 'World city formation: an agenda for research and action', *International Journal of Urban and Regional Research*, **6**(3): 309–344.

Glaeser, E. (2011), *Triumph of the City: How Our Greatest Invention makes us Richer, Smarter, Greener, Healthier, and Happier*, London: Pen Books.

Golubchikov, O. (2010), 'World-city-entrepreneurialism: globalist imaginaries, neoliberal geographies, and the production of new St Petersburg', *Environment and Planning A*, **42**(3): 626–643.

Grant, R. and J. Nijman (2002), 'Globalization and the corporate geography of cities in the less-developed world', *Annals of the Association of American Geographers*, **92**(2): 320–340.

Grosfoguel, R. (1995), 'Global logics in the Caribbean city system: the case of Miami', in: Knox, P.L. and Taylor, P.J. (eds), *World Cities in a World System*, Cambridge: Cambridge University Press, pp. 156–170.

Hamnett, C. (1994), 'Social polarisation in global cities: theory and evidence', *Urban Studies*, **31**(3): 401–424.

Hoyler, M. and J. Harrison (2018), 'Global cities research and urban theory making', *Environment and Planning A*, in press.

Hoyler, M. and A. Watson (2013), 'Global media cities in transnational media networks', *Tijdschrift Voor Economische En Sociale Geografie*, **104**(1): 90–108.

Hymer, S.H. (1972), 'The multinational corporation and the law of uneven development', in: Bhagwati, J. (ed.), *Economics and World Order from the 1970s to the 1990s*, London: Collier-Macmillan, pp. 128–152.

Keil, R., X. Ren and N. Brenner (2017), *The Globalizing Cities Reader*, London: Routledge.

Kleibert, J. (2017), 'On the global city map, but not in command? Probing Manila's position in the world city network', *Environment and Planning A*, **49**(12), 2897–2915.

Kloosterman, R.C. and S. Musterd (2001), 'The polycentric urban region: towards a research agenda', *Urban Studies*, **38**(4): 623–634.

Knox, P.L. and P.J. Taylor (1995), *World Cities in a World-System*, Cambridge: Cambridge University Press.

Krätke, S. (2014), 'How manufacturing industries connect cities across the world: extending research on "multiple globalisation"', *Global Networks*, **14**(2): 121–147.

Krijnen, M., D. Bassens and M. Van Meeteren (2017), 'Manning circuits of value: Lebanese professionals and expatriate world-city formation in Beirut', *Environment and Planning A*, **49**(12), 2878–2896.

Kuhn, T. (1970), *The Structure of Scientific Revolutions* (2nd edn), Chicago: University of Chicago Press.

Lai, K. (2012), 'Differentiated markets: Shanghai, Beijing and Hong Kong in China's financial centre network', *Urban Studies*, **49**(6): 1275–1296.

Mahutga, M.C., X. Ma and D.A. Smith (2010), 'Economic globalisation and the structure of the world city system: the case of airline passenger data', *Urban Studies*, **47**(9): 1925–1947.

Malecki, E.J. (2012), 'Internet networks of world cities: agglomeration and dispersion', in: Derudder, B., Hoyler, M., Taylor, P.J. and Witlox, F. (eds), *International Handbook of Globalization and World Cities*, Cheltenham, UK and Northampton, MA, USA: Edward Elgar Publishing, pp. 117–125.

Martinus, K., T.J. Sigler and G. Searle (2015), 'Strategic globalizing centers and sub-network geometries: a social network analysis of multi-scalar energy networks', *Geoforum*, **64**: 78–89.

Massey, D. (2007), *World City*, Cambridge: Polity Press.

McCann, E. and K. Ward (2011), *Mobile Urbanism: Cities and Policymaking in the Global Age*, Minneapolis: University of Minnesota Press.

Mingione, E. (1996), *Urban Poverty and the Underclass*, Oxford: Blackwell Publishers.

Neal, Z. (2011), 'Differentiating centrality and power in the world city network', *Urban Studies*, **48**(13): 2733–2748.

Neal, Z. (2017), 'Well connected compared to what? Rethinking frames of reference in world city network research', *Environment and Planning A*, **49**(12), 2859–2877.

Nijman, J. (2011), *Miami: Mistress of the Americas*, Philadelphia: University of Pennsylvania Press.

Parnreiter, C. (2014), 'Network or hierarchical relations? A plea for redirecting attention to the control functions of global cities', *Tijdschrift Voor Economische En Sociale Geografie*, **105**(4): 398–411.

Robinson, J. (2002), 'Global and world cities: a view from off the map', *International Journal of Urban and Regional Research*, **26**(3): 531–554.

Roy, A. (2009), 'The 21st-century metropolis: new geographies of theory', *Regional Studies*, **43**(6): 819–830.

Saey, P. (2007), 'How cities scientifically (do not) exist: methodological appraisal of research on globalizing processes of intercity networking', in: Taylor, P.J., Derudder, B., Saey, P. and Witlox, F. (eds), *Cities in Globalization: Practices, Policies and Theories*, London: Routledge, pp. 298–313.

Sassen, S. (1991), *The Global City*, Princeton: Princeton University Press.

Sassen, S. (1995), 'On concentration and centrality in the global city', in: Knox, P.L. and Taylor, P.J. (eds), *World Cities in a World System*, Cambridge: Cambridge University Press, pp. 63–76.

Sassen, S. (2000), 'The global city: strategic site/new frontier', *American Studies*, **41**(2/3): 63–76.

Short, J.R., C. Breitbach and S. Buckman (2000), 'From world cities to gateway cities: extending the boundaries of globalization theory', *City*, **4**(3): 317–340.

Sigler, T.J. and K. Martinus (2017), 'Extending beyond "world cities", in World City Network (WCN) research: urban positionality and economic linkages through the Australia-based corporate network', *Environment and Planning A*, **49**(12), 2916–2937.

Sklair, L. (2001), *The Transnational Capitalist Class*, Oxford: Wiley-Blackwell Publishing.

Skórska, M.J. and R.C. Kloosterman (2012), 'Performing on the global stage: exploring the relationship between finance and arts in global cities'. *GaWC Research Bulletin 412*, http://www.lboro.ac.uk/gawc/rb/rb412.html (accessed 16 June 2018).

Smith, M.P. (1998), 'The global city-whose social construct is it anyway? A comment on White', *Urban Affairs Review*, **33**(4): 482–488.

Smith, R.G. (2003), 'World city topologies', *Progress in Human Geography*, **27**(5): 561–582.

Smith, R.G. and M.A. Doel (2011), 'Questioning the theoretical basis of current global-city research: structures, networks and actor-networks', *International Journal of Urban and Regional Research*, **35**(1): 24–39.

Spence, L. (2005), 'Country of birth and labour market outcomes in London: an analysis of labour force survey and census data', London: Greater London Authority.

Taylor, P.J. (2004), *World City Network: A Global Urban Analysis*, London: Routledge.

Taylor, P.J. and B. Derudder (2016), *World City Network: A Global Urban Analysis* (2nd edn), London: Routledge.

Taylor, P.J. and R.E. Lang (2005), 'US cities in the "World City Network"'. Metropolitan Policy Program Survey Series. Washington, DC: Brookings Institution.

Therbörn, G. (2011), 'End of a paradigm: the current crisis and the idea of stateless cities', *Environment and Planning A*, **43**(2): 272–285.

Timberlake, M., M.R. Sanderson and X. Ma (2012), 'Testing a global city hypothesis: an assessment of polarization across US cities', *City & Community*, **11**(1): 74–93.

Van der Waal, J. (2015), *The Global City Debate Reconsidered: Economic Globalization in Contemporary Dutch Cities*, Amsterdam: Amsterdam University Press.

Van der Waal, J. and Burgers (2009), 'Unravelling the global city debate on social inequality: a firm-level analysis of wage inequality in Amsterdam and Rotterdam', *Urban Studies*, **46**(13): 2715–2729.

Van Meeteren, M., B. Derudder and D. Bassens (2016), 'Can the straw man speak? An engagement with postcolonial critiques of "global cities research"', *Dialogues in Human Geography*, **6**(3): 247–267.

Wall, R.S. and G. Van der Knaap (2011), 'Sectoral differentiation and network structure within contemporary worldwide corporate networks', *Economic Geography*, **87**(3): 267–308.

Wallerstein, I. (1979), *The Capitalist World-Economy,* Cambridge: Cambridge University Press.

Watson, A. and J.V. Beaverstock (2014), 'World city network research at a theoretical impasse: on the need to re-establish qualitative approaches to understanding agency in world city networks', *Tijdschrift Voor Economische En Sociale Geografie*, **105**(4): 412–426.

Wills, J., K. Datta, Y. Evans, J. Herbert, J. May and C. McIlwaine (2010), *Global Cities at Work: New Migrant Divisions of Labour*, London and New York: Pluto Press.

Wójcik, D. (2013), 'The dark side of NY–LON: financial centres and the global financial crisis', *Urban Studies*, **50**(13): 2736–2752.

Wójcik, D. (2013), 'The dark side of NY–LON: financial centres and the global financial crisis', *Urban Studies*, **50**(13): 2736–2752.

27. Ports, cities and the global maritime infrastructure
Markus Hesse and Evan McDonough

CONCEPTUALIZING MARITIME GLOBALIZATIONS

Globalization in general is a process and phenomenon that encompasses a broad range of economic, social, cultural or environmental issues. Maritime globalizations can, hence, be conceptualized in different ways. From an economic exchange perspective, the development of world trade and the establishment of global production networks (GPN) appear to be the most significant changes with regards to maritime activities and infrastructures (Coe and Yeung, 2015). At centre-stage here is not only the sheer expansion of the spaces that are involved in economic exchange, but also the integration of different localities in the complex creation of value that has unfolded in the context of globalization and the establishment of GPNs. These processes require a supporting system of connecting infrastructures, technologies, and operations, which is provided by logistics and physical distribution (Coe, 2014; Hesse and Rodrigue, 2006). While transport has always been key to the maritime industries, logistics and physical distribution now comprise a wide variety of related services, such as information management, law, financing and insurance (O'Connor et al., 2016).

Moreover, the maritime system has always had, and still has, a strong relationship with concrete territories, creating its own geographies of spaces and flows (Comtois and Slack, 2009). On the one hand, this applies to shipping routes, which have always played a role as connectors between the locales of the supply of and the demand for commodities. Shipping routes have changed, according to the provision of infrastructures such as canals, for geo-political reasons, or as a consequence of trade integration. On the other hand, port cities are still crucial nodes and interfaces in the system of global flows. Port cities are not only the entry and exit points for flows of goods, but they also generate labour markets for workers in the shipping and handling activities, and constitute a gateway to critical masses of customers in port regions and in their hinterlands. Thus, they form a strategic location for staff from the various maritime industries and the related producer services. Port cities can, therefore, be seen as both nodes and main facilitators of maritime trade. In this latter respect they have gone through massive changes during the accelerated globalization of trade and transport.

Regarding maritime businesses that are associated with globalization in practice, we can distinguish a double meaning of these processes: first, the link between globalization and maritime activity views the latter as being derived from world trade, commerce and globalized production systems. The more extensive global economic exchange was becoming over the last decades, the higher the turnover of goods and thus the demand for transport has been (UNCTAD, 2016). The other element to be taken into account here is the proactive role and strategic behaviour of corporate actors to become embedded in the system of global flows, by exploring markets, inserting themselves in supply chains, or acquiring shares through direct investment (Frémont, 2007; McCalla et al., 2004). The activities of these actors thus foster the intensification of globalization processes.

In comparable ways, national governments and municipal decision makers have strived to become global players and thus have lobbied for ports, shipping lines or transshipment firms to be based or strengthened in their respective territories (cf. for Europe Verhoeven, 2009; Vieira et al., 2014). These developments have a local dimension that brings about certain infrastructure requirements. These have not only culminated at port cities, but have also actively been regulated by port cities, port authorities and governments at various spatial scales. This active regulation also applies to cases where shipbuilding and maritime trade have vanished and port functions abandoned, in order to redevelop converted waterfront areas for housing or office purposes. Since the late 1970s and early 1980s, and beginning with redevelopment in Baltimore, Maryland (USA) and the London Docklands (UK), the urban waterfront has become a key concept for urban policy and planning around the world – another case of globalization (re-)shaping (former areas of) maritime industries and port cities (Desfor et al., 2011; Brownill, 2013).

Useful accounts for conceptualizing the maritime geography of globalization have evolved from different disciplinary approaches. First of all, maritime trade and transport is now considered to be an articulation of logistics processes and the management of value chains (Robinson, 2002). Second, institutional analyses have emphasized the role of global actors in the maritime industries, particularly as terminal operators and shipping lines have built global alliances and networks (McCalla et al., 2004; Soppé et al., 2009). Third, transport geography and urban geography have always considered the port city to be the main interface between sea and land, between the city and the maritime world, providing opportunities and challenges for both (Hayuth, 1982; Hoyle, 1989). Fourthly, research on the development trajectories of port cities has revealed not only the shifting geographies of maritime trade and transport towards East Asia, but also identified different roles of these cities according to various maritime specializations, such as more centralized mainports, on the one hand, and, on the other, more decentralized hub port strategies, while seeing port cities in general as 'front-line soldiers of globalization' (Ducruet and Lee, 2006).

In order to provide an up-to-date picture of these events, we firstly apply a supply chain perspective regarding maritime and port geographies, since the movement of ocean vessels and the territorial or geographical impact of these flows cannot be separated from the underlying logistics networks, or their spatio-temporal performance and overarching political-economic power configuration (Olivier and Slack, 2006). Second, we employ a relational perspective with respect to ports, cities and their interaction against the background of globalized logistics networks (see e.g. Jacobs and Lagendijk, 2014). This relational approach is inspired by understanding cities and regions as increasingly dependent on, and co-constituted by, larger associations, networks, mobilities and identities, rather than conceiving these entities as fixed in territories within a bounded space, delineated by clear margins.

SPATIO-TEMPORAL DIMENSIONS OF GLOBAL MARITIME SPACES AND FLOWS

Maritime trade and infrastructures are particularly concentrated and thus highly visible in port cities, as these are the connectors between land and sea, shipping companies and service providers, their customers, corporations, workforce, regulatory bodies and other agents such as labour unions. Port cities have played a crucial role in urban history (see also Chapter 2 by Peer Vries in this *Handbook*). With rising volumes of trade and transport, the cities that hosted an extensive port (the 'mainport') are considered gateways that serve major hinterlands and thus constitute critical nodes for organizing related services. Traditionally, the role of the gateway has primarily been based on territorial assets, such as a proper interface between sea and land, and also on broad and deep urban labour markets, which have enabled economic growth and specialization of these cities. Nowadays, many port cities also seek to establish strategic positionality within global logistics chains and global maritime networks (Jacobs and Lagendijk, 2014). Such efforts are still connected to the port, to water access and maritime industries, but they can also be considered increasingly independent from these mainly physical or territorially bounded conditions. Land-side access, hinterland connectivity and the presence of the associated maritime services play a role here as well. As these services can also be offered from more remote locations and do not necessarily require face-to-face contact between provider and client on site, the service component of the maritime industry appears to be generally more mobile, and volatile (Hall and Jacobs, 2012; Verhetsel and Sel, 2009; Verhetsel et al., 2016). This also means that the traditional link between port (maritime industries) and port city has been fundamentally altered in recent times.

However, given that globally significant ports are still located in major, if not global cities, these hubs of global maritime flows create a specific global–local relationship within the process of urbanization (Boschken, 2013; O'Connor et al., 2016). This relationship initially evolved against the background of physical conditions (such as topography, nautical situation, urban-economic framework conditions), a critical mass of customers both locally and in the hinterland, and an increasing degree of inter-nationalization of commodity markets. More recently, this relationship has been transformed by technological changes, the emergence of new types of maritime services, and variegated responses from both corporate and policy actors. The study of global transport flows and their local, urban context thus lies at the intersection of transport geography and both human and urban geography (Ng et al., 2014), transcending the conceptualization of a locality as a mere port-city interface (Hoyle, 1989). Hence contemporary port cities are considered to be co-constituted by both global flows and by networks at the local, regional and national scales, which makes them a template case of relationality. This co-constitution tends to be an outcome of a reciprocal, rather than a linear, causal relationship. In this context, it seems difficult to identify distinct causal mechanisms, drivers and outcomes of the related processes. In port cities, it all comes together: while the global flows are part of the global restructuring of production and consumption patterns, local strategies aim to position the respective ports within such larger flows and processes (Hall and Hesse, 2013).

In addition to this complex spatial context, maritime globalization also takes place within a unique trajectory of technological and regulatory change, which has displayed many significant ruptures throughout its distinct history. One that can be considered most significant with regard to the globalization of ocean trade in the course of the twentieth century is the invention of the shipping container concept in 1956 by an American entrepreneur who realized that shipping cargo in a truck-sized container would drastically reduce loading time and labour costs. Subsequently, the container itself would become a main enabler of global maritime flows and the 'work horse of international trade' (Rodrigue and Slack, 1998–2017). Containerization has since become an overall driver of world trade, global production and associated transport services, as it helped standardizing processes, hence added the desired capacity for expanding world trade and provided much cheaper transport rates. As a consequence, intermodal transshipment quickly revolutionized global shipping, production and consumption patterns around the world by the 1980s and 1990s, with significant implications for existing port cities. Containerized maritime shipping was gradually introduced in the Caribbean, Latin America, the Middle East and South Asia in the 1970s and 1980s, and has remained the dominant physical transport format of global trade since the 1990s (Guerrero and Rodrigue, 2014).

> Within containerized China, the Yangtze River Delta and Pearl River Delta are two of the largest geographical concentrations of manufacturing factories and logistics activities in the world. These regions include two major ports of Shanghai and Ningbo of the Yangtze River and the three major ports of Hong Kong, Shenzhen, and Guangzhou of the Pearl River Delta. (Wang, 2014: 22)

Concurrently, a network of logistic firms located in global cities have emerged, profiting from the advantages of agglomeration, and thus becoming the 'basing points' for the global restructuring and re-organizing of production systems and supply chains at the local, regional and global levels around the world (O'Connor et al., 2016).

International trade was also encouraged by the global trend of deregulation and privatization beginning in the 1980s (see the following section), as the establishment of free trade agreements helped to relax border customs procedures, enabling the increasingly unconstrained flow of goods between countries. Infrastructure policy and technical improvements also allowed containerization to reach deep inland and become fully integrated with existing domestic rail and road networks, as well as the development of secondary, satellite and/or inland nodes in the supply chain (Rodrigue and Slack, 1998–2017). The development of these inland nodes led them to be complementary to, and at times in competition with, the established mainports. As the world's dominant manufacturing platforms shifted from Europe and North America to Asia, and to China in particular during a period of rapid industrialization there, so did the main shipping flows. The mechanization of transport flows as provided by containerization played a central role in the emergence of the contemporary global regime of logistics and distribution (Ducruet and Notteboom, 2012).

In order to cope with the growing demand for freight transport in general, and specifically container transshipments, port and terminal operators were increasingly forced to adapt to the demand while simultaneously limiting costs. Containerization is highly capital-intensive, as it requires the provision of fully automated terminal operations. The associated growth led to the interrelated phenomena of building large-scale container carriers, so-called megaships, and providing sufficient handling capacity at container ports, so-called mega-hubs. The underlying economies of scale, which are inherent in current globalized maritime shipping and containerization, have also stimulated mergers, acquisitions, and global alliances in the shipping industries. These alliances reflect a need of shipping companies to become more resilient with respect to the rising volatility of the markets (i.e. given the global overcapacity of ocean ships and the falling freight rates in post-2008 developments), and also reflect the generally high demand for capital that is required for operating an ocean carrier fleet.

Beginning in the early 2000s, the continued rise of container shipments globally and in East Asia in particular (the 'China effect') led to the introduction of mega cargo ships capable of holding over 10 000 shipping containers (usually measured in twenty-foot equivalent units, or TEUs). Today the largest of these ships has a capacity of 18 000 TEUs (Rodrigue 1998–2017a), and as these vessels become ever larger, it is increasingly difficult, if not impossible, to integrate the required large-scale port infrastructures with the existing local environment of port cities.

While the seemingly unlimited growth of maritime trade is a vital part of the standard narrative of the shipping industry across the globe in the late twentieth and early twenty-first centuries, the constraints to these interrelated processes of growth have become rather obvious as well. Traditional port cities often lack space for port expansion, and their local nautical conditions are rarely viable for such deep-draft megaships. Also, the increasing demand for housing or offices at attractive waterfront settings in close proximity to port-industrial activity calls into question the compatibility and legitimacy of a working port in core urban areas. The related pressure works in two different directions simultaneously: competition and economies of scale call for expansion, while land-use conflicts exacerbate an at times adversarial relationship between the port and the city. This constellation is particularly evident in the cases of European mainports, such as Rotterdam in the Netherlands (see e.g. Priemus, 2001), Antwerp in Belgium (see van Hamme and Strale, 2012), or Hamburg, Germany (see Hesse, 2006). In contrast to these European cases, port expansion and the development of mainports and mega-hubs in China has taken place at a much faster pace. This is not only a consequence of the rapid industrialization of regions such as the Chinese Pacific coast or the Singapore Strait, but also a result of the completely different approaches to governing related developments and managing land-use conflicts in these countries. As urbanization may continue to proceed at an unprecedented level there compared to the West, so too may the development of new ports and harbours (Wang, 2014).

ORGANIZATIONAL AND GOVERNANCE DIMENSIONS

The political regulation of maritime industries and, hence, of seaports have undergone significant changes. This was related to globalization and the emergence of international markets, and also due to the associated deregulation and decentralization of public services. At the local level, this is most notable with port authorities, which, although in different organizational configurations, play a strategic role in the management of ports. In so doing, they are also vital for the development of port cities as such (Juhel, 2001). Port authorities have long been, and still are, powerful entities that are in charge of operating ports, infrastructure maintenance, and, more specifically, of planning land use and real estate development in port areas to increase efficiency (Doig, 2001). Initially, in many cases they represented a municipal institution. Beginning in the 1980s, and in line with the dominant neoliberal view on governance in general and the role of the state more in particular, public port administrative bodies were increasingly transformed into hybrid port authorities with various combinations of private, public or semi-public ownership and responsibility. 'Over the last two to three decades (port authorities) have undergone a reform from rather task oriented organizations to more autonomous and commercially acting organizations' (Van der Lugt et al., 2013: 103).

Through these institutional changes attempts were made to improve the strategic capabilities of the previously public authorities, and to attract global investors for financing further port growth and expansion. However, there seems to be no uniform pattern of operation and organization of port governance. There is, then, not just one trajectory of port reform and politico-administrative changes. Instead, there is a broad range of models, ranging from the fully privatized ports in the UK, a rather diverse situation in Continental Europe, to the case of the US, where port administration and policy remains primarily a municipal responsibility (Ng and Pallis, 2010). The governance of Latin American ports, for instance, gives a good indication of the mismatch between formal devolution and privatization processes on the one hand, and what Wilmsmeier and Sanchez (2017: 171, studying the case of Chile) have called a 'sclerotic' institutional situation, where a rigid policy and planning regime undermines the benefits of port reform. These internal issues currently add to difficult market conditions in the course of the economic crisis. Different countries have demonstrated rather unique patterns in this regard. For example, in South Africa, where successive waves of privatization shifted the ownership of ports from a state's department to a state-owned company, divided into distinct segments (Havenga et al., 2017). A separate account is provided by Asian ports, particularly by Chinese ports, which were essential for the country's emergence as the global manufacturing plant. The governance of ports in China was changed by various national port reforms since 2004, with a more recent focus on more integrative developments, including seaport collaboration, hinterland integration and new geo-economic initiatives (Notteboom and Yang, 2017).

Viewed from governments' perspectives, it looks as if port reforms have provided certain improvements, while overall claims to foster efficiency have not yet been brought to fruition (Brooks et al., 2017). At a higher scale, we do see some convergent patterns which can be considered a more direct outcome of globalization, as a certain internationalization and homogenization of regulation, policy making and knowledge production has taken place in the global maritime environment. The introduction of so-called 'flags of convenience', once initiated by ship-owners in order to independently register their vessels outside their country of origin (e.g. by choosing registry in countries such as Panama or Liberia), mainly to avoid labour and environmental regulations became widespread in the post-war era. More recent patterns of a related (re-)regulation of maritime and shipping industries include labour issues, market access or environmental security. These practices lead to a related convergence, which can be attributed to the work done by global regulatory bodies, most notably the International Maritime Organization (IMO) and the World Trade Organization (WTO).

Moreover, it is worth reflecting on the impact of authorities that have been supporting particular forms of port governance through policy discourses and particular narratives of 'reform'. The World Bank has been one of the main drivers of port reform, and thus of globalization in both developing and emerging economies, as well as in the industrialized world. The World Bank's 'port reform toolkit' (2007), for example, is a set of recommendations intended to provide direction to ports within the changing economic environment, by developing competitive strategies for ports and their governance, finding appropriate models for function and organization, and also for shared responsibility between public and private agents. The Bank developed a taxonomy of governance models that included four different forms of port organization:

> At the two extremes are the public service port and the wholly private port, both characterized by very little sharing of responsibility between public and private actors. In the third category, the tool port, the public sector is dominant as it owns the land, the infrastructure and the equipment, and private sector activity is limited to specific operations, most commonly cargo handling performed using equipment owned by the public authority. The fourth category, the landlord port, is one in which the public authority owns the land and the infrastructure and leases these to private operators as a concession, with equipment and operations in the hands of the private sector. (Debrie et al., 2013: 56–57)

The latter case still seems to be the most common model for the organization of port authorities. However, authors such as Debrie et al. (2013) are critical of the blueprint character of the Bank's partly mystified model, which tends to neglect the specificities of different regions or, particularly, different port cities.

In the same vein, the Organisation for Economic Co-operation and Development (OECD) has engaged port-city issues in the context of a series of global case studies of port cities. The studies emphasized issues of competitiveness, but also pointed to the difficulties of localizing the economic benefits of ports in an increasingly interconnected world (Merk, 2013). While mainports still turned out to be rather influential in achieving their goals, city governments have often struggled to achieve a balance between the port and other urban development pressures. Thus power relationships are generally most critical in terms of port policy and port-urban development, not least since the different forces that set the framework conditions for global maritime businesses and their variegated geographies are rather complex and consist of a variety of often conflicted economic actors: (1) shipping lines that capture much of the business, so they can put pressure on ports to host their flows at relatively low rates; (2) firms that trigger the demand for shipping and distribution (equipment manufacturers, retail chains increasingly including large Internet firms such as Amazon.com), thus putting pressure on shipping lines to offer lower freight rates; (3) local authorities in charge of large ports that are competing against neighbouring ports in order to remain the selected port of call of major shipping lines, thus putting pressure on local places to accommodate ever larger container ships; and also (4) national governments which seek to maintain or improve the competitiveness of their maritime industries and mainports by offering favourable fiscal and regulatory conditions (e.g. easy and cheap ship registry, less strict working conditions, tax reductions).

Since 2008, the financial crisis has added to the quandaries and political battles between the different actors in port-maritime and port-urban affairs, for four reasons. First, the crisis has revealed how vulnerable and volatile the trajectories of trade and thus of shipping markets are (Ng and Liu, 2010). Second, the once highly attractive field of ship financing turned out to be, instead, very risky for investors, who were often burdened with mounting losses and debts. Third, as port expansion plans turned out to be rather contested in many port cities, it becomes extremely difficult to balance the different needs and interests that are on display in port cities. Fourth, in the latter context, climate change is about to pose new challenges for port operations and port governance, as extreme weather conditions and particularly rising sea levels may disproportionately affect port cities (Ng et al., 2016).

CASE STUDY: THE PORT OF ROTTERDAM AND VENLO (THE NETHERLANDS)

Once a global colonial power, the Netherlands seems to have retained a distinct aptitude for trade and commerce, and for positioning local and regional places and businesses within global flows by strategically attracting and generating commerce through the development of successful maritime nodes at the local and regional scale (Levelt, 2010: 11). The growth of shipping container terminal operations at both the 'mainport' of Rotterdam, Europe's dominant global maritime port, and in Venlo, an emerging secondary, inland port, each comprise case studies of local maritime industries 'going global' by strategically and successfully inserting their local operations into global shipping networks.

Rotterdam clearly represents a template case of a highly dynamic mainport development that perfectly illustrates the changes that have taken place at port sites and in port cities as a consequence of both technological and organizational changes and the overarching process of globalization altering the spatial distribution of production and consumption. All major seaports have been confronted in recent years with pressures arising from the growth of global trade and the associated increase of container shipments. At the same time, their specific function as gateways for global flows has partly shifted to their hinterlands, which may be more conducive to organizing physical flows, in a larger spatial-organizational context.

Between the fifteenth and nineteenth centuries, Rotterdam was transformed from a fishing port to a global commercial trading hub. In the industrial era Rotterdam became a significant gateway for industrial activities in its hinterland, notably the Ruhr area (Loyen et al., 2003; Rodrigue, 1998–2017b). Since the 1980s, the Port of Rotterdam has benefited from the long-term mainport development strategy to prioritize the port (in addition to Amsterdam Schiphol Airport) as the Netherlands' primary transport hub. Approximately 175 000 people are employed by Rotterdam's port-related activities, processing 466 million tonnes of cargo in 2015, more than both the second and third busiest European ports (Antwerp and Hamburg) combined (Port of Rotterdam, 2017).

Like many other large-scale seaports in the West, the port continues to pursue a long-term strategy to operate as an international hub for logistics services through expansion, by extending the spaces of operation further downstream along the *Nieuwe Waterweg*, away from the city centre and towards the North Sea, first with the 1960s *Europoort* oil and petrochemical complex built on reclaimed land, and more recently with the massive *Maasvlakte 2* deep-sea extension some 50 kilometres from the centre of the city (Rodrigue, 1998–2017b). Subsequently, an ambitious, post-industrial urban waterfront conversion strategy was implemented in Rotterdam's original working port area, the *Kop van Zuid*, and more recently the adjacent *Stadshavens* project (Doucet, 2013). These developments took place against the background of changes in governance structures that were targeting Rotterdam as well, most notably as to the shift from the municipal port authority to the corporatized, business-like entity, the 'Port of Rotterdam' (De Langen and Heij, 2014; Dooms et al., 2013). Recently, there have been calls to reconsider the 'big is beautiful' mainport strategy of the national government and to opt for diversification of the port activities instead (RLI, 2016).

Speaking in more geographical terms, mainports such as Rotterdam are increasingly dependent on efficient transport corridors to their hinterland in order to cope with the growth of container shipments and to pass the transport units efficiently and rapidly through the port (port-city) bottlenecks to meet the requirements of just-in-time delivery. In this context, the *Betuwelijn* was planned to provide a dedicated freight-rail link between the Port of Rotterdam and the Ruhr area, with potential connections further hinterland. The A73 motorway, for example has already developed into a significant corridor that has attracted many logistics firms between Venlo and Nijmegen (Hesse, 2008: 51).

The small city of Venlo in Limburg, in the south-east of the Netherlands now serves as an 'extended gate' to major deep-sea gateways such as Rotterdam and Antwerp (Rodrigue and Notteboom, 2009: 177). Upstream from Rotterdam on the Meuse, Venlo has quickly become an important 'extended gate' between the Port of Rotterdam, the Port of Antwerp, and markets located across the nearby (open) border with Germany and elsewhere in Continental Europe. Its main function is to compensate for the shortage of land and accessibility at the core areas of the mainports. Inland ports such as Venlo are a more recent and unforeseen form of globalized logistics operations. These inland ports are becoming key nodes and independent structures within shipping corridors, and a distinct phenomenon presenting unique port-city challenges (Witte et al., 2014).

The pivotal moment in Venlo's trajectory occurred in 1992, when European Container Terminals (ECT) decided to expand its existing operations in Rotterdam by establishing a major logistics hub there, TCT Venlo, its first inland terminal (Rodrigue and Notteboom, 2009). ECT's decision was based on the scarcity of land and handling capacity in Rotterdam at the time, and the availability of easily developable and relatively cheap municipal land in Venlo as well as by the already well-developed infrastructural networks (the road, rail and waterways) around the inland city (Raimbault et al., 2015). A coalition of actors in the public and private sectors emerged to continue to attract the development of logistics companies and infrastructure (Raimbault et al., 2015). Across the province of Limburg economic and infrastructural development strategies prioritize logistics and external corporate investments (Hesse,

2015). The city's population rose from 64 392 at the beginning of the 1990s (before the arrival of ECT in Venlo), to 100 371 by 2015 (CBS – Statistics Netherlands, 2017).

Given the importance of foreign trade in general and the Dutch logistics industry in particular relative to the small size of the Netherlands itself, maritime logistics flows here exhibit a pattern of spilling over from major ports such as Rotterdam's, and into smaller cities such as Venlo. This 'extended gate' seems emblematic of the general phenomenon of 'port regionalization' (Notteboom and Rodrigue, 2005) emerging in the environment of most major contemporary seaports. It also demonstrates how local places have been successfully positioned between the global shipping flows in the deep-sea Ports of Rotterdam, and Antwerp, and supply chains deeper in the Continent.

CONCLUSION AND RESEARCH OUTLOOK

Over the last decades, if not the entire twentieth century, ocean transport has become intimately related to the geographies of globalization. Maritime industries have been operating at an increasingly global scale, deeply intertwined with the growth of world trade, the increasing dispersal and unbundling of production, the liberalization of markets, and also rapidly growing consumer base in emerging markets. These developments are, inevitably, reflected in the spatial layout and functioning of port cities.

On the demand side, it is obvious that further growth rates in world trade, an associated rise in the demand for goods movements and the related demand for the construction of new ships may become increasingly unpredictable industries. Also, maritime shipping is quite prone to cyclical patterns of growth and decline, indicating the increasing volatility of the business, which poses a major challenge to capital-intensive infrastructure investments (Hübner, 2017). Future prospects of globalization might be situated between different alternative, or complementary, scenarios (cf. Credit Suisse Research Institute, 2016: 20). Globalization marching on in the same direction as before could also be accompanied (or followed) by either an emerging multi-polar world with different patterns of trade or a possible re-nationalization of economic policies. The latter is obviously looming on the horizon as a consequence of *Brexit* and *Trump-ism*. This would have a significant impact on the role of port cities as major interfaces in the system of global flows.

On the supply side, the particular geographies of maritime globalization may continue to shift as well. For instance, new or extended corridors for shipping lines will develop, such as the Northwest Passage, the extended Panama Canal and the Nicaragua Canal, or the 'One Belt, One Road' initiative of the Chinese government. Closely related to the routing and scheduling decisions made by shipping lines and their alliances (although not exclusively determined by these), the role of mainports, secondary ports and their respective hinterland may change as well. This will then be important for the very geographical and urban dimensions that the port-city interface will play out in the future.

The rising interest of human geographers in maritime issues in general and in port-urban affairs in particular (see Ng et al., 2014) has already materialized in a variety of conceptual and empirical contributions to this subject matter. In light of the changes named above and given the contested nature of port-city relations in general, the issues that appear most challenging and relevant here include the possible limits to the growth for ports (and thus also for mega ships), the call for a more environmentally sensitive path of development for maritime operations, and a balanced development of maritime and urban issues within port cities.

REFERENCES

Boschken, H.L. (2013), 'Global cities are coastal cities too: paradox in sustainability?', *Urban Studies*, **50**(9), 1760–1778.

Brooks, M.R., K.P. Cullinane and A.A. Pallis (2017), 'Revisiting port governance and port reform: a multi-country examination', *Research in Transportation Business & Management*, **22**, 1–10.

Brownill, S. (2013), 'Just add water. Waterfront regeneration as a global phenomenon', in M.E. Leary and J. McCarthy (eds) *The Routledge Companion to Urban Regeneration*, London, New York: Routledge, pp. 44–55.

CBS – Statistics Netherlands (2017), *Population dynamics; birth, death and migration per region*, http://statline.cbs.nl/Statweb/publication/?DM=SLEN&PA=37259eng&D1=0,22-24&D2=0&D3=1049&D4=0,10,20,30,40,(1-1)-l&LA=EN&VW=T (accessed on 28 March 2017).

Coe, N.M. (2014), 'Missing links: logistics, governance and upgrading in a shifting global economy', *Review of International Political Economy*, **21**(1), 224–256.

Coe, N.M. and H.W.C. Yeung (2015), *Global Production Networks: Theorizing Economic Development in an Interconnected World*, Oxford: Oxford University Press.

Comtois, C. and B. Slack (2009), *The Geography of Transport Systems*, London: Routledge.

Credit Suisse Research Institute (2016), *Getting Over Globalization*, Zurich: Credit Suisse AG.

De Langen, P.W. and C. Heij (2014), 'Corporatisation and performance: a literature review and an analysis of the performance effects of the corporatisation of Port of Rotterdam Authority', *Transport Reviews*, **34**(3), 396–414.

Debrie, J., V. Lavaud-Letilleul and F. Parola (2013), 'Shaping port governance: the territorial trajectories of reform', *Journal of Transport Geography*, **27**, 55–65.

Desfor, G., J. Laidley, Q. Stevens and D. Schubert (eds) (2011), *Transforming Urban Waterfronts: Fixity and Flows*, London: Routledge.

Doig, J.W. (2001), *Empire on the Hudson: Entrepreneurial Vision and Political Power at the Port of New York Authority*, New York: Columbia University Press.

Dooms, M., L. Van der Lugt and P.W. de Langen (2013), 'International strategies of port authorities: the case of the Port of Rotterdam Authority', *Research in Transportation Business & Management*, **8**, 148–157.

Doucet, B. (2013), 'Variations of the entrepreneurial city: goals, roles and visions in Rotterdam's Kop van Zuid and the Glasgow harbour megaprojects', *International Journal of Urban and Regional Research*, **37**(6), 2035–2051.

Ducruet, C. and S.W. Lee (2006), 'Frontline soldiers of globalisation: port-city evolution and regional competition', *Geojournal*, **67**(2), 107–122.

Ducruet, C. and T. Notteboom (2012), 'The worldwide maritime network of container shipping: spatial structure and regional dynamics', *Global Networks*, **12**(3), 395–423.

Frémont, A. (2007), 'Global maritime networks: the case of Maersk', *Journal of Transport Geography*, **15**(6), 431–442.

Guerrero, D. and J.-P. Rodrigue (2014), 'The waves of containerization: shifts in global maritime transportation', *Journal of Transport Geography*, **34**, 151–164.

Hall, P.V. and M. Hesse (eds) (2013), *Cities, Regions and Flows*, Abingdon: Routledge.

Hall, P.V. and W. Jacobs (2012), 'Why are maritime ports (still) urban, and why should policy-makers care?', *Maritime Policy & Management*, **39**(2), 189–206.

Havenga, J., Z. Simpson and L. Goedhals-Gerber (2017), 'International trade logistics costs in South Africa: informing the port reform agenda', *Research in Transportation Business & Management*, **22**, 263–275.

Hayuth, Y. (1982), 'The port-urban interface: an area in transition', *Area*, **14**(3), 219–224.

Hesse, M. (2006), 'Global chain, local pain: the regional implications of global production and distribution networks in the German North Range', *Growth and Change*, **32**(4), 570–596.

Hesse, M. (2008), *The City as a Terminal. The Urban Context of Logistics and Freight Distribution*, Aldershot: Ashgate Publishing.

Hesse, M. (2015), 'Selling the region as a hub: the promises, beliefs, and contradictions of economic development strategies attracting logistics and flows', in: Julie Cidell and David Prytherch (eds), *Transport, Mobility and the Production of Urban Space*, New York and Milton Park, UK: Routledge, pp. 207–227.

Hesse, M. and J.-P. Rodrigue (2006), 'Global production networks and the role of logistics and transportation', *Growth and Change*, **37**(4), 499–509.

Hoyle, B.S. (1989), 'The port-city interface: trends, problems and examples', *Geoforum*, **20**, 33–42.

Hübner, J.-H. (2017), 'Shipping markets and their economic drivers', in: M.G. Kavussanos and I.D. Visvikis (eds), *The International Handbook of Shipping Finance. Theory and Practice*, London: Palgrave Macmillan, pp. 1–39.

Jacobs, W. and A. Lagendijk (2014), 'Strategic coupling as capacity: how seaports connect to global flows of containerized transport', *Global Networks*, **14**(1), 44–62.

Juhel, M.H. (2001), 'Globalisation, privatisation and restructuring of ports', *International Journal of Maritime Economics*, **3**(2), 139–174.

Levelt, M. (2010), *Global Trade and the Dutch Hub. Understanding Variegated Forms of Embeddedness of International Trade in the Netherlands. Clothing, Flowers, and High-tech Products*, Oisterwijk, Netherlands: Utgeverij BOX Press.

Loyen, R., E. Buyst and G. Devos (eds) (2003), *Struggling for Leadership: Antwerp-Rotterdam Port Competition between 1870–2000*, Heidelberg: Springer Science & Business Media.

McCalla, R.J., B. Slack and C. Comtois (2004), 'Dealing with globalisation at the regional and local level: the case of contemporary containerization', *The Canadian Geographer/Le Géographe canadien*, **48**(4), 473–487.

Merk, O. (2013), *The Competitiveness of Global Port-Cities: Synthesis Report*, Paris: OECD.

Ng, A.K.Y. and J.J. Liu (2010), 'The port and maritime industries in the post-2008 world: challenges and opportunities', *Research in Transportation Economics*, **27**, 1–3.

Ng, A.K.Y. and A.A. Pallis (2010), 'Port governance reforms in diversified institutional frameworks: generic solutions, implementation asymmetries', *Environment and Planning A*, **42**(9), 2147–2167.

Ng, A.K.Y., A. Becker, S. Cahoon, S.-L. Chen, P. Earl, Z. Yan (eds) (2016), *Climate Change and Adaptation Planning for Ports*, London: Routledge.

Ng, A.K.Y., C. Ducruet, W. Jacobs, J. Monios, T. Notteboom, J.-P. Rodrigue, B. Slack, K. Tam and G. Wilmsmeier (2014), 'Port geography at the crossroads with human geography: between flows and spaces', *Journal of Transport Geography*, **41**, 84–96.

Notteboom, T. and J.P. Rodrigue (2005), 'Port regionalization: towards a new phase in port development', *Maritime Policy & Management*, **32**(3), 297–313.

Notteboom, T. and Z. Yang (2017), 'Port governance in China since 2004: institutional layering and the growing impact of broader policies', *Research in Transportation Business & Management*, **22**, 184–200.

O'Connor, K., B. Derudder and F. Witlox (2016), 'Logistics services: global functions and global cities', *Growth and Change*, **47**(4), 481–496.

Olivier, D. and B. Slack (2006), 'Rethinking the port', *Environment and Planning A*, **38**(8), 1409–1427.

Port of Rotterdam (2017), *Facts and figures*, https://www.portofrotterdam.com/en/the-port/port-facts-and-figures/other-ports (accessed on 28 March 2017).

Priemus, H. (2001), 'Mainports as integrators of passenger, freight and information networks: from transport nodes to business generators; the Dutch case', *European Journal of Transport and Infrastructure Research (EJTIR)*, **1**(2), 143–167.

Raimbault, N., W. Jacobs and F. van Dongen (2015), 'Port regionalization from a relational perspective: the rise of Venlo as a Dutch international logistics hub', *Tijdschrift voor Economische en Sociale Geografie*, **107**(1), 16–32.

RLI (2016), *Mainport voorbij [Beyond Mainports]* http://www.rli.nl/sites/default/files/advies_mainports_voorbij_voor_website.pdf (accessed on 27 March 2017).

Robinson, R. (2002), 'Ports as elements in value-driven chain systems: the new paradigm', *Maritime Policy and Management*, **29**(3), 241–255.

Rodrigue, J.-P. (1998–2017a), *Freight transportation and value chains*, https://people.hofstra.edu/geotrans/eng/ch5en/conc5en/ch5c3en.html (accessed on 29 March 2017).

Rodrigue, J.-P. (1998–2017b), *Evolution of the Port of Rotterdam, 1400–2030*, https://people.hofstra.edu/geotrans/eng/ch4en/conc4en/evolrotterdam.html (accessed on 29 March 2017).

Rodrigue, J.-P. and T. Notteboom (2009), 'The terminalization of supply chains: reassessing the role of terminals in port/hinterland logistical relationships', *Maritime Policy and Management*, **36**(2), 165–183.

Rodrigue, J.-P. and B. Slack (1998–2017), *Intermodal transportation and containerization*, https://people.hofstra.edu/geotrans/eng/ch3en/conc3en/ch3c6en.html (accessed on 29 March 2017).

Soppé, M., Parola, F. and A. Frémont (2009), 'Emerging inter-industry partnerships between shipping lines and stevedores: from rivalry to cooperation?', *Journal of Transport Geography*, **17**(1), 10–20.

UNCTAD (2016), *Review of Maritime Transport 2016*, Geneva: UNCTAD.

Van der Lugt, L., M. Dooms and F. Parola (2013), 'Strategy making by hybrid organizations: the case of the port authority', *Research in Transportation Business & Management*, **8**, 103–113.

van Hamme, G. and M. Strale (2012), 'Port gateways in globalization: the case of Antwerp', *Regional Science Policy & Practice*, **4**(1), 83–96.

Verhetsel, A. and S. Sel (2009), 'World maritime cities: from which cities do container shipping companies make decisions?', *Transport Policy*, **16**, 240–250.

Verhetsel, A., T. Vanelslander and M. Balliauw (2016), 'Maritime world cities: development of the global maritime management network', *International Journal of Shipping and Transport Logistics*, **8**(3), 294–317.

Verhoeven, P. (2009), 'European ports policy: meeting contemporary governance challenges', *Maritime Policy & Management*, **36**(1), 79–101.

Vieira, G.B.B., F.J. Kliemann Neto and F.G. Amaral (2014), 'Governance, governance models and port performance: a systematic review', *Transport Reviews*, **34**(5), 645–662.

Wang, J.J. (2014), *Port-City Interplays in China*, Farnham: Ashgate.

Wilmsmeier, G. and R.J. Sanchez (2017), 'Evolution of national port governance and interport competition in Chile', *Research in Transportation Business & Management*, **22**, 171–183.

Witte, P., B. Wiegmans, F. van Oort and T. Spit (2014), 'Governing inland ports: a multi-dimensional approach to addressing inland port–city challenges in European transport corridors', *Journal of Transport Geography*, **36**, 42–52.

World Bank (2007), *The Evolution of Ports in a Competitive World. World Bank Port Reform Toolkit* (2nd edn), Module 2. Washington, DC: The World Bank.

PART V

GEOGRAPHIES OF GOVERNANCE

28. Global governance, human rights and humanitarianism

Barbara Oomen

INTRODUCTION

Human rights have been characterized as the 'lingua franca of global moral thought' (Ignatieff, 1999: 12). This illustrates the power, the spread and the degree of institutionalization of the basic idea that all human beings are born with universal, inalienable and indivisible rights. The Universal Declaration of Human Rights (UDHR), adopted by the United Nations in 1948, has 370 translations, more than any other document in the world. If humanitarianism is understood as working to advance these rights, there is a wealth of organizations and individuals active in this regard all over the world. They work within an ever-expanding framework of global governance, defined by Finkelstein as the governing of relationships of a variety of actors that transcend national borders without sovereign authority (Finkelstein, 1995: 369).

Yet, in spite of this general global rise of rights, their two central points of departure – that human rights are universal, and that states should be key actors in protecting them – have been subject to a variety of contestations over place and time. This chapter sets out the way in which the globalization of human rights has been conceptualized (i) both over time and place, and (ii) how it has been institutionalized. Two examples, concerning the rise of the Responsibility to Protect and of Human Rights Cities, illustrate the main point to be made about human rights in the context of globalization – that human rights theoretically and legally rely on nation states for their protection and promotion, but from an empirical perspective can only be fully realized by the combined forces of global governance and local adaption and protection.

CONCEPTUALIZATION

How is globalization perceived within the field of human rights? In order to answer this question, it is important to first distinguish a number of disciplinary perspectives upon human rights. Philosophy, law and the social sciences, including human geography, all offer – mutually supportive – understandings of what human rights and humanitarianism are and should be, what they are today and how they work within a context of global governance.

Human rights, after all, first of all concerns the *philosophical* idea that, to quote the first article of the UDHR, 'All human beings are born free and equal in dignity and rights' and that they are 'endowed with reason and conscience and should act towards one another in a spirit of brotherhood'. This idea has deep roots in the history of philosophy. The Hammurabi code already set rules for rulers to obey, and the Greek philosopher Aristotle formulated a concept of natural law, a law with universal validity. Enlightenment philosophers like Thomas Hobbes set out how the happiness of man could be secured via respect for natural rights, which was to be in the hands of the sovereign. John Locke added to this the insight that if rulers did not respect these rights, people were justified in rebelling against them, providing the legitimation for the French and the American revolutions. All these philosophers coupled their understanding of rights to their protection by a sovereign ruler. The move towards universalism was made by Immanuel Kant, who put forward the categorical imperative, the idea that rational men should only follow those rules which all other rational men would follow as well, and thus constituting a universal law.

The actual codification of such universal rights, after the atrocities of the Second World War, built on all these philosophical foundations. After the War, philosophers continued to either support or question the idea of human rights, or to inquire their nature – as with John Rawls, in his famous *Theory of Justice* (Rawls, 2009). Hannah Arendt, for instance, critically examined the value of human rights in the absence of a political authority standing above sovereign nation states – where were those people with nothing but their 'naked humanity' to turn to secure their human rights? (Arendt, 2004). Giorgio Agamben built on this critique with his notion of the 'homo sacer', like a refugee, who has much less rights than citizens of nation states (Agamben and Heller-Roazen, 1998). A different strand of philosophical critiques of human rights take issue with the Western Enlightenment roots of the concept, arguing for the need to include different epistemologies in its grounding and further development (Baxi, 2012; De Sousa Santos, 2015). More generally, the universality of human rights remains a subject of debate amongst philosophers (Donnelly, 2003), with some arguing for an approach based on human needs and capabilities, instead of on rights (Sen, 1999). This whilst other, more utilitarian discussions focus on how much there is in today's world to show that human rights have failed to deliver on their promise, arguing that in our multipolar world the heydays of human rights and global governance could well be over (Hopwood, 2013).

Whilst philosophers struggled, and continue to do so, to provide the philosophical foundations for human rights, global governance and humanitarianism, lawyers consistently developed the relevant fields of international law – international humanitarian law, international human rights law, international public law in general and, increasingly, international criminal law. International humanitarian law, for one, is the legal framework applicable during situations of armed conflict, codified largely in the Geneva Conventions of 1949, ratified by all UN member states (Crawford and Pert, 2015). International human rights law, in turn, is laid down in an ever-expanding number of international and regional treaties, with their implementation by the states who ratified them monitored by specialized bodies but also by national and regional courts. The International Bill of Rights, which consists of the International Covenant on Civil and Political Rights and the International Covenant on Economic, Social and Cultural Rights have been complemented by a variety of specialized treaties (like the Convention against Torture or the Convention on the Elimination of all forms of Racial Discrimination) or treaties specifying the relevance of the UDHR for specific groups, like women, children and people living with disabilities. Together, they set out a wide range of civil, political, economic, social and cultural rights applicable to every human being. In ratifying these treaties, states commit to respecting, protecting and fulfilling these rights (De Schutter, 2014).

Another ever-expanding legal field is that of international criminal law. The Genocide Convention, defining genocide and turning it into an international crime was adopted the day before the UDHR in 1948. The notion of international crimes, so heinous that they are of concern to the whole international community and must not go unpunished even if nations are unwilling or unable to prosecute them has received important legal recognition with the formation of the International Criminal Court (Schabas, 2001). This first 'world court', based in The Hague but recognized by 124 countries, prosecutes individuals for international crimes even when their countries of origin do not, thus forming another palpable example of the increase in global governance. For all the developments on the international plane, however, much of the most impactful activity in the field of human rights and humanitarianism is governed regionally. Courts like the Inter-American, African and European Court of Human Rights have large regional impact, not only because of the amount of cases put before them, but also because their case law becomes part of the law of the members states, and is subsequently applied by national judges and policy-makers on a daily basis.

It is this spread of human rights that has spurred disciplinary fields other than philosophy and law to take an interest in the topic: over the past years more and more social scientists have added their disciplinary perspective to the study of human rights. Anthropologists, first, were firmly opposed to the very notion of human rights in the 1940s, expressing the fear that the Universal Declaration would become a statement of the values prevalent in Western societies only, and not show enough regard for cultural diversity (AAA, 1947). In recent decades, however, anthropologists have not only embraced the notion of human rights but also produced a wealth of evidence in the past years on how norms travel between the global and the local (Goodale and Merry, 2007). One key focus is the interplay between cultural traditions and universal norms, and the legal pluralism that results from it (Oomen, 2014a). Another important emphasis is the role of 'translators' in translating human rights into the vernacular (Merry, 2006). Sociologists, in turn, have focused on the role of civil society in human rights activism, and on setting out why certain rights issues receive public attention and others not (Brunsma et al., 2015). Adding to this, a wealth of recent political science research has demonstrated not only the impact of human rights, but also the mechanisms, actors and processes behind such impact (Haglund and Stryker, 2015; Risse et al., 2009). In line with this more empirical approach towards human rights, social scientists have begun to develop human rights indicators, to measure progress in the field of civil, political, social, economic and cultural rights (Green, 2001). Geographers also have a key role to play in setting out the importance of place in the actual realization of human rights, and spatial inequality in access to rights (Carmalt, 2007).

Each of these disciplines, of course, provides a different conceptualization of the relationship between human rights, global governance and globalization. Philosophers emphasize the notion of common humanity. Lawyers discuss whether strengthening of human rights is needed, for instance via a world court. Social scientists emphasize the intricate relationship between globalization and human rights. Drawing together these insights, the following section will consider how global governance, human rights and humanitarianism have developed over place and time.

SPATIOTEMPORAL DIMENSIONS

When Eleanor Roosevelt told the United Nations General Assembly, on 10 December 1948, that the UDHR it had just adopted could well become 'the Magna Carta of all mankind', this was just one marking moment in a long process of formulating rights and assigning responsibility for their realization. Another such moment, in 1863, was the formation of the Red Cross, based on the idea of international cooperation to ensure care for all wounded during warfare. The rise in importance of nation states as the building blocks of international organization, in the nineteenth century, also strengthened the need for international cooperation, as marked by the Peace Conferences in the Hague and the League of Nations. However, the atrocities of the Second World War provided the push towards a new world order – with a key role for organizations like the United Nations and the Council of Europe, and human rights as a shared moral foundation. In the decades after the war, however, the Cold War kept human rights from realizing their double potential as a universal language concerning social justice, and a mechanism to limit state sovereignty in the interests of the individuals involved.

This started to change in the 1960s, with the formation of human rights NGOs like Amnesty International. The historian Moyn has plausibly dated the birth of human rights as both an ideology and a movement in the 1970s, the period in which many Western powers started – often upon prompting by their population – to include human rights in their foreign policies (Moyn, 2010). It was the end of the Cold War, with the fall of the Berlin Wall in 1989, however, that further pushed rights talk to global prominence. This period was characterized by states and international organizations setting human rights as a condition for international cooperation, and even taking over governance in countries like Kosovo and East Timor, but also by citizens from Argentina to New Zealand framing their claims for social justice on the basis of human rights. At the Vienna World Conference on Human Rights of 1993 all nations in the world reaffirmed their commitment to the purpose and principles of the UN and the UDHR. This apparent triumph of global governance, human rights and humanitarianism, however, substantially sobered after the turn of the century – with the rise of nationalism, the tenacity of religious and cultural identity and the 'War on Terror'. Protest against, for instance, torture by the United States does also show the degree to which human rights promotion has become a global endeavour, with Southern countries holding those in the West accountable to the norms that these Western countries once regarded mostly as export products and foreign policy concerns (Merry, 2009; Oomen, 2014b).

All this underscores the central paradox of human rights that are put to law and depend on protection by states that are also often their greatest violators. It is not surprising, therefore, that it is individuals and civil society organizations that have played a key role in the human rights movement. If it was not for civil society, it is unlikely that promoting and encouraging respect for human rights would have been included as one of the UN purposes, that the UDHR or any subsequent treaty, whether concerning children, women or people living with disabilities would have been adopted, and that institutions like the International Criminal Court (ICC) would have been set up (Glasius, 2006; Neier, 2012). The same applies for the actual realization of human rights. In writing about Latin America, Risse et al. (1999) have described the 'human rights spiral' of cooperation between local and international civil society that brings about change in a given state.

What, against this background, can be said about the global spread of human rights? From a formal point of view, this is remarkable. All countries in the world signed the UDHR, and virtually all countries in the world have ratified conventions like the Convention of the Rights of the Child (with the exception of the US) and the International Covenant on Civil and Political Rights (with the exception of China). Human rights also cover more and more issues – civil, political, economic, social and cultural – with explicit recognition of more and more groups – women, children, migrant workers, people with disabilities and lesbian, gay, bisexual and transgender (LGBT) people. All over the world, from Argentina to New Zealand, people conceive of and claim social justice within the framework of human rights. Social media offer unprecedented opportunities to bring human rights violations to the global attention, which sometimes leads to action. At the same time, human rights continue to be violated on a large scale all over the world, with violations ranging from mass killings and torture to day to day discrimination, and rights are under increased threat with the rise of populism, nationalism and religious fundamentalism (Human Rights Watch, 2015). Here, identity politics, both at state and at subnational levels, lead to a questioning of the notion of universal human rights.

ORGANIZATIONAL DIMENSIONS

The degree to which human rights deliver upon their promise strongly depends on the institutionalization of the concept, and power relations at play in its implementation.

Here, over the past decades, the global organizational framework to secure human rights and to monitor their implementation has considerably expanded. Under the auspices of the United Nations Economic and Social Council, specialized agencies like United Nations High Commissioner for Refugees (UNHCR), Children's fund (UNICEF), UN Women and the United Nations Development Programme (UNDP) all work on rights-related themes. In 2006, the UN created the UNHRC, the Human Rights Council, as a specialized intergovernmental body to promote and protect human rights. It does so, amongst others, via a process of Universal Periodic Review (UPR) in which all member states are questioned on their human rights record once every four years, and through focusing on thematic issues, often informed by the High Commissioner on Human Rights and Special Rapporteurs. In addition to this, every one of the ten UN core human rights treaties has its own monitoring mechanism. The Women's Convention, to give one example, is monitored by a committee to which countries submit four-yearly reports, and individuals can bring their complaints. These monitoring agreements, however, fall under the category of soft law, as monitoring mechanisms have very little means to enforce their decisions and recommendations.

Much real human rights enforcement, however, does not take place at the international, but rather at the regional and national level. Here, courts like the Inter-American, African and European Court of Human Rights play a key role. The case law of the European Court of Human Rights, for instance, has shaped policies, been quoted by national courts, and affected the lives of virtually everyone in the 47 Council of Europe member states that it unites. Within the European Union, staying in the same part of the world, fundamental rights have strongly gained importance inside Europe, and in its external relations. Nationally, all branches of government are held to the human rights agreements entered into, with national human rights institutes monitoring what they do in over 100 countries. To cite an example from the field of international criminal law: the ICC might be important in prosecuting the masterminds behind international crimes like torture, but national courts will prosecute the many other perpetrators, both within and outside of the countries concerned.

Securing human rights, however, is far more than an affair of states and their organs. The key role of civil society organizations has received increased recognition at the international level, for instance in allowing for more participation in UN proceedings. In addition, there is more and more attention for the role of other non-state actors, with regards to human rights protection, as illustrated by the adoption of guiding principles on business and human rights (Ruggie, 2007). In practice, furthering any given human rights issue depends on the collaboration between a variety of actors – state and non-state, local and global. Putting women's rights on the international agenda and working towards their implementation, for instance, is a combined effort of states, international organizations, civil society, the media, employers and other actors, amongst which women themselves (Hellum and Aasen, 2013). In seeking recognition for indigenous rights, to give another example, the Saramaka people of Surinam enlisted the support of NGOs, international lawyers and other actors to support their successful claim for recognition of their land rights by the Inter-American Court of Human Rights (Price, 2011).

Here, power dynamics like access to funds and information crucially determine which rights issues get to occupy centre stage. The Western dominance in providing philosophical foundation for, and actually formulating human rights, long lead to more global attention for civil and political rights, like the freedom of expression, than for economic, social or cultural rights. This has changed with the rise to global power of countries like Brazil, South Africa and India, in which civil society and courts have substantially strengthened the actual value of social and economic rights (Gauri and Brinks, 2008). An ongoing example of 'donor-driven justice' is the field of transitional justice; the way in which justice after violent conflict gets conceptualized and implemented, often bearing the stamp of those who pay for it more than the interests of victims (Oomen, 2005). The world's top-10 human rights organizations, headed by Amnesty International, Global Rights, Human Rights Watch, the International Committee for the Red Cross and the International Federation for Human Rights, might all operate globally but originate and have their headquarters and main funders in the Western hemisphere, selecting priorities accordingly. Small Southern organizations, in contrast, are often understaffed and lack funding to combat their priorities, like poverty. An additional problem is the way in which authoritarian regimes bar civil society and the spread of information, making that some of the worst human rights violations in today's world, as in North Korea and Ethiopia, lack solid coverage.

The structures to secure human rights, in sum, might be firmly in place, but in practice some human beings are still treated much more equally than others. One key explanation here lies in the fact that securing human rights is still primarily a state responsibility, but not always in the state interest.

TWO EXAMPLES

Let us consider two examples of the interplay between human rights and globalization that illustrate how the human rights movement questions classic notions of state sovereignty, and operates in a permanent to-and-fro between global and local forces.

The Responsibility to Protect

One of the classical challenges to the very idea of human rights, as set out by philosophers like Arendt and made painfully clear by mass murder in countries like Cambodia, Rwanda and Syria is the question what can be done the moment that states do not protect, but actually violate the human rights of their own citizens. Should the international community step in, and what grounds are there to do so? The UN Charter prohibits the use of force against other states, but some member states, especially after the Kosovo crisis, have argued in favour of the possibility of a humanitarian intervention. The UN Security Council, as the only UN body with the right to authorize collective action to maintain international peace and security, however, has been very hesitant to accept such a notion. A closely related principle, however, rose to prominence after the turn of the century. In 2005, against the background of the atrocities in Rwanda and Srebrenica a decade before, the United Nations endorsed the principle of the Responsibility to Protect (R2P) (Evans, 2009). This principle is based on the notion that all states have the responsibility to protect all population from genocide and mass atrocities, and was worked out by the International Commission on Intervention and State Sovereignty on the basis of existing international law (ICISS, 2001). If states refrain from protecting their citizens against international crimes, the UN Security Council can, as a measure of last resort, sanction the use of force under the R2P principle. This R2P is coupled to two other principles, the responsibility to prevent, and the responsibility to rebuild.

Since its adoption, the Security Council has invoked the R2P in cases like the Ivory Coast, Yemen, Mali, Sudan and South Sudan. An important invocation was in the case of Libya, in 2011, the only instance in which the R2P was used to authorize a military intervention. The systematic attacks by Muammar Gaddafi on the Libyan population, in combination with his threats of imminent mass murder prompted a Security Council Resolution, referring to the R2P. This formed the basis for NATO planes bombing Gaddafi's forces in March 2011 (Hehir, 2013). The international community left as soon as it came in, and was criticized, afterwards, for not basing the intervention upon the right premises (Gaddafi's intentions towards mass murder were contested) and for not staying long enough to give meaning to the third element of the R2P – the responsibility to rebuild (Bellamy, 2014). Clearly, securing human rights calls for long-term commitment and a key role for local actors.

Human Rights Cities

The invocation of the R2P shows the breaches in the sanctity of state sovereignty made in cases of the grossest human rights violations, as does the role of the ICC in today's world. The centrality of the state, however, is also increasingly put to task where it concerns a much wider range of human rights. One interesting example here is the rise of human rights cities, local authorities that explicitly base their policies on international human rights law (Oomen et al., 2016). Gwangju in South Korea, for instance, building on the tradition of urban resistance that goes back to the 1980 Gwangju uprising, seeks to both develop democracy and human rights in the local community and to spread them widely, and hosts the World Human Rights Cities forum every year. San Francisco, to quote another example, has passed an ordinance to base part of its policy on the UN Women's Convention, an international treaty that the United States has not even ratified. Cities worldwide, additionally, have incorporated Lefebvre's classic notion of the right to the city into urban legislation – Mexico City forms one example of the many cities to take the Right to the City as a stepping stone for local decision-making (Oomen et al., 2016).

The degree to which local authorities are stepping up to take responsibility for the realization of international human rights is not surprising. Economic and social rights have become more and more prominent over the years, and the role of local authorities in delivering services like housing, education and access to work has increased as a result of decentralization policies worldwide. In addition, more people live in cities than ever, and their highly diverse background strengthens the need for a common normative framework. In line with the networked, multi-layered character of governance in this day and age, such cities organize themselves in global city networks. The European cities that signed the European Charter for the Safeguarding of the Human Rights in the City is one example, as are the cities worldwide brought together by the NGO PDHRE (People's Decade on Human Rights Education).

One factor that motivates cities to engage with both human rights law and discourse is to distance themselves from more conservative and exclusionary national policies. The – over 300 – sanctuary cities in the United States, for instance, protect undocumented migrants by not prosecuting them and offering other types of services (Lippert and Rehaag, 2012). Similarly, cities of sanctuary in the UK explicitly welcome refugees, with greater respect for their human rights than is called for by national policies. During the 2015 refugee crisis cities like Barcelona, Athens and Amsterdam took the initiative in refugee welcome. Networks seeking to facilitate such processes, like Solidacities and Eurocities, show how local authorities have become yet another actor in the global governance of human rights.

CONCLUSION AND RESEARCH AGENDA

The appeal, spread and institutionalization of human rights relates to globalization in a variety of ways. It is one of its drivers, with an increased strengthening and explication of common norms and the people and institutions committed to securing them (Cushman, 2015). It is, however, also a product of the other forces of globalization –

the increase in spread and integration of information, economies and peoples enables exchange on, and strengthening of human rights. Finally, the recognition of the equality and dignity of every human being can serve as an important countervailing vision against some of the vicissitudes of, for instance, economic globalization or the encroachment upon privacy that comes with advances in the field of information technology (Brysk, 2002).

It has become clear that, over the past decades, human rights have been institutionalized to an unprecedented degree, with every state in the world pledging adherence to them on paper. For human rights to deliver upon their promise of global justice, however, more is needed. First, there is a need to build upon, and to engage, a variety of worldviews, religions and cultures worldwide in both human rights agenda-setting and their implementation. Next, the move from formulation to implementation of human rights can only take place successfully if driven forward by local actors, with state support and the benefits of international networks and all that global governance and globalization have to offer.

This then, should be the future research agenda for human rights in relation to globalization. It ranges from the epistemologies of human rights to the practice of their implementation by and for those actors and individuals that stand to benefit most. Human rights, after all, might have become a global moral lingua franca, but do need to listen, speak and relate to the other languages of the world, in order to move – to a much larger extent than is the case today – from talk to action.

REFERENCES

AAA. (1947), 'Statement on human rights', *American Anthropologist*, **49**(4), 539–543.
Agamben, G. and D. Heller-Roazen (1998), *Homo Sacer: Bare Life and Sovereign Power* (trans. Daniel Heller-Roaze), Stanford: Stanford University Press.
Arendt, H. (2004), *The Origins of Totalitarianism; introduction by Samantha Power*, originally published 1951: New York: Harcourt.
Baxi, U. (2012), *The Future of Human Rights*, Oxford: Oxford University Press.
Bellamy, A.J. (2014), 'From Tripoli to Damascus? Lesson learning and the implementation of the Responsibility to Protect', *International Politics*, **51**(1), 23–44.
Brunsma, D.L., K.E.I. Smith, and B.K. Gran (2015), *Handbook of Sociology and Human Rights*, Boulder and London: Paradigm Publishers.
Brysk, A. (2002) *Globalization and Human Rights*, Berkeley: University of California Press.
Carmalt, J.C. (2007), 'Rights and place: using geography in human rights work', *Human Rights Quarterly*, **29**(1), 68–85.
Crawford, E. and A. Pert (2015), *International Humanitarian Law*, Cambridge: Cambridge University Press.
Cushman, T. (2015), 'The globalization of human rights' in: B.S. Turner and R.J. Holton (eds), *The Routledge International Handbook of Globalization Studies*, Abingdon: Routledge, 589–603.
De Schutter, O. (2014), *International Human Rights Law: Cases, Materials, Commentaries* (2nd edn), Cambridge: Cambridge University Press.
De Sousa Santos, B. (2015), *If God Were a Human Rights Activist*, Stanford: Stanford University Press.
Donnelly, J. (2003), *Universal Human Rights in Theory and Practice*, Ithaca: Cornell University Press.
Evans, G. (2009), *The Responsibility To Protect: Ending Mass Atrocity Crimes Once and For All*, *Responsibility to Protect*, Washington: Brookings Institution Press.
Finkelstein, L.S. (1995), 'What is global governance?', *Global Governance*, **1**(3), 367–372.
Gauri, V. and D.M. Brinks (2008), *Courting Social Justice: Judicial Enforcement of Social and Economic Rights in the Developing World*, Cambridge: Cambridge University Press.

Glasius, M. (2006), *The International Criminal Court: A Global Civil Society Achievement*, Oxford and New York: Routledge.

Goodale, M. and S.E Merry (2007), *The Practice of Human Rights: Tracking Law between the Global and the Local*, Cambridge: Cambridge University Press.

Green, M. (2001), 'What we talk about when we talk about indicators: current approaches to human rights measurement', *Human Rights Quarterly*, **23**(4), 1062–1097.

Haglund, L., and R. Stryker (2015), *Closing the Rights Gap: From Human Rights to Social Transformation*, Oakland: University of California Press.

Hehir, A. (2013), 'The permanence of inconsistency: Libya, the Security Council, and the Responsibility to Protect', *International Security*, **38**(1), 137–159.

Hellum, A., and H.S. Aasen (2013), *Women's Human Rights: CEDAW in International, Regional and National Law* (Vol. 3), Cambridge: Cambridge University Press.

Hopwood, S. (2013), *The Endtimes of Human Rights*, Cornell: Cornell University Press.

Human Rights Watch (2015), *World Report 2015*, https://www.hrw.org/sites/default/files/wr2015_web.pdf (accessed 16 June 2018).

ICISS (2001), *The Responsibility to Protect*, http://www.dfait-maeci.gc.ca/iciss-ciise/pdf/Commission-Report.pdf (accessed 16 June 2018).

Ignatieff, M. (1999), *Whose Universal Values? The Crisis in Human Rights*, Amsterdam: Praemium Erasmianum Foundation.

Lippert, R., and S. Rehaag (2012), *Sanctuary Practices in International Perspectives: Migration, Citizenship and Social Movements*, New York: Routledge.

Merry, S.E. (2006), *Human Rights and Gender Violence: Translating International Law into Local Justice*, Chicago: University of Chicago Press.

Merry, S.E. (2009), 'Human rights in the imperial heartland', in: J. Maskovsky and I. Susser (eds), *Rethinking America: The Imperial Homeland in the 21st Century*, Herndon: Paradigm Publishers, 49–65.

Moyn, S. (2010), *The Last Utopia: Human Rights in History*, Boston: Harvard University Press.

Neier, A. (2012), *The International Human Rights Movement: A History*, Princeton: Princeton University Press.

Oomen, B. (2005), 'Donor-driven justice and its discontents: the case of Rwanda', *Development & Change*, **36**(5), 887–910.

Oomen, B. (2014a), 'The application of socio-legal theories of legal pluralism to understanding the implementation and integration of human rights law', *European Journal of Human Rights*, **4**, 471–495.

Oomen, B. (2014b), *Rights for Others: The Slow Home-Coming of Human Rights in the Netherlands*, Cambridge: Cambridge University Press.

Oomen, B., M. Davis and M. Grigolo (2016), *Global Urban Justice: The Rise of Human Rights Cities*, Cambridge: Cambridge University Press.

Price, R. (2011), *Rainforest Warriors: Human Rights on Trial*, Philadelphia: University of Pennsylvania Press.

Rawls, J. (2009), *A Theory of Justice*, Boston: Harvard University Press, originally published 1971.

Risse, T., S.C. Ropp and K.A. Sikkink (1999), *The Power of Human Rights: International Norms and Domestic Change*, Cambridge: Cambridge University Press.

Ruggie, J.G. (2007), 'Business and human rights: the evolving international agenda', *The American Journal of International Law*, **101**(4), 819–840.

Schabas, W.A. (2001), *An Introduction to the International Criminal Court*, Cambridge: Cambridge University Press.

Sen, A. (1999), *Development as Freedom*, New York: Anchor Books.

29. States, globalizing tendencies and processes of supranational governance

Alun Jones

GEOGRAPHICAL CONTRIBUTIONS TO THE STUDY OF EU INTEGRATION AND SUPRANATIONALISM

Globalization has unfolded together with, and in some ways fostered, macro-regionalization among states. Across the globe, there are numerous examples of macro-regionalization be it in Southeast Asia, North and South America, or South Africa. However, it is the European Union (EU) as a supranational project that represents the closest, though still incomplete, expression of macro-regionalization. Pinning down the nature and shape of its identity has occupied academic researchers since its creation in the years that followed Europe's brutal twentieth-century's wars. For some the EU is on a journey to an unknown destination, is only 'partially formed' and perennially suffers from capability and expectations deficits. The journey has been marked by frequent setbacks, political reversals, blocking, and regular outbreaks of nationalism as state sovereignty forcefully butts up against political integration agendas. At a time when the need for a united response to the challenges of globalization has never been greater – be it combating financial instability and climate change, and easing transnational flows of labour and migration, or tackling the changed geopolitical global landscape and the rise of new geo-economic powers – the EU's capacity to act is often in disarray. Most problems stem from the increasingly vexed relations between the EU and its member states over the political management of closer economic integration. From the wellsprings of Europe's diversity, political unity has taken on rather more mercurial qualities.

Academic conceptualization of the supranational project in Europe post-1945 follows the contours of its political development. Emphasis at particular moments has been placed on approaches that attempt to capture its politico-economic dynamics and their spatial consequences. In broad terms one can periodize the EU into six key phases in its development, and within each a mixture of integration efforts, often supported by binding legislation, and vocalized conflicts over their depth, speed and consequences for state sovereignty. Geographers have also long been interested in the European project, and their research activities have responded in important ways to these politico-economic developments. Territory and territoriality have been critical vectors for them in investigating the European integration process (Mamadouh and Van der Wusten, 2008). Thus, they have examined the growth and development of territory under expansive supranational governance, the territorial expression of supranational integration across political landscapes, and the territorial challenges and responses to economic globalization. Geographers have also in turn placed particular emphasis upon the spatialities of integration, the scaling of territoriality and the simultaneous transfer of state power upwards to supranational agency. In research terms, this has furnished endless opportunity for the scrutiny and critique of the unfinished adventure that is European integration (Jones and Clark, 2010).

Over time, geographers have increasingly treated the EU as a bounded socio-spatial entity that engages globally through a distinctive mode of geopolitical behaviour underpinned by a set of norms and values (Bialasiewicz, 2011). Its member states subscribe to these normative dimensions and prospective members are forced to do likewise. European integration therefore invokes particular imagined geographies, of both itself and of the world. Consequently, political geographers have not only focused upon the multi-scalar (re-)configuration of European governance but also addressed supranationalism as a 'bordering' project linking what might be described as Europeanization processes with territory (Clark and Jones, 2008). Economic geographers, meanwhile, have tended to focus on the consequences of a changing Europe for its, broadly interpreted, geo-economic power. Here, they have emphasized how new architectures of supranational governance impact upon European competitiveness, economic outlook, regional growth and inequality. Particularly, they have examined the role of supranational institutions and their policies in addressing European cohesion, promoting regional growth and social inclusion, and stimulating cross-border exchange, innovation and cooperation (Crescenzi and Rodríguez-Pose, 2011). Their work has not only charted enlargement of the supranational project and its economic consequences, for example, recent economic transitions in the eastern enlargement of the EU, but also examined the increasing openness and interpenetration of national economies and sovereign states (Smith, 2013). In this they have also re-asserted the region as a critical locus of economic order and as a potent foundation of competitive advantage. Consequently, both economic and political geographers have begun to tackle how supranationalism has led to growing regional self-consciousness in socio-political and economic terms and spurred efforts to promote and mobilize Europe globally.

From the late 1980s the creation of a single market, supported by a common currency, and a commitment on the part of the EU to social cohesion and economic and financial solidarity have become the edifice of supranationalism in Europe. As with political geography research, EU enlargement has stimulated much interest in supranationalism among economic geographers. The release of market-driven dynamics across Europe has highlighted the tension between integration and transition. It has put into sharp relief the considerable socio-economic differences within the expanded EU and the political difficulties of addressing these (Pickles and Smith, 2005). Keeping uneven development in the EU within tolerable limits has proved untenable. The crisis of 2007/2008 and the aftershocks have acutely challenged the EU's commitment to cohesion and solidarity as core EU economies have been reluctantly forced to play the role of debt guarantors to 'peripheral' economies, whilst also imposing severe structural adjustments. The crisis also highlights the mismatch between an EU of monetary integration and the much more fragmented commitment to political union, highlighting the tension of governing a monetary union in the absence of deep-rooted, universally accepted political union (Martin and Pollard, 2017).

PERIODIZING SUPRANATIONAL GOVERNANCE IN EUROPE

There are a number of ways in which the phases of supranational governance in Europe might be identified, either drawing upon landmark events such as enlargement or treaty changes, or crisis moments where state resistance to European integration efforts is rekindled. Here, we set out a longitudinal examination of the evolution of supranational governance in Europe based upon six key phases in its development. In each phase the emerging architecture of the 'European project' is briefly outlined to expose the fragility and precariousness of the state-supranational governance relation.

1945–1957: Formative Years of Supranationalism: Trade and Economic Cooperation

In the early years following the end of the Second World War the mainstream advocates of European integration were inspired by federalism. However, key to integrationist thinking among political elites at this time was the avoidance of war and a particular fear in France of Germany's political and economic resurgence. Moreover, there was wide acknowledgement of the significance of coal and steel industries in the economic and political DNA of both countries. Integrating West Germany into a supranational project on the European continent without exacerbating France's mistrust of its neighbour was therefore essential. Jean Monnet, who was in charge of plans for revitalizing the post-war French economy, drew up proposals for a European Coal and Steel Community (ECSC) that would pool coal and steel resources under the executive coordination of a Higher Authority. Underlying this was a view that supranational integration could not be achieved all at once, nor in a single framework, but, rather, through concrete measures that would bring about solidarity in one limited but decisive economic sector. In doing so, war between France and (West) Germany would not only become unthinkable but impossible. In short, European integration could be built only through practical achievements (Archer, 1990).

The ECSC, established by a treaty signed in Paris in 1951, brought together not only West Germany and France but also Italy, the Netherlands, Belgium and Luxembourg. It marked the first stage of supranational governance and first attempt at containment of the nation-state in post-war Europe. Additionally, it set out a potential roadmap for further integration based upon key economic sectors, the creation of a High Authority above the nation-state that would oversee integration and enable it to sponsor further integration efforts, as well as progressively champion the entanglement of national economies. This approach to supranational governance was termed neo-functionalism; an incremental project for integration that would see loyalty transferred from nation-states to a European authority, an economic integration that would necessitate further European interventions and supranational institutionalization. For example, a coal and steel community would require cooperation over transport policy at a European level. Crucially, economic integration would spur political cooperation in Europe, put differently; the economic would 'spillover' into the political. Neo-functionalism was thus a logic of European integration that was driven by both the widening and deepening of supranational governance and fear of the spread of communism across Europe.

Between 1950 and 1957, discussions took place over the creation of a common market among the six ECSC member states. This would seek to abolish quotas and tariffs on trade between them, establish a common external tariff, unify trade policy towards the rest of the world, draw up common policies for a number of key sectors, and organize an internal market. Signed in Rome in March 1957, the European Economic Community (EEC) promised ever closer union between its member states, their peoples, economies and regions. Its creation was the most significant event for the economic, political, and social geography of Europe in the second half of the twentieth century. The intended free movement of goods, labour, services and capital across its members was vital to the economic resurgence of Western Europe, though of course this came with challenges to the autonomy of the state. Thus, state-imposed limits to European supranationalism drove much of the negotiations.

The consequence of these state fears of European encroachment was a hybrid EEC between supranationalism and intergovernmentalism reflected in institutions such as the European Commission (which replaced the High Authority) charged with the advocacy and protection of European integration, a Council of Ministers that would legislate, a Court of Justice that would adjudicate and a European Assembly (now Parliament) that would offer opinion. Tensions over the (dis)benefits of this integration approach were never far from the surface and focused upon the loss of sovereignty, national independence and national identity, a perceived reduction in the powers of national governments to deal with increased competition, loss of jobs and other economic threats caused by the dismantling of tariff barriers. Supporters of integration, on the other hand, emphasized the peace and security engendered by integration, the opportunities presented for trade in a bigger geographical market, the benefits from free movement of people between countries and the prospects for the EEC to act as a 'global (soft) power' that is, as a counterweight to the US–USSR military-based power dynamic. Indeed, the EEC's Treaty of Rome made specific provision for trading links and closer association with non-member states both inside and outside Europe, thus placing itself potentially as an attractor node in the web of global trading relationships.

Geographically then, the EEC was a Western European project determined by national requirements; a cooperative venture in which states with their own geography, history, language, traditions and institutions were the preeminent actors. Put bluntly, the EEC was something made in (Western) Europe for Europe.

1957–1973: Building Supranationalism by Institutionalizing European Integration

Bringing six countries into a common market required an institutional and regulatory framework. The Treaty of Rome intended this to be achieved over a twelve-year period during which there would be a staged removal of trade barriers. By 1968 the EEC had agreed a common external tariff on imports from outside the Community and had also set up a Common Agricultural Policy (CAP) with the aim of increasing farm output, incomes and ensuring market stability for a range of crop and livestock sectors. The CAP came to symbolize the integration project, attracting not only most of the EEC's budgetary funds but also large amounts of criticism. During the 1960s trade levels within the EEC grew three times faster than trade with countries in the rest of the world and per capita income levels grew at over 4.5 per cent; a figure that can in part be attributed to the 'EEC effect'. What was also significant was the ability of the EEC to negotiate as a bloc on trade agreements with other countries both in Western Europe and in Africa. This enthusiasm about the EEC was, however, tempered by serious disputes about the limits of supranationalism especially the growing power of the European Commission, and its proposals for decisions to be made on qualified bases within the Council of Ministers. Creeping supranationalism driven by the European Commission particularly angered the French government which boycotted the Council in 1965. The 'empty chair' crisis, as it became known, represented a major challenge to the emergent Community. Its resolution, by allowing member states to veto legislation if national interests were threatened, set back the Community; the spirit of ever closer union was thus confronted by more pragmatic and nationally centric priorities. The political, economic and institutional interleaving of supranationalism and its emergent nationalisms was symbolized by the Community's inability to agree on the city location for the European project with EEC institutions located in Brussels, Luxembourg and Strasbourg.

By 1970 the idea of European unity had been considerably fashioned by French interests – from a productivist CAP, closer links with former French colonies in Africa, distancing from the US and Atlanticism, and resistance to overtures from UK governments to join the Community. From a conceptual standpoint, the appearance of French nationalism in the EEC during the 1960s compromised the explanatory potential of neo-functionalist accounts with their integrationist logics based on technocratic rationales. In essence, intergovernmentalism began to outweigh supranationalism as an explanans of integration in Western Europe. Member state governments and not Community institutions were thus seen to determine the speed and depth of European integration.

1973–1986: Tensions of Supranationalism: Enlargement, Conflict and Growing Nationalism

Economic enmeshment with the European integration project was the primary motive for successive UK applications to join the EEC. Changes in French political leadership and less hostility to the potential disruptive tendencies of the UK in the European project, paved the way for full British membership in 1973 (along with that of Denmark and Ireland). This first enlargement of the EEC coincided with global political uncertainty, the Middle East conflict and a quadrupling of oil prices that unleashed a rapid downturn in economic growth in Europe. Consequences for manufacturing industries prompted increased unrest in the UK and inevitably calls grew for a referendum on continued membership of the supranational project. The ambivalence of British attitudes to supranationalism in Europe became increasingly apparent, despite the result of the 1975 referendum committing the UK to the project which, in its wake, left most of the principal political parties in the UK divided on the issue. Political and public anxieties arose because European integration was considered as compromising the power of parliament and government. The key question was the degree to which British sovereignty could be modified in a globalized interdependent world system to enable pragmatic relations with its European partners. A number of inducements were given by the EEC to the UK to shore up its support for the supranational project, not least the allocation of funds to tackle growing regional disparities. By the early 1980s, the Thatcher government in the UK began to reveal a deep-seated prejudice towards supranationalism and what she regarded as its undermining of the nation-state. In her view, European integration should not extend beyond the removal of barriers to trade and investment (particularly services) and the coordination of European economic policies based on the paramountcy of intergovernmental relations. Particularly galling to her was the unfairness of the UK's contribution to the EEC budget, which, because of the mechanics of the CAP and the UK's large volume of imports from outside Europe (which were penalized through common external tariffs imposed by the EEC), the UK found itself a net contributor to a European project for which it had only lukewarm enthusiasm. Resolving this perceived injustice dominated the Thatcher government's approach to supranationalism (that is, full support for open economic competition and a free market though with limited bureaucratic interference from the EEC) in the early 1980s and, as a consequence, retarded European political integration efforts throughout this period.

By the mid-1980s the European integration project had in many commentators' eyes stalled. It badly needed an injection of political purpose and commitment. The Single Market Programme launched by the European Commission in 1985 provided this renewed ambition (Laffan et al., 2000). Its principal goal (and thus one the UK Thatcher government could endorse) was to remove all quantitative and qualitative barriers to the free movement of goods and services across the EC in order to create a market place of over 250 million people, and in doing so resuscitating European growth and competitiveness internationally (European Community, 1988). However, it would require changes to the governance of the EC – speedier decision making, increased surveillance responsibilities for the European Commission, and pressures for more deeply integrated economic and monetary policies. The transformation of the European project was thus launched and, with it, new rounds of political wrangling between states over the nature and speed of integration in a Community of now 12 states with the accession of Greece (1981) and Spain and Portugal (both 1986).

1986–2000: Deepening Supranationalism for a Competitive EU in a Globalizing World

Lifting the European Community from its stagnant to its dynamic phase required improving decision making (including recognition of the European Court of Justice in the defence and promotion of EC integration), enhancing administrative efficiency, promoting market liberalization, achieving socio-economic cohesion and projecting Europe as a global actor. These goals necessitated an extension of the supranational project into hitherto restricted areas such as economic and fiscal measures, foreign and security policies, and institutional re-organization. Revitalizing supranationalism would thus require treaty changes endorsed by all the member states. The Single European Act of 1986 paved the way for the Single Market Programme. This, it was argued would enable Europe, with its market of over 350 million people (with the accession of Spain and Portugal in 1986), to compete favourably in the global economy. Geographically, of course, it would increase the volume of trade within the EC, offer companies opportunities to operate in a larger market and better exploit economies of scale and secure competitiveness. However, the dismantling of barriers would also expose regional and sectoral vulnerabilities and aggravate, potentially, uneven regional development. To address this, the EC agreed on a package of funding measures (known as Structural Funds) that would be targeted at the development problems of the poorest regions of the Community.

The Single Market Programme also spurred a fresh round of negotiations on economic, monetary and political union. Coinciding with the collapse of communism and the unification of Germany in 1990, talks began on how to usher in a New Europe in the post-Cold War era based upon closer political cooperation and underpinned by economic and monetary union. In the UK, the conservative government and mainstream press unequivocally opposed economic and monetary union, seeing it as an unacceptable abrogation of national sovereignty, and a blatant attempt by the forces of supranationalism to aggrandize power in Brussels.

By the early 1990s therefore Europe was in the throes of momentous political, economic and institutional change. Events in Poland and Hungary, the reunification of Germany, the dissolution of the Soviet Union in 1991, and moves to deepen political-economic and monetary union in the EU presaged a decade of historic change in Europe (Jones, 1995). The Treaty on European Union (TEU), signed in Maastricht in February 1992, and entering into force in November 1993 marked a key stage in the development of the EU by setting out a political direction for Europe, involving economic and monetary union, EU citizenship, social rights, enhanced democracy and roles for EU institutions, and a common foreign and security policy. The Treaty of Maastricht thus enabled the EU to go clearly beyond its original economic objective, that is, the creation of a common market, and set out its political ambitions. However, not all member states in the EU shared the enthusiasm for deeper European integration, giving rise to a situation of 'multiple Europes' moving at different speeds reflecting varying degrees of commitment to the integration project (Graham, 1998).

The UK, for example, did not support membership of a European currency union, the 'Eurozone' and steadfastly protected its own currency. Other member states signed up to the timetable for monetary union which involved the free movement of capital, the mutual convergence of their economic policies, and the creation of a single currency (Euro) and establishment of a European Central Bank (ECB). Deepening political integration also involved transferring powers to the EU's institutions, strengthening their democratic legitimacy and, at the same time, seeking to improve their effectiveness.

Throughout the 1990s further reforms to the EU were put in place and a first round of post-Cold War enlargement was completed in 1995 with countries that were economically and politically liberal democracies with market economies, although geopolitically not committed to the Western camp (Finland, Sweden, and Austria). The Treaty of Amsterdam, signed by EU states in 1997, sought to establish an area of freedom, security and justice for EU citizens. The Schengen Area agreement (EU states except UK and Ireland) led to the removal of internal border controls and a common visa policy, operating very much like a single state for international travel purposes with external border controls for travellers entering and exiting the area. Switzerland,

Norway and Iceland, while not members of the EU, are participants in the Schengen Area agreement.

The Treaty of Amsterdam also set out new principles and responsibilities for the EU in the field of common foreign and security policy, with the emphasis on projecting the EU's values to the outside world, protecting its interests and reforming its modes of action. Importantly, the Treaty introduced a High Representative for the Common Foreign and Security Policy, effectively a 'name and a face' on EU policy to the outside world. Although the Amsterdam Treaty did not provide for a common defence policy, it did increase the EU's responsibilities for peacekeeping and humanitarian work. Through these changes the EU sought to translate its economic muscle into political power and agency on the global stage (Dinan, 1999).

2000–2017: The Supranationalism–Globalization Conundrum

Preparations, negotiations and financial assistance for the big bang enlargement of the EU to include ten states in Central and Eastern Europe (as well as Cyprus and Malta) took place from 1998 to 2004. Making this 'New Europe' by aligning Eastern European economies, institutions and policies to those operating in the EU required liberalizing economic and agricultural sectors; reforming the judicial system; combating corruption; applying the rules on food safety; upholding minority rights; and improving and protecting the environment. The 2004 enlargement increased the number of EU member states by 66 per cent, the population of the EU by approximately 20 per cent, while only raising economic output of the EU by a mere 5 per cent. The accession of Bulgaria and Romania to the EU in 2007 added a further 30 million citizens (a 6 per cent increase in EU population) yet only contributed a 1 per cent increase to the EU's GDP. Re-joining the 'European family' as the negotiations were portrayed was often acrimonious, and the applicant states were deeply critical of the heavy handed, hierarchical and condescending nature of the EU side.

There are a number of serious issues originating in Europe that now threaten macro-regionalism in Europe. These may be grouped in the following way: a growing detachment of the EU from its citizens (in effect an elite, rather than a citizen project); certain EU regions not seeing their interests best served in current state-supranationalism arrangements; the waning of the appeal of the EU as a result of economic recession and austerity measures; and a growing anti-EU sentiment in core EU states symbolized in right-wing movements against perceived large influxes of migrants.

The relationship between macro-regional integration in Europe and globalization is naturally dynamic. The global financial crisis post-2008 has had serious repercussions for the supranational project in Europe (Serricchio et al., 2013). Critically, the turbulence of globalizing capitalism has undermined confidence in the ability of the EU to tackle growing social and economic inequalities. As a result, the political and financial architecture of the European supranational project has been shaken by state crises linked to global downturn. In such austere contexts, as in Greece, the unwilling-ness of richer EU states to commit extra resources to achieve regional economic convergence across the EU has been fully exposed.

The supranational project in Europe has now reached an important crossroads. From earlier concerns with achieving balanced economic development across Europe, and greater social and political cooperation and convergence within it, an increasing strategic concern of the EU project is now focused upon dealing with the global economy. Arguments centre on the capacity of the EU to protect itself from the vagaries of global capitalism, compete globally against other macro-regional groupings such as North America, and the Asia-Pacific Economic Cooperation Zone, and respond and adjust to the emerging desires of certain member states, notably the UK, to chart its own politico-economic destiny outside of perceived restrictive and intrusive macro-regional regulatory structures. Without doubt, the Brexit vote in 2016 to leave the EU has been the single biggest shock to supranationalism in Europe and marks a critical stage in the long-term future of the EU project (Jessop, 2017). It is the fullest expression of the desire of the state to wrestle from macro-regionalism in order to retain its own sovereignty for control of money, markets and people.

CONCLUDING REMARKS

Supranationalism in Europe has been sponsored by the EU and its constituent states to varying degrees. Economic cooperation in Europe was encouraged and facilitated by the member states as a means of generating trade, economic growth and employment. Achieving economic integration has, though, required greater political and financial regulation in Europe as the complexities of managing open borders and markets have increased with the EU expansion to include more member states. Having started with six members in 1957 it has grown to twenty-eight states by 2017, and will fall to twenty-seven states in March 2019.

The relationship between macro-regionalism in Europe and globalization processes is complex. How far globalization has provoked rounds of EU integration, or how such integration has affected global trends is at the centre of these debates. What is clear is that the EU has been forced to respond to globalization by intensifying its internal market (including removing barriers to the free movement of goods, people, capital and services), protecting its external borders, and ensuring it negotiates with one voice in global trade discussions.

These responses have both widened and deepened the political-institutional management of the EU project and, in turn, highlighted varying state positions on this process. Without doubt, and despite its aspirations to portray itself in unified ways beyond its shores, the EU is politically, economically and culturally divided. As a supranational entity in a globalizing world, it has established wide-ranging politico-economic and cultural interconnections that have spawned all manner of trade and aid relations, partnerships and formal associations. Yet, despite the rhetoric of an outward-facing EU, exclusion, protection and division are the mobilizing vocabulary for the EU.

Within the supranational project itself, regionalist and independence struggles fuelled by either perceived state neglect or exploitation, combined with declared cultural difference and economic ambition, are on the increase. The relationship between people, place and political organization, be it at local, national or supranational levels, is changing. Ironically, the allure of the EU may be diminishing for its own citizens yet, simultaneously, on the increase for those migrants from sub-Saharan Africa, the Middle East and Asia risking their lives to enter it. The EU supranational project remains characterized by a continuous tension between integration and disintegration, offering an uncertain trajectory as it renegotiates and remakes itself into the globalizing future.

REFERENCES

Archer, C. (1990), *Organizing Western Europe*, Routledge: New York.
Bialasiewicz, L. (ed.) (2011), *Europe in the World: EU Geopolitics and the Making of European Space*, Ashgate Publishing: Farnham.
Clark, J. and A. Jones (2008), 'The spatialities of Europeanisation: territory, government and power in "Europe"', *Transactions of the Institute of British Geographers*, **33**(3), 300–318.
Crescenzi, R. and A. Rodríguez-Pose (2011), *Innovation and Regional Growth in the European Union*, Springer: Berlin.
Dinan, D. (1999), *Ever Closer Union*, Lynne Rienner: Boulder.
European Community (1988), *Europe 1992: The Overall Challenge* (summary of the Cecchini Report), EU Commission – SEC Document: Brussels.
Graham, B. (ed.) (1998), *Modern Europe: Place, Culture and Identity*, Arnold: London.
Jessop, B. (2017), 'The organic crisis of the British state: putting Brexit in its place', *Globalizations,* **14**(1), 133–141.
Jones, A. (1995), *The New Germany*, Wiley: London.
Jones, A. and J. Clark (2010), *The Spatialities of Europeanisation*, Routledge: New York.
Laffan, B., R. O'Donnell and M. Smith (2000), *Europe's Experimental Union: Rethinking Integration*, Routledge: London.
Mamadouh, V. and H. Van der Wusten (2008), 'The European level in EU governance: territory, authority and trans-scalar networks', *GeoJournal*, **72**(1–2), 19–31.
Martin, R. and J. Pollard (eds) (2017), *Handbook on the Geographies of Money and Finance*, Edward Elgar Publishing: Cheltenham, UK and Northampton, MA, USA.
Pickles, J. and A. Smith (2005), *Theorizing Transition: The Political Economy of Post-Communist Transformation*, Routledge: London.
Serricchio, F., M. Tsakatika and L. Quaglia (2013), 'Euroscepticism and the Global Financial Crisis', *Journal of Common Market Studies*, **51**(1), 51–64.
Smith, A. (2013), 'Europe and an inter-dependent world: uneven geo-economic and geo-political developments', *European Urban and Regional Studies*, **20**(1), 3–13.

30. Maritime trade and geopolitics: the Indian Ocean as Japan's sea lane
Takashi Yamazaki

INTRODUCTION

This chapter explores the historical relationship between globalization, maritime trade and geopolitics. It focuses on the case of Japan which recognizes the Indian Ocean as its 'vital' sea lane or SLOC (sea line of communication) for the state economy. As maritime trade has developed globally through the expansion of capitalism (and colonialism) from Western Europe, its importance for the state economy has prompted foreign and security policies that focus on the security of sea lanes to sustain international trade. Therefore maritime security policies have been constructed on the premise that maritime states (both littoral and archipelagic ones) need to secure the control of the seas and the sea lanes for their political economic development through diplomatic and military actions.

Maritime security policies cannot be divorced from geographical factors mainly because these factors directly or indirectly constrain and inform such policies (Germond, 2015: 138). The (material) importance of sea lanes for maritime trade is often conditioned by the geographical configurations of seas and sea channels as seen typically in chokepoints (like a strait). In this sense, we can call such geographically constrained and informed maritime policies 'maritime geopolitics' which often takes the form of classical geopolitics on sea powers (maritime states) as exemplified by Mahan (1890 [1987]). Drawing on critical geopolitical perspectives (Ó Tuathail, 1996; Germond, 2013), it can also be said that the geographical location of a state and the fixity of its position on the globe, condition maritime policy-makers' identification of external 'others' as threats. The discursive justification of their policy and the reasoning of maritime geopolitics for sea-lane security can therefore constitute the subject of critical geopolitical research (see Germond, 2013).

This chapter first looks at the historical development of maritime trade as a fundamental aspect of globalization and the interconnectedness of the world. It then examines how maritime geopolitics has been constructed under such global dynamics. In the second half, it uses the case of Japan to illustrate this point by considering its geopolitical codes pertaining to the Indian Ocean as its sea lane and to demonstrate how such geopolitical codes have remained unchanged and have repeatedly been employed to justify Japan's postwar maritime geopolitics with – and against – neighboring states.

MARITIME TRADE AND GLOBALIZATION

The global expansion of maritime trade is one of the fundamental aspects of economic globalization. In the contemporary globalizing world, a state cannot survive without conducting trade or exchanging goods with other states within its regions and in other regions. Maritime trade is a core activity of state agencies and private corporations to connect different places in the world.

Historically, due to the limited size of the regional economies and the low level of technology, trade was in large part carried out on land and/or on internal waters. During the fifteenth and sixteenth centuries, however, advances in ship design and navigation technology led to the Europeans' discovery of the Americas, the opening-up of new trade routes to Asia around Africa, and Magellan's circumnavigation of the globe (WTO, 2013: 46). Figure 30.1 shows the main Portuguese and Spanish maritime trade routes in the sixteenth century, which resulted out of the exploration during the Age of Discovery. These sixteenth-century routes already include the Straits of Florida between Florida and Cuba, the Strait of Malacca between the Malay Peninsula and Sumatra, and other chokepoints which have been heavily used for maritime trade ever since.

This outline of the world economy was already laid out in the seventeenth and eighteenth centuries. The Industrial Revolution in the early 1800s prompted the massive expansion of trade, capital and technology flows, the explosion of migration and communications, and the time–space compression of the world economy (WTO, 2013: 46). The construction of the Suez Canal in 1869 and the Panama Canal in 1914 created significant shortcuts for inter-oceanic navigation. The map in Figure 30.2 shows the present-day density of commercial shipping in the world's seas, and indicates the highly dense shipping routes around the southern fringe of the Eurasian Continent and the broadly spreading routes in the Atlantic and Pacific Oceans.

A comparison between the two maps shows that our world has been increasingly interconnected since the Age of Discovery and that sea routes, canals, ports, and vessels for maritime trade have become the basic infrastructures for modern capitalism, colonialism, and contemporary globalization. Globalizing maritime trade also promoted the establishment of the shipbuilding industries, transnational shipping agents, banks, and maritime insurance companies. In order to sustain the development of the state economies, international trade of a large amount of resources, food, and manufactured goods has become inevitable. As a mode of international trade, cargo transportation by ship has been more cost-effective than by road, railway or aircraft. Hence maritime trade has played an essential part in the expansion of the modern world economy.

Another important dimension in the development of maritime trade infrastructures is how to secure trade routes using naval forces, which will be discussed in the next section.

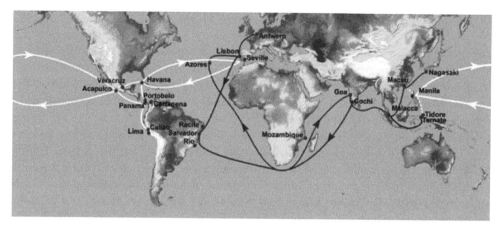

Source: https://commons.wikimedia.org/wiki/File:16th_century_Portuguese_Spanish_trade_routes.png (extracted May 27, 2017).

Figure 30.1   *Trade routes of Spain (white) and Portugal (dark) in the sixteenth century*

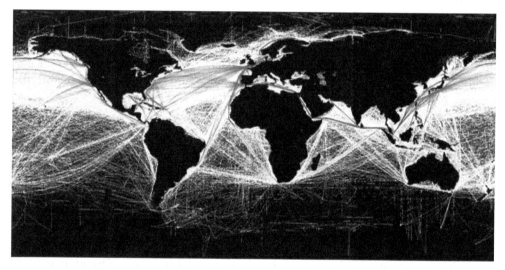

Source: https://en.wikipedia.org/wiki/Ship_transport#/media/File:Shipping_routes_red_black.png (extracted May 27, 2017).

Figure 30.2   Shipping routes of the world

THE GEOPOLITICS OF SEAS

As maritime trade was globalized, the security of trade routes and governance over oceans became important on the diplomatic/military agendas of imperial states, leading to the rise of classical (maritime) geopolitics and the institutionalization of seas.

The Rise of Classical Geopolitics

According to Wallerstein (1979), modern capitalism began in sixteenth-century Western Europe and spread towards the rest of the world through exploration and colonization by sea powers such as the Netherlands and the United Kingdom. What made this possible was open sea sailing equipped with advanced navigation and shipbuilding technologies. The hegemonic rise of these states was promoted through the technological innovations of the time. After the Industrial Revolution in the eighteenth century, the introduction of the steam boat accelerated maritime trade and colonization with the support of strong naval forces. Maritime trade became an essential aspect of imperialism in the nineteenth century and sustained the re-ordering of the world led by the United Kingdom and other Western powers (Flint and Taylor, 2007; Flint, 2011). Intensified competition for colonies among imperial powers, among other things, finally led to the First World War (1914–1918).

Major works of classical geopolitics appeared in that period of time. 'Classical geopolitics' originally refers to a system of geographical knowledge aimed at prescribing foreign and military policies for a selected imperialist state. It became state-centric and geographically deterministic because such prescribed policies were often shaped according to the geographical location and configurations of a particular state. In other words, the political economic future or survival of a state tended to be described according to its geographical position (i.e. as a continental or an oceanic state).

Halford Mackinder (1861–1947), a British geographer and parliament member, was well-known as a classical geopolitical thinker. He observed that the major axis of conflict was between land- and sea-based powers (Flint, 2011: 6–8). His famous 'Heartland Theory' (Mackinder, 1904) emphasized the geostrategic advantage of land powers occupying the center of the Eurasian landmass while designating the function of sea powers as containing land powers. His geopolitical framing of a world order significantly influenced subsequent geopolitical thinkers such as Karl Haushofer (1869–1946) and Nicholas Spykman (1893–1943). Even after the Second World War, Mackinder became an intellectual inspiration for Cold War strategists and proponents of the North Atlantic Treaty Organization in the second half of the twentieth century (Flint, 2011: 8).

The intellectual who influenced Mackinder was Alfred Thayer Mahan (1840–1914), a United States naval historian and admiral. Mahan contended that with a command of the sea, even if local and temporary, naval operations in support of land forces could be of decisive importance (Sumida, 1999). He also believed that naval supremacy could be exercised by a transnational consortium acting in defense of a multinational system of free trade (Sumida, 1999). Compared to Mackinder, Mahan explicitly framed the importance of sea powers in terms of trade for economic wealth.

In the modern world, resource-rich countries make use of a monopoly power to pursue their international interests while resource-scarce countries prioritize the pursuit of resource security in their foreign policies (WTO, 2013: 167). Therefore, the geopolitics of maritime trade has been a focal point of international relations and has had a decisive impact on underlying technological and structural trends of international trade for the past centuries (WTO, 2013: 55).

The Institutionalization of Seas

As Mahan argued one and a half centuries ago, the relationship between maritime trade and the capability of sea powers has been regarded as a crucial factor to secure the political economic development of a state. As the scope of maritime trade became global in the seventeenth century, how such trade could be secured became a matter of multilateral negotiation over the principle of 'freedom of the seas' (Grotius, 1609 [2004]). This principle has been repeatedly discussed and challenged among European thinkers and within the international political arena.

For instance, the UK's naval supremacy in the nineteenth century ensured that the world sea lanes, being so important for the global economy, remained open, not just to British trade but also to the commerce of the world. By the turn of the century, however, the consolidation and expansion of European colonial empires in Africa and Asia were a clear sign that the British 'imperialism of free trade' was already waning (WTO, 2013: 50–51).

Laws and regulations regarding the sea are called 'the international law of the sea'. It provides a framework for contested issues such as the breadth of territorial waters, resource exploitation on the continental shelf, and the use of the high seas. The international law of the sea has a long history from the Age of Discovery in which Spain and Portugal defended a *mare clausum* (closed sea) principle. *Mare clausum* is a Latin term used in international law to refer to the sea under the jurisdiction of a particular state and not accessible to other states. To counter this principle, Hugo Grotius, a Dutch jurist, propagated a *mare liberum* (free sea) principle in the early seventeenth century. By claiming a free sea, Grotius provided an ideological justification for the Dutch assault on existing trade monopolies through its strong naval power and for establishing its own monopoly thereafter (Saha, 2010: 73).

Discussions on the high seas cannot be separated from those on territorial waters. This separation became another issue in the discussion on 'freedom of the seas' in the eighteenth century, leading to the international recognition of 'narrow territorial waters' and 'broad high seas'. The latter, however, allowed imperial powers (with the naval force to use these broad high seas) to pursue their colonization overseas. One of the focuses of imperial rivalry in the late nineteenth and early twentieth centuries was placed on control and security over maritime routes for trade and military deployment. Therefore, if the high seas are broad and free enough for ships of any registries to pass, then they do not become any source of conflict.

After having been swinging between the principles of closed and free seas, the institutionalization of the use of open seas finally led to the conclusion of United Nations Convention on the Law of the Sea (UNCLOS) in 1982. UNCLOS is a territorial (zonal) allocation of the degree of exclusive/free use of the seas depending on the distance from the coastal line of states' land territory. Article 87(1) provides: 'high seas are open to all states, whether coastal or land-locked'.

Such abstract legal spaces, however, need to be reconfigured to see how they work in concrete geographical settings. The actual shapes of maritime trade routes are constricted bundles of curves, especially in the seas connected by chokepoints (see the map in Figure 30.2). The geographical configurations of open seas include several chokepoints or sea passages (straits or canals) of strategic importance which have often caused geopolitical tensions among the respective states until today. The (in)security of these chokepoints can greatly influence the global flow of trade items as well as naval deployment, which has constituted geopolitics over sea lanes.

THE SECURITY OF SEA LANES FOR JAPAN

As mentioned above, globalizing maritime trade has (re)constructed maritime geopolitics. In order to illustrate this process this section examines Japan's postwar maritime geopolitics that has focused on sea-lane security in the Indian Ocean and demonstrates how and why geopolitical codes of the Indian Ocean have changed/are unchanged in the shifting geopolitical context of Asia.

Premises of Japan's Sea-Lane Security

As the development of maritime trade became crucial for the wealth of states, the security of maritime trade routes came to have important geostrategic implications, especially for littoral and archipelagic states. Such geostrategic routes are called sea lanes, sea roads, or sea lines of communication (SLOCs). Sea lanes are the key maritime passageways that facilitate heavy shipping traffic and host the transportation of key maritime trade goods such as crude oil. Sea lanes often include chokepoints such as sea straits and channels, the disruption of which can have an adverse political economic effect on the states and the corporations using them (Khalid, 2012).

The globalization of maritime trade and its intensification, therefore, have inevitably (re)constructed geopolitics over sea lanes for their user states. In order to understand how sea lanes have connected maritime trade and maritime geopolitics, this section investigates how sea lane security is incorporated into its user state's maritime geopolitics. The case chosen is Japan because it is an archipelagic state in Asia with numerous past military confrontations with neighboring and Western states, and because it has been dependent for its national and imperial development on maritime trade since the late nineteenth century. Postwar maritime trade, in particular, is one of the elements for Japan's success in becoming a global power regardless of its scarcity of natural resources (Graham, 2006: 10–11). Currently Japan ranks sixth regarding the size of its territorial waters and of its Exclusive Economic Zones (EEZs)[1] in the world. In this sense, Japan can be seen as a maritime state seeking resources, markets, and places of investment overseas.

Unlike the prewar period, however, Japan's military capabilities have been restrained according to the new Constitution, and its security policies have been closely tied to US foreign/security policies toward the Western Pacific. Thus Japan's sea-lane security policies have been framed to combine security alliance with the US and dependence on global maritime trade for its economic prosperity.

---

[1]   An Exclusive Economic Zone (EEZ) is a sea zone over which a coastal state assumes jurisdiction regarding the exploration and use of marine resources from its coast out to 200 nautical miles offshore. This concept was adopted at the Third United Nations Conference on the Law of the Sea (1982).

In the 1970s, at the peak of Japan's economic growth, the Ministry of Defense of Japan began to publish an annual white report entitled *Defense of Japan* (*Bouei-hakusho* 防衛白書 hereafter *DOJ*) which describes Japan's security policy towards sea lanes. During the 1980s the increasing dependence on the Middle East for the import of petroleum made *DOJ* focus on the Indian Ocean as a vital sea lane for Japan which connects the Persian Gulf and the East China Sea. A content analysis of successive *DOJs* should reveal how Japan has perceived the Indian Ocean from a geopolitical point of view during and after the Cold War.

Such a geopolitical perception of a state is called a 'geopolitical code'. Geopolitical codes can be described as 'more or less coherent perspectives on global geopolitics and/or the position and interests of one particular state' and have become the basis 'of prescribed or pursued foreign policies' (Nijman, 1994). A geopolitical code consists of several calculations such as the specification of allies and enemies, the way to maintain good links with allies, the way to counter enemies, and the justification of such calculations (Flint, 2011: 43–44). Therefore, an investigation of Japan's geopolitical codes on the Indian Ocean could illuminate how Japan's sea-lane security policies have been framed in relation to globalizing maritime trade and shifting geopolitical contexts in Asia.

Japan's Geopolitical Codes on the Indian Ocean

DOJ is a comprehensive report on Japan's security policies and its perception of international security affairs. It can be used as a source of information that contains Japan's geopolitical codes on the Indian Ocean and their temporal shifts. The full-textual data for *DOJ* from 1970 to 2016 are available and downloadable at the Ministry's website (Ministry of Defense, 1970–2016). There is also a search engine for the contents of *DOJs* at the website so that particular terms can be searched and located. This search engine was used to find and locate terms such as 'Indian Ocean (*indoyō* インド洋)' and 'sea lanes (*shī rēn* シーレーン)'. *DOJ* has hierarchal contents consisting of Part, Chapter, Section, and one or two sub-section levels.

There are 458 sub-sections containing the term 'Indian Ocean' against 92 for 'sea lanes'. As shown in Figure 30.3, there are recognizable peaks in the appearance of 'Indian Ocean' and 'sea lanes' which are not proportional to a general increase in the number of pages of *DOJ*. For each appearance, a paragraph (or a group of paragraphs) that contains a searched term was examined and classified by five criteria: term location (where it appeared in *DOJ*), key player country/region, key player administrative/military unit, subject(s) of the description, and geopolitical context of the description.

The content analysis of *DOJs* shows several interesting tendencies. First, descriptions on the Indian Ocean had appeared only in the first part of *DOJ* before 2001, meaning that originally the Indian Ocean had been described as a place *external* to Japan and that after 2001 the Ocean began to be referred to as more than an external place for Japan. Descriptions on the Ocean appeared in other parts of *DOJ* after 9/11 took place and promoted Japan's involvement in the War on Terror in 2001.

Second, before the end of the Cold War, key player countries related to the Indian Ocean were rather limited to a few countries such as the US and the USSR, suggesting that *DOJ* saw the geopolitical context of the Indian Ocean in light of the confrontation

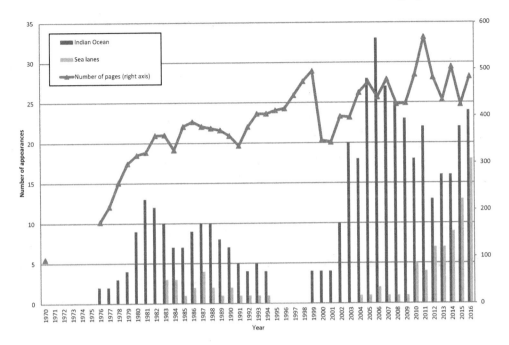

Note: There is no comparable data for the periods between 1971 and 1975 and between 1995 and 1998.

Figure 30.3   The contents of DOJ, *1970–2016*

between the two global powers. After the demise of the USSR, *DOJ* almost ceased to refer to the former USSR or to Russia as a key player in the region.

Third, given Japan's postwar dependence on US military presence, it is not difficult to understand that *DOJ* continues to pay attention to US military posture toward the Indian Ocean. For Japan, the deployment of US military forces in the Indian Ocean as well as in the West Pacific has been considered crucial for its resource import and maritime trade. More specifically, the military deployment of the USSR towards the Indian Ocean and Afghanistan in the 1970s and 1980s directed Japan's attention to US military posture to counter such a challenge.

Fourth, even after the end of the Cold War, *DOJ* continued to mention the military role of the US in the Indian Ocean. This is mainly because Japan (has) depended on US military deployment in the West Pacific and Indian Ocean for its military and resource security according to the Japan–US security alliance. *DOJ* states the role of the US after the end of the Cold War as follows:

> The U.S. has so far deployed the Pacific Command in the Asia-Pacific region as the Joint Force of the Amy, Navy, Air Force, and Marine and carried out policies to prevent conflicts in this region and protect interests of the U.S. and its allies by concluding security agreements with Japan and several countries in the region. ... The U.S. Pacific Command

located in Hawaii responds to unexpected contingencies swiftly and flexibly and forward-deploys the forces consisting mainly of its own navy and air force units in the Pacific and Indian Ocean. (*DOJ 1992*)[2]

Finally, *DOJ*'s descriptions on the Indian Ocean began to mention other key players in the region after 2001, including Japan itself. The appearance of 'Indian Ocean' decreased in the first part and moved to other parts of *DOJ*, meaning that the Indian Ocean began to be mentioned in the parts on counterterrorism measures in the Indian Ocean and humanitarian and reconstruction support in Iraq. The *DOJ* has clearly different geopolitical codes of the Indian Ocean, representing the shift from a bipolar to multi-polar structure in which Japan could play a certain role.

THE INDIAN OCEAN AS JAPAN'S SEA LANE

DOJ often refers to the Indian Ocean as an important maritime transportation route but does not necessarily use the term 'sea lane(s)'. As shown in Figure 30.3, 'sea lanes' appear from the early 1980s to the early 1990s and from the mid-2000s to present. From 1983 to 1990, 'sea lanes' were referred to as a key word for the Japan–US defense cooperation in *DOJ*. 'Sea lanes' were used by *DOJ* to express one of the policy objectives of such defense cooperation.

The discussion about sea lanes between Japan and the US began in the late 1970s and was finally laid down in the 'Guidelines for Defense Cooperation between Japan and the U.S.' in 1978. The Guidelines stipulated that Japan's Self Defense Forces (SDFs) had the responsibility for the defense of Japan's territory and the surrounding sea and air areas, while the US forces supplement functions beyond the abilities of the SDFs. Following the USSR military intervention in Afghanistan in 1979, Japanese Prime Minister Suzuki and US President Reagan defined their bilateral relationship as an 'alliance' and agreed on an appropriate 'division of roles' for security issues in 1981. With regards to this division of roles, Suzuki stated that Japan would defend sea lanes within 1 000 nautical miles from its territory, while the US 7th Fleet would handle the security of the Persian Gulf (Kotani, 2006: 195–196). *DOJ* began to refer to these sea lanes in 1983, when the Japan–US joint research on the defense of sea lanes started (it ended in 1986).

Although Japan was to protect sea lanes in its surrounding areas according to the above-mentioned guidelines, *DOJ* continued to express Japan's (SDFs') inability to defend itself from nuclear threats from the USSR. *DOJ* recognized that the SDFs' capabilities were not sufficient to carry out the geopolitical division of roles with US military forces.

The appearance of 'sea lanes' from the mid-2000s has different nuances from the previous decades. 'Sea lanes' appeared again in 2004 in the section titled 'U.S. Forces in the Asia-Pacific Region'. The section explains the roles of US 7th Fleet as 'an operational unit with a mission to protect the territory, nation, *sea lanes*, allies, and

---

[2]   http://www.clearing.mod.go.jp/hakusho_data/1992/w1992_01.html (accessed May 27, 2017), my translation into English.

other vital national interests of the U.S.' (*DOJ 2004:* 73, emphasis added).[3] The term continues to be used in this description till present, suggesting that *DOJ* regards one of the roles of US 7th Fleet as the protection of sea lanes and allies for the US.

In 2006, the above-mentioned description on US 7th Fleet began to include the 'Indian Ocean'. These two descriptions were formerly separated but combined in 2004 to indicate both action areas and missions for the 7th Fleet. Here *DOJ* came to clearly describe in a single description the roles of the Fleet as the protection of sea lanes, allies, and the Indian Ocean. *DOJ 2006* also contains the declaration of 'The Japan–U.S. Alliance in the New Century' which was published jointly by Japan's Prime Minister Koizumi and US President Bush in 2006.[4] It states in the first part that:

> The United States and Japan share interests in: winning the war on terrorism; maintaining regional stability and prosperity; promoting free market ideals and institutions; upholding human rights; securing freedom of navigation and commerce, including *sea lanes*; and enhancing global energy security. It is these common values and common interests that form the basis for U.S.–Japan regional and global cooperation. (emphasis added)

'Sea lanes' has become a term mentioned in the (geopolitical) context of the Japan–US alliance in which Japan basically depends on and cooperates with the US for its own security from the Indian Ocean to the West Pacific. All descriptions on sea lanes in *DOC 2011* also appear in the context of the Japan–US defense cooperation.

However, *DOJ 2010* has different features on the description of sea lanes in two respects. First, it begins to explain Japan–India defense cooperation and exchange by stating that 'India is located in the center of *sea lanes* which connect Japan with the Middle East and Africa, making it an extremely important country for Japan in a geopolitical sense, which relies on maritime transportation for most of its trade' (*DOJ 2010*, my translation, emphasis added).[5] In this sub-section, *DOJ 2010* emphasizes that Japan and India share fundamental values such as democracy, the rule of law, respect for human rights, and capitalist economies and that these countries have established a strategic global partnership for the peace, stability and prosperity of Asia and the world.

From 2006 to 2008 *DOJ* also paid attention to the movement of China in the Indian Ocean. This is mainly because China became interested and intensified military activities in the Indian Ocean. Compared to Japan's strong dependence on US 7th Fleet for the security of remote sea lanes, these descriptions imply a more independent Japanese effort to be involved in the protection of such sea lanes.

Second, *DOJ 2010* includes the summary of the report titled *Japan's Visions for Future Security and Defense Capabilities in the New Era: Toward a Peace-Creating Nation* (*Japan's Visions*). The report was made by the Council on Security and Defense Capabilities in the New Era. In the 2009 Lower House election, the coalition government led by the Liberal Democratic Party (LDP) was replaced by a coalition led by the Democratic Party of Japan. In that context the report showed new directions of

---

3   http://www.clearing.mod.go.jp/hakusho_data/2004/2004/pdf/16130000.pdf (accessed May 27, 2017).

4   http://www.mofa.go.jp/region/n-america/us/summit0606.html (accessed May 27, 2017).

5   http://www.clearing.mod.go.jp/hakusho_data/2010/2010/html/m3322300.html (accessed May 27, 2017).

Japan's security policies promoted by the new government. In the original report 'sea lanes' (and 'SLOCs' in the English version) appear several times as shown below:

> The scarcity of resources and energy in Japan makes *SLOCs* and their surroundings an important security issue. Japan relies for most of its energy supply on maritime transportation across *the Indian Ocean*. Thus, the security of *the SLOCs* that run from the Persian Gulf, *the Indian Ocean*, the Strait of Malacca, the South China Sea, the Bashi Channel, and the east coast of Taiwan to Japan's vicinity and the stability of the coastal countries around the *sea lines* are of crucial importance for Japan. This will not change in the future. (*Japan's Visions*: 2010, 12, emphasis added)[6]

In sum, with regards to sea-lane security in the Indian Ocean, it can be said that Japan aims at maintaining the alliance with the US as a global power and seeks cooperation with India as a regional power by enhancing the SDFs' capability for such purposes. In order to achieve these objectives, Japan would need to be more actively involved in the security of the Indian Ocean.

THE SHIFTING IMPORTANCE OF THE INDIAN OCEAN FOR JAPAN

Until 2001 almost all descriptions on the Indian Ocean had appeared in the first part of *DOJ*, indicating that the Indian Ocean had never constituted any national security issue for postwar Japan. This is mainly because the Japanese Constitution prohibits the use of military forces outside Japan's territory. However, 9/11 has completely changed these geopolitical codes regarding the Indian Ocean.

During the Gulf War in 1991, Japan did not dispatch any military forces according to the Japanese Constitution but provided financial support for the US and relevant countries after the war. Japan's 'passive' attitude towards such an international crisis became controversial within Japan and prompted the LDP-led government to dispatch the SDFs for UN Peace Keeping Operations in the 1990s. Since then Japan has become more active in pursuing a 'military' contribution in the international arena. Immediately after 9/11, the Koizumi cabinet decided to dispatch the SDFs to the Indian Ocean and to Pakistan to provide logistic support for the US-led military operation in Afghanistan. During the Iraq War, the Koizumi cabinet also sent the SDFs to Iraq to provide humanitarian and reconstruction assistance.

9/11 facilitated international military cooperation (i.e. 'War on Terror') initiated by the US and it contributed to the construction of a new geopolitical context in East Asia that allowed Japan to become more active in the international security arena. Accordingly, Japan has been more proactive in dispatching the SDFs for international disaster-relief activities such as the tsunami relief in Indonesia. Japan's interest and involvement in the Indian Ocean region should be seen in this context.

After 9/11, the Indian Ocean began to be mentioned in the second and latter parts of *DOJ*. *DOJ 2002* and *2003* referred to the Ocean in the part on responses to security

---

6   http://www.kantei.go.jp/jp/singi/shin-ampobouei2010/houkokusyo_e.pdf (accessed May 27, 2017).

emergency after 9/11 and SDF dispatch to the Ocean. Then *DOJ* increased the number of descriptions on the Indian Ocean in the part on international cooperation for the War on Terror, the subsequent humanitarian intervention in Iraq, and the recent involvement in disaster relief in Indonesia and other countries.

An important implication of this shift is a completely new definition of the Indian Ocean as a 'space of engagement' (Cox, 1998) for the SDFs. According to the Antiterrorism Special Measures Law (2001–2007) and the Law Concerning the Special Measures on Humanitarian and Reconstruction Assistance in Iraq (2003–2009), the basic plans to implement these laws designated the Ocean as an area for SDFs' activities. The Indian Ocean is no longer a remote space for Japan's security policy.

There has also been a new trend in which Japan is seeking bilateral defense cooperation and exchanges with countries other than the US such as Australia and India. In addition to the emerging description of China as a challenger to the Indian Ocean, we see an attempt to build such a regional security network that will definitely be continued in the future.

JAPAN'S GEOPOLITICAL SELF-IMAGE AS A MARITIME STATE

Japan's security policy has been consistently premised on its geography of isolation by the seas (Graham, 2006: 36) and reinforced by basic dependence on foreign imports of minerals and fossil fuels (Graham, 2006: 11) since the end of the Second World War. These factors have constituted Japan's geopolitical self-vision as a maritime state in *DOJs*. As shown in Figure 30.3, descriptions on sea lanes have been increasing in number since 2010. An important turning point is the formulation of 'The National Security Strategy (Kokka anzen hoshō senryaku 国家安全保障戦略)' in 2013 (hereafter *NSS 2013*).[7] *NSS 2013* re-defines Japan as follows:

> Surrounded by the sea on all sides and blessed with an immense exclusive economic zone and an extensive coastline, Japan as a maritime state has achieved economic growth through maritime trade and development of marine resources, and has pursued 'Open and Stable Seas'. (*NSS 2013:* 2)

Based on this premise and to ensure maritime security it states:

> As a maritime state, Japan will play a leading role, through close cooperation with other countries, in maintaining and developing 'Open and Stable Seas', which are upheld by maritime order based upon such fundamental principles as the rule of law, ensuring the freedom and safety of navigation and overflight, and peaceful settlement of disputes in accordance with relevant international law. More concretely, Japan will take necessary measures to address various threats in *sea lanes of communication*, including anti-piracy operations to ensure safe maritime transport and promote maritime security cooperation with other countries. (*NSS 2013:* 16, emphasis added)

Then the document specifies sea lanes that are crucial for Japan:

---

7   http://www.cas.go.jp/jp/siryou/131217anzenhoshou/nss-e.pdf (accessed May 27, 2017).

In particular, *sea lanes of communication*, stretching from the Persian Gulf, the Strait of Hormuz, the Red Sea and the Gulf of Aden to the surrounding waters of Japan, passing through *the Indian Ocean*, the Straits of Malacca, and the South China Sea, are critical to Japan due to its dependence on the maritime transport of natural and energy resources from the Middle East. In this regard, Japan will provide assistance to those coastal states alongside the *sea lanes of communication* and other states in enhancing their maritime law enforcement capabilities, and strengthen cooperation with partners on the *sea lanes* who share strategic interests with Japan. (*NSS 2013*: 17, emphasis added)

Accordingly, *DOJ 2014* contains the descriptions regarding *NSS 2013* and adds a new sub-section on China's maritime activities, indicating that Japan formally acknowledges China as a challenger to the security of Japan's sea lanes. *DOJ 2014* (p. 44) points out that China has been intensifying its activities in the South China Sea in which China has territorial disputes over the Spratly and Paracel Islands[8] with its neighboring countries. It also refers to the improvement of Chinese Navy's capabilities that allows its advancement into the Indian Ocean. Such a perception of China is implicitly reflected in Japan's geopolitical self-vision described in *NSS 2013*.

Geostrategic prescriptions against China's challenge, as mentioned in the previous section, reinforce the Japan–US security alliance and the promotion of bilateral defense cooperation with India and with Australia. *DOJ 2015* and *2016* devote pages to these issues, in particular, to the role of Australia for sea lane security while *DOJ 2016* (p. 59) provides more descriptions as to China's maritime activities in remote seas including the Indian Ocean (to protect its own sea lanes as the 'String of Pearls' strategy). As dependence on maritime trade increases in a globalizing world, Japan's maritime geopolitics continues to single out new challengers to its sea lanes and to reinforce its geopolitical self-image as a vulnerable maritime state in the West Pacific, in a fashion that follows the deterministic tradition of classical geopolitics.

CONCLUSION

As discussed above, the historical development of maritime trade is a fundamental aspect of globalization and the interconnectedness of the world. Under such global dynamics, the maritime security policies of sea powers have been shaped and reshaped over time. This is because the availability of maritime trade routes for free trade in the high/open seas inevitably requires infrastructures to sustain it, including the deployment of naval forces. It can thus be said that maritime geopolitics over sea lanes has been a political outcome of globalizing maritime trade.

This chapter, using the case of Japan and its sea-lane security policies as described in *DOJ*s, analyzed Japan's geopolitical codes on the Indian Ocean. The results of the analysis showed that the Indian Ocean has been constantly coded as a sea lane vital for Japan's maritime trade and that such coding has been repeatedly employed to justify

---

[8]   The Spratly Islands are called *Nánshā Qúndǎo* 南沙群岛 in Chinese, *Kapuluan ng Kalayaan* in Tagalog, and *Quần đảo Trường Sa* in Vietnamese. The Paracel Islands are called *Xisha Qúndǎo* 南沙群岛 in Chinese and *Quần đảo Hoàng Sa* in Vietnamese.

Japan's military alliance with the US to protect the sea lane against emerging challengers.

Despite the deepening globalization and interconnectedness of the world, Japan's maritime geopolitics has stemmed from ontological anxieties about the disruption of its sea lanes by their challengers. Japan's geopolitical self-image has been that of a vulnerable maritime state located in the West Pacific. Based on such a self-image, Japan's sea-lane security policies have been shaped to build security networks with the US and other friendly states along the sea lanes. It is clear that behind this image, there has been an increasing fear of being disconnected from the external world to which Japan feels increasingly connected. Just as deterministic classical geopolitics, so Japan's maritime geopolitics has been founded on the imagination of an inescapable geographical destiny.

REFERENCES

Cox, K. (1998), 'Spaces of dependence, spaces of engagement and the politics of scale, or: looking for local politics', *Political Geography*, **17**(1), 1–23.

Flint, C. (2011), *Introduction to Geopolitics*, New York: Routledge.

Flint, C. and P. Taylor (2007), *Political Geography*, Essex: Pearson Education.

Germond, B. (2013), 'The European Union at the Horn of Africa: the contribution of critical geopolitics to piracy studies', *Global Policy*, **4**(1), 80–85.

Germond, B. (2015), 'The geographical dimension of maritime security', *Marine Policy*, **54**, 137–142.

Graham, E. (2006), *Japan's Sea Lane Security, 1940–2004*, London: Routledge.

Grotius, H. (trans. by Hakluyt, R.) (1609 [2004]), *The Free Sea (Mare Liberum)*, Indianapolis: Liberty Fund.

Khalid, N. (2012), 'Sea lines under strain: the way forward in managing sea lines of communication', *The IUP Journal of International Relations*, **6**(2), 57–66.

Kotani, T. (2006), 'Defense of sea lanes: development and limits of "people-to-people cooperation" in the Japan–US alliance (Shī rēn bōei: nichibei dōmei niokeru "hito to hito no kyōryoku" no tenkai to sono genkai)', *Dōshisha hōgaku*, **58**(4), 179–207.

Mackinder, H.J. (1904), 'The geographical pivot of history', *The Geographical Journal*, **23**(4), 421–437.

Mahan, A.T. (1890 [1987]), *The Influence of Sea Power upon History, 1660–1783*, New York: Dover Publications.

Ministry of Defense (1970–2016), *Defense of Japan (Bouei-hakusho* 防衛白書), Tokyo: Ministry of Defense of Japan. http://www.mod.go.jp/j/publication/wp/ (accessed May 27, 2017).

Nijman, J. (1994), 'Geopolitical codes', in: O'Loughlin, J. (ed.), *Dictionary of Geopolitics*, pp. 89–90, Westport: Greenwood Press.

Ó Tuathail, G. (1996), *Critical Geopolitics*, London: Routledge.

Saha, T.K. (2010), *Textbook on Legal Methods, Legal Systems and Research*, New Delhi: Universal Law Publishing.

Sumida, J.T. (1999), *Inventing Grand Strategy and Teaching Command*, Washington, DC: The Woodrow Wilson Center Press.

Wallerstein, I. (1979), *The Capitalist World-Economy*, Cambridge: Cambridge University Press.

WTO (2013), *World Trade Report 2013*, Geneva: World Trade Organization.

31. Alter-globalization movements and alternative projects of globalization
Byron Miller

INTRODUCTION

In 2010, the 93-year-old former French Resistance fighter, Nazi concentration camp survivor, French diplomat, and co-author of the United Nations Universal Declaration of Human Rights, Stephane Hessel, published a short book – *Indignez-vous!* – which was quickly translated into more than thirty languages, including the English version, *Time for Outrage!* Selling more than three million copies in short order, it became a spark (certainly not the only one) for a new wave of resistance to neoliberal globalization, including the Spanish Indignados movement and the global Occupy movement. Taking aim at globalization, these contemporary movements aligned with Hessel's central concerns: (1) the 'immense gap between the very poor and the very rich, which never ceases to expand', and (2) 'human rights and the state of the planet' (Hessel, 2011: 17). What differentiates these movements from many of the preceding movements targeting globalization is not so much the specific grievances – increasing inequality, employment insecurity and job loss, the dismantling of social safety nets, and despoilment of the environment have all been significant issues since the 1970s – but the solutions proposed. Rather than disengage, close borders, and retreat into the nation state, the alter-globalization movements (also called 'global justice movements') seek to redress the ravages of neoliberal globalization by replacing it with an entirely different kind of globalization. As Hessel reminds us in *Indignez-vous!*, revulsion to the horrors of Nazism and Fascism led to the creation of a new post-Second World War global order that aimed to ensure 'universal rights' around the world, not 'international rights'. The Universal Declaration of Human Rights, which was adopted by the United Nations in December, 1948, represented the brightest post-war vision of a just world order, declaring that 'everyone, as a member of society, has the right to social security and is entitled to realization, through national effort and international cooperation and in accordance with the organization and resources of each State, of the economic, social and cultural rights indispensable for his [sic] dignity and the free development of his [sic] personality' (article 22). Moreover, 'everyone is entitled to all the rights and freedoms set forth in this Declaration … no distinction shall be made on the basis of the political, jurisdictional or international status of the country or territory to which a person belongs' (article 2). Clearly, the initial post-war vision of a just and globally-integrated world was quite different from the present-day market-fundamentalist reality. Indeed, the values of alter-globalization movements bear a strong affinity to this mid-twentieth century vision, before nation states and the institutions of global governance were captured by powerful economic actors prosecuting a neoliberal agenda (see Stiglitz, 2002).

What might constitute a just global order is by no means a simple question, however. It is, nonetheless, a critical question to ask if we are to understand what goals alter-globalization movements might reasonably pursue and what strategies might be suitable to their achievement.

JUSTICE IN A GLOBALIZING WORLD

While a comprehensive survey of concepts of justice is far beyond the scope of this chapter, a brief overview is needed because justice movements face special problems in the context of globalization. Concepts of justice have a long history. Among contemporary theorists, John Rawls' (1971) *A Theory of Justice* is often considered foundational. For Rawls, questions of justice start with 'the distributive aspects of the basic structure of society' (Rawls, 1971: 9). For Rawls and his followers, justice is ultimately a question of the distribution of goods and benefits in society and most treatments of justice represent variations on this theme. In *Seeking Spatial Justice* (2010), Ed Soja, for example, argues strongly that spatial distributions of resources, access, services, and so on constitute a fundamental component of justice. Soja does, however, go beyond a strict distributional framework to consider Lefebvre's notion of 'the right to the city', which above all rests on the notion of a 'right to inhabit space', that is, the right to benefit from the use values of particular spaces. The Lefebvrian perspective also entails a recognition of how the exchange value logic of uneven development can enhance the qualities and opportunities present in some places and diminish those of others. In other words, questions of justice are innately spatial.

In *Justice, Nature, and the Geography of Difference* (1996), David Harvey takes a particularly nuanced and process-oriented view of questions of justice. Not only does he argue that all questions of justice must be evaluated in their social and geographical context, he argues that the very production of geographical differences is a question of social justice:

> Concerns about social justice (and how to understand and operationalize foundational beliefs about that contested term) ... intertwine with the question of how to understand foundational geographical concepts ... At the intersection of all these arguments lies the question of the *just production of geographical differences*. We need critical ways to think about how differences in ecological, cultural, economic, political, and social conditions get produced (particularly through those human activities that we are in a position in principle to modify or control) and we also need ways to evaluate the justice/injustice of the differences so produced. (Harvey, 1996: 5)

Harvey makes two important points here: (1) conceptions of justice are fundamentally geographical, and (2) justice is not merely a question of distribution of social goods in a particular geographical context, but also a question of the processes that produce geographical differences and distributions in the first place. Although it often goes unspoken, alter-globalization movements, by focusing on questions of justice at a global scale, are pursuing new conceptions of, and strategies for, achieving justice that are not confined to the scale of the nation state, but rather respond to the new relational geographies being produced by neoliberal globalization. In doing so they are breaking down the traditional framings of questions of justice.

One of the most wide-ranging and systematic thinkers on questions of justice has been Nancy Fraser. Drawing from both critical theory and feminist theory, Fraser has developed highly innovative and integrative frameworks for thinking about justice. In 'From Redistribution to Recognition? Dilemmas of Justice in a "Post-Socialist" Age', Fraser (1995) argues that demands for 'recognition of difference' have become a major theme in contemporary struggles, but that 'struggles for recognition occur in a world of exacerbated material inequality ... [and] material inequality is on the rise in most of the world's countries' (Fraser, 1995: 68). Fraser asserts that 'justice today requires *both* redistribution *and* recognition' (Fraser, 1995: 69) as well as understanding of the relation between them. In a move that parallels Harvey's argument, she proposes we consider both remedies within status quo relationships, as well as remedies that transform relationships (affirmation remedies and transformation remedies, respectively). Conceptualizing these concepts in a two by two table, Fraser identifies four approaches to addressing justice issues: (1) affirmation remedies to redistribution issues occur through the *liberal welfare state*, which provides 'surface reallocations of existing goods to existing groups' (Fraser, 1995: 87), (2) affirmation remedies to recognition issues occur through *mainstream multiculturalism*, which supports 'surface reallocations of respect to existing identities of existing groups; supports group differentiation' (Fraser, 1995: 87), (3) transformation remedies to redistribution issues require *socialism*, which pursues a 'deep restructuring of relations of production, blurs group differentiation, [and] can help remedy some forms of misrecognition' (Fraser, 1995: 87), and (4) transformation remedies to recognition issues may occur through *deconstruction*, which involves 'deep restructuring of relations of recognition [and may blur] group differentiation' (Fraser, 1995: 87). Fraser considers two pairs of remedies more promising than the others: pairing the affirmative redistribution politics of the welfare state with the recognition politics of multiculturalism has the advantage of promoting group differentiation and welfare; even more attractive, Fraser argues, is pairing the transformative redistribution program of socialism with the transformative recognition of deconstruction – rigid group differentiation is undermined and a more egalitarian political economy is created. Fraser's arguments for transformative remedies to both redistribution and recognition problems resonate with the goals of many alter-globalization movements (more on this below), but rely on the geographical assumption of the nation state as the arena of remedy. This nation state framework is obviously at odds with the issues that concern alter-globalization movements, which deal with issues of justice beyond the nation state.

That virtually all frameworks for understanding and remedying injustice presume the Westphalian nation state is a topic Fraser directly addresses in 'Reframing Justice in a Globalizing World' (2005). While drawing from her previous work, Fraser argues that in a globalizing world:

> [I]t has ceased to be axiomatic that the modern territorial state is the appropriate unit for thinking about issues of justice, and that the citizens of such states are the pertinent subjects of reference. The effect is to destabilize the previous structure of political claims-making – and therefore to change the way we argue about social justice. (Fraser, 2005: 71)

That globalization radically changes how we must think about social justice is at the heart of the concerns and actions of alter-globalization movements. Certainly, questions of redistribution (economic inequality) and recognition (what constitutes respect and public standing) remain primary substantive issues, but a new meta-level question arises: what is the appropriate geographical frame for resolving these issues? Rather than approaching justice as a two-dimensional problem – redistributive/economic questions and recognition/cultural questions – a third dimension encompassing the politics of representation must be added. Traditionally the nation state has been taken for granted as the appropriate geographical frame for seeking justice, but this frame can produce forms of misrepresentation when it comes to global issues and processes. Incorrect territorial boundary-setting can produce a significant form of injustice, 'wrongly exclud[ing] some people from the chance to participate *at all* in its authorized contests over justice' (Fraser, 2005: 76). The nation state frame 'is a powerful instrument of injustice, which gerrymanders political space at the expense of the poor and despised' (Fraser, 2005: 78) – a critical concern of alter-globalization movements that seek to redress forms of injustice stemming from relationships that transcend the nation state.

As with issues of redistribution and recognition, Fraser sees the politics of spatial frame-setting taking both affirmative and transformative forms. In an affirmative framing politics, the boundaries of the nation state are contested and perhaps redrawn, while still accepting that 'the territorial state is the appropriate unit within which to pose and resolve disputes about justice' (Fraser, 2005: 80). In a transformative framing politics, the state-territorial 'grammar is [challenged as] out of synch with the structural causes of many injustices in a globalizing world, which are not territorial in character' (Fraser, 2005: 81). Global forces of injustice are not only non-territorial, they frequently are highly mobile as well. These characteristics necessitate the abandonment of Westphalian frame-setting processes and their replacement with framing processes that include all affected claimants, regardless of location. In addition to addressing issues of redistribution and recognition across nation state borders, then, alter-globalization movements must often focus on creating new transnational public spheres and democratic arenas where transnational elites cannot monopolize agendas and deny standing to those with grievances.

A Diversity of Alter-Globalization Movements

Given the diversity of global justice issues – redistribution, recognition, and representation – as well as political strategies – affirmative and transformative – it is not surprising that there is a diversity of alter-globalization movements. While what counts as an alter-globalization movement is in large measure a function of how we choose to classify social movements, there is general consensus around what the most influential contemporary alter-globalization movements have been. In one of the most comprehensive accounts of alter-globalization, Pleyers (2010) identifies a series of protests and events that gave birth to the key streams of the alter-globalization movement: (1) The October 1993 march in Bangalore, India, at which over a half million Indian farmers protested proposed changes to the General Agreement on Tariffs and Trade (GATT). This gave rise to Via Campesina, a global network of small farmers eventually spanning fifty-six countries and 100 million members; (2) The seizing of control of seven towns in Chiapas, Mexico, by indigenous activists on the same day in 1994 that the North American Free Trade Agreement (NAFTA) took effect. Known as the Zapatistas, these activists organized a major international gathering in 1996 that formed the People's Global Action network; (3) The first major protest against a G-8 summit in May 1998 in Birmingham, England. In alliance with Jubilee 2000, more than 70 000 people protested, calling for third world debt relief; (4) 'The Battle in Seattle' in December 1999, at which labor unions, environmental organizations, women's rights organizations, human rights organizations, students and academics converged to analyze and protest the World Trade Organization's (WTO) 'globalization from above' vision and replace it with a vision of 'globalization from below'. The protests resulted in the failure of the Seattle round of WTO negotiations and have inspired massive protests at most global elite gatherings since; and (5) The first World Social Forum, held in Porto Alegre, Brazil, in January 2001, with more than 12 000 activists from almost 100 countries in attendance. Establishment of the World Social Forum represented not only a direct response to the global elites' World Economic Forum, it represented an explicit focus on the construction of alternatives to neoliberal globalization.

To the events and movements identified by Pleyers should be added at least two more: (6) The Climate Justice movement, which is actually an amalgamation of diverse streams of climate change mobilization seeking 'democratic accountability and participation, ecological sustainability and social justice ... [in] solutions to climate change' (Chatterton et al., 2013: 606). The Climate Justice movement first emerged in the early 2000s and has been particularly active at United Nations COP (Conference of the Parties) climate change summits, strongly opposing ecological modernization approaches to climate change that do not address the problems wrought by the growth dynamics of unrestrained capitalism; and (7) The so-called 'Movement of the Squares', encompassing the Greek anti-austerity protests starting in May 2010, the Arab Spring starting in Tunisia in December 2010, the Spanish 15-M Movement/Indignados/Take the Square protests beginning in May 2011, and the Occupy Movement beginning in September 2011, originally directed at the Wall Street financial district in New York and eventually spreading to hundreds of cities around the world. All of these movements developed in reaction to the global financial crisis and authoritarian austerity measures instituted by numerous national governments which bowed to the pressures and policies of supra-national governance institutions such as the International Monetary Fund and the European Union (della Porta, 2015).

In alter-globalization events and movements Pleyers sees two different logics: 'resisting through subjectivity' and 'resisting through reason'. Resisting through subjectivity is rooted not in economic calculation but rather authentic non-commodified living, emphasizing the 'will to think and to act by oneself, to develop and express one's own creativity, to construct one's own existence ... Lived experience, the assertion of subjectivity, identity and creativity are set up against the triumph of market utilitarianism' (Pleyers, 2010: 36). Perhaps most emblematic of this approach is the Zapatista movement. In contrast to approaches emphasizing subjectivity, resisting through reason draws on technical expertise and abstract knowledge to challenge the neoliberal global order, above all to delegitimize the market rationality that produces inequality and exploitation and to lay sound foundations for humane alternatives to neoliberal globalization. Exemplifying this latter approach is ATTAC (the Association for the Taxation of Financial Transactions for the Aid of Citizens), a network of radical intellectuals founded in France, now spanning 40 countries. ATTAC identifies itself as an alter-globalization organization:

> [O]ppos[ing] neo-liberal globalization and develop[ing] social, ecological, and democratic alternatives so as to guarantee fundamental rights for all. [ATTAC fights for] the regulation of financial markets, the closure of tax havens, the introduction of global taxes to finance public goods, the cancellation of the debt of developing countries, fair trade, and the implementation of limits to free trade and capital flows. (ATTAC, 2018)

These two approaches – resisting through subjectivity and resisting through reason – bear a strong affinity to Fraser's (1995) politics of recognition and politics of redistribution. Important to emphasize, however, is the fact that while different movements may have different starting points – concern with the need for recognition of diverse everyday lived experiences and concern with the mechanisms and processes that create unjust distributions of wealth and opportunity – they inevitably weave both logics together.

All forms of alter-globalization oppose market-fundamentalist neoliberal governance. Market fundamentalism assumes that market actors are the only type of actors that should count and that markets represent the best and most efficient form of social action coordination. These assumptions ignore the value of diverse subjectivities and non-market-based forms of action coordination. In other words, neoliberalism undermines governance rooted in communication – the very essence of democracy. Not surprisingly, then, the heart of the alter-globalization agenda focuses on radical democratization and commoning, goals that advance both a politics of recognition and a politics of redistribution, while challenging the structures of political representation found in the contemporary Westphalian state system (Fraser, 2005) by emphasizing 'globalization from below' (della Porta et al., 2006). While efforts to enhance the representation of diverse voices are common to virtually all alter-globalization movements, the World Social Forums are probably the best-known examples of global deliberative bodies that acknowledge the standing of citizens regardless of national origin.

Alter-Globalization Movements: Collective Identities and Mobilization Strategies

Alter-globalization movements face particularly difficult challenges as they attempt to mobilize peoples around the globe to address issues of justice that span nation states, greatly varying economic and social conditions, and diverse cultures of meaning and communication. As difficult as collective action problems can be within a shared culture and territory, they are dramatically more challenging on a global scale. In an alter-globalization movement one expects collective action from a great diversity of people around the world, but participation is always uneven.

All forms of collective action have a common problem: how to get people to participate. Mancur Olson (1965) identified free-riding as the central issue of collective action: a rational self-interested person could reap the benefits of other people's collective work while doing nothing, so why participate? Olson believed the solution lay in offering selective incentives, but research on collective action shows that this is almost never the case. The 'glue' of collective action is, rather, collective identity. Most people hold multiple collective identities and do not view the world from a purely self-interested perspective.

But which collective identities? It is virtually impossible to create 'master frames' that will resonate globally with all people. Most people's collective identities are not global but built through everyday experiences and interactions that can vary greatly, both by geographic scale and frequency of interaction (Miller, 2004). *Cosmopolitan* identities are common among leading alter-globalization activists – people who frequently travel around the world and interact with many people in diverse parts of the world. The daily lives of cosmopolitans often take place in global circuits of interaction and encounters are often with the relatively well-educated and well-to-do, although not exclusively so. Many other activists, including many alter-globalization activists, are strongly rooted in place. These activists interact frequently with others on a local scale and become members of *tightly-knit place-based* communities. Their collective identities, and perspectives on globalization processes, are from a local context, looking outward. Still others may have very little direct interaction with others at any scale, but think of themselves as members of a broader national or regional community, based on information they consume from both mainstream and social media. While this form of collective identity may represent an *imagined community*, it can nonetheless be an effective basis for collective action. And there are isolated individuals – people who interact very little with others at any scale – whose identities are rooted in *individualism*; these are people who are unlikely to get involved in collective action or, if they do, are likely to do so in response to selective incentives. How does a chaotic mish-mash of different collective identities and allegiances get forged into a coherent global social movement? One answer is that it does not: global social movements are always messy and fluid alliances that are constantly evolving, falling apart, and being reconstituted.

A common narrative of global mobilization relies on the notion of the rise of 'the network society' associated with the information technology revolution – based first and foremost on communication and networking opportunities created through the Internet. Most prominent among exponents of technology-driven explanations of new global social movements is Manuel Castells. In *Networks of Outrage and Hope: Social Movements in the Internet Age*, Castells (2012: 9) asserts that 'in the network society communicative autonomy is primarily constructed in the Internet networks and in the platforms of wireless communication'. In this narrative, the mobilization of the Movement of the Squares 'began on the Internet social networks, as these are spaces of autonomy, largely beyond the control of governments and corporations that had monopolized the channels of communication as the foundation of their power, throughout history' (Castells, 2012: 2). Castells' account dramatically downplays the roles and messiness of pre-existing organizations, communication and cooperation among organizations, locally-based communities and their networks, forums, public protests, and a host of other long-standing means of building networks, sharing information, and fostering solidarity. Instead, his notion of global social movements strongly emphasizes the fluid construction of international linkages built among individuals using the Internet. In this regard, his account shares much with Hardt and Negri's (2004) notion of a globally networked 'multitude' arising to challenge global institutions and systems of power through non-hierarchical forms of action coordination that will ultimately produce global democratic governance.

While global democratic governance may be a goal of some alter-globalization movements, it does not seem to be a goal of most, and indeed many reject a strategy of 'discontents of the world, unite!' (Hale and Slaughter, 2005). Instead, much alter-globalization mobilization takes place at scales other than the global and focuses on solutions that are not global. Important in any analysis of alter-globalization mobilization is the recognition that social movement networks are not only topological but are also territorial, rooted in place, and scaled (Miller, 2013). In other words, alter-globalization movements are not always global themselves. Routledge et al. (2013) provide a number of examples of alter-globalization movements negotiating the tensions between horizontal and vertical forms of organization, and between more localized and transnational forms of networking and decision-making. They argue that a 'focus on processes necessitates an understanding of network growth and contraction strategies. This must take into account the fact that while networks of resistance operate transnationally, the struggles and the identities of resistance are often born locally through activists' sense and experience of place' (Routledge et al., 2013: 274–275). Often 'brokers' or 'imagineers' are needed to build lines of communication among activists operating in different geographical realms. The need for such bridging is not surprising given the need to frame messages that will resonate with 'a multitude of culture- and place-specific norms, values, and issues' (Miller, 2004: 237). And part of framing messages that resonate with different groups of people consists of proposing solutions to neoliberal globalization that are consistent with people's lived experiences, experiences that may be highly localized.

CONCLUSION: BUILDING A WORLD OF ALTER-GLOBALIZATION

Drawing on Raymond Williams' notion of 'militant particularism', David Harvey (1996) has argued that most social movements begin 'with place-specific experience, but [go] far beyond the bounds of place as ideals forged out of the affirmative experience of solidarities in one place get generalized and universalized' (Harvey, 1996: 32). This is true for most social movements, including alter-globalization movements. While some movements may try to build global alliances to reform global institutions (see Stiglitz, 2006), most responses stay closer to home with a focus on local, regional, national, or world-regional responses to neoliberal globalization (see Gibson-Graham, 2006; Fuller et al., 2010). The ravages of neoliberal globalization always touch down somewhere, and those locations can be settings for response, whether it is through the adoption of municipal alternative currencies, the building of regional networks of industrial cooperatives, supporting indigenous rights and economies, or working toward fair trade and debt forgiveness on a global scale. Alter-globalization movements pose alternatives to neoliberal globalization, alternatives that need not be global themselves. Their common thread is the desire to build a just world, one that produces a just distribution of resources and wealth and recognizes the voices of all people. It is an extraordinarily ambitious goal and one that will likely be achieved piecemeal, in fits and starts. Nonetheless, as the World Social Forum proclaims, 'another world is possible … .'

REFERENCES

ATTAC (2018), 'Overview'. https://www.attac.org/en/overview (accessed on March 2, 2018).
Castells, M. (2012), *Networks of Outrage and Hope: Social Movements in the Internet Age*, Cambridge, UK and Malden, MA: Polity.
Chatterton, P., D. Featherstone and P. Routledge (2013), 'Articulating Climate Justice in Copenhagen: Antagonism, the Commons, and Solidarity', *Antipode*, **45**(3), 602–620.
della Porta, D. (2015), *Social Movements in Times of Austerity*, Cambridge, UK and Malden, MA: Polity.
della Porta, D., M. Andretta, L. Mosca and H. Reiter (eds) (2006), *Globalization from Below*, Minneapolis: University of Minnesota Press.
Fraser, N. (1995), 'From Redistribution to Recognition? Dilemmas of Justice in a "Post-Socialist" Age', *New Left Review*, **212** (July–August), 68–93.
Fraser, N. (2005), 'Reframing Justice in a Globalizing World', *New Left Review*, **36** (November–December), 69–88.
Fuller, D., A. Jonas and R. Lee (eds) (2010), *Interrogating Alterity: Alternative Economic and Political Spaces*, Farnham: Ashgate Publishing.
Gibson-Graham, J.K. (2006), *A Postcapitalist Politics*, Minneapolis: University of Minnesota Press.
Hale, T. and A.-M. Slaughter (2005), 'Hardt & Negri's "Multitude": The Worst of Both Worlds'. *openDemocracy* https://www.opendemocracy.net/globalization-vision_reflections/marx_2549.jsp (accessed on March 8, 2018).
Hardt, M. and A. Negri (2004), *Multitude: War and Democracy in the Age of Empire*, New York: Penguin Press.
Harvey, D. (1996), *Justice, Nature, and the Geography of Difference*, Cambridge, MA and Oxford: Blackwell Publishers.
Hessel, S. (2010), *Indignez-vous!*, Paris: Indigène éditions.
Hessel, S. (2011), Indignez-vous!, *The Nation*, March 7/14, 15–19.
Miller, B. (2004), 'Spaces of Mobilization: Transnational Social Movements', in: C. Barnett and M. Low (eds), *Spaces of Democracy*, London and Thousand Oaks: Sage, pp. 223–246.
Miller, B. (2013), 'Spatialities of Mobilization: Building and Breaking Relationships', in: W. Nicholls, B. Miller, and J. Beaumont (eds), *Spaces of Contention: Spatialities and Social Movements*, Farnham: Ashgate Publishing, pp. 285–298.
Olson, M. (1965), *The Logic of Collective Action*, Cambridge, MA: Harvard University Press.
Pleyers, G. (2010), *Alter-Globalization*, Cambridge, UK and Malden, MA: Polity.
Rawls, J. (1971), *A Theory of Justice*, Cambridge, MA: Belknap Press.
Routledge, P., A. Cumbers and C. Nativel (2013), 'Global Justice Networks: Operational Logics, Imagineers and Grassrooting Vectors', in: W. Nicholls, B. Miller, and J. Beaumont (eds), *Spaces of Contention: Spatialities and Social Movements*, Farnham: Ashgate Publishing, pp. 261–284.
Soja, E. (2010), *Seeking Spatial Justice*, Minneapolis: University of Minnesota Press.
Stiglitz, J. (2002), *Globalization and Its Discontents*, New York and London: W.W. Norton and Company.
Stiglitz, J. (2006), *Making Globalization Work*, New York and London: W.W. Norton and Company.
United Nations (1948), *Universal Declaration of Human Rights*. http://www.ohchr.org/EN/UDHR/Documents/UDHR_Translations/eng.pdf (accessed on February 23, 2018).

PART VI

RESEARCHING AND TEACHING GEOGRAPHIES OF GLOBALIZATION

32. Multi-sited fieldwork in a connected world
Valentina Mazzucato and Lauren Wagner

INTRODUCTION: WHEN, WHERE AND WHY 'MULTI-SITED'

Over twenty years ago, anthropologist George Marcus published the seminal and frequently referenced article arguing for multi-sited ethnographic research in an age of globalization (Marcus, 1995). Building on a world systems theory perspective (Wallerstein, 1974), he emphasizes that, while an ethnographic subject may be framed within a specific place, it is also researchable as interconnected to many places – maybe even *necessarily* researchable through interconnectedness as a 'local' perspective in a globalized world becomes impossibly narrow. Though he develops this approach as emerging from ethnographies that preceded this article, he is nevertheless widely acknowledged as putting a name to a new methodology for investigating globalization in a 'multi-sited' way.

Part of the popularity of this article lies in how Marcus outlines, and provides examples of approaches to multi-sited ethnography. He suggests several modes through a trope of 'following' – follow the people, the thing, the metaphor, the story, the biography and the conflict – by which a researcher might maintain focus on a core topic while engaging with the multiple sites where that topic can be empirically observed. Many of these modes have been taken up by scholars to research topics in globalization, using ethnography as well as other methods, which have produced insightful investigations of how global phenomena manifest in and link dispersed localities.

Yet multi-sited research also presents significant challenges. While phenomena may be 'global' and even occur 'simultaneously' (Levitt and Glick Schiller, 2004) in different locations, researchers are constrained by the physics of time and space, and limited financial and temporal resources in their ability to engage with multiple sites of activity. While some topics seem to occur everywhere simultaneously, it is not possible to focus methodological attention on everything at once. Likewise, some phenomena engage in multiple places across different or extended time periods, which can equally present a challenge to a researcher trying to 'follow the story' as it unfolds without a full conclusion.

In this chapter, we discuss some parameters and possibilities for doing multi-sited research as a methodology that connects very closely with the theoretical problems of globalization. We first highlight some different modes for conceptualizing globalization as a theoretical problem, in order to reflect on how those theories relate to different possible methodological constructions. Next, we consider what constitutes a multi-sited research design that investigates a single phenomenon, as opposed to other research design models that may also include many sites but treat their core phenomenon in a different way. Then, we give some examples of what a multi-sited design looks like, drawing on our own research and that of others. Finally, we discuss some of the advantages and disadvantages of these designs, and what they are able to contribute to current research on globalization.

WHAT CONNECTS TO WHAT

As the preceding chapters of this book demonstrate, geographies of globalization range through many different fields of study, lifeworlds, systems, contexts and phenomena. Across all these factors, however, certain key elements to 'globalization' mark it as a distinctive framework for research. In order to envisage how to use multi-sited methods to study globalization topics, we first need to identify how globalization frames our research subject in a specific and unique way. We focus here on two parameters that often characterize what is challenging for researchers about studying globalization – namely *simultaneity* and *complexity*. These two parameters (independently or combined) can be methodologically difficult to observe in a single site. By addressing them, we will discuss how a multi-sited approach to researching globalization can produce different possibilities of analysis than focusing on a single site.

Imagining how different places are connected *simultaneously* (Levitt and Glick Schiller, 2004) requires recognizing how ongoing relationships are actively maintained when people are not sharing a localized, physical space at the same time. While researchers in the past – especially ethnographers – have often focused on the observable 'here and now', using globalization as a paradigm necessitates observations of what is going on 'elsewhere' at the same time that it may be materially interconnected to what is going on here. In other words, it requires understanding 'the stretching and deepening of social relations across national borders so that everyday activities are more influenced by events at greater distances' (Smart and Smart, 2003: 265).

These kinds of simultaneous relationships crossing distant spaces have been discussed through concepts like 'transnational' (Vertovec, 1999, 2001; Levitt, 2001; Portes et al., 1999), 'translocal' (Greiner and Sakdapolrak, 2013; Conradson and McKay, 2007), and 'commodity or value chains' (Baird, 2008) often in reference to migrant lives or commodities that connect different places through the simultaneous presence and activities of people connected through family, work, religion, or production and consumption processes. Rather than understanding 'community' as a unit bounded by physical geography – such as a firm, city or a neighbourhood – these terms consider that a community may be occurring simultaneously in very dispersed places. While new communications technologies are not always embedded in these contexts, they are often part of what facilitates simultaneity, through increasingly instantaneous communicative networks between places (Vertovec, 2001, 2004). But simultaneity is not limited to human communities; in a globalization framework, one can also take into account simultaneous proliferation of ideas (Levitt, 2001), movement of goods or finances (Haugen, 2017), or political change (Guarnizo et al., 2003). While any one of these may be empirically observable in a specific locality – for example the Occupy Movement in New York City – our analysis of them becomes more complete and insightful by observing their *simultaneous* occurrence in multiple spaces.

Analyses of globalization, likewise, generally employ a perspective of *complexity* in their approach to different phenomena. That is, different domains of social life – work, family, belonging, politics and so on – are understood to be interconnected, interrelated, and inseparably embedded within each other. While this perspective is not exclusive to studies in globalization, it is part of what makes these studies challenging to execute. In many branches of social science, part of the methodological approach is to delimit the site and refine variables in order to identify *causality* as a two-part relationship – a cause produces an effect. Research in globalization approaches causality as a complex relationship, where much of the researcher's work lies in the problem of identifying which elements are exerting influence on the topic at hand. A good example of the task of finding the 'right' variables is found in the work of Nicky Gregson, Mike Crang and their colleagues (Gregson ct al., 2016, 2017) researching the global circulation of freight cargo and recycling. When politics and discourses in the global North promote – and even enforce – 'resource recovery' of waste, they create effects in the global South, where much of the 'dirty work' of transforming waste into usable material has immense consequences for those hired to do that labour. This relationship is, however, not simply a 'cause-and-effect' one: many interwoven actors, sites, and flows enable this global chain to function, and its effects are felt across political, economic and health spheres in Europe and in Asia.

Studying people who may never meet face to face; objects/ideas that are being formulated and shaped in many places at the same time; one thing that is interrelated with many other things – all these require a multi-sited conceptualization of connected space. These spaces echo what geographer Doreen Massey called 'a global sense of place', that what occurs in a single place 'can be imagined as articulated moments in networks of social relations and understandings, but where a large proportion of those relations, experiences and understandings are constructed on a far larger scale than what we happen to define for that moment as the place itself, whether that be a street, or a region or even a continent' (Massey, 1994: 154). Such spaces may be simultaneously active or dormant, or sometimes disjunctive in their interconnected activity (Collins, 2009): they develop as networks of interconnection that persist and evolve, with many potential ways to intersect and influence one another.

While a first wave of multi-sited studies tended to focus on the connectivities created by labour or undocumented migrants, more recent studies have extended to non-migrant populations, the middle classes and elites. A burgeoning branch of geographical studies, taking up calls for studying up or sideways (Ortner, 2010), are turning from a focus on predominantly poor or marginalized sectors of society to studying the superrich (Ley, 2003). Such people are often embedded in global relationships of production and consumption, making it necessary for researchers to understand processes at multiple levels, often entailing multiple sites. Saskia Sassen, for example, argues that global financial capitals like Tokyo, London and New York are more connected between themselves than with neighbourhoods in the same cities (Sassen, 2001), as mobility of ideas, capital and elites circulate between them without engaging, to a great extent, with the 'locality' of the city that hosts them.

Similarly, David Conradson and Alan Latham (2005) argue for attention to 'middling transnationalism' as part of the lives of members of the middle classes. Their argument highlights how interconnections of some actors across distant spaces may appear 'ordinary' and innocuous while they parallel interconnections of others – such as labour migrants or other less advantaged groups, who are often the subject of enquiry – frequently characterized as difficult or extraordinary (Burawoy et al., 2000). These different layers of transnational engagement encourage us to think about simultaneity occurring 'from below' (Brecher et al., 2000) as well as from the 'middle' (Leinonen and Pellander, 2013; Wagner, 2014).

These examples demonstrate that multi-sited engagement with globalization is not predicated on class; it affects those who are elite and impoverished, urban and rural, mobile and immobile. Thus, for researchers investigating a wide variety of issues, a multi-sited methodology can be significant in identifying the different factors that influence their topic and developing an analysis that furthers scientific understanding of it. To do so, we argue, requires careful construction of a multi-sited research design.

WHAT IS A MULTI-SITED RESEARCH DESIGN AND WHAT IS IT NOT

It is important to clearly define and understand what multi-sited research designs are and what they are not, as they have at times been mis-specified in the literature. A key mis-characterization is when comparative international research is considered to be multi-sited. More than simply moving across borders, we argue that multi-sitedness involves conceptualizing how experiences and events in different places contribute to a multifaceted understanding of a single phenomenon.

Much like triangulation, doing multi-sited research entails collecting data that addresses the *same problem through different lenses*. While many research designs involve collecting data from multiple sources about the same issue, we contend that multi-sited research signifies a configuration where actors in at least two sites are connected to each other by concrete linkages, created and maintained through the flow of people, goods, money, and/or communication on an observable basis, over time. Such sites are not necessarily in different countries, but many multi-sited research designs are indeed located across political or natural borders. In short, multi-sited research defines as its objective 'the study of social phenomena that cannot be accounted for by focusing on a single site' (Falzon, 2009: 1).

Other research designs may employ different sites to look at the same problem comparatively; this approach is distinct from a multi-sited design. For example, a study designed to investigate Romanian female care workers in Italy and the UK may involve collecting data in multiple locations in order to compare these two groups. Alternately, the care workers may be themselves linked to each other, for example through kinship bonds or a common migrant organization, and the objective of the study incorporates the role of these bonds or organizations. While the first design incorporates variables of difference between two places, it does not account for the complex influence of connectivity between those places. The second design includes a dimension of simultaneity and complexity, through how these individuals may be interconnected via a social network. We contend that while the first design may be illustrative of many important factors, the second represents a multi-sited research design. If the essence of a multi-sited research design is to 'follow' – whether people, connections, associations or relationships (Marcus, 2011) – then there must be a 'followable' connection between geographically distant spaces to constitute a multi-sited design.

Increasingly, different formulations of 'site' are being recognized as useful in 'following' networks and flows. Beyond a doubt, the most significant is the ever-evolving landscape of mediated connectivity, whether through older technologies like radio and television, through mobile phone usage, Internet connectivity, and into new and as-yet-unknown modes through which individuals can connect to distant people, places and things. To demonstrate with the above example, a network of care workers may take place simultaneously in-person in different geographical locations, as well as online via forums, chats, websites, and other modes of mediated interaction (Brown, 2016). For many researchers, these are multiple sites, all of which are indispensable to building an analytical understanding of globalization.

It is important to note that multi-sited research designs do not imply using a specific methodology, although most multi-sited research designs have employed qualitative methods such as ethnography, semi-structured interviews, and network data collection because they lend themselves well to following the linkages that connect actors across localities and they help to address some of the challenges of multi-sited research designs, which will be discussed below.

Surveys and other quantitative methods can be employed, although it is difficult to obtain large samples that are connected to each other across multiple sites. In research on migration, only a few have succeeded in employing such methods (Mazzucato, 2009b; Osili, 2007) and even then, the samples are small. Beauchemin (2014) has outlined the main challenges to doing multi-sited research on a large scale, such as gaining correct contact details of people overseas, the double consent needed from people on each side of a connection, and the fact that people are connected to many people in multiple locations, greatly increasing the costs of tracing and interviewing all relevant persons. These are all factors to take into account when considering whether to engage in multi-sited research.

WHAT DO MULTI-SITED RESEARCH DESIGNS LOOK LIKE

To execute multi-sited research, several approaches and formats might be used. In this section, we detail three different possible practical approaches, followed by some issues to consider when choosing a multi-sited design.

Sequential Visits to Multiple Sites

Most multi-sited research has implemented Marcus' 'follow' trope referenced above, by having one researcher move physically from site to site in order to study the way the connections of a particular phenomenon under study work. As an example of 'following the relationship', Schmaltzbauer (2004) was interested in understanding how Guatemalan families continue to operate as families across great distances. She observed and interviewed migrant parents in the US and then went to Guatemala to follow some of the interviewed parents' children and their respective caregivers.

'Following the thing' can be illustrated by Kopytoff's famous example of a used car (Kopytoff, 1986: 1). He followed a second-hand car as it travelled from Europe to Africa to study the different meanings it takes on in the different contexts it passes through. These are just a few examples from the many where researchers 'followed' with participants over sequential visits to multiple sites, in order to understand global social phenomena from transnational families and civil society organizations to urban and political elites and sex workers as they move across borders and create and maintain social ties with people and organizations in the different locations they pass through.

Simultaneous Matched Samples

A less frequently used design is one in which teams of researchers are located in the different sites, and follow people who are connected to each other through some kind of relationship, simultaneously. For example, Mazzucato et al. (2006) followed one event, a funeral, across different sites by each being based in one of the sites where members of the same extended family were located. By recording the preparations, the event itself and the post-funeral arrangements as they took place at the same time and across multiple sites, they investigated how an important cultural ceremony took on new meanings and entailed new and modified practices through its transnational character. Poeze et al. (2016), similar to Schmaltzbauer above, were interested in understanding how people 'do family' across great distances, yet each researcher was based in a different site where the family members were located in order to observe their interactions in real time. For example, this design allowed the researchers to witness an international phone call by being with both the caller and the receiver of the call and to observe what immediate effect the call had on both ends. By sharing notes about what was happening in the lives of the caller and the called, the researchers were able to detect not only what was communicated between the parties, but also the strategies that they employed for not communicating certain information. This illustrated the work that goes on behind maintaining transnational relationships, showing the invisible (such as non-communication), and small everyday effects (such as not going to work in order to cater to a migrant's request) that are not easy to evoke on recall.

Multi-Sitedness Through Technological Connection

One of the challenges of doing research on people, things and ideas that move is precisely their mobility. A researcher, or even a team of researchers, may have difficulty being in the right place at the right time. This has led to the conundrum for researchers of having to sacrifice depth or breadth (Rutten, 2007). One way to partially overcome the difficulty in following people on the move, is through following them virtually, either by interacting with them online (Schapendonk, 2009) or by following their interactions with others (Madianou and Miller, 2011). Schapendonk, for example, met with various African migrants in North Africa, and then again on their routes to Europe via Greece and Turkey. However, to be able to follow up with respondents in the different locations where they ended up, they communicated with them via social media. Madianou and Miller observed how Filipino mothers kept an eye on their teenage children in the Philippines, by keeping webcam communication open at all times, while going about their daily lives in their homes in the UK. In all these examples, researchers did not physically travel to speak with or observe their respondents. In some cases they communicated with them via ICTs and in others they observed their respondents' behaviours by observing their interactions through ICTs. Both types of studies show how ICT is both a subject of study and a tool for enabling multi-sited research.

What to Consider When Deciding Whether to Use Multi-Sited Designs

The main advantage of using a multi-sited design is that knowing what is happening in one site enables the researcher to ask better questions or notice more details in another site. For example, money sent by a migrant to family at home can have different meanings on the sending and receiving end (Mazzucato, 2009a). These different stories show how people are differently positioned in a transnational social field (Levitt and Glick Schiller, 2004), and would not be observable by researching in only one site. By working in teams in linked field sites, researchers can observe the immediate effects of people's simultaneous engagement over these sites – such as what happens when money is sent, or when money is requested and not sent. Recognizing the architectural styles of houses migrants see in the US helps a researcher to recognize architectural similarities with the houses they construct in their countries of origin (Lopez, 2014). All of these help researchers gain a more complete understanding of phenomena that are situated in multiple, simultaneous places, connected in globalization.

By participating in movement, researchers can gain more insight into their respondent's lived realities. Schapendonk (2009) experienced the exhaustion of migrants' trips; Wagner (2017) showed how diasporic Moroccans share the experience of the journey towards their vacation in Morocco while in transit. By taking into account multiple sites as well as the 'off-time', 'transition' or movement between them, researchers can gain access to important information that enables better understanding of the 'sites' themselves. For example, in sharing details with each other about their respective linked field sites, Poeze et al. (2016), witnessed that not communicating certain information is a technique that both migrant parents and their children's caregivers in the home country use as a strategy to reduce tensions in the family. The effects of transition (Schapendonk, 2009; Wagner, 2017) or of absence (Poeze et al., 2016) only become visible through observing movements – whether physical or technological – between multiple sites.

Yet doing multi-sited research comes with its challenges that are also important to consider before embarking on such an adventure. As mentioned at the start of this chapter, there is always the challenge of combining depth with breadth. Researchers have developed ways to do this through working with teams, or using virtual methods, but there will always be some concessions between the two. Furthermore, having multiple field sites also means needing time to develop networks in and familiarity with multiple sites. This means having to build relationships of trust in multiple contexts, understanding different cultural and linguistic contexts and possibly having to find multiple institutional supports. All of this takes time, and sometimes much time.

There are ethical considerations particular to multi-sited research as well. Multi-sited research leads to asymmetrical information between respondents and researchers where the latter can come to know more about a respondent's family for example, than the respondent knows him or herself (Mazzucato, 2009a). In such cases, these researchers have opted for not divulging any information from respondents in one site to respondents in another site, as part of the confidentiality agreements made to all respondents. This also meant publishing using pseudonyms for names of people and places that would reveal people's identities. Researchers will need to reflect on how to respond in an ethically appropriate way in such situations, and on what they promise their respondents.

Finally, especially multi-sited research conducted in teams requires a different research praxis than solo research such as data sharing, co-publishing and adaptation to accommodate for different researchers' ideas and schedules. For example, for a multi-sited team research to have its advantages, researchers need to share data, including field observations, interview transcripts and the like. This requires particular methods for data sharing, from the development of forms for sharing data as the research unfolds, to techniques for transferring data from one computer to another. All this requires trust and adaptability between researchers and for it to work, it must lead to co-authored publications. Furthermore, working with teams of local scientists in the different locations means negotiating diverse institutional cultures and recognizing that in certain countries, where material circumstances for researchers are fewer, this will mean that such researchers will fulfill different roles in the research, but that all roles are necessary for the successful completion of a research project and should be recognized through co-authorship of publications. Social scientists are more often trained as solo researchers where things like good practices around data sharing and skills for working in teams are not taught and co-authorship is still seen as a second-best solution.

CONCLUSION

Despite these challenges, multi-sited research designs have emerged as an answer to the necessity of studying flows in a globalized world. They are part of a reassessment of the importance of fieldwork, fieldtrips and other visits to the field of geographical inquiries and a necessary move as a concrete way to study the interconnections that are the 'stuff' that globalization is made of. Multi-sited research designs allow following the movement of people, things, ideas, discourses, or money as they connect sites across the world, and also seeing the disconnections, or new places of marginality by showing who or what is connected and who or what is not or no longer. They are not only a way to study how cheap travel, greater transport and communication infrastructure and the availability of an increasing range of ICTs have connected people and places in the world, but these same features of globalization are what has made multi-sited research possible.

As connected sites proliferate, and as the types of sites diversify such as through virtual or augmented reality, multi-sited research designs can be carried forward to include more and new sites. Team research will increasingly be indispensable to incorporate such a diversity of sites in a way that does not compromise the breadth required for doing global research but also caters to the depth that is needed for quality research. Researchers thus need to develop different praxes for working in more and new types of sites, in teams, sharing data and co-publishing, also with colleagues located in different national and institutional contexts.

REFERENCES

Baird, J. (2008), *Frontiers of Commodity Chain Research*, Stanford: Stanford University Press.
Beauchemin, C. (2014), 'A Manifesto for Quantitative Multi-sited Approaches to International Migration', *International Migration Review*, **48**(4), 921–938.
Brecher, J., T. Costello and B. Smith (2000), *Globalization from Below: The Power of Solidarity*, Cambridge: South End Press.
Brown, R. (2016), 'Multiple Modes of Care: Internet and Migrant Caregiver Networks in Israel', *Global Networks*, **16**(2), 237–256.
Burawoy, M., J. Blum, S. George, Z. Gille, T. Gowan, L. Haney, M. Klawiter, S. Lopez, Ó Riain and M. Thayer (2000), *Global Ethnography*, Berkeley: University of California Press.
Collins, F.L. (2009), 'Transnationalism Unbound: Detailing New Subjects, Registers and Spatialities of Cross-Border Lives', *Geography Compass*, **3**(1), 434–458.
Conradson, D. and A. Latham (2005), 'Transnational Urbanism: Attending to Everyday Practices and Mobilities', *Journal of Ethnic and Migration Studies*, **31**(2), 227–233.
Conradson, D. and D. McKay (2007), 'Translocal Subjectivities: Mobility, Connection, Emotion', *Mobilities*, **2**(2), 167–174.
Falzon, M.-A. (2009), *Multi-sited Ethnography: Theory, Praxis and Locality in Contemporary Research*, Surrey: Ashgate Publishing.
Gregson, N., M. Crang, and C.N. Antonopoulos (2017), 'Holding Together Logistical Worlds: Friction, Seams and Circulation in the Emerging "Global Warehouse"', *Environment and Planning D: Society and Space*, **35**(3), 381–398.
Gregson, N., M. Crang, J. Botticello, M. Calestani and A. Krzywoszynska (2016), 'Doing the "Dirty Work" of the Green Economy: Resource Recovery and Migrant Labour in the EU', *European Urban and Regional Studies*, **23**(4), 541–555.
Greiner, C. and P. Sakdapolrak (2013), 'Translocality: Concepts, Applications and Emerging Research Perspectives', *Geography Compass*, **7**, 373–384.
Guarnizo, L.E., A. Portes and W. Haller (2003), 'Assimilation and Transnationalism: Determinants of Transnational Political Action among Contemporary Migrants', *American Journal of Sociology*, **108**(6), 1211–1248.
Haugen, H.Ø. (2017), 'Petty Commodities, Serious Business: The Governance of Fashion Jewellery Chains between China and Ghana', *Global Networks*, doi:10.1111/glob.12164.
Kopytoff, I. (1986), 'The Cultural Biography of Things: Commoditization as a Process', in: Appadurai, A. (ed.), *The Social Life of Things: Commodities in Cultural Perspective*, Cambridge: Cambridge University Press, 64–93.
Leinonen, J. and S. Pellander (2014), 'Court Decisions over Marriage Migration in Finland: A Problem with Transnational Family Ties', *Journal of Ethnic and Migration Studies*, **40**(9), 1488–1506.
Levitt, P. (2001), *The Transnational Villagers*, Oakland: University of California Press.
Levitt, P. and N. Glick Schiller (2004), 'Conceptualizing Simultaneity: A Transnational Social Field Perspective on Society', *International Migration Review*, **38**(3): 1002–1039.
Ley, D. (2003), 'Artists, Aestheticisation and the Field of Gentrification', *Urban Studies*, **40**(12), 2527–2544.
Lopez, S.L. (2014), *The Remittance Landscape: Spaces of Migration in Rural Mexico and Urban USA*, Chicago: University of Chicago Press.

Madianou, M. and D. Miller (2011), 'Mobile Phone Parenting: Reconfiguring Relationships Between Filipina Migrant Mothers and Their Left-Behind Children', *New Media & Society*, **13**(3), 457–470.

Marcus, G.E. (1995), 'Ethnography in/of the World System: The Emergence of Multi-Sited Ethnography', *Annual Review of Anthropology*, **24**(1), 95–117.

Marcus, G.E. (2011), 'Multi-sited Ethnography: Five or Six Things I Know About It Now', in: Coleman, S. and P. von Hellerman (eds), *Multi-sited Ethnography: Problems and Possibilities in the Translocation of Research Methods*, New York, London: Routledge, 16–32.

Massey, D. (1994), 'A Global Sense of Place', in: D. Massey, *Space, Place and Gender*, Cambridge: Polity Press, 146–156.

Mazzucato, V. (2009a), 'Bridging Boundaries With a Transnational Research Approach: A Simultaneous Matched Sample Methodology', in: M.A. Falzon (ed.), *Multi-sited Ethnography: Theory, Praxis and Locality in Contemporary Social Research*, Hampshire: Ashgate, 215–232.

Mazzucato, V. (2009b), 'Informal Insurance Arrangements in Ghanaian Migrants' Transnational Networks: The Role of Reverse Remittances and Geographic Proximity', *World Development*, **37**(6), 1105–1115.

Mazzucato, V., M. Kabki and L. Smith (2006), 'Transnational Migration and the Economy of Funerals: Changing Practices in Ghana', *Development and Change*, **37**(5), 1047–1072.

Ortner, S.B. (2010), 'Access: Reflections on Studying Up in Hollywood', *Ethnography*, **11**(2), 211–233.

Osili, U. (2007), 'Remittances and Savings from International Migration: Theory and Evidence Using a Matched Sample', *Journal of Development Economics*, **83**(2007), 446–465.

Poeze, M., E. Dankyi and V. Mazzucato (2016), 'Navigating Transnational Childcare Relationships: Migrant Parents and Their Children's Caregivers in the Origin Country', *Global Networks*, **17**(1), 111–129.

Portes, A., L. Guarnizo and P. Landolt (1999), 'The Study of Transnationalism: Pitfalls and Promise of an Emergent Research Field', *Ethnic and Racial Studies*, **22**(2), 217–237.

Rutten, M. (2007), '*Leuke vakantie gehad?' Verhalen over antropologisch veldwerk* ['Had a Nice Vacation?' Stories about Anthropological Fieldwork], Amsterdam: Aksant.

Sassen, S. (2001), *The Global City: New York, London, Tokyo*, Princeton: Princeton University Press.

Schapendonk, J. (2009), 'Staying Put In Moving Sands: The Stepwise Migration Process Of Sub-Saharan African Migrants Heading North', *Respacing Africa*, **2007**(4), 113–138.

Schmaltzbauer, L. (2004), 'Searching for Wages and Mothering from Afar: The Case of Honduran Transnational Families', *Journal of Marriage and Family*, **66**(5), 1317–1331.

Smart, A. and J. Smart (2003), 'Urbanization and the Global Perspective', *Annual Review of Anthropology*, **32**(1), 263–285.

Vertovec, S. (1999), 'Conceiving and Researching Transnationalism', *Ethnic and Racial Studies*, **2**(2), 447–462.

Vertovec, S. (2001), 'Transnationalism and Identity', *Journal of Ethnic and Migration Studies*, **27**(4), 573–582.

Vertovec, S. (2004), 'Cheap Calls: The Social Glue of Migrant Transnationalism', *Global Networks*, **4**, 219–224.

Wagner, L.B. (2014), 'Rhythms of a Transnational Marriage: Temporal Topologies of Borders in a Knowledge Migrant Family', *Etnofoor*, **26**(1), 81–105.

Wagner, L.B. (2017), *Becoming Diasporically Moroccan: Linguistic and Embodied Practices for Negotiating Belonging*, Bristol: Channel View Publications.

Wallerstein, I. (1974), *The Modern World-System I: Capitalist Agriculture and the Origins of the European World-Economy in the Sixteenth Century*, New York: Academic Press.

33. Teaching globalisations

Matthew Sparke

INTRODUCTION

Both of the letters 's' in 'globalisations' are critical when it comes to teaching. The first stands instead of the American 'z', indexing non-American perspectives and experiences that diverge from dominant discourses that align capital 'G' Globalization with McDonaldization, Coca-Cola-nization, and Americanization more generally. The second 's' at the end of 'globalisations' points in turn towards the plurality and heterogeneity of these other perspectives and experiences of global integration. A major pedagogic challenge, nevertheless, is to acknowledge this plurality and heterogeneity while also attending simultaneously to the singularizing effects and integrative imperatives that are represented by the dominant discourse of Globalization. To do this, and to debunk all the associated myths about Globalization being new, inevitable and flattening, a major argument of this chapter is that teaching globalisations both demands and enables ongoing efforts to denaturalize the market fundamentalist policies and norms of neoliberalism. At the same time, I also want to argue that we need to work as teachers to help students think, learn, research, and engage beyond the discursive and disciplinary bounds of neoliberal Globalization.

Denaturalization of the dominant discourse is demanded because neoliberal policies and norms are continually legitimized around the world by mythic depictions of Globalization as an unavoidable golden dawn of hyper-competitive but enriching flat-world integration. 'I feel about globalization a lot like I feel about the dawn,' declared the *New York Times* Globalization booster Thomas Friedman in a typical early example of the naturalizing discourse that paradigmatically went on to recommend the 'golden straitjacket' of pro-market 'golden rules' needed to succeed under the rising golden sun of market integration (Friedman, 2000: 103). This kind of naturalizing-turned-normative discourse has now been so widely used to justify the rules of pro-market policy-making that in many students' minds Globalization and neoliberalism are the same thing or at least confusingly fused. It is in turn against this confusing discursive and disciplinary fusion that teaching globalisations becomes so important as a way of introducing analysis and evidence that denaturalizes the normative neoliberal common sense. By bringing into view the non-natural emergence of neoliberalism as a hegemonic regime, and by highlighting all the experiences of market-rule that, as it were, fall off the edge of the 'flat-world' pro-market vision of Globalization, I am suggesting, in what follows, that we can also help students imagine alternative globalisations too.

Other geographers have already mapped these critical teaching opportunities starting from situated knowledge of other globalisations, including an inspiring antipodean textbook that begins with a deliberately 'out of place' insistence that 'Globalization Is

Not Spelt with a Zed' (Murray and Overton, 2015: xviii). Refusing what Gillian Hart once critiqued (from another perspective partly located in the Global South) as the 'impact model' of Globalization (Hart, 2003), and denaturalizing dispossession associated with the militarized as well as economized deployment of the dominant discourse (Hart, 2005), such teaching can draw on the academic progress that has been made in charting the uneven political geographies of dominance, governance and resistance shaping actually existing globalisations (for preliminary progress reports see Sparke, 2004, 2006 and 2008; and for some subsequent surveys, Barkan, 2011; Klein et al., 2014; Sidaway, 2012; Silvey and Rankin, 2010). Historical research can likewise be used pedagogically to highlight the heterogeneity of historical-geographies embodied in (but obscured by) 'borderless-world' type arguments about Globalization 'making geography history' (Domosh, 2010, 2013; Nally, 2014; Smith, 2005). And with attention to embodiment at a much more personal scale, feminist geographies of the 'global intimate' have underlined the associated diversity of experiences in everyday life (Katz, 2001; Mountz and Hyndman, 2006; Pratt and Rosner, 2012).

What all this critical work by geographers underlines is that diverse non-normative, non-dominant globalisations can be better acknowledged and analyzed once the dominant discourse of Globalization is denaturalized. It is in this same spirit that my own textbook *Introducing Globalization* highlights alternatives to neoliberal Globalization alongside an historical-geographical account of its emergence, expansion and entrenchment (Sparke, 2013). The pedagogic aim in this respect is to foster what I refer to as student response abilities in the face of the mind-numbing neoliberal emphasis on personal responsibility. Resources for this kind of counter-hegemonic pedagogy are being increasingly diversified by efforts to advance teaching and action research that contests neoliberalism, taking back both global and local economies from the straitjacket of Globalization (e.g. Darian-Smith and McCarty, 2017; Gibson-Graham et al., 2013; Houston and Lange, 2017; Roy et al., 2016). Here, in the rest of this chapter I seek to add my own further reflections on these teaching opportunities by focusing on three main themes.

First I seek to conceptualize how we can teach students that more monolithic and hegemonic discourses about Globalization (with a capital 'G', a 'zee', and no 's's) are at once both distinct from and influential upon the many diverse ways in which people actually experience globalisations (with two 's's and a lower case 'g'). Second I describe how we can teach against the myth-making about Globalization being new, inevitable and levelling by tracing the spatio-temporal dimensions of actually existing globalisations. And third, by teaching about this uneven experience of globalisations in ways that remain attentive to the globalization of neoliberal governmentality, I argue we can also contribute to students' meta-cognitive awareness about their own experiences of market discipline (and disobedience) in the contemporary corporatizing university. It is through such meta-learning that I argue that we can help students start to turn neoliberal governmentality's trademark emphasis on personal responsibility into new and more enabling forms of response ability for life in a globalising world.

CONCEPTUALIZATION

My own approach to teaching globalisations begins by telling students that it is best to conceptualize a double definition (see Sparke, 2013, Chapter 1). In this way I argue that globalisation is both a *descriptive term* used to name processes of increasing global integration, and a *prescriptive discourse* that is often instrumentalized politically to persuade people about the way the world should be integrated – most commonly thereby prescribing the neoliberal policies of market liberalization, deregulation, and privatization by arguing that such market-friendly reforms are required for successful Globalization. Such prescription works ideologically, I argue, by depicting Globalization as a novel, non-stop juggernaut of global change against which resistance is useless, and for which, in former British prime minister Margaret Thatcher's famous axiom, There Is No Alternative. Teaching against this type of TINA-tout discourse it is useful to explore with students the various myths that are used to promote and prop-up the big G Globalization story: including most repeatedly, that it is new, that it is inevitable and that it is a flattening or levelling or somehow equalizing force. Noticing these myths helps students start to see themselves as decoders of dominant political and media narratives (although some resist the critiques and see themselves instead as defenders of neoliberal commonsense). Wherever students stand, the work of de-mythologization further helps to set the scene for evidence-based analysis of the actual spatio-temporal complexity of real world globalisations. But it is equally important to stress that they will not always find the myths of Globalization reproduced verbatim in exactly the same way and style by all politicians and pundits. Globalization as prescriptive discourse instead moves with a mobile army of metaphors, metonyms and global-morphisms, and it has also increasingly taken diversions, shifting course over time and space in response to the far-from-flat contexts through which it has passed.

In the 1990s, when Globalization was still being consolidated as a buzz-word of political speech, the dominant pro-market discourse often substituted an array of synonyms including 'global market', 'world economy', and 'global competition'. The all too familiar prescriptive conclusion was still that countries and individuals needed to adopt neoliberal policies and practices in order to survive and thrive. And Globalization was thereby metaphorized as a kind of non-stop market-based competition with everyone, everywhere on the globe. Here, for example, was the US conservative Newt Gingrich arguing for the neoliberal 'Contract With America' on the basis of this naturalizing logic back in 1994:

> [W]e need to recognize the objective reality of the world market, to realize that we create American jobs through world sales and that we need to make a conscious national decision that we want to have the highest value added jobs on the planet with greatest productivity so we can have the highest take-home pay and the greatest range of choices of lifestyles. In order to do that we have to literally rethink the assumptions that grew up in a self-indulgent national economy and we have to recognize that litigation, taxation, regulation, welfare, education, the very structure of government, the structure of health all those things have to be reexamined from the standpoint of what will make us the most competitive society on the planet, the most desirable place to invest to create job the place with the best trained and most entrepreneurial workforce. (quoted in Fisher et al., 1994: 188)

More recently, as the failings of this sort of Washington Consensus have become more obvious to more people around the world, the discourse of Globalization is still in use but now performs a rather more contradictory role that demands special care to interpret. On the one side it has become a punching bag for hard-right reactionary nationalists who rail against 'Globalists' and Globalization, even as they go on promoting neoliberal reforms and the interests of transnational corporations and global finance. 'Hillary Clinton is the chief emissary for Globalism', claimed Donald Trump in a resonant populist line of attack in his successful 2016 presidential campaign, before proceeding as US president to push in classic neoliberal style for massive tax cuts, business deregulation, banking deregulation, and the rollback of health care for the poor (Bessner and Sparke, 2017). On the other side, in tandem with the rise of reactionary hyper-nationalists such as Trump, Globalization continues as the rallying cry for global business elites who still want to argue that it represents the golden dawn of a more prosperous world, but who also now acknowledge that more must be done for those who have been burned by the rising sun. 'So while globalization has been glorious for billions of people, for some, it hasn't been glorious at all,' equivocated Hans-Paul Bürkner, the chairman of *The Boston Consulting Group*, one of the most influential business consultancies in the world in January 2017. Warning fellow elites about the spreading reactionary backlash in the aftermath of Trump's electoral triumph, Bürkner continued '… the world risks going into reverse. This must not be allowed to happen. It would be an utter disaster if globalization was stopped in its tracks' (Bürkner, 2017).

Faced with all the prescriptive political announcements, denouncements and re-pronouncements about big G Globalization, teachers need to help students evaluate claims about the successes of neoliberalism by asking 'success for who?' An especially useful teaching text in this respect – both because it comes from the Global South and because it offers a compellingly graphic global historical-geography – is subversively entitled *How to Succeed at Globalization: A Primer for The Roadside Vendor* (Fisgón, 2004). But whether a book like this is used or not, it remains vital to teach students about the political grievances of all those left on the ravaged roadsides after the Globalization juggernaut of market-managed interdependency has sped through. This enables them to start to come to terms with why there has been so much resistance to neoliberal Globalization from global justice movements, and why such resistance may in turn be understood to be 'anti-Globalization' but also simultaneously supportive of alter-globalisations – following French ideas of 'alter-mondialisme' and their ongoing diversification in global justice organizing venues such as the World Social Forum (Caruso, 2017; Sparke et al., 2005).

Another benefit of denaturalizing the ties between Globalization and neoliberalism is that it prepares the way for students to explore empirically the vast variety of actual globalisations. As Paul Amar (2013) argues, not all of these experiences can be accommodated in a singular narrative of global neoliberalization, and certainly not one that narrativizes neoliberalism as being just about the rise of market rule and globalized consumerism. Instead, Amar's examples of evolving forms of security and sexuality usefully remind students that other arrangements of governmentality and subjectivity are emerging in conjunction with global integration. But to begin to come to terms with such post-neoliberal formations, it is equally important to address the ways in which

even the market-dominated process of capitalist globalization has itself been characterized by complexities that explode the simplifying Globalization myths of newness,
inevitability and flat-world levelling. It is to these intertwined histories, governmentalities and spatialities that we now turn.

SPATIO-TEMPORAL COMPLEXITIES

Still the most systematic social science approach to tracing the spatio-temporal
complexities of actually existing globalisations is the now classic text *Global Transformations* (Held et al., 1999). This historically organized 'transformationalist'
approach usefully attempts to document the changing extensity, intensity, velocity and
impact propensity of global interconnections over time, but as a teaching text it remains
limited by its attempts to square the circle of political debate over neoliberalization
(Sparke, 2001). Setting up an evidence-based middle way between the exaggerations of
globalization hype and the denialism of globalization scepticism, the approach nevertheless fails to address the fundamentally political agonistics over neoliberalism that are
at stake in the tension between these alternative positions (see however, Held and
McGrew, 2002). Rather than acknowledge that the hyperglobalists' hype mainly stems
from efforts to prescribe neoliberal norms, and rather than explore why those
expressing scepticism do so in part because they seek to defend non-neoliberal national
economic governance, the tranformationalist approach takes the focus off the ideological and discursive struggles altogether. For the same reason, I think, it also fails to
connect pedagogically with what students are deliberating when they come to study
globalization and associated debates over the necessity (or not) of pro-market reforms.
To be sure, the rigorous scholarly focus on tracing uneven historical-geographical
transformations still works well as an antidote to the simplifications of the big G
Globalization meta-narrative. But my own approach has been to complement these
transformationalist lessons by using them to counter point directly the myth-making
that suggests that Globalization is new, inevitable and levelling.

Against the myth of newness attention to 'extensity' under earlier forms of global
interconnection also helps to highlight a number of significant precedents that actively
shape some of the power geometries of today's globalisations too. Histories of
colonialism and imperialism are important to teach about in this way not just because
they illustrate extensive global networking in prior centuries but also because of how
those older networks have left legacies shaping everything from the contemporary
prominence of older global cities (such as London) to patterns of post-colonial
resistance sovereignty and its disciplining by debt (such as illustrated by the case of
Haiti from the 1791 revolution onwards). In this way, I explain to students that
announcements of Globalization's novelty also obscure historical geographies of
connectivity and adverse incorporation that remain consequential today. From the
transatlantic slave trade, to the imperial oil extraction networks that tie the Middle East
to the West, to the global ecological extensity of the Anthropocene-*cum*-Capitalocene, I
use examples in my textbook and teaching that are aimed in this way to bring into view
histories of violence and geographies of global exploitation about which sun-dazed
sightings of Globalization's new dawn are blind.

At the same time, adapting another lesson from the transformationalist approach I also seek to teach students that in terms of intensity, velocity and impact propensity, today's global interdependencies are wider reaching, faster and more impactful on more people than globalisations of the past. These are the grains of truth that Globalization myth-making incorporates into its narrativization of the necessity of neoliberalism. The challenge, I therefore suggest, is to acknowledge these new globalisation developments without making a new world order argument for pro-market reform (a challenge that is further heightened by the common student complaint that teaching against neoliberalism is 'biased'). Transformationalist sensitivity to global shifts in political-economy through the twentieth century is especially useful for denaturalizing this linkage between neoliberalism and Globalization. Reciprocally, once this is no longer seen as a natural linkage it becomes possible instead to explain the articulations with actual evidence of changing global conditions. In this way, the increasing intensity, velocity and impact of global economic interdependencies from the 1950s to the 1970s can be used to explain to students the conjoined rise of both big G Globalization discourse and the systematic prescription of neoliberalism. In short, I teach in this way that post-war global economic re-integration led eventually, by the 1970s, to the breakdown of Fordist efforts to balance national mass production with national mass consumption (Sparke, 2013, Chapters 3, 4, 5). As corporations started to rely increasingly on global markets and then also on global production networks, their reasons for investing in national welfare-state liberalism and tolerating active government economic management diminished. And consequently, as David Harvey has argued in ways that students find especially useful, corporations could then radically escalate their struggles against national labour movements in concert with the more general promotion of neoliberal ideals (Harvey, 2005, 2016).

Other valuable teaching texts usefully detail how neoliberal ideals had an ideational pre-history before the 1970s – not least of all in the work of the *Mont Pelerin Society* and *Chicago School of Economics* – and how they have also continued to evolve as ideas subsequently through the work of think tanks, academics, and various 'falling forward' crisis experiments in policy-making (Brown, 2015; Mirowski, 2014; Peck, 2010). These kinds of readings help students come to terms in turn with the central paradox of the inevitability myth surrounding Globalization. They thus underline that it has taken a combination of ideational work *and* active political implementation to force through the expansion and entrenchment of pro-market reforms. The globalized and globalizing reform process was in fact never inevitable – even if many students tend to assume it is based on unexamined teleological ideals about the price of progress, modernity and so on. It instead required endless appeals to inevitability by boosters in order to progress. And politically it also required all sorts of state action and extra-state involvement by businesses and NGOs to bring the neoliberal 'anti-state state' into being (Sparke, 2013, Chapter 7).

The vast variety of neoliberalizing dynamics also contributes in turn to the complexity of actually existing globalisations, with all sorts of path dependence and contextual conditions shaping the expansion of neoliberal norms in particular countries and communities. Most famously there were the Thatcher and Reagan revolutions of the 1980s in the UK and US, both of which evolved out of the stagflationary crises of the 1970s and the subsequent electoral losses faced by the Labour party and the Carter administration respectively. Elsewhere it was military violence rather than electoral losses that led to rapid pro-market reforms, including most notably, the one in Chile in 1973 when the *coup d'état* led by Augusto Pinochet enabled a forced experiment in overnight neoliberalization led by Chicago trained economists (Klein, 2007). Differently again, in other more privileged times and places neoliberal reforms were developed as a so-called Third Way or New Labour policy by Western political leaders pulling traditionally left-leaning and centrist parties in more conservative pro-market directions with a wide array of policy innovations in education, welfare reform and free trade. Meanwhile, in many poor countries in the 1980s and 1990s, neoliberalism was instead imposed from the outside through the so-called Washington Consensus conditionalities of World Bank and IMF structural adjustment programmes. Then, more recently there has been talk of a 'Bcijing Consensus' emerging alongside various other 'aidez-faire' innovations in neoliberalization that conjoin communist party management with the market remaking of everything from regional planning to patriotism and sexuality (Rofel, 2007; Zhang and Ong, 2008; Zhang and Peck, 2014). A key lesson for students reviewing these complex and geographically variegated scenes of neoliberalization is that there has been nothing natural about the global expansion and entrenchment of pro-market policy, and nor has there in turn been anything inevitable about the resulting intensification of global market interdependencies.

It is the resulting growth in the intensity, velocity and impacts of globalized integration that has nevertheless been mythologized with metaphors of global flattening. Teaching against this mythic meta-narrative was perhaps more challenging in the initial years of Globalization hype. Around the year 2000 even critical writings on the new world order tended to recycle the mytheme of global levelling, exemplified perhaps most influentially by the 'smooth space' of globalization evoked in the academic bestseller *Empire* (Hardt and Negri, 2001). The limits of this flat-world view, with its notably decentred depiction of a capitalist empire without an imperial hegemon, soon became clear in the context of America's global war on terror after 2001 (Sparke et al., 2005, Chapter 5). But more recently, as examples of growing economic inequalities, political instabilities and international asymmetries and conflicts have multiplied, it has become almost impossible for students to buy into the vision of the flat world. In the classroom this means that attention to evolving historical hegemonies, path dependencies and the influence of contextual conditions on crisis-management can further help nuance explanations of why the world is not flat.

Even as it becomes easier to teach against flat-world myths, it remains a challenge to tease out what can be explained in terms of neoliberalization and what involves other influences. Much of the intensification of in-country economic inequality can be explained as an outcome of pro-market policy-making and all the resulting patterns of labour flexibilization, deunionization, outsourcing, offshoring, as well as the domestic neoliberal rollbacks in welfare, housing, education and health services in particular countries. At the same time, neoliberalization has also obviously benefited global elites, heightening inequality by enabling value capture by the transnational business class, along with their increasing tax avoidance through globalizing strategies such as inversions, transfer pricing and the use of tax havens. The result is that the world's wealthiest people still fly fast above the borders, immiseration and precarity facing ordinary people, thereby also continuing to be able to take comfort in the flat-world visions provided by their orbital world view, and doing so even as they enclave themselves off in gated communities of fellow neoliberal believers when they disembark. Still, as contradictory and complex as it is, the global expansion of neoliberalization only explains so much of this intensified inequality. It does not address effectively all the other factors – factors that range from inter-state violence to racist violence to intimate partner violence – that come together to codetermine suffering in what affluent Americans dismissingly refer to as 'fly-over country'. Reflecting on these additional aspects of uneven development in the classroom certainly helps critique flat-world discourse. But by showing that the causes of inequality, instability and uneven development are complex and geo-historically over-determined, it also provides a critical context for classroom conversations over what is to be done in response. It is to these kinds of conversations that we turn next.

INSTITUTIONAL OPPORTUNITIES FOR META-LEARNING IN THE CORPORATIZING UNIVERSITY

Students in today's universities are all too familiar with the market metrics, tools, and assumptions influencing their educational experiences. They compare universities and degree options through a vast array of competitive global ranking systems and Return on Investment (ROI) ratings. They find themselves repeatedly pressured by family members and peers (not to mention lenders) to justify their educational self-investment in terms of future economic returns. And, as they approach graduation, many feel obliged to turn their self-capitalization into new forms of self-commodification, marketing themselves for an ever-more competitive and globalized job market. Of course, they may not see themselves as neoliberal market subjects or entrepreneurs of their educational selves, and most continue to find the terminology of neoliberalism confusing, especially in the US where the word 'liberalism' is widely assumed – partly due to the hegemony of neoliberal ideas – to refer only to ideals of welfare-state redistribution and liberal human rights protection. But most students still experience higher education in ways that are profoundly shaped by the market-made and market-making rationalities of neoliberalism. While many would prefer for this to be left unexamined (and while even some progressive educators pre-emptively profess a preference for post-neoliberalism), neoliberal student experiences remain the basis of a huge meta-learning opportunity: an opportunity to teach about the connections between Globalization and neoliberalism, denaturalizing the linkage, and enabling students to think more critically about how their own global lives are conditioned by neoliberalization – even and, indeed, especially as some of them seek to imagine alternatives to Globalization.

It is in turn no small irony that many of us who seek to teach these students critically about globalisations do so in institutions that are also being radically restructured by the market metrics, tools and assumptions of neoliberalism (Castree and Sparke, 2000; Larner, 2015; Morrissey, 2013; Newfield, 2016; Rhoads and Torres, 2006). Teaching in corporatizing universities, we face increasing challenges created both directly by for-profit interests as well as by the wider imperatives of neoliberal market rationalities. There are country-specific variations, of course, but signs of such corporatization are now seen around the world: whether in the pressure to teach ever-larger numbers of students with fewer resources; or in the emphasis on high demand vocational degrees in Business and STEM (Science, Technology, Engineering and Math); or in the re-setting of research and faculty recruitment priorities around what wealthy donors demand; or in the competitive benchmarking of faculty performance with output and impact targets; or in the efforts to commercialize and thereby privatize the results of research funded by public grants; or just in the relentless branding of universities and all the logo-covered gear that now increasingly replaces books in university bookshops. This tableau of corporatization is thus also another backcloth against which we can (and must) teach critical lessons about the expanding reach of market forces. Once students can see the connections between these developments and their own experiences, and especially their own enlistment into being market subjects making investments in themselves, such teaching can be deeply meaningful and radically transformative (even if it is also at times disillusioning and discomforting). Ultimately this is how I think we can turn the neoliberal emphasis on personal responsibilization into a much more radical set of response abilities enabling the contestation and reimagination of Globalization (see also Sparke, 2013, Chapter 10).

Many other scholars of globalisations have elsewhere described useful examples of critical teaching that challenges market responsibilization and corporatization in the contemporary university (e.g. Darian-Smith and McCarty, 2017; Houston and Lange, 2017; Routledge and Derickson, 2015; Roy et al., 2016). Much of this work is experimental and iterative, cultivating new student responses, fostering meta-learning about these responses, and thereby also opening up possibilities for yet further experimentation. To provide an example of this kind of evolutionary process, I will conclude here with a brief synopsis of some of the meta-learning and responses I have seen in my own students.

Over the years I have been inspired by students who have developed a wide range of response abilities: from learning how to campaign for *United Students Against Sweatshops* (http://usas.org/) and *Universities Allied for Essential Medicines* (http://uaem.org/), to becoming involved in university apparel licensing committees and pressuring administrations to stop contracting with anti-union vendors. I have also witnessed students becoming articulate analysts of how global market forces influence their own lives, whether in terms of consumer choices or student debt or future plans for life after graduation. Sobering as these meta-learning moments often are, they also usefully add new nuance to arguments about neoliberal responsibilization and corporatization that go beyond simplistic theories of totalizing determination by global markets. For example, I have seen students develop sophisticated analyses of the limitations imposed by market rationalities on their well-intentioned service learning endeavours both locally and globally, thereby providing powerful insight into the increasingly influential brand of neoliberal humanitarianism that Katharyne Mitchell and I have called the 'New Washington Consensus' (Chahin, 2013; Mitchell and Sparke, 2016). In addition, what has also been educational about these student responses is their critically cosmopolitan composition. The related meta-learning made possible by the diverse global perspectives represented in our increasingly globalized classes is what inspired me in turn to develop a so-called Massive Open Online Course (MOOC) and it is with this example that I will close.

Entitled *Globalization and You* in its first two iterations on *Coursera*, I hoped at the outset that my MOOC would create an opportunity for globally inclusive online conversations about globalisations in the plural as well as critical analysis of neoliberal Globalization at the same time. The '*and You*' in the title was designed in this way to convey both a concern for students' situated knowledges and a critical focus on the market remaking of subjectivity. But I was also aware from the start that my critiques of Globalization as new, inevitable and levelling on a platform that was being widely hyped as the new, inevitable and levelling future of global education had a huge and problematic irony. My hope, nevertheless, was to turn this irony into a teaching opportunity. Inspired in part by the sharply ironic 'open letter' to *Coursera* co-founder Daphne Koller from Robert Meister (2013), I thought this might actually create a globalised meta-learning experiment in which participants could reflect on the processes of corporatization and student self-responsibilization reshaping global higher education. I have since described the results of this experiment at length in another article (Sparke, 2017). I must simply end here by noting that they were mixed and compromised.

The MOOC did enable me to offer my critical lectures about Globalization on a global scale (to about 75,000 registrants in the end). It showed how online courses can foster global discussions among participants from diverse countries, including many in the Global South (see also Spiegel et al., 2017). Participants were in turn able to share diverse perspectives on neoliberalization and their various alignments and non-alignments with associated forms of personal responsibilization. For a few this even led to embodied responses too, such as the small group of people who met through the MOOC's discussion threads and then came together in person at a protest against climate change in New York. But in terms of teaching globalisations, I do not think the MOOC led to many radical projects of anti-neoliberal resistance, and meanwhile it gave *Coursera* just one more batch of 'learner' analytics to boast about to the venture capitalists that are funding the platform (in the hopes of doing to university instructors what they have already done to journalists, retailers, and travel agents globally). For these reasons, I came away from my MOOC experiment concluding that the master's teaching tools will not dismantle the master's global classroom, however critical the lessons about Globalization may be. And for the same reasons, I want to return myself to focusing on working with students in smaller but still experimental efforts at teaching globalisations.

ACKNOWLEDGEMENTS

Many thanks to Mónica Farías and Tiffany Grobelski for their valuable comments on an earlier draft.

REFERENCES

Amar, P. (2013), *The Security Archipelago: Human-security States, Sexuality Politics, and the End of Neoliberalism*, Durham: Duke University Press.

Barkan, J. (2011), 'Law and the Geographic Analysis of Economic Globalization', *Progress in Human Geography*, **35**(5), 589–607.

Bessner, D. and M. Sparke (2017), 'Nazism, Neoliberalism and the Trumpist Challenge to Democracy', *Environment and Planning A*, **49**(6), 1214–1223.

Brown, W. (2015), *Undoing the Demos: Neoliberalism's Stealth Revolution*, Cambridge, MA: MIT Press.

Bürkner, H-P. (2017), 'What Leaders Must Do Now', *Handelsblatt Global*, accessed on 17 January 2017, https://global.handelsblatt.com/opinion/what-leaders-must-do-now-683899.

Caruso, G. (2017), 'Open Cosmopolitanism and the World Social Forum: Global Resistance, Emancipation, and the Activists' Vision of a Better World', *Globalizations*, **14**(4), 504–518.

Castree, N. and M. Sparke (2000), 'Professional Geography and the Corporatization of the University: Experiences, Evaluations and Engagements', *Antipode*, **32**(3), 222–229.

Chahin, D. (2013), 'From Good Intentions to Praxis: Learning from the Successes and Failures of the Critical Development Forum', Transcript of talk presented at Engineering, Social Justice, and Peace (ESJP) Conference in Troy, NY, accessed on 7 May 2017, https://static1.squarespace.com/static/5259ce6ee4b05804955c2799/t/52912e03e4b0a535daf07358/1385246211400/Chahim_ESJP2013_Talk.pdf.

Darian-Smith, E. and P.C. McCarty (2017), *The Global Turn: Theories, Research Designs, and Methods for Global Studies*, Berkeley: University of California Press.

Domosh, M. (2010), 'The World was Never Flat: Early Global Encounters and the Messiness of Empire', *Progress in Human Geography*, **34**(4), 419–435.

Domosh, M. (2013), 'Geoeconomic Imaginations and Economic Geography in the early 20th Century', *Annals of the Association of American Geographers*, **103**(4), 944–966.

Fisgón, E. (2004), *How to Succeed at Globalization: A Primer for the Roadside Vendor*, New York: Metropolitan.

Fisher, M., T. Gillespie and B. Schellhas (eds) (1994), *Contract with America: The Bold Plan by Rep. Newt Gingrich, Rep. Dick Armey and the House Republicans to Change the Nation*, New York: Times Books.

Friedman, T. (2000), *The Lexus and the Olive Tree: Understanding Globalization*, New York: Farrar, Straus and Giroux.

Gibson-Graham, J.K., S. Healy and J. Cameron (2013), *Take Back the Economy: An Ethical Guide for Transforming Our Communities*, Minneapolis: University of Minnesota Press.

Hardt, M. and A. Negri (2001), *Empire*, Cambridge: Harvard University Press.

Hart, G. (2003), *Disabling Globalization: Places of Power in Post-apartheid South Africa*, Berkeley: University of California Press.

Hart, G. (2005), 'Denaturalizing Dispossession: Critical Ethnography in the Age of Resurgent Imperialism', in: *From Local Processes to Global Forces*, Centre for Civil Society, University of KwaZulu-Natal Research Reports volume 1.

Harvey, D. (2005), *A Brief History of Neoliberalism*, Oxford: Oxford University Press.

Harvey, D. (2016), 'Neoliberalism is a Political Project', *Jacobin*, accessed on 7 May 2017, https://www.jacobinmag.com/2016/07/david-harvey-neoliberalism-capitalism-labor-crisis-resistance/.

Held, D. and A.G. McGrew (2002), *Globalization/Anti-globalization*, Cambridge, UK: Polity.

Held, D., A.G. McGrew, D. Goldblatt and J. Perraton (1999), *Global Transformations: Politics, Economics and Culture*, Stanford: Stanford University Press.

Houston, S. and K. Lange (2017), '"Global/Local" Community Engagement: Advancing Integrative Learning and Situated Solidarity', *Journal of Geography in Higher Education*, **42**(1), 44–60.

Katz, C. (2001), 'On the Grounds of Globalization: A Topography for Feminist Political Engagement', *Signs*, **26**, 1213–1234.

Klein, N. (2007), *The Shock Doctrine: The Rise of Disaster Capitalism*, New York: Metropolitan Books/Henry Holt.

Klein, P., E. Pawson, M. Solem and W. Ray (2014), 'Geography Education for "An Attainable Global Perspective"', *Journal of Geography in Higher Education*, **38**(1), 17–27.

Larner, W. (2015), 'Globalising Knowledge Networks: Universities, Diaspora Strategies, and Academic Intermediaries', *Geoforum*, **59**, 197–205.

Meister, R. (2013), 'An Open Letter to a Founder of Coursera', *The Chronicle of Higher Education*, accessed on 21 May 2017, http://www.chronicle.com/blogs/conversation/2013/05/21/can-venture-capital-deliver-on-the-promise-of-the-public-university/.

Mirowski, P. (2014), *Never Let a Serious Crisis Go to Waste: How Neoliberalism Survived the Financial Meltdown*, New York: Verso.

Mitchell, K. and M. Sparke (2016), 'The New Washington Consensus: Millennial Philanthropy and the Making of Global Market Subjects', *Antipode*, **48**(3), 724–749.

Morrissey, J. (2013), 'Regimes of Performance: Practices of the Normalized Self in the Neoliberal University', *British Journal of Sociology of Education*, **36**(4), 614–634.

Mountz, A. and J. Hyndman (2006), 'Feminist Approaches to the Global Intimate', *Women's Studies Quarterly*, **34**(1–2), 446–463.

Murray, W. and J. Overton (2015), *Geographies of Globalization* (2nd edn), New York: Routledge.

Nally, D. (2014), 'Globalization', in: John Morrissey, David Nally, Ulf Strohmayer and Yvonne Whelan (eds), *Key Concepts in Historical Geography*, Los Angeles: Sage, pp. 225–237.

Newfield, C. (2016), *The Great Mistake: How We Wrecked Public Universities and How We Can Fix Them*, Baltimore: Johns Hopkins University Press.

Peck, J. (2010), *Constructions of Neoliberal Reason*, Oxford: Oxford University Press.

Pratt, G. and V. Rosner (eds) (2012), *The Global and the Intimate: Feminism in Our Time*, New York: Columbia University Press.

Rhoads, R. and C.A. Torres (2006), *The University, State, and Market: The Political Economy of Globalization in the Americas*, Stanford: Stanford University Press.

Rofel, L. (2007), *Desiring China: Experiments in Neoliberalism, Sexuality, and Public Culture*, Durham: Duke University Press.

Routledge, P. and K. Derickson (2015), 'Situated Solidarities and the Practice of Scholaractivism', *Environment and Planning D: Society and Space*, **33**, 391–407.

Roy, A., G. Negrón-Gonzales, K. Opoku-Agyemang and C. Talwalker (eds) (2016), *Encountering Poverty: Thinking and Acting in an Unequal World*, Berkeley: University of California Press, Poverty, Interrupted Series.

Sidaway, J.D. (2012), 'Geographies of Development: New Maps, New Visions?', *The Professional Geographer*, **64**(1), 49–62.

Silvey, R. and K. Rankin (2010), 'Development Geography: Critical Development Studies and Political Geographic Imaginaries', *Progress in Human Geography*, **35**(5) 696–704.

Smith, N. (2005), *The Endgame of Globalization*, New York: Routledge.

Sparke, M. (2001), 'Networking Globalization: A Tapestry of Introductions', *Global Networks*, **1**(2), 171–179.

Sparke, M. (2004), 'Political Geographies of Globalization: (1) Dominance', *Progress in Human Geography*, **28**(6), 777–794.

Sparke, M. (2006), 'Political Geographies of Globalization: (2) Governance', *Progress in Human Geography*, **30**(2), 1–16.

Sparke, M. (2008), 'Political Geographies of Globalization: (3) Resistance', *Progress in Human Geography*, **32**(1), 1–18.

Sparke, M. (2013), *Introducing Globalization: Ties, Tensions and Uneven Integration*, Oxford, UK: Wiley-Blackwell.

Sparke, M. (2017), 'Situated Cyborg Knowledge In Not So Borderless Online Global Education: Mapping the Geosocial Landscape of a MOOC', *Geopolitics*, **22**(1), 51–72.

Sparke, M., E. Brown, D. Corva, H. Day, C. Faria, T. Sparks and K. Varg (2005), 'The World Social Forum and the Lessons for Economic Geography', *Economic Geography*, **81**(4), 359–380.

Spiegel, S., H. Gray, B. Bompani, K. Bardosh and J. Smith (2017), 'Decolonising Online Development Studies? Emancipatory Aspirations and Critical Reflections – A Case Study', *Third World Quarterly*, **38**(2), 270–290.

Zhang, J. and J. Peck (2014), 'Variegated Capitalism, Chinese Style: Regional Models, Multi-scalar Constructions', *Regional Studies*, **50**(1), 52–78.

Zhang, L. and A. Ong (2008), *Privatizing China: Socialism from Afar*, Ithaca: Cornell University Press.

Index